AF738685

Compact Objects in the Universe

Eleftherios Papantonopoulos ·
Nikolaos Mavromatos
Editors

Compact Objects in the Universe

Springer

Editors
Eleftherios Papantonopoulos
Physics Division, School of Applied
Mathematical and Physical Sciences
National Technical University of Athens
Athens, Greece

Nikolaos Mavromatos
Physics Division, School of Applied
Mathematical and Physical Sciences
National Technical University of Athens
Athens, Greece

Physics Department
King's College London
London, UK

ISBN 978-3-031-55097-3 ISBN 978-3-031-55098-0 (eBook)
https://doi.org/10.1007/978-3-031-55098-0

© The Editor(s) (if applicable) and The Author(s), under exclusive license to Springer Nature Switzerland AG 2024, corrected publication 2024

This work is subject to copyright. All rights are solely and exclusively licensed by the Publisher, whether the whole or part of the material is concerned, specifically the rights of translation, reprinting, reuse of illustrations, recitation, broadcasting, reproduction on microfilms or in any other physical way, and transmission or information storage and retrieval, electronic adaptation, computer software, or by similar or dissimilar methodology now known or hereafter developed.
The use of general descriptive names, registered names, trademarks, service marks, etc. in this publication does not imply, even in the absence of a specific statement, that such names are exempt from the relevant protective laws and regulations and therefore free for general use.
The publisher, the authors and the editors are safe to assume that the advice and information in this book are believed to be true and accurate at the date of publication. Neither the publisher nor the authors or the editors give a warranty, expressed or implied, with respect to the material contained herein or for any errors or omissions that may have been made. The publisher remains neutral with regard to jurisdictional claims in published maps and institutional affiliations.

This Springer imprint is published by the registered company Springer Nature Switzerland AG
The registered company address is: Gewerbestrasse 11, 6330 Cham, Switzerland

If disposing of this product, please recycle the paper.

Preface

During the past two decades we have witnessed extraordinary progress in precision instrumentation for astrophysical and cosmological observations, which placed Cosmology and Astrophysics in the map of high precision scientific fields. Moreover, the discovery in 2015 of gravitational waves (GW) by the LIGO-VIRGO Collaborations, using extremely high precision interferometric devices, created a new field for astronomical and cosmological searches. The discovery of GW, one of the main predictions of the theory of General Relativity, which we now believe with great confidence that are produced by the merger of spinning black holes (like in the event of 2015), or other compact objects like neutron stars, as well as a plethora of phase transitions in the early Universe, opens up the way for potential experimental measurements, or falsification, of theoretical models beyond general relativity (GR), given the differences in the profiles of the GW produced from mergers of spinning compact objects in such theories as compared to the GR case.

So far, the latest GW detections by ground-based interferometers and the discovery of a binary pulsar whose orbital period changes in accordance with the predicted GW emission, verified the predictions of the Einstein's theory. However, there are many theoretical reasons pointing to the fact that we need to go beyond GR, one of the most compelling ones being the need for the construction of an ultraviolet complete consistent quantum theory of the gravitational interaction, which most likely will only encompass GR as its low-energy limit (see, for instance, the example of string theory). In this respect, the experimental detection of GW opened up an intensive activity of understanding all aspects of the GWs and the implications of this discovery to alternative theories of Gravities and their cosmological implications.

The modified gravity theories can give us important information on the structure and properties of the compact objects predicted by these theories and also the observational signatures which they can introduce. It is expected that future ground and space-borne detectors will improve our perception of gravitational interactions in particular will shed light into the existence of exotic compact objects, like hairy black holes and wormholes. These compact objects provide a window into the microscopic structure of spacetime in quantum gravity and the hope is to understand the nature of their interior, potential applications to quantum cosmology, and the development of further connections to quantum information and laboratory quantum simulation. The formation of these compact objects may also trigger possible holographic phase transitions to other compact objects.

The aim of this volume is to collect the presentations of a selected group of plenary speakers in the 11th Aegean Summer School, which was held in Ermoupolis of the island of Syros (Greece), from 5 to 15 of September 2022. The topic of the school was to discuss the recent developments in theories of Gravity and Cosmology, and in observations discussing several state-of-the-art experiments. In the theory part, topics discussed include high curvature theories and scalar-tensor theories and also the physics of black holes, their formation, their stability and their connection to astrophysical objects. Models of early and late cosmological evolution have been presented and their connection with dark matter and dark energy have been discussed. There were also talks on quantum gravity and holography. In the experimental part, talks on Gravitational Waves and their observational signatures, especially in the context of the LIGO, VIRGO, LISA and BINGO experiments.

The contents of this volume consist of a selected group of invited talks that are representative of the above topics. Specifically,

Eugeny Babichev, Christos Charmousis, Nicolas Lecoeur, discussed "Exact Black Hole Solutions in Higher Order Scalar Tensor Theories". In this chapter explicit black hole solutions in higher-order scalar-tensor theories were discussed. First the so-called stealth solutions present in theories with parity and shift symmetry were analyzed. Using stealth metrics they constructed an analytic stationary solution of scalar-tensor theory which is called disformed Kerr metric. Applying a Kaluza-Klein reduction of a higher-dimensional Lovelock theory they considered scalar-tensor theories. Using this reduction procedure, they obtained non-stealth black holes with interesting properties and they analyzed the obtained solutions with respect to the geometry of the internal space according to their Kaluza Klein origin.

Panagiota Kanti discussed "Compact Objects in EsGB Theory and Beyond", analyzing novel and physically interesting spherically-symmetric solutions that describe compact objects in EsGB theories with a general coupling function between the scalar field of the theory and the quadratic Gauss-Bonnet term and in more general scalar-tensor Horndeski theories. A variety of spherically-symmetric solutions for several different forms of the coupling function and their main features are discussed. Also wormhole solutions are presented and it was demonstrated that the EsGB theory naturally supports traversable wormholes without the need for exotic matter. In a more extended scalar-tensor Horndeski theory, solutions were discussed containing black holes and novel regular scalarised particle-like solutions, as well as traversable wormholes obtained by appropriate (disformal) transformations applied to black hole solutions.

Daniela D. Doneva, Stoytcho S. Yazadjiev review the "Dynamics of Scalarization". In General Relativity solutions describing compact objects such as neutron stars and black holes endowed with scalar field have been constructed in a number of modified theories of gravity. The mechanisms of the spontaneous scalarization can be different and the dynamics of such processes is developing at a slower speed due to the high complexity of the problem especially in the nonlinear fully coupled regime. The advance in this topic is crucial, though, for studying a number of astrophysical implications. In this chapter some of the developments in this area focusing especially on the scalar-Gauss-Bonnet gravity as a representative example, are discussed.

Cristián Erices, Simón Riquelme, Nicolás Zalaquett presented an "EFT Approach to Black Hole Scalarization and Its Compatibility with Cosmic Evolution" and examined its compatibility with cosmic inflation and big bang cosmology from an effective field theory (EFT) point of view. They considered how higher-order theories of gravity, up to cubic operators in Riemann curvature, fit within this context. It was found that there is a non-trivial interplay between the Wilson coefficients constraining the parameter space where scalarization may actually occur. They claimed that the EFT does exhibit black hole scalarization, remaining compatible with the inflationary paradigm, and admitting General Relativity as a cosmological attractor.

Rodrigo Olea, discusses "Conformal Renormalization and Energy Functionals in Anti-De Sitter Gravity". He explores, within an appropriate holographic framework implemented by the AdS geometry, the physical consequences of embedding Einstein-AdS gravity in Conformal Gravity in four and six dimensions.

Kyriakos Destounis, Francisco Duque study "Black-Hole Spectroscopy: Quasinormal Modes, Ringdown Stability and the Pseudospectrum". Black hole spectroscopy is a powerful tool to probe the Kerr nature of astrophysical compact objects and their environment. They review the black hole spectroscopy program and its challenges regarding quasinormal mode detection, the overtone status and the recent evidence that supports the existence of nonlinearities in the spectrum of black holes. They discuss the pseudospectrum: a mathematical notion that can shed light on the spectral stability of quasinormal modes, and discuss its novel applications in black holes and exotic horizonless compact objects and also potential ringdown imprints and waveform stability issues in current and future gravitational-wave detectors.

Filipe Moura, João Rodrigues, described the "WKB Method and Quasinormal Modes of String-Theoretical D-Dimensional Black Holes". They first give a brief introduction to quasinormal modes in dissipative systems, and then they review the WKB formalism in the context of the analytical calculation of quasinormal frequencies. They calculate the quasinormal frequencies associated with gravitational perturbations of d-dimensional spherically-symmetric black holes with string corrections.

Lucas G. Collodel, in his article discusses "Spacetime Geometry of Rotating Boson Stars and KBHsSH". He gives an overview of the proprieties of rotating boson stars and Kerr black holes with scalar hair. A fluid description of the scalar matter is provided, and by adopting a frame co-moving with the fluid he finds a structural connection between both objects. He shows that the unusual distribution of the scalar matter warps spacetime in a very unique way, such that some metric functions on the equatorial plane behave nonmonotonically.

Giancarlo Cella, gave an overview of the physics results during the "Seven Years of LIGO-Virgo Science". He discussed several new detected signals, including peculiar events not seen before such as coalescences with a large mass or large mass asymmetry, and coalescences of a neutron star-black hole pair. The analysis of these signals can provide information about stellar compact object populations, the local expansion rate of the universe, the nature of black holes and the validity of General Relativity. The article also discusses other searches, which constrain the emission of isolated sources and the possible background of gravitational waves from unresolved sources of astrophysical or cosmological nature.

BINGO Group The BINGO Group presents two chapters: "Astrophysics of Radio Sources: Analysis of 21 cm Post-reionization Signals and Modeling of FRBs as PBHs with Magnetic Fields" and the "BINGO-ABDUS: A Radiotelescope to Unveil the Dark Sector of the Universe". In these chapters, different types of astrophysical information are described that are observed when a radiotelescope operates in a 21-cm intensity mapping on large scales and how one can handle the images to extract the hidden 21-cm signal. They provide a new candidate to describe Fast Radio Bursts and their magnetic characteristic in addition to one most currently accepted and likely magnetar nature. In the second chapter, there will be a discussion for understanding the recent Universe dynamics that have shed light on the existence of the Dark Sector and claim that BINGO will explore the southern celestial Hemisphere to observe neutral hydrogen signals from when the Universe was between 8.969–12.112 billion years and identifying individually fast radio sources.

Nikolaos Stergioulas discusses "Machine Learning Applications in Gravitational Wave Astronomy". Gravitational wave astronomy has emerged as a new branch of observational astronomy, since the first detection of gravitational waves in 2015. New computational detection algorithms are needed in this direction and application of machine-learning techniques have been developed to accelerate the detection and parameter estimation of gravitational wave sources. In this chapter, several different applications are summarized, including the application of artificial neural networks and autoencoders in accelerating the computation of surrogate models, deep residual networks in achieving rapid detections with high sensitivity as well as artificial neural networks for accelerating the construction of neutron star models in an alternative theory of gravity.

Last, but not least, **Ioannis Dalianis** discusses the "Features in the Inflaton Potential and the Spectrum of Cosmological Perturbations". It is known that small-scale perturbations are generated during the later stages of inflation, when possible strong features in the inflaton potential can break scale invariance and leave characteristic imprints on the spectrum. He focuses on and reviews three types of features: inflection points and steep steps in the potential, as well as sharp turns in the inflationary trajectory in field space. The purpose of his study is to show that such features induce a strong enhancement of the curvature spectrum within a certain wavenumber range and they also generate characteristic oscillatory patterns that are transferred in the spectrum of secondary GWs, potentially observable by operating or designed experiments.

The contents of this volume are addressed to both experts and beginners in the field, given that the articles are quite informative on the state-of-the-art developments in their respective subjects, but also written in a pedagogical way, without sacrificing though mathematical rigor, when necessary. We hope the readers will enjoy the volume!

Athens, Greece Eleftherios Papantonopoulos
London, UK Nikolaos Mavromatos

The original version of the book has been revised: The Book authors affiliations have been updated. A correction to this book can be found at https://doi.org/10.1007/978-3-031-55098-0_14

Contents

Modifications of General Relativity: Scalar Tensor Theories

Exact Black Hole Solutions in Higher Order Scalar Tensor Theories

Eugeny Babichev, Christos Charmousis, and Nicolas Lecoeur

Abstract In this chapter we discuss explicit black hole solutions in higher-order scalar-tensor theories. After a brief recap of no hair theorems, we start our discussion by so-called stealth solutions present in theories with parity and shift symmetry. Stealth solutions are such that their metric are Ricci flat General Relativity solutions, but they are accompanied by a non-trivial scalar field, in both spherically-symmetric and rotating cases. The stealth metrics then enable us to construct an analytic stationary solution of scalar tensor theory which is called disformed Kerr metric. This solution constitutes a measurable departure from the usual Kerr geometry of GR. We discuss within parity and shift symmetric theories several non stealth solutions. We then consider scalar-tensor theories stemming from a Kaluza-Klein reduction of a higher-dimensional Lovelock theory. These theories encompass all Horndeski functionals and hence go beyond parity and shift symmetry. Reduction and singular limits allow one to obtain non-stealth black holes with differing interesting properties which are not Ricci flat metrics. We analyse the solutions obtained and classify them with respect to the geometry of the internal space according to their Kaluza Klein origin.

1 Black Holes and Scalar Tensor Theories

General Relativity (GR) is a highly successful theory of gravity as it has predicted multiple gravitational phenomena and successfully passed multiple observational tests at different energy and length scales [1]. It is an effective field theory (EFT) depending on a unique curvature scale. The puzzle of dark energy and dark matter,

E. Babichev · C. Charmousis (✉) · N. Lecoeur
Université Paris-Saclay, CNRS/IN2P3, IJCLab, 91405 Orsay, France
e-mail: christos.charmousis@ijclab.in2p3.fr

E. Babichev
e-mail: eugeny.babichev@ijclab.in2p3.fr

N. Lecoeur
e-mail: nicolas.lecoeur@ijclab.in2p3.fr

© The Author(s), under exclusive license to Springer Nature Switzerland AG 2024

E. Papantonopoulos and N. Mavromatos (eds.), *Compact Objects in the Universe*,
https://doi.org/10.1007/978-3-031-55098-0_1

occurring at classical scales (IR) as well as the expected breakdown of GR at ultraviolet scales (UV) question the interval of validity of GR on both ends, at the IR and UV scales. It is important therefore to seek theories which consist measurable departures from the realm of GR and which are again effective theories (in the absence of a quantum theory of gravity). Gravity wave (GW) detections have put a very strong constraint on the speed of gravitons [2], whose applicability however, for cosmological dark energy and higher-order scalar tensor theories, is questionable [3]. Future GW detectors such as LISA [4] or the Einstein telescope [5] will give better insight on constraints involving dark energy, and the cosmological puzzles remain open and are even enhanced as precision tests of cosmology improve. In the domain of strong gravity, black holes stand out as an extraordinary laboratory of strong gravity, and numerous ongoing observations such as gravitational waves of compact binaries [2], the very large telescope interferometer of GRAVITY [6] or again the Event Horizon Telescope [7], will help to measure the validity and possible observational limits of GR.

Many directions can be considered for modifying gravity (see [8] for a review), but geometric modifications all amount to altering the number of degrees of freedom as compared to the two metric degrees of freedom in GR. This follows from Lovelock's theorem [9] which states that any four-dimensional action depending on the metric only and yielding second-order conserved field equations is the Einstein-Hilbert action with cosmological constant. The consequence is twofold: first, the most simple modification is to add a single degree of freedom, represented by a scalar field, second, many other modified gravity theories can be cast into a scalar-tensor form (see for instance [10]) or have a scalar tensor theory limit [11, 12]. The focus of the present chapter regards exact[1] BH solutions in four-dimensional scalar-tensor theories with a single scalar field ϕ, with action $S\left[g_{\mu\nu},\phi\right]=\int \mathrm{d}^4x\sqrt{-g}\mathcal{L}$ (see [13] for previous reviews). Also, our attention will be restricted to asymptotically flat BHs in vacuum without a cosmological constant in the action,[2] and more precisely to BHs which behave asymptotically at leading order like the GR Schwarzschild/Kerr solutions, in order for them to display the expected Newtonian limit [15].

The simplest theories are scalar-tensor Lagrangians which have at most two derivatives, however, as we will now see, no-hair theorems greatly restrict the existence of non GR solutions for such Lagrangians. Since we are interested in modified gravity, we will consider a nonminimal coupling of the scalar to gravity,[3] and to fix this

[1] We will focus on explicit solutions throughout this chapter. Numerical solutions will be treated elsewhere.

[2] Solutions with (a)dS asymptotics are in general less constrained than asymptotically flat solutions and generically easier to obtain: some scalar-tensor actions admit solutions with (effective or not) cosmological constant but do not allow asymptotically flat solutions, see e.g. [14].

[3] The case of minimally coupled scalar field will not be dealt with in this chapter. BH solutions can be constructed with such minimal coupling by reverse engineering the scalar field potential: one first imposes the solution, then finds the potential supporting it, see e.g. [16].

frame we shall suppose that any matter sources couple minimally to gravity. For nonminimal coupling and at most two derivatives, the theories are of a generalized Brans-Dicke form (see for example [17]),

$$\mathcal{L} = \phi R - \frac{\omega(\phi)}{\phi}(\partial\phi)^2 - V(\phi). \tag{1}$$

where R is the Ricci scalar of the metric $g_{\mu\nu}$ and $\omega(\phi)$ measures the strength of the scalar coupling, infinity representing the GR limit. No-scalar-hair theorems exist for this kind of Lagrangians [18]. They greatly restrict the existing BH solutions to be GR solutions with a trivial scalar field. There is a particular Lagrangian with a non-minimally conformally coupled scalar ϕ, and second-order in derivatives, which gives rise to an asymptotically flat BH. It is usually written as

$$\mathcal{L} = R - \frac{1}{2}(\partial\phi)^2 - \frac{1}{12}R\phi^2 \tag{2}$$

(this is of the form (1) up to redefinition of the scalar field), and its static, spherically-symmetric solution is known as the Bocharova-Bronnikov-Melnikov-Bekenstein (BBMB) black hole [19],

$$\mathrm{d}s^2 = -f\mathrm{d}t^2 + \frac{\mathrm{d}r^2}{f} + r^2\mathrm{d}\Omega^2, \quad f(r) = \left(1 - \frac{M}{r}\right)^2, \quad \phi(r) = \frac{2\sqrt{3}M}{r-M}, \tag{3}$$

where $\mathrm{d}\Omega^2$ is the metric of the unit two-sphere.[4] This describes an extremal BH with double horizon at $r = M$ where M is the mass of the BH. The scalar field diverges at the horizon, bringing about one of the pathologies of the BBMB solution.

In a word, scalar-tensor Lagrangians which are second order in derivatives allow very few asymptotically flat BH solutions. A way to circumvent no-hair theorems is to consider higher-order scalar-tensor theories. After all, if we are interested in UV corrections to gravity, such terms are bound to play an important role. This chapter reviews exact asymptotically flat BH solutions in four-dimensional, higher-order scalar-tensor theories. These theories are first described in Sect. 2. Section 3 then presents the so-called stealth solutions, where a non-trivial scalar field supports a GR metric. On the other hand, non-stealth BHs, for which the metric differs from GR, are studied in Sect. 4. Finally, a summary and discussions appear in the last section.

---

[4] The generalization of the BBMB solution to (a)dS asymptotics is the Martinez-Troncoso-Zanelli (MTZ) BH [20], and requires to include in the BBMB Lagrangian (2) not only a cosmological constant Λ, but also a self-interacting potential $\lambda\phi^4$, with $\lambda = \Lambda/72$. The planar AdS version requires the addition of axionic fields [21]. In the presence of a cosmological constant the scalar singularity can be hidden by the event horizon.

2 Horndeski and Beyond

2.1 Horndeski Theories

As advocated in the introduction, higher-order scalar-tensor theories enable one to evade no-scalar-hair theorems and to obtain BH solutions. In general, however, higher derivative Langrangians lead to field equations with time derivatives of order higher than two, which lead (in general again) to the propagation of unphysical degrees of freedom with an associated Hamiltonian unbounded from below. This is the Ostrogradsky instability, and the unphysical degrees of freedom are called Ostrogradsky ghosts (see [22] and references therein). Of course, a sufficient condition for preventing the Ostrogradsky instability is for the field equations to be of second order. It turns out that the most general four-dimensional scalar-tensor action with second order field equations was described as early as in 1974 by Horndeski [23]. His work went largely unnoticed until its rediscovery by [24] in 2011. Fortunately, independent progress on scalar-tensor theories was accomplished during those almost forty years following differing paths and arriving to the same conclusions. Starting from the Galileon [25] (most general scalar action in flat spacetime with invariance under *Galilean boost* $\phi \to \phi + b^\mu x_\mu + c$ with constant b^μ and c), 'covariantization' [26] and several steps led to the generalized Galileon [27], whose action is simply written in terms of four arbitrary functions G_2, G_3, G_4, G_5. It was later shown [28] that this simple generalized Galileon action is equivalent to the Horndeski action. As a consequence, the Horndeski action is now conventionally written in its generalized Galileon form, namely,

$$S\left[g_{\mu\nu}, \phi\right] = \int \mathrm{d}^4x\sqrt{-g}\left\{\mathcal{L}_2 + \mathcal{L}_3 + \mathcal{L}_4 + \mathcal{L}_5\right\}, \tag{4}$$

$$\mathcal{L}_2 = G_2, \quad \mathcal{L}_3 = -G_3\Box\phi, \quad \mathcal{L}_4 = G_4 R + G_{4X}\left[(\Box\phi)^2 - \left(\phi_{\mu\nu}\right)^2\right], \tag{5}$$

$$\mathcal{L}_5 = G_5\, G_{\mu\nu}\phi^{\mu\nu} - \frac{1}{6}G_{5X}\left((\Box\phi)^3 - 3\Box\phi\left(\phi_{\mu\nu}\right)^2 + 2\phi_{\mu\nu}\phi^{\nu\rho}\phi^\mu_\rho\right), \tag{6}$$

where G_2, G_3, G_4, G_5 are arbitrary functions of the scalar field ϕ and its kinetic term $X \equiv -\left(\partial\phi\right)^2/2$: $G_2 = G_2\left(\phi, X\right)$, etc. A subscript X means derivation with respect to X. $G_{\mu\nu}$ is the Einstein tensor of the metric $g_{\mu\nu}$, and the following notations are used for brevity: $\phi_\mu = \partial_\mu\phi$, $\phi_{\mu\nu} = \nabla_\mu\nabla_\nu\phi$, $\left(\phi_{\mu\nu}\right)^2 = \phi_{\mu\nu}\phi^{\mu\nu}$. Note that, if all functions G_2, G_3, G_4, G_5 depend only on the kinetic term X, the Horndeski action is shift-symmetric, i.e. unchanged under shifts of the scalar $\phi \to \phi + \text{const}$. This continuous symmetry brings about a conserved Noether current,

$$J^\mu = -\frac{1}{\sqrt{-g}}\frac{\delta S}{\delta\left(\partial_\mu\phi\right)}, \tag{7}$$

and the scalar field equation becomes $0 = \nabla_\mu J^\mu$. As a result the shift-symmetric version of Horndeski is accordingly easier to deal with. In addition, if $G_3 = G_5 = 0$, a shift-symmetric Horndeski action acquires the discrete parity symmetry $\phi \to -\phi$, which turns out to simplify things as well. Last but not least, by definition of Horndeski theory, every scalar-tensor action whose field equations are second order can be cast into the form (4), modulo a number of (obvious or less obvious) integrations by parts. For instance, the derivative coupling with the Einstein tensor, $G^{\mu\nu}\partial_\mu\phi\,\partial_\nu\phi$, corresponds to a galieon $G_4 = X$, while a coupling $\phi\mathcal{G}$ with the Gauss-Bonnet invariant $\mathcal{G} \equiv R^2 - 4R_{\mu\nu}R^{\mu\nu} + R_{\mu\nu\rho\sigma}R^{\mu\nu\rho\sigma}$ coincides with $G_5 = -4\log|X|$, see [28, 29]. More details on Horndeski theories can be found in the review [30].

2.2 *Degenerate Higher-Order Scalar-Tensor (DHOST) Theories*

Second order field equations are a sufficient condition for avoiding Ostrogradsky instability, but are by no means necessary, at least for scalar tensor theories.[5] Indeed, with field equations whose time derivatives are of order higher than two, the Ostrogradsky ghost emerges as an extra degree of freedom associated with the need of additional initial conditions. However, degeneracy of the kinetic matrix of the system eliminates this extra degree of freedom, yielding theories without Ostrogradsky instability despite having field equations of order higher than two. Exploiting this fact in the context of scalar-tensor theories leads to the Degenerate Higher-Order Scalar-Tensor (DHOST) theories (see [32] for a review), which evidently comprise the Horndeski theory. They were first constructed for Lagrangians quadratic in the second derivatives [33], then extended to cubic order [34]. We will not need the full expression of the DHOST Lagrangian, and will rather write it down partly when needed. From our point of view which focuses on solutions rather than theories themselves, it is more interesting to note that DHOST theories admit a quite simple interpretation in terms of *conformal-disformal transformations* of Horndeski theory [35]. A conformal-disformal transformation [36] depends on two functions of the scalar and its kinetic term, the conformal factor $C(\phi, X)$ and the disformal factor $D(\phi, X)$, mapping a metric $g_{\mu\nu}$ to a new metric $\widetilde{g}_{\mu\nu}$ according to

$$\widetilde{g}_{\mu\nu} = C(\phi, X)\, g_{\mu\nu} + D(\phi, X)\,\partial_\mu\phi\,\partial_\nu\phi. \tag{8}$$

Given an initial Horndeski action $S\left[g_{\mu\nu}, \phi\right]$, it is possible to express it in terms of the metric $\widetilde{g}_{\mu\nu}$,

$$S\left[g_{\mu\nu}, \phi\right] = \widetilde{S}\left[\widetilde{g}_{\mu\nu}, \phi\right], \tag{9}$$

[5] Going beyond scalar tensor theories (vector/tensor, bigravity theories, etc.), it turns out that avoiding instabilities is far more complex [31].

and it turns out that the resulting action $\widetilde{S}$ is a DHOST theory. More precisely, three cases can be distinguished. If the two factors depend only on the scalar field, $C = C(\phi)$ and $D = D(\phi)$, then the new action also belongs to the Horndeski class [37], (8) is in this case an internal map. If $C = C(\phi)$, but the disformal factor depends on X, $D = D(\phi, X)$, then the new action belongs to the so-called beyond Horndeski class [38], which, in addition to the four Horndeski terms (4), admit two additional contributions,

$$S_{\text{bH}} = \int \mathrm{d}^4x\sqrt{-g}\Big\{\mathcal{L}_{4\text{bH}} + \mathcal{L}_{5\text{bH}}\Big\}, \tag{10}$$

$$\mathcal{L}_{4\text{bH}} = F_4\,\varepsilon^{\mu\nu\rho\sigma}\,\varepsilon^{\alpha\beta\gamma}{}_{\sigma}\,\phi_\mu\phi_\alpha\phi_{\nu\beta}\phi_{\rho\gamma}, \tag{11}$$

$$\mathcal{L}_{5\text{bH}} = F_5\,\varepsilon^{\mu\nu\rho\sigma}\,\varepsilon^{\alpha\beta\gamma\delta}\phi_\mu\phi_\alpha\phi_{\nu\beta}\phi_{\rho\gamma}\phi_{\sigma\delta}, \tag{12}$$

with functions[6] $F_4 = F_4(\phi, X)$ and $F_5 = F_5(\phi, X)$, and $\varepsilon_{\mu\nu\rho\sigma}$ is the Levi-Civita tensor. Finally, if $C = C(\phi, X)$, the resulting action is a DHOST action, but not of the beyond Horndeski type [35].

More generally, DHOST theories are stable under conformal-disformal transformations: if the initial action $S\left[g_{\mu\nu}, \phi\right]$ is a DHOST action, then after the conformal-disformal transformation (8), the resulting action $\widetilde{S}\left[\widetilde{g}_{\mu\nu}, \phi\right]$ given by (9) also belongs to the DHOST family [35], or in other words, degeneracy of the kinetic matrix remains unaffected by a conformal-disformal transformation. This stability enables *generation of solutions* [40] in the following way: starting from a *seed solution* $\left(g_{\mu\nu}, \phi\right)$ to the initial action S, one gets a solution $\left(\widetilde{g}_{\mu\nu}, \phi\right)$ to the new action $\widetilde{S}$ since

$$\widetilde{\mathcal{E}}_{\mu\nu} = \frac{\sqrt{-g}}{\sqrt{-\widetilde{g}}}\frac{\delta g^{\rho\sigma}}{\delta \widetilde{g}^{\mu\nu}}\mathcal{E}_{\rho\sigma}, \quad \widetilde{\mathcal{E}}_{\phi} = \frac{\sqrt{-g}}{\sqrt{-\widetilde{g}}}\mathcal{E}_{\phi}, \tag{13}$$

where $\mathcal{E}_{\mu\nu} = (-g)^{-1/2}\,\delta S/\delta g^{\mu\nu}$, etc., are the field equations. This useful fact will be exploited in the next section.

3 Stealth Black Holes and Their Disformal and Conformal Transformations

As we saw, no-hair theorems establish that, under certain assumptions, a scalar-tensor theory admit no other BH solution than a GR metric (Schwarzschild or Kerr) with a trivial scalar field. Hence, finding a GR BH dressed with a non-trivial scalar field already represents a step forward. Writing the field equations as

[6] Considering the full beyond Horndeski Lagrangian, i.e. the pure Horndeski contribution (4) and the pure beyond Horndeski part (10), the degeneracy of the kinetic matrix imposes the relation $XG_{5X}F_4 = 3F_5\left(G_4 - 2XG_{4X}\right)$ [39] leaving us at the end with one additional degree of freedom equivalent to the disformal function D.

$$G_{\mu\nu} = T_{\mu\nu} \tag{14}$$

where $T_{\mu\nu}$ is the energy-momentum tensor associated to the scalar field, such solutions require $T_{\mu\nu} = 0$, so that the scalar field, although non-trivial, does not modify the GR field equations.[7] These solutions are therefore known as *stealth solutions*, as opposed to non-stealth solutions, whose metric is not a GR metric and which will be studied in later sections. For stealth solutions, the non-trivial scalar field still enters the perturbation theory and may therefore modify, for instance, the emission of GW. Furthermore, one may, using stealth solutions, construct non stealth images by using conformal/disformal transformations. This latter possibility may be very useful as the field equations for stationary black holes are extremely complex for analytic study. The current section first presents Schwarzschild and Kerr stealth solutions, before moving on to the generation of a *disformed Kerr metric* by a disformal transformation, as well as conformal cosmological versions.

3.1 Generic Stealth Schwarzschild in Horndeski Theories

The first asymptotically flat BH in higher-order scalar-tensor theories was constructed in [43]. The action is

$$S = \int \mathrm{d}^4x\sqrt{-g}\Big\{R + \beta G^{\mu\nu}\partial_\mu\phi\,\partial_\nu\phi\Big\}, \tag{15}$$

which belongs to the Horndeski class with a unique nonvanishing function, $G_4 = 1 + \beta X$, with β a coupling. The action is shift-symmetric, and the authors are looking for a static, spherically-symmetric solution,

$$\mathrm{d}s^2 = -h\,(r)\,\mathrm{d}t^2 + \frac{\mathrm{d}r^2}{f\,(r)} + r^2\mathrm{d}\Omega^2. \tag{16}$$

Under this setting, there exists a no-hair argument [44], which requires the additional assumption that $\phi = \phi\,(r)$. However, following [45], the authors of [43] realized that such a radial profile for the scalar field is not the most general compatible with the symmetries of the spacetime. Indeed, the action depends on the scalar field only via its derivatives, so do the field equations, therefore, a linear time dependence of the scalar is allowed,

$$\phi = qt + \psi\,(r)\,, \tag{17}$$

with q a constant. This form of the scalar is then found to support a stealth Schwarzschild solution,

---

[7] One can slightly modify the above relation to allow for de Sitter stealth solutions, i.e. Einstein metric solutions with a non trivial scalar field. Such solutions have self tuning properties for the cosmological constant, see for example [41, 42].

$$h(r) = f(r) = 1 - \frac{2M}{r}, \tag{18}$$

with the usual mass integration constant M, and the radial part of the scalar field reads

$$\psi'(r) = \pm q \frac{\sqrt{2Mr}}{r - 2M}. \tag{19}$$

The apparent logarithmic divergence of $\psi(r)$ at the Schwarzschild radius $r = 2M$ stems from the poor behaviour of the coordinates (t, r, θ, φ) at the horizon. Changing for horizon-crossing coordinates [15], (v, r, θ, φ) or (u, r, θ, φ), where $v = t + r^\star$ and $u = t - r^\star$ are the usual advanced and retarded time in terms of the tortoise coordinate $r^\star = \int \mathrm{d}r/\sqrt{h(r) f(r)}$, the Schwarzschild metric is well-behaved and the scalar field is regular at the future horizon, $\phi(v, r = 2M) = qv + \text{const.}$, if the $+$ sign is chosen in (19), or at the past horizon, $\phi(u, r = 2M) = qu + \text{const.}$, if the $-$ sign is chosen in (19). Note that the kinetic term of the solution is constant,

$$X = -\frac{1}{2}(\partial\phi)^2 = \frac{1}{2}\left(\frac{q^2}{h(r)} - f(r)\psi'(r)^2\right) = \frac{q^2}{2}. \tag{20}$$

This property was exploited in [46] to construct other solutions, or more precisely, to find more general forms of function G_4 supporting the existence of the stealth Schwarzschild solution.[8] Stealth solutions prove to be a good starting point for our solution quest, however, they have at least two shortcomings : first they are too much GR-like by definition, and second their construction imposes the absence of the scalar field kinetic term in the action which brings about perturbative problems. The former problem was shown to lead to non hyperbolic scalar field perturbations [47] akin to strong coupling. This is unlike the de Sitter stealth versions, which do include a kinetic term [48]. Finally, in [42] stealth solutions were presented in beyond Horndeski theories, while Refs. [49, 50] found conditions for the existence of stealth Schwarzschild solutions in DHOST theories, Ref. [49] studying in particular the non-shift symmetric case.

3.2 A Stealth Kerr Black Hole in a GR Like DHOST Theory

The authors of [51] found a way of generalising the previous construction to a stealth Kerr BH in DHOST theory. Note that this construction is not possible in Horndeski theory; in fact it turns out that a stealth stationary metric is a lot harder to obtain than

---

[8] Generalizations to stealth Schwarzschild-de-Sitter are presented in both [43, 46], in which case a bare cosmological constant Λ is added into the action, as well as a nontrivial kinetic term. The effective cosmological constant of the BH is then a combination of the bare one Λ, and of the G_2 and G_4 terms as evaluated at the constant value of X. The solutions have self tuning properties [41, 42].

a stealth static metric which appears quite generically in shift and parity symmetric Horndeski theories. We now use the usual notations of DHOST theory [33], with in particular a kinetic term $X \equiv (\partial\phi)^2$, which also turns out to be more clear for the following. Reference [51] considers the most general shift-symmetric quadratic DHOST theory with speed of GW equal to the speed of light [52],

$$S = \int d^4x \Big\{ f(X) R + P(X) + Q(X) \Box\phi + A_3(X) \phi^\mu \phi_{\mu\nu} \phi^\nu \Box\phi \\ + A_4(X) \phi^\mu \phi_{\mu\nu} \phi^{\nu\rho} \phi_\rho + A_5(X) \left(\phi^\mu \phi_{\mu\nu} \phi^\nu\right)^2 \Big\}, \quad (21)$$

with four free functions f, P, Q, A_3, while degeneracy of the kinetic matrix yields A_4 and A_5 [33],

$$A_4 = \frac{1}{8f}\left(48 f_X^2 - 8(f - X f_X) A_3 - X^2 A_3^2\right), \quad (22)$$

$$A_5 = \frac{A_3}{2f}(4 f_X + X A_3). \quad (23)$$

It turns out that only within this theory, which has graviton speed identical to GR, it is seemingly possible to obtain a stealth Kerr solution. Taking inspiration from the stealth Schwarzschild solution, the authors of [51] are looking for a Kerr solution with constant kinetic term $X = X_0 \equiv -q^2$. This greatly simplifies the DHOST field equations, since the A_4 and A_5 terms do not contribute for constant X, and the Kerr metric implies a vanishing Ricci tensor. It is then found that such a stealth Kerr solution is allowed provided the DHOST functions verify[9]

$$P(X_0) = P_X(X_0) = Q_X(X_0) = A_3(X_0) = 0. \quad (24)$$

It thus remains to solve the requirement of constant kinetic term, $X = X_0 = -q^2$ for the Kerr geometry. This can be linked to the mass-shell equation of a point-particle with mass q,

$$-q^2 = (\partial\phi)^2 = g^{\mu\nu} \partial_\mu \phi \, \partial_\nu \phi \equiv g^{\mu\nu} p_\mu p_\nu, \quad (25)$$

where p^μ is the four-momentum of the point-particle, $p^\mu = dx^\mu / d\lambda$, with λ an affine parameter along the geodesic followed by the point-particle. Accordingly, ϕ is a *Hamilton-Jacobi functional* for the geodesic trajectory, i.e., its exterior derivative is the momentum one-form of the geodesic, $d\phi = \partial_\mu \phi \, dx^\mu = p_\mu dx^\mu$. It is then only a matter of using Carter's integration of Kerr geodesics [53] to find the form of $p_\mu dx^\mu$ and therefore of ϕ. The final form of the regular scalar field, with $X = -q^2$, is[10]

---

[9] Reference [51] constructs more generally a stealth Kerr-dS solution, in which case conditions (24) are replaced by analogous ones involving the cosmological constant Λ.

[10] In the stealth Kerr-dS case, regularity of the scalar field at both the event and the cosmological horizons requires it to depend also on the colatitude angle θ, $\phi = \phi(t, r, \theta)$.

$$\phi = q\left(t \pm \int \frac{\sqrt{2Mr\left(r^2+a^2\right)}}{\Delta}\mathrm{d}r\right), \tag{26}$$

where as usual, M and a are the mass and angular momentum per unit mass of the Kerr spacetime, and $\Delta = r^2 + a^2 - 2Mr$. All this is written in the usual Boyer-Lindquist coordinates (t, r, θ, φ), where the Kerr metric is [15]

$$\begin{aligned} \mathrm{d}s^2 = & -\left(1-\frac{2Mr}{\Sigma}\right)\mathrm{d}t^2 + \frac{\Sigma}{\Delta}\mathrm{d}r^2 + \Sigma \mathrm{d}\theta^2 - \frac{4Mar\sin^2\theta}{\Sigma}\mathrm{d}t\mathrm{d}\varphi \\ & + \frac{\sin^2\theta}{\Sigma}\left[\left(r^2+a^2\right)^2 - a^2\Delta\sin^2\theta\right]\mathrm{d}\varphi^2, \end{aligned} \tag{27}$$

with $\Sigma = r^2 + a^2\cos^2\theta$. The expression (26) reduces to the stealth Schwarzschild scalar (17)–(19) for vanishing rotation, and just as in this case, regularity at the horizons $\Delta = 0$ can be verified by changing to horizon-crossing coordinates for the Kerr metric. Other interesting works include Ref. [54], which undertook a systematic study of the most general quadratic DHOST theory in order to determine conditions for it to admit the stealth Kerr solution.[11] The linear metric perturbations of the stealth Kerr solution were studied in [55] and they were found to have a modified Teukolsky form. Regarding the stability of stealth solutions in DHOST theories, odd-parity perturbations around static stealth solutions were discussed in [56], while Refs. [57, 58] extended the analysis to even-parity perturbations, stating that they were strongly coupled. In fact, this strong coupling problem also exists for the stealth Kerr solution [57] and is most probably due to the fact that the kinetic term is absent for this solution which entails the non hyperbolic character of the scalar perturbation equation. Discussions on the significance of strong coupling in stealth solutions and ways to circumvent the problem have been recently discussed in [59].

3.3 Disformal Kerr Metric

As we pointed out earlier, see discussion around Eq. (13), a conformal-disformal transformation (8) maps a seed solution $\left(g_{\mu\nu}, \phi\right)$ of an initial DHOST theory to another solution $\left(\tilde{g}_{\mu\nu}, \phi\right)$ for a new DHOST theory. Let us first consider the case of a purely disformal transformation with, $C(\phi, X) = 1$. For a start a disformal transformation depending solely on X, of a stealth Schwarzschild solution turns out to be a coordinate transformed stealth Schwarzschild solution with a rescaled mass parameter [48]. Therefore a disformal map in this case simply carries the same Ricci flat geometry from Horndeski to beyond Horndeski theory (essentially because X is a constant). The disformal transformation of the seed stealth Kerr solution turns out

[11] And, more generally, the Kerr-Newman-dS solution.

to be far more interesting and was undertaken in [60, 61] with a constant disformal factor $D(X) = \text{const.}$, since again X is constant. Note in passing that a dependence of D on the scalar field (26) would introduce an explicit time-dependence in the metric and thus spoil the spacetime symmetries. We will come back to that in the next section. The disformed Kerr metric is written as

$$\widetilde{g}_{\mu\nu} = g_{\mu\nu} - \frac{D}{q^2}\partial_\mu\phi\,\partial_\nu\phi, \tag{28}$$

with D a dimensionless constant, and $g_{\mu\nu}$ is the Kerr metric (27) with mass M and angular momentum per unit mass a. The DHOST theory admitting the disformal Kerr metric as a solution can be found in [61]. Its black hole properties were analysed in [60] which we now turn to. Eq. (28) produces the following line element,

$$\begin{aligned} \mathrm{d}\widetilde{s}^2 = & -\left(1 - \frac{2\widetilde{M}r}{\Sigma}\right)\mathrm{d}t^2 - \frac{4\sqrt{1+D}\widetilde{M}ar\sin^2\theta}{\Sigma}\mathrm{d}t\mathrm{d}\varphi + \Sigma\mathrm{d}\theta^2 \\ & + \frac{\sin^2\theta}{\Sigma}\left[\left(r^2+a^2\right)^2 - a^2\Delta\sin^2\theta\right]\mathrm{d}\varphi^2 - 2D\frac{\sqrt{2\widetilde{M}r\left(r^2+a^2\right)}}{\Delta}\mathrm{d}t\mathrm{d}r \\ & + \frac{\Sigma\Delta - 2D(1+D)\widetilde{M}r\left(r^2+a^2\right)}{\Delta^2}\mathrm{d}r^2, \end{aligned} \tag{29}$$

with $\widetilde{M} = M/(1+D)$, while Σ and Δ are as in the seed Kerr metric, in particular Δ still features the parameter M and not $\widetilde{M}$. $\Sigma = 0$ corresponds to the ring singularity, just like for Kerr [15]. Regarding the conserved charges, the disformed Kerr metric has mass $\widetilde{M}$, while its angular momentum per unit mass is $\widetilde{a} \equiv a\sqrt{1+D}$. The disformal transformation thus rescales the mass and angular momentum, $J = aM \to \tilde{J} = J/\sqrt{1+D}$. In fact, for vanishing rotation, $a = 0$, that is to say, if the seed metric is stealth Schwarzschild, then a change of coordinates removes the $\mathrm{d}t\mathrm{d}r$ term of (29) and the disformed metric is again Schwarzschild, with a mere rescaling of the mass, see also [40, 48]. The disformal transformation turns out to be non-trivial only if D and a are both non zero as (29) is a non Ricci flat metric. Indeed, if (29) were the Kerr metric in different coordinates, then it would also share its property of circularity, which for the Kerr metric manifests itself by the invariance under $(t, \varphi) \to (-t, -\varphi)$. The disformed Kerr metric as written in (29) does not possess this invariance because of the $\mathrm{d}t\mathrm{d}r$ term, and the authors of [60] prove that this is not due to the choice of coordinates, by referring to the more fundamental definition of circularity, which means integrability of the 2-submanifold orthogonal to the two Killing vector fields ∂_t and ∂_φ of an axisymmetric spacetime.

Non-circular spacetimes have a richer causal structure than circular ones [62]. For disformed Kerr just as for Kerr, there is an ergosphere or static limit, where ∂_t is null, and inside[12] which static observers, i.e. observers with constant r, θ and φ, cease to exist. Inside the ergosphere, the stationary limit is defined as the hypersurface inside

[12] 'Inside' here means 'for decreasing r'.

which stationary observers, that is, with constant r and θ, cease to exist. Equivalently, this stationary limit is a Killing horizon: there exists a linear combination $\partial_t + \Omega \partial_\varphi$ which is null, with Ω then defined as the angular velocity of the BH. For circular spacetimes like Kerr, this Killing horizon coincides with the event horizon (this is a particular case of the rigidity theorem, see e.g. [63]), however, when the spacetime is non-circular, the stationary limit is no more the event horizon, which must be looked for as a null hypersurface inside the Killing horizon.

Concerning the particular case of the disformed Kerr metric [60], the static limit or ergosphere lies at

$$r(\theta) = \widetilde{M} + \sqrt{\widetilde{M}^2 - a^2 \cos^2\theta}, \tag{30}$$

just as for Kerr up to the rescaling of the mass, while the stationary limit or Killing horizon is located at $r = R_0(\theta)$ such that

$$P(R_0(\theta), \theta) = 0 \tag{31}$$

where

$$P(r, \theta) \equiv r^2 + a^2 - 2\widetilde{M}r + \frac{2D\widetilde{M}ra^2 \sin^2\theta}{\Sigma}, \tag{32}$$

yielding a fourth-order algebraic equation for $R_0(\theta)$. Finally, the event horizon $r = R(\theta)$ is given by a nonlinear ordinary differential equation,

$$\left(\frac{\mathrm{d}R(\theta)}{\mathrm{d}\theta}\right)^2 + P(R(\theta), \theta) = 0. \tag{33}$$

Reference [60] investigated numerically the parameter space allowing solution to this equation, finding in particular that for a nonvanishing disformal parameter D, this candidate event horizon exists only for an angular momentum per unit mass $\widetilde{a} < \widetilde{a}_c < 1$, with the upper bound depending on the value of D, $\widetilde{a}_c = \widetilde{a}_c(D)$. Of course, the parameter D could be constrained by observations, since it deforms the shadow with respect to the one of a Kerr BH [64], or modifies the orbit of stars around the BH [65].

3.4 Conformal Kerr Metric: Cosmological Black Holes

In the previous subsection a purely disformal transformation was applied to obtain new stationary solutions distinct from the Kerr metric. There we took a trivial conformal factor $C = 1$ in order to retain asymptotically flat solutions. In this section we instead assume a non-trivial conformal factor depending explicitly on the scalar field, $C = C(\phi)$, while we set the disformal factor to zero, $D = 0$. The causal structure is preserved under such a transformation, i.e. starting with the seed solution of a theory with the speed of gravity equals unity, one obtains a theory with the same property. Having in mind an asymptotically flat seed metric with a non-trivial asymptotic dependence of the scalar field on time, such a method naturally gives rise to a time-dependent black hole metric. In particular, rotating growing black holes embedded in Friedmann-Lemaître-Robertson-Walker (FLRW) universe can be constructed [66]. Indeed, if the gradient of the scalar field ϕ is time-like, one may choose ϕ to be (proportional to) a time coordinate, so that the new metric $C(\phi)g_{\mu\nu}$ asymptotically has a form $C(\phi)g_{\mu\nu}^{(\text{flat})}$ with ϕ being the conformal time, i.e. FLRW metric in conformal time is recovered asymptotically.

A suitable case of the seed solution is a stealth Kerr black hole, which has been already described in detail above. It has the correct flat metric asymptotic with a non-trivial time-dependent scalar field. The metric that is obtained by conformal transformation is neither spherically symmetric nor stationary. To uncover the properties of such a conformal metric one defines the trapping (or apparent) horizons, which use the notion of 'expansion', related to the definition of null geodesic congruences [67]. While in the general case of non-zero rotation it is quite a non-trivial task to find the trapping horizons, the problem is greatly simplified when the rotation is set to zero. In this case, the metric for the resulting conformal solution has the Schwarzschild form, written in Painlevé-Gullstrand coordinates, multiplied by the conformal factor that depends on the Painlevé-Gullstrand time:

$$d\tilde{s}^2 = A^2(\tau)\left\{-\left(1-\frac{2M}{r}\right)d\tau^2 + 2\sqrt{\frac{2M}{r}}d\tau dr + dr^2 + r^2 d\Omega^2\right\}. \tag{34}$$

Clearly, at large r the metric becomes the FLRW line element in conformal time τ. The scale factor of the universe $A(\tau)$ is related to the choice of the conformal factor $C(\phi)$ and is defined by the DHOST theory, for which the conformal solution is written. The obtained metric is non-singular everywhere apart from the Big Bang singularity $A = 0$ and the curvature black hole singularity $r = 0$. The metric (34) has been discussed in the context of GR as an *ad-hoc* line element in [68], while its properties were recently studied [69] for particular choices of the scale factor $A(\tau)$.

The analysis of the conformal spacetime (34) reveals two trapping horizons,

$$\tau_+ = -\alpha r\left(1-\sqrt{\frac{2M}{r}}\right)^{-1}, \quad \tau_- = \alpha r\left(1+\sqrt{\frac{2M}{r}}\right)^{-1}, \tag{35}$$

where the '+' sign corresponds to the black hole horizon and '-' sign corresponds to the cosmological horizon, while α is the exponent in the scale factor, $A(\tau) \propto |\tau|^\alpha$. One should note that r is not a physical distance but a comoving coordinate, therefore a relevant measure for the physical radius is $R_{phys} = A(\tau)\,r$. The case of non-zero rotation can be treated similarly, however because of the absence of spherical symmetry the analysis is considerably more complicated, see the details in [66]. As a further generalisation of the method presented above, the stealth Kerr-de-Sitter solution of a DHOST theory can be considered as a seed metric. In this case, the scale factor of the asymptotically FLRW line element is composed from two parts, the original de Sitter expansion and the extra expansion due to the conformal factor $C(\phi)$, see details in Ref. [66].

4 Black Holes Beyond Ricci Flat Metrics

Stealth BHs turn out to be a good starting point and interesting as seed solutions, nevertheless their Ricci flat geometry makes them very much alike GR albeit the problems associated with the scalar perturbations. It is therefore paramount to move on and try to seek solutions which differ from the Schwarzschild/Kerr metrics predicted by GR.[13]

4.1 Parity and Shift-Symmetric Theories

As explained when introducing the Horndeski action (4), the field equations simplify under the assumption that all Horndeski functions G_2, G_3, G_4, G_5, and also the beyond Horndeski functions F_4 and F_5, depend only on the kinetic term X and not on the scalar field ϕ, because this implies invariance of the action under shifts $\phi \to \phi + \text{const.}$, and therefore a conserved Noether current J^μ, see Eq. (7). In addition, the terms G_3, G_5 and F_5 have long resisted the exact resolution of field equations: before 2020, no exact asymptotically flat solution with Newtonian asymptotics could be found when including those terms. We thus begin with a presentation of the non-stealth solutions for the following subclass of beyond Horndeski theories,

$$S = \int \mathrm{d}^4x \sqrt{-g} \Big\{ G_2(X) + G_4(X) R + G_{4X} \left[(\Box\phi)^2 - \phi_{\mu\nu}\phi^{\mu\nu} \right] + F_4(X)\, \epsilon^{\mu\nu\rho\sigma} \epsilon^{\alpha\beta\gamma}{}_{\sigma} \phi_\mu \phi_\alpha \phi_{\nu\beta} \phi_{\rho\gamma} \Big\} \tag{36}$$

which is not only shift-symmetric but also parity-symmetric, $\phi \to -\phi$, thanks to the absence of G_3, G_5 and F_5. It is important to note for the exact solutions presented here that scalar quantities associated to ϕ, like X or the norm $J_\mu J^\mu$ of the Noether current, are finite everywhere for $r \geq r_h$, where r_h is the horizon radius. This can be important concerning the regularity of the solutions as advocated in [44]. In fact, all solutions presented here have $J_\mu J^\mu = 0$ everywhere, in other words $J^r = 0$ and as such have one less integration constant associated to the scalar field. Moreover, they are all static and spherically-symmetric,

[13] Note that every asymptotically flat BH presented from now on admits a direct generalization to (a)dS asymptotics by simply introducing a cosmological constant Λ into the action.

$$\mathrm{d}s^2 = -h\left(r\right)\mathrm{d}t^2 + \frac{\mathrm{d}r^2}{f\left(r\right)} + r^2\mathrm{d}\Omega^2. \tag{37}$$

and this is also true for $\phi = \phi(r)$ although one could take a linear time dependence for the scalar.

Pure Horndeski Black Hole

The first such asymptotically flat BH solution to action (36) was found in [70]. Given the symmetries of the spacetime, looking for a scalar field $\phi = \phi\left(r\right)$, the only nonvanishing component of the current is J^r. The idea of [70] was to inspect carefully the form of J^r, in order to identify which Horndeski functionals enabled a non-trivial scalar field from the equation $J^r = 0$ *for spherical symmetry*. This results in the following pure Horndeski theory with two coupling constants η (dimensionless) and β (dimension length),

$$G_2 = \eta X, \quad G_4 = 1 + \beta\sqrt{-X}, \quad F_4 = 0, \tag{38}$$

where the G_2 term is a canonical kinetic term, and the constant 1 in G_4 corresponds to a pure Einstein-Hilbert term. The BH is homogeneous, $h = f$ in the metric (37), with potential

$$f\left(r\right) = 1 - \frac{2M}{r} - \frac{\beta^2}{2\eta r^2}, \tag{39}$$

while the scalar field reads

$$\phi'\left(r\right) = \pm\frac{\sqrt{2}\beta}{\eta r^2\sqrt{f\left(r\right)}}. \tag{40}$$

The previous is solution only if η and β have the same sign. In the metric, M is an integration constant and is the Arnowitt-Deser-Misner (ADM) mass of the spacetime [71], and also its thermodynamic mass [72]. The spacetime is clearly asymptotically Schwarzschild at leading order, but differs from the GR result for finite r. It presents a curvature singularity at $r = 0$. If $\eta > 0$, there is a unique horizon, while for $\eta < 0$, there are no horizons if $M < \left|\beta\right|/\sqrt{-2\eta}$ (naked singularity) and two horizons if $M > \left|\beta\right|/\sqrt{-2\eta}$. As regards the scalar field, it is real and finite for any $r \geq r_h$, while it becomes imaginary when $f\left(r\right) < 0$. Its kinetic term is finite apart from the central singularity,

$$X = -\frac{\beta^2}{\eta^2 r^4}, \tag{41}$$

and is negative, consistently with the $\sqrt{-X}$ appearing in G_4. Note that when we set $M = 0$ we do not have flat spacetime but rather a non trivial spacetime. The linear perturbations around this BH have been recently discussed in [73].

Beyond Horndeski Black Hole

More recently, the authors of [74] undertook a systematic study of the shift-symmetric beyond Horndeski field equations in spherical symmetry, rewriting them in a way facilitating their integration, in particular for the parity-symmetric case (36), which permits to identify theories admitting solutions. For instance, the following theory was obtained, displaying again a canonical kinetic term, but also a nonzero beyond Horndeski F_4 term,

$$G_2 = \frac{8\eta\beta^2}{\lambda^2}X, \quad G_4 = 1 + 4\eta\beta\left(\sqrt{-X} + \beta X\right), \quad F_4 = -\frac{\eta\beta^2}{X}, \tag{42}$$

with three couplings η (dimensionless), β, λ, with both β and λ positive and of dimension length. The metric solution is homogeneous, $h = f$ in (37), where

$$f(r) = 1 - \frac{2M}{r} + \eta\frac{\arctan(r/\lambda) - \pi/2}{r/\lambda}. \tag{43}$$

Again, M is an integration constant and the mass of the solution, which behaves asymptotically as

$$f(r) = 1 - \frac{2M}{r} - \frac{\eta\lambda^2}{r^2} + \mathcal{O}\left(\frac{1}{r^4}\right). \tag{44}$$

The scalar field supporting the solution is $\phi = \phi(r)$ where

$$\phi'(r) = \pm\frac{1}{\beta\sqrt{2f(r)}\left(1 + (r/\lambda)^2\right)}, \tag{45}$$

and is seen to be real and finite for any r greater than the horizon, $r \geq r_h$, and imaginary when $f(r) < 0$, just as for the previous case. The kinetic term is well-defined everywhere, including now at $r = 0$,

$$X = -\frac{\lambda^4}{4\beta^2\left(r^2 + \lambda^2\right)^2}. \tag{46}$$

Concerning the horizons, their presence or absence can be understood by looking at the behaviour of the metric near $r = 0$,

$$f(r) = 1 + \eta - \frac{2M + \pi\eta\lambda/2}{r} - \frac{\eta r^2}{3\lambda^2} + \mathcal{O}\left(r^4\right), \tag{47}$$

where appears a threshold mass $M_0 = -\pi\eta\lambda/4$. For $M > M_0$, there is a unique horizon (this is in particular the case for $\eta > 0$); for $M < M_0$, there are two horizons if M is not too small, and zero horizon (naked singularity) for small masses; while for the limit case $M = M_0$, there is zero or one horizon according to $\eta \gtrless -1$, and $f(r)$ does not diverge at $r = 0$ but the spacetime remains singular there because $f(0) \neq 1$.

Non-homogeneous Black Hole

Furthermore, the formalism developed in [74] makes it possible to construct non-homogeneous BHs, with $f \neq h$ in (37). This article presents the following example theory admitting such a solution,

$$G_2 = \frac{8\eta\beta^2}{\lambda^2}\frac{X}{1-\xi^2 X}, \quad G_4 = \frac{1 + 4\eta\beta\left(\sqrt{-X} + \beta X\right)}{1-\xi^2 X}, \tag{48}$$

$$F_4 = \frac{\xi^6 X^2 - \left(3\xi^2 + 4\eta\beta^2\right)\xi^2 X - 8\eta\beta\xi^2\sqrt{-X} - 4\eta\beta^2}{4X\left(1-\xi^2 X\right)^2}. \tag{49}$$

As compared to the previously considered theory (42), there is an additional coupling ξ of dimension length. Although the expression of F_4 is largely modified, those of G_2 and G_4 are only changed by a factor $1/\left(1-\xi^2 X\right)$, which is in fact directly linked to the 'inhomogeneity' of the black hole: the metric function $h(r)$ which multiplies $-dt^2$ is as before,

$$h(r) = 1 - \frac{2M}{r} + \eta\frac{\arctan(r/\lambda) - \pi/2}{r/\lambda}, \tag{50}$$

see Eq. (43), while the metric function $f(r)$ which divides dr^2 becomes

$$f(r) = \frac{h(r)}{\left(1-\xi^2 X\right)^2}. \tag{51}$$

The scalar field reads

$$\phi'(r) = \pm\frac{1}{\beta\sqrt{2f(r)}\left(1+(r/\lambda)^2\right)}, \tag{52}$$

which is the same expression in terms of $f(r)$ as above, Eq. (45), but $f(r)$ is now different. Hence, the kinetic term $X = -f(r)\phi'(r)^2/2$ is exactly identical to (46), with the behaviour $X = \mathcal{O}\left(1/r^4\right)$ at infinity, which renders the theory (48)–(49) indistinguishable from the previous one (42) in this limit, with a canonical kinetic term and Einstein-Hilbert term. The inhomogeneity (51) is thus asymptotically suppressed with a factor $1/r^4$, but grows significant as r becomes comparable to the coupling λ. Importantly, the negative sign of X ensures that the denominator in (51) is always greater than 1, so in particular never vanishes. In addition, X is bounded everywhere, see (46), therefore, $f(r)$ vanishes only when $h(r)$ does, and because $h(r)$ has the same expression as the previous metric potential (43), the inhomogeneous spacetime has the same horizons as the previous homogeneous spacetime, with again a singularity at $r = 0$.

4.2 Non-parity Symmetric Theories

Construction of the Four-Dimensional Einstein-Gauss-Bonnet Gravity

Exact solutions in theories without parity symmetry remained unknown for a long time. This was quite frustrating as interesting numerical solutions had been known for a long time [75] with interesting beyond GR phenomenology. Also for certain parity breaking theories there were found interesting physical effects such as BH scalarisation [76, 77] (for an updated review see [78]). This changed in 2020, in relation with the quest for a four-dimensional Einstein-Gauss-Bonnet theory of gravity (see [79] for a review). The present section presents a concise overview of this search, while the reader mostly interested in the resulting theory and its BH solutions can skip directly to Eq. (67).

In order to understand the relevance of the Gauss-Bonnet curvature invariant, one must have in mind the Lovelock theorem [9] which in essence states: in D-dimensional spacetime, the most general action in terms of the metric only and yielding second-order conserved field equations is

$$S = \int \mathrm{d}^D x \sum_{k=0}^{\lfloor \frac{D-1}{2} \rfloor} \alpha_k \mathcal{R}^{(k)}, \tag{53}$$

where the upper limit of the sum is the floor of $(D-1)/2$. The α_ks are coupling constants, while $\mathcal{R}^{(k)}$ is a curvature invariant of order k, given by

$$\mathcal{R}^{(k)} = (2k)!\, \delta^{\mu_1}_{[\alpha_1} \delta^{\nu_1}_{\beta_1} \cdots \delta^{\mu_k}_{\alpha_k} \delta^{\nu_k}_{\beta_k]} \prod_{i=1}^{k} R^{\alpha_i \beta_i}{}_{\mu_i \nu_i}, \tag{54}$$

where the brackets denote the usual antisymmetrization. One can normalize the couplings so that $\alpha_1 = 1$, and call $\alpha_0 = -2\Lambda$, then the sum up to $k = 2$ reads

$$\sum_{k=0}^{2} \alpha_k \mathcal{R}^{(k)} = -2\Lambda + R + \alpha_2 \mathcal{G}, \tag{55}$$

where $\mathcal{G} \equiv \mathcal{R}^{(2)} = R^2 - 4R_{\mu\nu}R^{\mu\nu} + R_{\mu\nu\rho\sigma}R^{\mu\nu\rho\sigma}$ is known as the Gauss-Bonnet invariant. The Gauss-Bonnet term appears as the first higher-order correction to the Einstein-Hilbert action, and (55) is usually called Einstein-Gauss-Bonnet gravity. Nevertheless, in $D = 4$ dimensions, the sum in (53) stops at $k = 1$ and one recovers GR, because for any $k \geq 3$, $\mathcal{R}^{(k)}$ vanishes identically, while $\mathcal{R}^{(2)} = \mathcal{G}$ does not vanish in general, but is a boundary term which does not contribute to the field equations by virtue of the Chern-Gauss-Bonnet theorem.[14] Only from $D = 5$ dimensions and higher does the Gauss-Bonnet term give a non-trivial contribution to Lovelock's action. This is precisely why Lovelock's theorem implies, as argued in the introduction to this chapter, that any modification of gravity in four dimensions must include other degrees of freedom than the two degrees of freedom of the metric. Importantly however, in the framework of $D = 4$ scalar-tensor theories, the Gauss-Bonnet term gives non-trivial contributions when non-minimally coupled to the scalar field ϕ, in actions such as

$$\int \mathrm{d}^4x\sqrt{-g}\Big\{R + f(\phi)\,\mathcal{G} + \cdots\Big\} \tag{56}$$

where $f(\phi)$ is a function of the scalar field.[15] In addition to Lovelock's mathematical result, the Gauss-Bonnet invariant is physically motivated by many considerations from string theory, where for instance, 1-loop corrected heterotic string effective action presents terms as $\mathrm{e}^{\phi}\mathcal{G}$ where ϕ is the dilaton [84].

In this context, one can wonder what is the natural non-trivial generalization of Einstein-Gauss-Bonnet gravity in four dimensions. A first attempt was made by [85] through a diagonal Kaluza-Klein dimensional reduction [86]. This consists in starting from the D-dimensional Einstein-Gauss-Bonnet gravity,

$$\int \mathrm{d}^Dx\sqrt{-g_{(D)}}\Big\{R_{(D)} + \hat{\alpha}\mathcal{G}_{(D)}\Big\}, \tag{57}$$

with coupling $\hat{\alpha}$ and the subscript (D) stands for D-dimensional quantities. Then, the D-dimensional metric $\mathrm{d}l_{(D)}^2$ splits into a four-dimensional *spacetime* $\mathrm{d}s_{(4)}^2$ and an $n = D - 4$-dimensional *internal space* $\mathrm{d}\widetilde{s}_{(n)}^2$,

$$\mathrm{d}l_{(D)}^2 = \mathrm{d}s_{(4)}^2 + \mathrm{e}^{-2\phi}\mathrm{d}\widetilde{s}_{(n)}^2, \tag{58}$$

and the scalar field ϕ appears in the conformal factor of the internal space and depends only on the spacetime coordinates. The authors of [85] then compute that, up to integration by parts,

---

[14] The Chern-Gauss-Bonnet theorem [80] applied to a four-dimensional manifold $\mathcal{M}$ relates its Euler characteristic $\chi(\mathcal{M})$ to the integral of the Gauss-Bonnet curvature, $\chi(\mathcal{M}) = \int_{\mathcal{M}}\mathcal{G}\,\mathrm{d}\mu/\left(8\pi^2\right)$, where $\mathrm{d}\mu$ is the volume form.

[15] As regards BH physics, such scalar-tensor Gauss-Bonnet models were known since 1995 to allow for hairy BH solutions [75], giving then rise to numerous other static [81] or spinning [82] solutions, all requiring numerical analysis or power series expansions [83]. These theories also lead to spontaneous scalarization, again for static [76] or rotating [77] BHs.

$$\int \mathrm{d}^D x \sqrt{-g_{(D)}} R_{(D)} = T \int \mathrm{d}^4 x \sqrt{-g}\, \mathrm{e}^{-n\phi} \Big\{ R + \widetilde{R} \mathrm{e}^{2\phi} + n\,(n-1)\,(\partial\phi)^2 \Big\}, \tag{59}$$

and

$$\begin{aligned} \int \mathrm{d}^D x \sqrt{-g_{(D)}} \mathcal{G}_{(D)} = T \int \mathrm{d}^4 x \sqrt{-g}\, \mathrm{e}^{-n\phi} \Big\{ & \mathcal{G} + \widetilde{\mathcal{G}} \mathrm{e}^{4\phi} \\ & + 2\widetilde{R} \mathrm{e}^{2\phi} \Big[R + (n-2)\,(n-3)\,(\partial\phi)^2 \Big] - 4n\,(n-1)\, G^{\mu\nu} \phi_\mu \phi_\nu \\ & + 2n\,(n-1)\,(n-2)\,\Box\phi\,(\partial\phi)^2 - n\,(n-1)^2\,(n-2)\,(\partial\phi)^4 \Big\}, \end{aligned} \tag{60}$$

where spacetime quantities of $\mathrm{d}s^2_{(4)}$ are as usual $R, \mathcal{G}$, while quantities from the internal space $\mathrm{d}\widetilde{s}^2_{(n)}$ are with tildes, $\widetilde{R}, \widetilde{\mathcal{G}}$. One can forget about the global proportionality factor T, which is the volume of the internal space, $T = \int \mathrm{d}^n x \sqrt{\widetilde{g}}$. The resulting four-dimensional action is therefore

$$\begin{aligned} S = \int \mathrm{d}^4 x \sqrt{-g}\, \mathrm{e}^{-n\phi} \Big\{ & R + \widetilde{R} \mathrm{e}^{2\phi} + n\,(n-1)\,(\partial\phi)^2 \\ & + \hat{\alpha} \Big[\mathcal{G} + \widetilde{\mathcal{G}} \mathrm{e}^{4\phi} + 2\widetilde{R} \mathrm{e}^{2\phi} \Big(R + (n-2)\,(n-3)\,(\partial\phi)^2 \Big) - 4n\,(n-1)\, G^{\mu\nu} \phi_\mu \phi_\nu \\ & + 2n\,(n-1)\,(n-2)\,\Box\phi\,(\partial\phi)^2 - n\,(n-1)^2\,(n-2)\,(\partial\phi)^4 \Big] \Big\}, \end{aligned} \tag{61}$$

and in this action, one can now analytically continue n to be a real parameter. The authors of [85] were interested in getting a four-dimensional solution descending directly from a D-dimensional solution of (57), and found that a sufficient condition for this was to take the internal space $\mathrm{d}\widetilde{s}^2_{(n)}$ to be a product of $n/2$ two-spheres of same radius ρ [87], yielding

$$\widetilde{R} = n/\rho^2, \quad \widetilde{\mathcal{G}} = n\,(n-2)\,/\rho^4. \tag{62}$$

This procedure led them to scalar-tensor solutions to action (61) for any n, nevertheless, the obtained metric, of the spherically-symmetric form (37) with $h = f$, behaves asymptotically as $f\,(r) \approx 1/\,(n+1)\left(1 - 2M/r^{n+1}\right)$, which is not a Newtonian behaviour unless $n = 0$, but in this case (61) reduces to pure GR. The way out of this deadlock was first described[16] by [90], which proposed the following procedure: take $n \to 0$ and the coupling $\hat{\alpha} \to \infty$, while maintaining the product $n\hat{\alpha} = \text{const.} \equiv \alpha$, with α a new coupling. In [90], the authors apply this procedure to a maximally-symmetric internal space of curvature γ, $\widetilde{R}_{abcd} = \gamma\,(g_{ac}g_{bd} - g_{ad}g_{bc})$, implying

$$\widetilde{R} = \gamma n\,(n-1)\,, \quad \widetilde{\mathcal{G}} = \gamma^2 n\,(n-1)\,(n-2)\,(n-3)\,, \tag{63}$$

[16] To be precise the renewed interest for constructing a four-dimensional Einstein-Gauss-Bonnet gravity originates from the adventurous and interesting attempt of [88], despite its shortcomings [89].

but their prescription in fact applies to any internal space such that $\widetilde{R} \propto n$ and $\widetilde{\mathcal{G}} \propto n$, like (62), because these terms, multiplied by $\hat{\alpha}$ in (61), then possess a regular limit. The only term of (61) which remains ambiguous in this limit is $e^{-n\phi}\hat{\alpha}\mathcal{G}$, and it is schematically regularized by expanding the exponential as

$$e^{-n\phi}\hat{\alpha}\mathcal{G} = \underbrace{\hat{\alpha}\mathcal{G}}_{\text{BT}} - n\hat{\alpha}\phi\mathcal{G} + \mathcal{O}\left(n^2\right) \underset{n\to 0}{\longrightarrow} -\alpha\phi\mathcal{G}, \tag{64}$$

where in the intermediate step, $\hat{\alpha}\mathcal{G}$ drops out since it is a boundary term (BT) in four dimensions. Very generally, the procedure of [90] can thus be applied for any internal space for which the following regularized curvature invariants are well-defined,

$$\widetilde{R}_{\text{reg}} = \lim_{n\to 0}\frac{\widetilde{R}}{n}, \quad \widetilde{\mathcal{G}}_{\text{reg}} = \lim_{n\to 0}\frac{\widetilde{\mathcal{G}}}{n}, \tag{65}$$

bringing about the regularized Kaluza-Klein action,

$$S = \int d^4x\sqrt{-g}\Big\{R + \alpha\Big[-\phi\mathcal{G} + \widetilde{\mathcal{G}}_{\text{reg}}e^{4\phi} + 2\widetilde{R}_{\text{reg}}e^{2\phi}\left(R + 6\left(\partial\phi\right)^2\right) + 4G^{\mu\nu}\phi_\mu\phi_\nu + 4\Box\phi\left(\partial\phi\right)^2 + 2\left(\partial\phi\right)^4\Big]\Big\}. \tag{66}$$

Shortly after the publication of [90], two other articles, Refs. [91, 92], came in almost simultaneously with another proposal for regularizing Einstein-Gauss-Bonnet gravity in four dimensions. In this case, the Gauss-Bonnet piece in four dimensions is defined by first considering the D-dimensional piece $\int d^Dx\sqrt{-g}\hat{\alpha}\mathcal{G}$, second, substracting the identical action but for a conformally related metric $\widetilde{g}_{\mu\nu} = e^{2\phi}g_{\mu\nu}$, then redefining the coupling $\hat{\alpha} = \alpha/\left(D-4\right)$ and finally computing the limit when $D \to 4$. The obtained action is nothing but (66) with $\widetilde{R}_{\text{reg}} = \widetilde{\mathcal{G}}_{\text{reg}} = 0$, which can thus be seen as a Kaluza-Klein along a flat internal space.

Last but not least, another complementary point of view leads to the regularized Kaluza-Klein action (66): in [93], the author is looking for the most general four-dimensional scalar-tensor action with second order field equations and such that the scalar field equation of motion has conformal symmetry, that is to say, is invariant under $g_{\mu\nu} \to e^{2\sigma}g_{\mu\nu}$, $\phi \to \phi - \sigma$, with σ an arbitrary function on spacetime. Requiring such a symmetry is not so strange at it might first seem. Indeed, the BBMB action, briefly presented in the introduction, see (2), is usually said to enjoy such a conformal invariance,[17] but strictly speaking, this holds true only for the scalar-tensor part $\sqrt{-g}\left(\left(\partial\phi\right)^2/2 + R\phi^2/12\right)$, while the pure Einstein-Hilbert term breaks this invariance. Hence, in the BBMB case, the conformal invariance holds only at the level

[17] More precisely, with the notations of (2), the invariance is under $g_{\mu\nu} \to \Omega^2 g_{\mu\nu}$, $\phi \to \Omega^{-1}\phi$, which is equivalent upon redefining the scalar field $\phi \to e^{\phi}$.

of the scalar field equation of motion, and [93] wonders what is the most general scalar-tensor action with this property.[18] The result is

$$S = \int d^4x\sqrt{-g}\Big\{R - 2\lambda e^{4\phi} - \beta e^{2\phi}\Big[R + 6\left(\partial\phi\right)^2\Big] + \alpha\Big[-\phi\mathcal{G} + 4G^{\mu\nu}\phi_\mu\phi_\nu + 4\Box\phi\left(\partial\phi\right)^2 + 2\left(\partial\phi\right)^4\Big]\Big\}, \tag{67}$$

with three independent couplings λ, β and α. This is the same action as (66) with

$$2\lambda = -\alpha\widetilde{\mathcal{G}}_{\text{reg}}, \quad \beta = -2\alpha\widetilde{R}_{\text{reg}}. \tag{68}$$

Due to its *generalized conformal invariance*, the field equations of action (67) combine in the following way,

$$2g^{\mu\nu}\frac{\delta S}{\delta g^{\mu\nu}} + \frac{\delta S}{\delta\phi} = R + \frac{\alpha}{2}\mathcal{G}. \tag{69}$$

Hence, any scalar-tensor solution to the field equations of (67) must satisfy the purely geometric equation $R + \alpha\mathcal{G}/2 = 0$. This reproduces exactly the trace of the field equations for the purely metric Einstein-Gauss-Bonnet gravity in D dimensions, $\int d^Dx\sqrt{-g}\Big\{R + \alpha\mathcal{G}\Big\}$. Given the previously detailed construction and properties of action (67), it can be called the four-dimensional Einstein-Gauss-Bonnet gravity. As advertised, it is a non-parity symmetric Horndeski theory (4), with functions

$$\begin{aligned} G_2 =& -2\lambda e^{4\phi} + 12\beta e^{2\phi}X + 8\alpha X^2, \quad G_3 = 8\alpha X, \\ G_4 =& \, 1 - \beta e^{2\phi} + 4\alpha X, \quad G_5 = 4\alpha\log|X|. \end{aligned} \tag{70}$$

The theory acquires shift-symmetry when $\lambda = \beta = 0$, the $\phi\mathcal{G}$ term being shift-symmetric because $\mathcal{G}$ is a boundary term. Note also that the $e^{2\phi}X$ term in G_2 can be transformed into a canonical kinetic term for the scalar field $\Phi \equiv e^\phi$, but we will continue working with ϕ. This is important as scalar field perturbations can be affected by the absence of the lowest order term in the action. Let us now present the various exact, asymptotically flat BH solutions to action (67), of the static, spherically symmetric, homogeneous form,

$$ds^2 = -f(r)\,dt^2 + \frac{dr^2}{f(r)} + r^2d\Omega^2. \tag{71}$$

[18] And, as already said, with second order field equations, otherwise, most general theories are obtained, see [94].

We will classify these solutions according to the *geometric characteristics of the internal space* as it appears in the Kaluza-Klein picture (66). Importantly, they all correspond to different relative values of the couplings λ, β and α, thus they are solutions to different theories.

Maximally-Symmetric Internal Space

The first solution was obtained by [90], following their Kaluza-Klein regularization procedure along a maximally-symmetric internal space (63). Using (65) and (68), this corresponds to $\lambda = 3\beta^2/(4\alpha)$. The metric is

$$f(r) = 1 + \frac{r^2}{2\alpha}\left(1 \pm \sqrt{1 + \frac{8\alpha M}{r^3}}\right). \tag{72}$$

Very interestingly, this profile can be seen as the continuation to $D \to 4$ of the spherically-symmetric Boulware-Deser solution of pure metric Einstein-Gauss-Bonnet gravity in $D \geq 5$ [95] (see also [88]). The $\pm$ in front of the square root comes from the quadratic order of the field equations and appears for all solutions to action (67), but only the $-$ branch is asymptotically flat and reduces to GR in the limit of vanishing coupling α, therefore, the $+$ branch is always omitted from now on. The integration constant M is the ADM and thermodynamic mass of the spacetime, which behaves asymptotically as

$$f(r) = 1 - \frac{2M}{r} + \frac{4\alpha M^2}{r^4} + \mathcal{O}\left(\frac{1}{r^7}\right). \tag{73}$$

The scalar field is radial and reads[19]

$$\phi(r) = \log\left(\frac{\sqrt{-2\alpha/\beta}}{r}\right) - \log\left(\sigma\left(c \pm \int \frac{\mathrm{d}r}{r\sqrt{|f(r)|}}\right)\right), \tag{74}$$

where c is an arbitrary integration constant and σ is a function whose expression depends on the sign of $f(r)$: if $f(r) \geq 0$, $\sigma = \cosh$, while if $f(r) < 0$, $\sigma = |\cos|$. Therefore the scalar field is well-defined when $f(r) \geq 0$ (in particular outside the event horizon), and even when $f(r) < 0$ apart on a set of null measure. The horizons of (72) depend on the sign of α. If $\alpha > 0$, there are two possible horizons at

$$r_\pm = M \pm \sqrt{M^2 - \alpha}, \tag{75}$$

[19] The divergence of $\phi(r)$ as $r \to \infty$ is irrelevant as it can be eliminated by field redefinition. Indeed the redefined scalar $\Phi = \mathrm{e}^\phi$ already mentioned between (70) and (71), vanishes as $r \to \infty$.

in other words, the spacetime has no horizons if $M < \sqrt{\alpha}$ and has two horizons otherwise. For $\alpha < 0$, the square root in (72) becomes ill-defined before reaching r_+ if the mass is too small (naked singularity), while if the mass is sufficiently large, there is indeed a horizon at r_+ but no inner horizon since the square root becomes ill-defined. Reference [96] proved that the expression (75) for the horizon would strongly constrain the negative values of α if there were a Birkhoff-like theorem for the considered theory. In this case, an atomic nucleus of radius R would produce the gravitational field (72), and is not a BH so verifies $r_+ < R$, yielding $-10^{-30}\,\mathrm{m}^2 < \alpha < 0$. When $\alpha > 0$, the metric is well-defined up to $r = 0$, where it behaves as

$$f(r) = 1 - \sqrt{\frac{2Mr}{\alpha}} + \frac{r^2}{2\alpha} + \mathcal{O}\left(r^{7/2}\right). \tag{76}$$

Although $f(r)$ is well-defined at $r = 0$, the curvature invariants diverge there (they would not if $f(r) = 1 + \mathcal{O}\left(r^2\right)$, which is the case only for $M = 0$ but this reduces to flat spacetime), so the spacetime is singular at $r = 0$.

Flat Internal Space

The second solution was presented by [92], after their regularization which corresponds to a flat internal space and therefore to $\lambda = \beta = 0$, see (68). The metric potential $f(r)$ is exactly the same as (72), while the radial scalar field is now given by

$$\phi(r) = \int \frac{\pm 1 - \sqrt{f(r)}}{r\sqrt{f(r)}} \mathrm{d}r \tag{77}$$

up to a global additive constant since this particular theory is shift-symmetric. The scalar field is well-defined for $r \geq r_+$ but becomes imaginary when $f(r) < 0$. The kinetic term itself, whose regularity is of interest since the theory is shift-symmetric, is imaginary for $f(r) < 0$,

$$X = -\frac{\left(\pm 1 - \sqrt{f(r)}\right)^2}{2r^2}. \tag{78}$$

This unpleasant feature was cured in [96], by adding a linear time dependence to the scalar field. It now reads

$$\phi = qt + \int \frac{\pm\sqrt{q^2r^2 + f(r)} - f(r)}{rf(r)} \mathrm{d}r, \tag{79}$$

while the metric remains unchanged and does not feel the influence of the scalar hair q, which is an arbitrary integration constant. As explained in the section regarding stealth solutions, the regularity of the scalar field at the horizon r_+ is manifest in horizon-crossing coordinates, since $\phi(v, r = r_+) = qv + \text{const.}$ or $\phi(u, r = r_+) =$

$qu +$ const. depending on the sign of q and the choice of $\pm$ in (79), where $v = t + r^\star$, $u = t - r^\star$ and $r^\star = \int \mathrm{d}r/f(r)$. Moreover, since $f(r)$ as given by (72) is bounded from below (with a minimum depending on the mass M), there always exists a range of the integration constant q such that the scalar field is well-defined in the whole spacetime, and so is the kinetic term,

$$X = \frac{\pm 2\sqrt{q^2 r^2 + f(r)} - 1 - f(r)}{2r^2}, \tag{80}$$

apart from the central singularity at $r = 0$. Note that X here is not constant unlike the stealth solutions we studied before. As a result disformal transformations of this solution can lead, in certain cases, to different spacetimes, such as traversable regular wormhole solutions [97]. This theory also involves interesting neutron star solutions with quite distinct properties from GR such as an absence of mass gap in between neutron stars and black holes [96].

Internal Space as a Product of Two Spheres

The third solution was described by [93], and in fact corresponds to the Kaluza-Klein regularization for an internal space which is a product of $n/2$ two-spheres of identical radius, see (62). Equations (65) and (68) imply that $\lambda = \beta^2/(4\alpha)$, and the metric is

$$f(r) = 1 + \frac{r^2}{2\alpha}\left(1 - \sqrt{1 + \frac{8\alpha M}{r^3} + \frac{8\alpha^2}{r^4}}\right), \tag{81}$$

M being again the mass. The asymptotic departure from Schwarzschild gets larger than in (73), since now,

$$f(r) = 1 - \frac{2M}{r} - \frac{2\alpha}{r^2} + \mathcal{O}\left(\frac{1}{r^4}\right). \tag{82}$$

In fact, this solution directly descends from a higher-dimensional solution [87] through the diagonal Kaluza-Klein reduction (58) [85], and the scalar field, whose exponential $\mathrm{e}^{-2\phi}$ multiplies the two-spheres of the internal space, has consistently a very simple expression,[20]

$$\phi(r) = \log\left(\frac{\sqrt{-2\alpha/\beta}}{r}\right). \tag{83}$$

[20] Footnote 19 explains how to get rid of the divergence of $\phi(r)$ as $r \to \infty$.

The candidate event horizon is now located at

$$r_+ = M + \sqrt{M^2 + \alpha}. \tag{84}$$

For $\alpha > 0$, there is always a unique horizon at r_+, while for $\alpha < 0$, the square root might become ill-defined: it turns out that there is no horizon for small masses, and a unique horizon at r_+ for sufficiently large masses. This was studied in detail in [98]. Also, the same kind of Birkhoff conjecture as explained between (75) and (76) now severely constrains the positive values of α for the present theory, $0 < \alpha < 10^{-30}\,\mathrm{m}^2$. When the metric reaches $r = 0$ with no problems in the square root, it behaves there as

$$f(r) = 1 - \operatorname{sgn}(\alpha)\sqrt{2} - \frac{Mr}{\sqrt{2}\,|\alpha|} + \mathcal{O}\left(r^2\right). \tag{85}$$

Again, $f(r)$ is well-defined at $r = 0$, but not sufficiently regular, so there is a curvature singularity at $r = 0$. A major difference with the previous spacetime (72) is that (81) does not reduce to flat spacetime when $M = 0$, but is rather a naked singularity at $r = 0$. This is due to the $8\alpha^2/r^4$ term in (81), which ultimately comes from the higher-dimensional origin of the solution with a horizon of non-trivial topology (product of two-spheres) [85, 87]. Note that disformal transformations of such solutions can lead to wormhole solutions [98].

As a summary, the four-dimensional Einstein-Gauss-Bonnet gravity (67) admits three distinct scalar-tensor BH solutions, for three distinct theories with respectively $\lambda = 3\beta^2/(4\alpha)$, $\lambda = \beta = 0$ and $\lambda = \beta^2/(4\alpha)$. Reference [93] in fact showed that these are the only ones[21] with the homogeneous[22] ansatz (71). Although each of the obtained theory can be mapped to the regularized Kaluza-Klein action (66) with different internal space (respectively maximally-symmetric, flat, and product of two-spheres), as regards the solutions, only the product of two-spheres case directly descends from a solution to Einstein-Gauss-Bonnet gravity in D dimensions through the diagonal Kaluza-Klein reduction (58), while a hypothetical Kaluza-Klein origin for the other two solutions remains unclear. Indeed, their metric function (72) generalizes the higher-dimensional Boulware-Deser solution [95], but their scalar field (74) or (77) does not fit into the diagonal Kaluza-Klein reduction picture (58).

In a nutshell, the search for a four-dimensional Einstein-Gauss-Bonnet gravity provided as a byproduct quite involved Horndeski theories, without parity nor shift symmetry in general and which, quite remarkably, admit exact BH solutions. Nevertheless, the formulation (67) of [93] shows that the resulting theory still admits a kind of symmetry, the generalized conformal invariance. The next paragraph pro-

---

[21] More precisely, Ref. [93] also finds a solution with constant scalar field, but we do not mention it because the field equations become partly degenerate at this constant value (strong coupling). This constant scalar even formally allows for a non-perturbative rotating solution [99], but degeneracy is manifest since this solution is parameterized by two arbitrary functions of θ which are not fixed by the field equations.

[22] Furthermore, for the shift-symmetric case $\lambda = \beta = 0$, unicity of the static BH was proven [100], in other words, non-homogeneous BHs do not exist in this case.

vides an example of a Horndeski theory generalizing (67) with terms breaking this generalized conformal invariance, and still admitting exact solutions.

A Horndeski Theory Without Any Symmetry

The authors of [101] constructed this theory by first considering an action similar to (67), but with generalized potentials in front of each term (for instance $\phi\mathcal{G} \to V(\phi)\,\mathcal{G}$, etc.), and then imposing that the resulting theory allow for a logarithmic scalar field similar to (83). The obtained action reads

$$\begin{aligned} S = \int \mathrm{d}^4x\sqrt{-g}\Big\{&R - 2\lambda_4 \mathrm{e}^{4\phi} - 2\lambda_5 \mathrm{e}^{5\phi} - \beta_4 \mathrm{e}^{2\phi}\left(R + 6\,(\partial\phi)^2\right) \\ &- \beta_5 \mathrm{e}^{3\phi}\left(R + 12\,(\partial\phi)^2\right) - \alpha_4\left(\phi\mathcal{G} - 4G^{\mu\nu}\phi_\mu\phi_\nu - 4\Box\phi\,(\partial\phi)^2 - 2\,(\partial\phi)^4\right) \\ &- \alpha_5 \mathrm{e}^{\phi}\left(\mathcal{G} - 8G^{\mu\nu}\phi_\mu\phi_\nu - 12\Box\phi\,(\partial\phi)^2 - 12\,(\partial\phi)^4\right)\Big\}, \end{aligned} \tag{86}$$

with six coupling constants λ_4, λ_5, β_4, β_5, α_4 and α_5. The part with λ_4, β_4 and α_4 is identical to the four-dimensional Einstein-Gauss-Bonnet gravity (67) with the generalized conformal invariance. On the other hand, the couplings with λ_5, β_5 and α_5 would correspond to a conformally-invariant scalar field in *five* dimensions [102]: the five-dimensional action

$$\begin{aligned} \int \mathrm{d}^5x\sqrt{-g}\Big\{&2\lambda_5 \mathrm{e}^{5\phi} + \beta_5 \mathrm{e}^{3\phi}\left(R + 12\,(\partial\phi)^2\right) \\ &+ \alpha_5 \mathrm{e}^{\phi}\left(\mathcal{G} - 8G^{\mu\nu}\phi_\mu\phi_\nu - 12\Box\phi\,(\partial\phi)^2 - 12\,(\partial\phi)^4\right)\Big\} \end{aligned} \tag{87}$$

is invariant under $g_{\mu\nu} \to \mathrm{e}^{2\sigma} g_{\mu\nu}$, $\phi \to \phi - \sigma$. Of course, this invariance does not hold in the four-dimensional action (86). It is hard to understand why a Lagrangian with interesting properties in five dimensions would fit in a four-dimensional action and give exact solutions (which we discuss in a moment). A possible interpretation, which was not given in [101], emerges by looking at the Kaluza-Klein reduction of the Gauss-Bonnet action into four dimensions, Eq. (60). In this expression, n initially corresponds to the dimension of the internal space, but is analytically continued to the real axis, and upon taking $n = -1$, the obtained terms are exactly the λ_5, β_5 and α_5 terms of (86), provided the curvature invariants of the internal space are related to the couplings as

$$2\lambda_5 = -\alpha_5\widetilde{\mathcal{G}}, \quad \beta_5 = -2\alpha_5\widetilde{R}. \tag{88}$$

This interpretation at least shows how this five-dimensional conformal scalar field can appear in four dimensions, but the negative dimension $n = -1$ of the internal space is mysterious, and it remains unclear why this dimension is special and gives exact BH solutions (which is not the case for other negative values of n). Action (86) belongs to the Horndeski class with complicated functions G_k, $2 \leq k \leq 5$, which can be found in [101]. There exists two solutions of the static, spherically-symmetric form (71), corresponding to different values of the couplings, and therefore to different theories. The first solution requires

$$\lambda_4 = \frac{\beta_4^2}{4\alpha_4}, \quad \lambda_5 = \frac{9\beta_5^2}{20\alpha_5}, \quad \frac{\beta_5}{\beta_4} = \frac{2\alpha_5}{3\alpha_4} \tag{89}$$

and reads[23]

$$f(r) = 1 + \frac{2\alpha_5\eta}{3\alpha_4 r} + \frac{r^2}{2\alpha_4}\left[1 - \sqrt{\left(1+\frac{4\alpha_5\eta}{3r^3}\right)^2 + 4\alpha_4\left(\frac{2M}{r^3}+\frac{2\alpha_4}{r^4}+\frac{8\alpha_5\eta}{5r^5}\right)}\right],$$
$$\phi = \ln\left(\frac{\eta}{r}\right), \quad \eta \equiv \sqrt{\frac{-2\alpha_4}{\beta_4}}. \tag{90}$$

This spacetime, with mass M, is a generalization of the above one (81), with deviation parameterized by the product $\alpha_5\eta$ and of order $\mathcal{O}\left(1/r^3\right)$ at infinity. As for this previous solution, the $M = 0$ case is not a flat spacetime. $r = 0$ is a curvature singularity, with either $f(0) = -1/5$ for $\alpha_5 > 0$ or $\lim_{r\to 0} f(r) = \infty$ if $\alpha_5 < 0$, but of course, there might be a curvature singularity before $r = 0$ if the square root becomes ill-defined.

The second solution exists for

$$\lambda_4 = \beta_4 = \alpha_4 = 0, \quad \lambda_5 = \frac{9\beta_5^2}{20\alpha_5}, \tag{91}$$

which corresponds to the pure $n = -1$ Kaluza-Klein described above, with a metric function and a scalar field given by

$$f(r) = \frac{1}{1+\frac{4\alpha_5\eta}{3r^3}}\left[1 - \frac{2M}{r} - \frac{4\alpha_5\eta}{15r^3}\right], \quad \phi = \ln\left(\frac{\eta}{r}\right), \quad \eta \equiv 2\sqrt{\frac{-\alpha_5}{3\beta_5}}. \tag{92}$$

It is a simple deviation from Schwarzschild, parameterized by the product $\alpha_5\eta$, and of order $\mathcal{O}\left(1/r^3\right)$ at infinity. If $\alpha_5 > 0$, the curvature singularity is at $r = 0$ with $f(0) = -1/5$, while if $\alpha_5 < 0$, the singularity is at $r = (-4\alpha_5\eta/3)^{1/3}$. Note that for both solutions (90) and (92), the candidate event horizon r_h is root of a cubic equation,

$$15r_h^3 - 30Mr_h^2 - 15\alpha_4 r_h - 4\alpha_5\eta = 0, \tag{93}$$

with $\alpha_4 = 0$ for the second case, and the precise existence or not of the horizon depends on the various signs of the couplings, see details in [101].

[23] With again a $\pm$ in front of the square root, where the $+$ branch is forgotten because not asymptotically flat.

5 Summary

Higher-order scalar-tensor theories admit a number of exact asymptotically flat BH solutions in four dimensions, of which this chapter has reviewed the most relevant ones. A first step was accomplished by showing the possibility for stealth solutions (Sect. 3), where the metric is Schwarzschild or Kerr, but the scalar field is non-trivial and thus modifies the perturbation theory around these GR backgrounds. The stealth Kerr solution even stands as an interesting starting point for generating, by a disformal transformation, the disformed Kerr metric (29), which displays new properties as compared to its GR counterpart, in particular due to its non-circularity. However, disformal transformation from a seed stealth solution is by no means the unique way of obtaining non-stealth solutions, that is to say, solutions for which the spacetime differs from the Schwarzschild or Kerr metrics predicted by GR: Sect. 4 presents a number of such non-stealth solutions, all of them with the static, spherically-symmetric form (37). The less difficult case of parity and shift-symmetric theories is dealt with in Sect. 4.1, with three different solutions, two of them being homogeneous (i.e. $g_{tt} = -g^{rr}$) and for which the associated (beyond) Horndeski theories admit a canonical kinetic term, while the third one is non-homogeneous. Then, Sect. 4.2 studies non-parity symmetric theories: a great progress was recently achieved in the course of the search for a four-dimensional Einstein-Gauss-Bonnet gravity, which appears as a Horndeski theory enjoying a generalized conformal invariance, see (67), in relation with the regularized Kaluza-Klein action of higher-dimensional Einstein-Gauss-Bonnet gravity, dimensionally reduced along an internal space of dimension $n \to 0$, see (66). This leads to three scalar-tensor BH solutions, depending on the internal space geometry (maximally-symmetric, flat, or product of two-spheres). Finally, a generalization (86) of (67), with additional terms breaking the generalized conformal invariance and possibly linked to a Kaluza-Klein procedure, completes the picture with two more solutions.

Apart from the solutions we have depicted here, DHOST theories have been shown to host exotic solutions such as regular black holes. These solutions are constructed without the need of fine-tuning, where the solution characteristics end up appearing in the coupling constants of the theory in question. These solutions have an inner and outer event horizon with a regular de Sitter central core [103] and host a large number of regular black hole geometries such as the Hayward solution [104] for example. Quite interestingly, these solutions appear to exist away from beyond Horndeski theories; i.e., only in pure DHOST theories. This may or not be due to the construction method, an extension of the well-known Kerr-Schild solution generating trick. Neutron star solutions have been found in the context of the four dimensional Einstein Gauss Bonnet theory which have interesting and distinct properties from GR [96]. For example there is no mass gap in between neutron star solutions and black holes and as such neutron stars can be more massive in accord to recent GW observables such as GW190814 [105]. Recently, smooth traversable wormholes have been found as purely vacuum modified gravity solutions in beyond Horndeski theories [97, 98]. These solutions are obtained starting from Horndeski theory and going via disformal transformation to a beyond Horndeski theory. Although there has been considerable recent progress in the field, the greatest challenge remains: finding explicit stationary solutions. We discussed the disformal Kerr metric and its black hole properties [60] which originates from a simple stealth Kerr metric. Very recently an intriguing stationary metric was also proposed [99], which originates from semi-classical gravity and is more akin to the theories we studied in Sect. 4 (rather than those of Sect. 3 involving stealth metrics). Although the origin of both solutions is distinctively different, both metrics exhibit non circularity, very seldom considered in numerical and ad hoc beyond Kerr constructions (see for example [62] and references within). It is paramount therefore to find such solutions and their properties and figure out what they signify on the observational front. These are some of the important next steps in the field of explicit compact object solutions beyond GR.

References

1. C.M. Will, Living Rev. Rel. **17**, 4 (2014). https://doi.org/10.12942/lrr-2014-4. arXiv:1403.7377 [gr-qc]]
2. B.P. Abbott et al. [LIGO Scientific and Virgo], Phys. Rev. Lett. **116**(6), 061102 (2016). https://doi.org/10.1103/PhysRevLett.116.061102. [arXiv:1602.03837 [gr-qc]]. B.P. Abbott et al. [LIGO Scientific and Virgo], Phys. Rev. Lett. **119**(16), 161101 (2017). https://doi.org/10.1103/PhysRevLett.119.161101[arXiv:1710.05832 [gr-qc]]
3. C. de Rham, S. Melville, Phys. Rev. Lett. **121**(22), 221101 (2018). https://doi.org/10.1103/PhysRevLett.121.221101. [arXiv:1806.09417 [hep-th]]
4. P. Amaro-Seoane et al., [LISA], [arXiv:1702.00786 [astro-ph.IM]]
5. M. Maggiore, C. Van Den Broeck, N. Bartolo, E. Belgacem, D. Bertacca, M.A. Bizouard, M. Branchesi, S. Clesse, S. Foffa, J. García-Bellido et al., JCAP **03**, 050 (2020). https://doi.org/10.1088/1475-7516/2020/03/050. ([arXiv:1912.02622 [astro-ph.CO]].)
6. R. Abuter et al., [GRAVITY], Astron. Astrophys. **615**, L15 (2018). https://doi.org/10.1051/0004-6361/201833718[arXiv:1807.09409 [astro-ph.GA]]. R. Abuter et al., [GRAVITY], Astron. Astrophys. **636**, L5 (2020). https://doi.org/10.1051/0004-6361/202037813[arXiv:2004.07187 [astro-ph.GA]]
7. D. Psaltis et al., Event horizon telescope. Phys. Rev. Lett. **125**(14), 141104 (2020). https://doi.org/10.1103/PhysRevLett.125.141104. [arXiv:2010.01055 [gr-qc]]
8. T. Clifton, P.G. Ferreira, A. Padilla, C. Skordis, Phys. Rept. **513**, 1–189 (2012). https://doi.org/10.1016/j.physrep.2012.01.001. [arXiv:1106.2476 [astro-ph.CO]]
9. D. Lovelock, J. Math. Phys. **12**, 498–501 (1971). https://doi.org/10.1063/1.1665613
10. T. Chiba, Phys. Lett. B **575**, 1–3 (2003). https://doi.org/10.1016/j.physletb.2003.09.033. [arXiv:astro-ph/0307338 [astro-ph]]
11. M.A. Luty, M. Porrati, R. Rattazzi, JHEP **09**, 029 (2003). https://doi.org/10.1088/1126-6708/2003/09/029. [arXiv:hep-th/0303116 [hep-th]]
12. C. de Rham, G. Gabadadze, Phys. Rev. D **82**, 044020 (2010). https://doi.org/10.1103/PhysRevD.82.044020. ([arXiv:1007.0443 [hep-th]].)
13. C.A.R. Herdeiro, E. Radu, Int. J. Mod. Phys. D **24**(09), 1542014 (2015). https://doi.org/10.1142/S0218271815420146. [arXiv:1504.08209 [gr-qc]]. A. Lehébel, [arXiv:1810.04434 [gr-qc]]
14. M. Rinaldi, Phys. Rev. D **86**, 084048 (2012). https://doi.org/10.1103/PhysRevD.89.084050. ([arXiv:1208.0103 [gr-qc]]. A. Anabalon, A. Cisterna, J. Oliva, Phys. Rev. D **89**, 084050 (2014). [arXiv:1312.3597 [gr-qc]]. M. Minamitsuji, Phys. Rev. D **89**, 064017

(2014). 10.1103/PhysRevD.89.064017 [arXiv:1312.3759 [gr-qc]]. https://doi.org/10.1103/PhysRevD.86.084048)
15. S.M. Carroll (Cambridge University Press, 2019). ISBN 978-0-8053-8732-2, 978-1-108-48839-6, 978-1-108-77555-7
16. O. Bechmann, O. Lechtenfeld, Class. Quant. Grav. **12**, 1473–1482 (1995). https://doi.org/10.1088/0264-9381/12/6/013[arXiv:gr-qc/9502011 [gr-qc]]. H. Dennhardt, O. Lechtenfeld, Int. J. Mod. Phys. A **13**, 741–764 (1998). https://doi.org/10.1142/S0217751X98000329[arXiv:gr-qc/9612062 [gr-qc]]. K.A. Bronnikov, J.C. Fabris, Phys. Rev. Lett. **96**, 251101 (2006). https://doi.org/10.1103/PhysRevLett.96.251101. [arXiv:gr-qc/0511109 [gr-qc]]. A. Anabalon, J. Oliva, Phys. Rev. D **86**, 107501 (2012). https://doi.org/10.1103/PhysRevD.86.107501[arXiv:1205.6012 [gr-qc]]. M. Cadoni, E. Franzin, Phys. Rev. D **91**(10), 104011 (2015). https://doi.org/10.1103/PhysRevD.91.104011[arXiv:1503.04734 [gr-qc]]. A. Bakopoulos, T. Nakas, Phys. Rev. D **107**(12), 124035 (2023). https://doi.org/10.1103/PhysRevD.107.124035[arXiv:2303.09116 [gr-qc]]. T. Karakasis, N.E. Mavromatos, E. Papantonopoulos, Phys. Rev. D **108**(2), 024001 (2023). https://doi.org/10.1103/PhysRevD.108.024001[arXiv:2305.00058 [gr-qc]]
17. T.P. Sotiriou, Lect. Notes Phys. **892**, 3–24 (2015). https://doi.org/10.1007/978-3-319-10070-8. [arXiv:1404.2955 [gr-qc]]
18. S.W. Hawking, Commun. Math. Phys. **25**, 167–171 (1972). https://doi.org/10.1007/BF01877518. J.D. Bekenstein, Phys. Rev. D **51**(12), R6608 (1995). https://doi.org/10.1103/PhysRevD.51.R6608. T.P. Sotiriou, V. Faraoni, Phys. Rev. Lett. **108**, 081103 (2012). https://doi.org/10.1103/PhysRevLett.108.081103[arXiv:1109.6324 [gr-qc]]
19. N.M. Bocharova, K.A. Bronnikov, V.N. Melnikov, Vestnik Moskov. Univ. Fizika **25**, 706 (1970). J.D. Bekenstein, Ann. Phys. **82**, 535 (1974)
20. C. Martinez, R. Troncoso, J. Zanelli, Phys. Rev. D **67**, 024008 (2003). ([arXiv:hep-th/0205319]. C. Martinez, J.P. Staforelli, R. Troncoso, Phys. Rev. D **74**, 044028 (2006) [arXiv:hep-th/0512022])
21. Y. Bardoux, M.M. Caldarelli, C. Charmousis, JHEP **09**, 008 (2012). https://doi.org/10.1007/JHEP09(2012)008. [arXiv:1205.4025 [hep-th]]
22. R.P. Woodard, Scholarpedia **10**(8), 32243 (2015). https://doi.org/10.4249/scholarpedia.32243. [arXiv:1506.02210 [hep-th]]
23. G.W. Horndeski, Int. J. Theor. Phys. **10**, 363–384 (1974). https://doi.org/10.1007/BF01807638
24. C. Charmousis, E.J. Copeland, A. Padilla, P.M. Saffin, Phys. Rev. Lett. **108**, 051101 (2012). https://doi.org/10.1103/PhysRevLett.108.051101. ([arXiv:1106.2000 [hep-th]].)
25. A. Nicolis, R. Rattazzi, E. Trincherini, Phys. Rev. D **79**, 064036 (2009). https://doi.org/10.1103/PhysRevD.79.064036. ([arXiv:0811.2197 [hep-th]].)
26. C. Deffayet, G. Esposito-Farese, A. Vikman, Phys. Rev. D **79**, 084003 (2009). https://doi.org/10.1103/PhysRevD.80.064015. ([arXiv:0901.1314 [hep-th]. C. Deffayet, S. Deser, G. Esposito-Farese, Phys. Rev. D **80**, 064015 (2009). [arXiv:0906.1967 [gr-qc]. https://doi.org/10.1103/PhysRevD.79.084003)
27. C. Deffayet, X. Gao, D.A. Steer, G. Zahariade, Phys. Rev. D **84**, 064039 (2011). https://doi.org/10.1103/PhysRevD.84.064039. ([arXiv:1103.3260 [hep-th]].)
28. T. Kobayashi, M. Yamaguchi, J. Yokoyama, Prog. Theor. Phys. **126**, 511–529 (2011). https://doi.org/10.1143/PTP.126.511. [arXiv:1105.5723 [hep-th]]
29. D. Langlois, K. Noui, H. Roussille, JCAP **09**, 019 (2022). https://doi.org/10.1088/1475-7516/2022/09/019. ([arXiv:2204.04107 [gr-qc]].)
30. T. Kobayashi, Rept. Prog. Phys. **82**(8), 086901 (2019). https://doi.org/10.1088/1361-6633/ab2429. [arXiv:1901.07183 [gr-qc]]
31. M. Crisostomi, K. Noui, C. Charmousis, D. Langlois, Phys. Rev. D **97**(4), 044034 (2018). https://doi.org/10.1103/PhysRevD.97.044034. [arXiv:1710.04531 [hep-th]]
32. D. Langlois, Int. J. Mod. Phys. D **28**(05), 1942006 (2019). https://doi.org/10.1142/S0218271819420069. [arXiv:1811.06271 [gr-qc]]

33. D. Langlois, K. Noui, JCAP **02**, 034 (2016). https://doi.org/10.1088/1475-7516/2016/02/034. ([arXiv:1510.06930 [gr-qc]].)
34. J. Ben Achour, M. Crisostomi, K. Koyama, D. Langlois, K. Noui, G. Tasinato, JHEP **12**, 100 (2016). https://doi.org/10.1007/JHEP12(2016)100. [arXiv:1608.08135 [hep-th]]
35. J. Ben Achour, D. Langlois, K. Noui, Phys. Rev. D **93**(12), 124005 (2016). https://doi.org/10.1103/PhysRevD.93.124005. [arXiv:1602.08398 [gr-qc]]
36. J.D. Bekenstein, Phys. Rev. D **48**, 3641–3647 (1993). https://doi.org/10.1103/PhysRevD.48.3641. [arXiv:gr-qc/9211017 [gr-qc]]
37. D. Bettoni, S. Liberati, Phys. Rev. D **88**, 084020 (2013). https://doi.org/10.1103/PhysRevD.88.084020. ([arXiv:1306.6724 [gr-qc]].)
38. J. Gleyzes, D. Langlois, F. Piazza, F. Vernizzi, Phys. Rev. Lett. **114**(21), 211101 (2015). https://doi.org/10.1103/PhysRevLett.114.211101. [arXiv:1404.6495 [hep-th]]. J. Gleyzes, D. Langlois, F. Piazza, F. Vernizzi, JCAP **02**, 018 (2015). https://doi.org/10.1088/1475-7516/2015/02/018. [arXiv:1408.1952 [astro-ph.CO]]. X. Gao, Phys. Rev. D **90**, 081501 (2014). https://doi.org/10.1103/PhysRevD.90.081501. [arXiv:1406.0822 [gr-qc]]. M. Zumalacárregui, J. García-Bellido, Phys. Rev. D **89**, 064046 (2014). https://doi.org/10.1103/PhysRevD.89.064046. [arXiv:1308.4685 [gr-qc]]
39. M. Crisostomi, M. Hull, K. Koyama, G. Tasinato, JCAP **03**, 038 (2016). https://doi.org/10.1088/1475-7516/2016/03/038. ([arXiv:1601.04658 [hep-th]].)
40. J. Ben Achour, H. Liu, S. Mukohyama, JCAP **02**, 023 (2020). https://doi.org/10.1088/1475-7516/2020/02/023. [arXiv:1910.11017 [gr-qc]]
41. C. Charmousis, D. Iosifidis, J. Phys. Conf. Ser. **600**, 012003 (2015). https://doi.org/10.1088/1742-6596/600/1/012003. ([arXiv:1501.05167 [gr-qc]].)
42. E. Babichev, G. Esposito-Farese, Phys. Rev. D **95**(2), 024020 (2017). https://doi.org/10.1103/PhysRevD.95.024020. [arXiv:1609.09798 [gr-qc]]
43. E. Babichev, C. Charmousis, JHEP **08**, 106 (2014). https://doi.org/10.1007/JHEP08(2014)106. ([arXiv:1312.3204 [gr-qc]].)
44. L. Hui, A. Nicolis, Phys. Rev. Lett. **110**, 241104 (2013). https://doi.org/10.1103/PhysRevLett.110.241104. ([arXiv:1202.1296 [hep-th]].)
45. E. Babichev, Phys. Rev. D **83**, 024008 (2011). https://doi.org/10.1103/PhysRevD.87.044032. ([arXiv:1009.2921 [hep-th]]. E. Babichev, G. Esposito-Farèse, Phys. Rev. D **87**, 044032 (2013). [arXiv:1212.1394 [gr-qc]]. https://doi.org/10.1103/PhysRevD.83.024008)
46. T. Kobayashi, N. Tanahashi, PTEP **2014**, 073E02 (2014). https://doi.org/10.1093/ptep/ptu096. [arXiv:1403.4364 [gr-qc]]
47. E. Babichev, C. Charmousis, G. Esposito-Farèse, A. Lehébel, Phys. Rev. D **98**(10), 104050 (2018). https://doi.org/10.1103/PhysRevD.98.104050. [arXiv:1803.11444 [gr-qc]]
48. E. Babichev, C. Charmousis, G. Esposito-Farèse, A. Lehébel, Phys. Rev. Lett. **120**(24), 241101 (2018). https://doi.org/10.1103/PhysRevLett.120.241101. [arXiv:1712.04398 [gr-qc]]
49. M. Minamitsuji, H. Motohashi, Phys. Rev. D **98**(8), 084027 (2018). https://doi.org/10.1103/PhysRevD.98.084027. [arXiv:1809.06611 [gr-qc]]. J. Ben Achour, H. Liu, Phys. Rev. D **99**(6), 064042 (2019). https://doi.org/10.1103/PhysRevD.99.064042. [arXiv:1811.05369 [gr-qc]]
50. H. Motohashi, M. Minamitsuji, Phys. Rev. D **99**(6), 064040 (2019). https://doi.org/10.1103/PhysRevD.99.064040. [arXiv:1901.04658 [gr-qc]]. M. Minamitsuji, J. Edholm, Phys. Rev. D **100**(4), 044053 (2019). https://doi.org/10.1103/PhysRevD.100.044053. [arXiv:1907.02072 [gr-qc]]
51. C. Charmousis, M. Crisostomi, R. Gregory, N. Stergioulas, Phys. Rev. D **100**(8), 084020 (2019). https://doi.org/10.1103/PhysRevD.100.084020. [arXiv:1903.05519 [hep-th]]
52. D. Langlois, R. Saito, D. Yamauchi, K. Noui, Phys. Rev. D **97**(6), 061501 (2018). https://doi.org/10.1103/PhysRevD.97.061501. [arXiv:1711.07403 [gr-qc]]
53. B. Carter, Phys. Rev. **174**, 1559–1571 (1968). https://doi.org/10.1103/PhysRev.174.1559
54. K. Takahashi, H. Motohashi, JCAP **06**, 034 (2020). https://doi.org/10.1088/1475-7516/2020/06/034. ([arXiv:2004.03883 [gr-qc]].)
55. C. Charmousis, M. Crisostomi, D. Langlois, K. Noui, Class. Quant. Grav. **36**(23), 235008 (2019). https://doi.org/10.1088/1361-6382/ab4fb1. [arXiv:1907.02924 [gr-qc]]

56. K. Takahashi, H. Motohashi, M. Minamitsuji, Phys. Rev. D **100**(2), 024041 (2019). https://doi.org/10.1103/PhysRevD.100.024041. [arXiv:1904.03554 [gr-qc]]
57. C. de Rham, J. Zhang, Phys. Rev. D **100**(12), 124023 (2019). https://doi.org/10.1103/PhysRevD.100.124023. [arXiv:1907.00699 [hep-th]]
58. K. Takahashi, H. Motohashi, JCAP **08**, 013 (2021). https://doi.org/10.1088/1475-7516/2021/08/013. ([arXiv:2106.07128 [gr-qc]].)
59. A. De Felice, S. Mukohyama, K. Takahashi, Phys. Rev. Lett. **129**(3), 031103 (2022). https://doi.org/10.1103/PhysRevLett.129.031103. [arXiv:2204.02032 [gr-qc]]
60. T. Anson, E. Babichev, C. Charmousis, M. Hassaine, JHEP **01**, 018 (2021). https://doi.org/10.1007/JHEP01(2021)018. ([arXiv:2006.06461 [gr-qc]].)
61. J. Ben Achour, H. Liu, H. Motohashi, S. Mukohyama, K. Noui, JCAP **11**, 001 (2020). https://doi.org/10.1088/1475-7516/2020/11/001. [arXiv:2006.07245 [gr-qc]]
62. T. Johannsen, Phys. Rev. D **87**(12), 124017 (2013). https://doi.org/10.1103/PhysRevD.87.124017. [arXiv:1304.7786 [gr-qc]]
63. S.W. Hawking, G.F.R. Ellis (Cambridge University Press, 1997). ISBN 978-1-00-925316-1-00-925315-4, 978-0-521-20016-5, 978-0-521-09906-6, 978-0-511-82630-6, 978-0-521-09906-6. V.P. Frolov, I.D. Novikov (2023). https://doi.org/10.1007/978-94-011-5139-9. https://doi.org/10.1017/9781009253161
64. F. Long, S. Chen, M. Wang, J. Jing, Eur. Phys. J. C **80**(12), 1180 (2020). https://doi.org/10.1140/epjc/s10052-020-08744-8. [arXiv:2009.07508 [gr-qc]]
65. T. Anson, E. Babichev, C. Charmousis, Phys. Rev. D **103**(12), 124035 (2021). https://doi.org/10.1103/PhysRevD.103.124035. [arXiv:2103.05490 [gr-qc]]
66. E. Babichev, C. Charmousis, N. Lecoeur, JCAP **08**, 022 (2023). https://doi.org/10.1088/1475-7516/2023/08/022. ([arXiv:2305.17129 [gr-qc]].)
67. V. Faraoni, Lect. Notes Phys. **907**, 1–199 (2015). ISBN 978-3-319-19239-0, 978-3-319-19240-6. https://doi.org/10.1007/978-3-319-19240-6. S.A. Hayward, Phys. Rev. D **49**, 6467–6474 (1994). https://doi.org/10.1103/PhysRevD.49.6467. R. Penrose, Phys. Rev. Lett. **14**, 57–59 (1965). https://doi.org/10.1103/PhysRevLett.14.57
68. H. Culetu, [arXiv:1202.2285 [gr-qc]]
69. T. Sato, H. Maeda, T. Harada, Class. Quant. Grav. **39**(21), 215011 (2022). [erratum: Class. Quant. Grav. **40**(7), 079501 (2023)]. https://doi.org/10.1088/1361-6382/ac902f. [arXiv:2206.10998 [gr-qc]]
70. E. Babichev, C. Charmousis, A. Lehébel, JCAP **04**, 027 (2017). https://doi.org/10.1088/1475-7516/2017/04/027. ([arXiv:1702.01938 [gr-qc]].)
71. R.L. Arnowitt, S. Deser, C.W. Misner, Phys. Rev. **122**, 997 (1961). https://doi.org/10.1103/PhysRev.122.997
72. G.W. Gibbons, S.W. Hawking, Phys. Rev. D **15**, 2752–2756 (1977). https://doi.org/10.1103/PhysRevD.15.2752
73. D. Langlois, K. Noui, H. Roussille, Phys. Rev. D **104**(12), 124044 (2021). https://doi.org/10.1103/PhysRevD.104.124044. [arXiv:2103.14750 [gr-qc]]. H. Roussille, [arXiv:2211.01103 [gr-qc]]
74. A. Bakopoulos, C. Charmousis, P. Kanti, N. Lecoeur, JHEP **08**, 055 (2022). https://doi.org/10.1007/JHEP08(2022)055. ([arXiv:2203.14595 [gr-qc]].)
75. P. Kanti, N.E. Mavromatos, J. Rizos, K. Tamvakis, E. Winstanley, Phys. Rev. D **54**, 5049–5058 (1996). https://doi.org/10.1103/PhysRevD.54.5049. [arXiv:hep-th/9511071 [hep-th]]
76. D.D. Doneva, S.S. Yazadjiev, Phys. Rev. Lett. **120**(13), 131103 (2018). https://doi.org/10.1103/PhysRevLett.120.131103. [arXiv:1711.01187 [gr-qc]]. H.O. Silva, J. Sakstein, L. Gualtieri, T.P. Sotiriou, E. Berti, Phys. Rev. Lett. **120**(13), 131104 (2018). https://doi.org/10.1103/PhysRevLett.120.131104. [arXiv:1711.02080 [gr-qc]]. G. Antoniou, A. Lehébel, G. Ventagli, T.P. Sotiriou, Phys. Rev. D **104**(4), 044002 (2021). https://doi.org/10.1103/PhysRevD.104.044002. [arXiv:2105.04479 [gr-qc]]
77. P. Cunha, V.P., C.A.R. Herdeiro, E. Radu, Phys. Rev. Lett. **123**(1), 011101 (2019). https://doi.org/10.1103/PhysRevLett.123.011101. [arXiv:1904.09997 [gr-qc]]. A. Dima, E. Barausse, N. Franchini, T.P. Sotiriou, Phys. Rev. Lett. **125**(23), 231101 (2020). https://doi.org/10.1103/

PhysRevLett.125.231101. [arXiv:2006.03095 [gr-qc]]. C.A.R. Herdeiro, E. Radu, H.O. Silva, T.P. Sotiriou, N. Yunes, Phys. Rev. Lett. **126**(1), 011103 (2021). https://doi.org/10.1103/PhysRevLett.126.011103. [arXiv:2009.03904 [gr-qc]]. E. Berti, L.G. Collodel, B. Kleihaus, J. Kunz, Phys. Rev. Lett. **126**(1), 011104 (2021). https://doi.org/10.1103/PhysRevLett.126.011104. [arXiv:2009.03905 [gr-qc]]
78. G. Antoniou, [arXiv:2308.03501 [gr-qc]]
79. P.G.S. Fernandes, P. Carrilho, T. Clifton, D.J. Mulryne, Class. Quant. Grav. **39**(6), 063001 (2022). https://doi.org/10.1088/1361-6382/ac500a. [arXiv:2202.13908 [gr-qc]]
80. S.-S. Chern, Ann. Math. **46**(4), 674–684 (1945). https://doi.org/10.2307/1969203
81. T.P. Sotiriou, S.Y. Zhou, Phys. Rev. Lett. **112**, 251102 (2014). https://doi.org/10.1103/PhysRevD.90.124063. ([arXiv:1312.3622 [gr-qc]]. T.P. Sotiriou, S.Y. Zhou, Phys. Rev. D **90**, 124063 (2014) [arXiv:1408.1698 [gr-qc]]. G. Antoniou, A. Bakopoulos, P. Kanti, Phys. Rev. D **97**(8), 084037 (2018). https://doi.org/10.1103/PhysRevD.97.084037. [arXiv:1711.07431 [hep-th]]. G. Antoniou, A. Bakopoulos, P. Kanti, Phys. Rev. Lett. **120**(13), 131102 (2018). https://doi.org/10.1103/PhysRevLett.120.131102. [arXiv:1711.03390 [hep-th]]. M. Saravani, T.P. Sotiriou, Phys. Rev. D **99**(12), 124004 (2019). https://doi.org/10.1103/PhysRevD.99.124004. [arXiv:1903.02055 [gr-qc]]. E. Babichev, W.T. Emond, S. Ramazanov, Phys. Rev. D **106**(6), 063524 (2022). https://doi.org/10.1103/PhysRevD.106.063524. [arXiv:2207.03944 [gr-qc]]. https://doi.org/10.1103/PhysRevLett.112.251102)
82. B. Kleihaus, J. Kunz, E. Radu, Phys. Rev. Lett. **106**, 151104 (2011). https://doi.org/10.1103/PhysRevD.90.044066. ([arXiv:1101.2868 [gr-qc]]. D. Ayzenberg, N. Yunes, Phys. Rev. D **90**, 044066 (2014). [Erratum: Phys. Rev. D **91**(6), 069905 (2015)]. [arXiv:1405.2133 [gr-qc]]. A. Maselli, P. Pani, L. Gualtieri, V. Ferrari, Phys. Rev. D **92**(8), 083014 (2015). https://doi.org/10.1103/PhysRevD.92.083014. [arXiv:1507.00680 [gr-qc]]. B. Kleihaus, J. Kunz, S. Mojica, E. Radu, Phys. Rev. D **93**(4), 044047 (2016). https://doi.org/10.1103/PhysRevD.93.044047. [arXiv:1511.05513 [gr-qc]]. L.G. Collodel, B. Kleihaus, J. Kunz, E. Berti, Class. Quant. Grav. **37**(7), 075018 (2020). https://doi.org/10.1088/1361-6382/ab74f9. [arXiv:1912.05382 [gr-qc]]. J.F.M. Delgado, C.A.R. Herdeiro, E. Radu, JHEP **04**, 180 (2020). https://doi.org/10.1007/JHEP04(2020)180. [arXiv:2002.05012 [gr-qc]]. https://doi.org/10.1103/PhysRevLett.106.151104)
83. F.L. Julié, E. Berti, Phys. Rev. D **100**(10), 104061 (2019). https://doi.org/10.1103/PhysRevD.100.104061. [arXiv:1909.05258 [gr-qc]]
84. B. Zwiebach, Phys. Lett. B **156**, 315–317 (1985). https://doi.org/10.1016/0370-2693(85)91616-8R.I. Nepomechie, Phys. Rev. D **32**, 3201 (1985). https://doi.org/10.1103/PhysRevD.32.3201. C.G. Callan, Jr., I.R. Klebanov, M.J. Perry, Nucl. Phys. B **278**, 78–90 (1986). https://doi.org/10.1016/0550-3213(86)90107-0. P. Candelas, G.T. Horowitz, A. Strominger, E. Witten, Nucl. Phys. B **258**, 46–74 (1985). https://doi.org/10.1016/0550-3213(85)90602-9. D.J. Gross, J.H. Sloan, Nucl. Phys. B **291**, 41–89 (1987). https://doi.org/10.1016/0550-3213(87)90465-2
85. C. Charmousis, B. Gouteraux, E. Kiritsis, JHEP **09**, 011 (2012). https://doi.org/10.1007/JHEP09(2012)011. ([arXiv:1206.1499 [hep-th]].)
86. J.M. Overduin, P.S. Wesson, Phys. Rept. **283**, 303–380 (1997). https://doi.org/10.1016/S0370-1573(96)00046-4. [arXiv:gr-qc/9805018 [gr-qc]]
87. G. Dotti, R.J. Gleiser, Phys. Lett. B **627**, 174–179 (2005). https://doi.org/10.1016/j.physletb.2005.08.110. [arXiv:hep-th/0508118 [hep-th]]. C. Bogdanos, C. Charmousis, B. Gouteraux, R. Zegers, JHEP **10**, 037 (2009). https://doi.org/10.1088/1126-6708/2009/10/037. [arXiv:0906.4953 [hep-th]]. H. Maeda, Phys. Rev. D **81**, 124007 (2010). https://doi.org/10.1103/PhysRevD.81.124007. [arXiv:1004.0917 [gr-qc]]
88. D. Glavan, C. Lin, Phys. Rev. Lett. **124**(8), 081301 (2020). https://doi.org/10.1103/PhysRevLett.124.081301. [arXiv:1905.03601 [gr-qc]]
89. M. Gürses, T. Ç. Şişman, B. Tekin, Eur. Phys. J. C **80**(7), 647 (2020). https://doi.org/10.1140/epjc/s10052-020-8200-7. [arXiv:2004.03390 [gr-qc]]. M. Gurses, T.Ç. Şişman, B. Tekin, Phys. Rev. Lett. **125**(14), 149001 (2020). https://doi.org/10.1103/PhysRevLett.125.149001. [arXiv:2009.13508 [gr-qc]]. J. Arrechea, A. Delhom, A. Jiménez-Cano, Chin. Phys. C

45(1), 013107 (2021). https://doi.org/10.1088/1674-1137/abc1d4. [arXiv:2004.12998 [gr-qc]]. J. Arrechea, A. Delhom, A. Jiménez-Cano, Phys. Rev. Lett. **125**(14), 149002 (2020). https://doi.org/10.1103/PhysRevLett.125.149002. [arXiv:2009.10715 [gr-qc]]. J. Bonifacio, K. Hinterbichler, L.A. Johnson, Phys. Rev. D **102**(2), 024029 (2020). https://doi.org/10.1103/PhysRevD.102.024029. [arXiv:2004.10716 [hep-th]]. W.Y. Ai, Commun. Theor. Phys. **72**(9), 095402 (2020). https://doi.org/10.1088/1572-9494/aba242. [arXiv:2004.02858 [gr-qc]]. S. Mahapatra, Eur. Phys. J. C **80**(10), 992 (2020). https://doi.org/10.1140/epjc/s10052-020-08568-6. [arXiv:2004.09214 [gr-qc]]. M. Hohmann, C. Pfeifer, N. Voicu, Eur. Phys. J. Plus **136**(2), 180 (2021). https://doi.org/10.1140/epjp/s13360-021-01153-0. [arXiv:2009.05459 [gr-qc]]. L.M. Cao, L.B. Wu, Eur. Phys. J. C **82**(2), 124 (2022). https://doi.org/10.1140/epjc/s10052-022-10079-5. [arXiv:2103.09612 [gr-qc]]
90. H. Lu, Y. Pang, Phys. Lett. B **809**, 135717 (2020). https://doi.org/10.1016/j.physletb.2020.135717. ([arXiv:2003.11552 [gr-qc]].)
91. P.G.S. Fernandes, P. Carrilho, T. Clifton, D.J. Mulryne, Phys. Rev. D **102**(2), 024025 (2020). https://doi.org/10.1103/PhysRevD.102.024025. [arXiv:2004.08362 [gr-qc]]
92. R.A. Hennigar, D. Kubizňák, R.B. Mann, C. Pollack, JHEP **07**, 027 (2020). https://doi.org/10.1007/JHEP07(2020)027. ([arXiv:2004.09472 [gr-qc]].)
93. P.G.S. Fernandes, Phys. Rev. D **103**(10), 104065 (2021). https://doi.org/10.1103/PhysRevD.103.104065. [arXiv:2105.04687 [gr-qc]]
94. E. Ayón-Beato, M. Hassaine, [arXiv:2305.09806 [hep-th]]
95. D.G. Boulware, S. Deser, Phys. Rev. Lett. **55**, 2656 (1985). https://doi.org/10.1103/PhysRevLett.55.2656
96. C. Charmousis, A. Lehébel, E. Smyrniotis, N. Stergioulas, JCAP **02**(02), 033 (2022). https://doi.org/10.1088/1475-7516/2022/02/033. [arXiv:2109.01149 [gr-qc]]
97. A. Bakopoulos, C. Charmousis, P. Kanti, JCAP **05**(05), 022 (2022). https://doi.org/10.1088/1475-7516/2022/05/022. [arXiv:2111.09857 [gr-qc]]
98. E. Babichev, C. Charmousis, M. Hassaine, N. Lecoeur, Phys. Rev. D **106**(6), 064039 (2022). https://doi.org/10.1103/PhysRevD.106.064039. [arXiv:2206.11013 [gr-qc]]
99. P.G.S. Fernandes, [arXiv:2305.10382 [gr-qc]]
100. P.G.S. Fernandes, P. Carrilho, T. Clifton, D.J. Mulryne, Phys. Rev. D **104**(4), 044029 (2021). https://doi.org/10.1103/PhysRevD.104.044029. [arXiv:2107.00046 [gr-qc]]
101. E. Babichev, C. Charmousis, M. Hassaine, N. Lecoeur, Phys. Rev. D **108**(2), 024019 (2023). https://doi.org/10.1103/PhysRevD.108.024019. [arXiv:2303.04126 [gr-qc]]
102. J. Oliva, S. Ray, Class. Quant. Grav. **29**, 205008 (2012). https://doi.org/10.1088/0264-9381/29/20/205008. ([arXiv:1112.4112 [gr-qc]].)
103. E. Babichev, C. Charmousis, A. Cisterna, M. Hassaine, JCAP **06**, 049 (2020). https://doi.org/10.1088/1475-7516/2020/06/049. [arXiv:2004.00597 [hep-th]]. O. Baake, C. Charmousis, M. Hassaine, M. San Juan, JCAP **06**, 021 (2021). https://doi.org/10.1088/1475-7516/2021/06/021. [arXiv:2104.08221 [hep-th]]
104. S.A. Hayward, Phys. Rev. Lett. **96**, 031103 (2006). https://doi.org/10.1103/PhysRevLett.96.031103. ([arXiv:gr-qc/0506126 [gr-qc]].)
105. R. Abbott et al., LIGO Scientific and Virgo. Astrophys. J. Lett. **896**(2), L44 (2020). https://doi.org/10.3847/2041-8213/ab960f. [arXiv:2006.12611 [astro-ph.HE]]

Compact Objects in EsGB Theory and Beyond

Panagiota Kanti

Abstract In the context of General Relativity, black holes are not allowed to possess scalar hair, wormholes are not traversable and particle-like solutions are irregular. Therefore, in order to derive novel and physically interesting solutions that describe compact objects one needs to address generalised gravitational theories. One popular class of such theories is the Einstein-scalar-Gauss-Bonnet (EsGB) theory with a general coupling function between the scalar field of the theory and the quadratic Gauss-Bonnet term. Starting from black holes, we present a variety of spherically-symmetric solutions for several different forms of the coupling function and discuss their main features. We then proceed to wormhole solutions and demonstrate that the EsGB theory naturally supports traversable wormholes without the need for exotic matter. Regular scalarised particle-like solutions also emerge in the context of the same theory which also possess interesting observable features such as photon rings and echoes. Moving beyond this class of theories, we then address the more extended scalar-tensor Horndeski theory, briefly mention the types of black-hole solutions that arise, and demonstrate that an appropriately constructed disformal transformation of a black-hole solution, such as the Lu-Pang solution, results into a traversable wormhole in the context of the beyond-Horndeski theory.

1 Introduction

In 1915, Albert Einstein formulated the General Theory of Relativity [1], a mathematically beautiful and, at the same time, a physically relevant theory which has so far passed all experimental tests. It is, in addition, a much more interesting theory compared to Newtonian Gravity since, not only does it describe the gravitational interactions between two massive bodies, but it also incorporates the equivalence

P. Kanti (✉)
Theory Division, Physics Department, University of Ioannina, GR 451 10 Ioannina, Greece
e-mail: pkanti@uoi.gr

© The Author(s), under exclusive license to Springer Nature Switzerland AG 2024

E. Papantonopoulos and N. Mavromatos (eds.), *Compact Objects in the Universe*,
https://doi.org/10.1007/978-3-031-55098-0_2

between mass and energy, and proceeds further to predict the existence of new gravitational solutions, such as black holes or wormholes, and new phenomena associated with them.

However, General Relativity (GR) is not a perfect theory—if such a thing ever exists. On the cosmological side, the Standard Cosmological Model, which has been formulated on the mathematical and physical basis provided by GR, has a number of open problems: the nature of dark matter and dark energy, the coincidence problem, the spacetime singularities, the right model for inflation, to mention a few. Also, the prospect of the unification of gravity with the other forces seems unlikely within the GR—the latter theory is a tensorial theory rather than a gauge field theory, and is not renormalizable. But even on the gravitational side, GR is a rather restricted theory predicting in fact the existence of a limited number of types of compact objects beyond stars, some of them with undesirable properties: black holes obey "no-hair" theorems, wormholes cannot keep their tunnels open or are plagued by internal singularities and gravitational particle-like solutions (solitons) simply do not exist.

Therefore, in order to find new black-hole solutions, traversable wormholes or particle-like solutions, going beyond General Relativity seems to be a one-way road. For this reason, a large number of generalised theories of gravity have been formulated by adding to the Einstein-Hilbert action new fields, new gravitational terms and couplings among them. The simplest extension of GR is the scalar-tensor theories [2–6], however, as it quite well known, one must consider non-minimally coupled scalar fields in order for physically-interesting solutions to emerge.

In this review, we will focus first on the Einstein-scalar-Gauss-Bonnet (EsGB) theory as an indicative example of a non-minimal-coupled scalar-tensor theory. It is a rather simple theory containing a single scalar degree of freedom coupled to the quadratic, gravitational Gauss-Bonnet term. This theory has been studied for decades, as it arises in the context of superstring effective theory at low energies [7–9] or in Kaluza-Klein compactifications of Lovelock's theory [10]. Due to the presence of the quadratic Gauss-Bonnet (GB) term, this generalised theory reduces to GR, with a trivial scalar field, in the limit of weak gravity but may lead to important modifications from GR in the strong-gravity regime. These modifications include novel gravitational solutions characterised by a non-trivial scalar field. Thus, in the context of the EsGB theory, we will search for new scalarised solutions describing various types of compact objects. As we will demonstrate, the configuration space of solutions of the EsGB theory is indeed a particularly rich one, with some areas of it still waiting to be explored.

The EsGB theory is in fact a subclass of Horndeski theory [11], which is a the most general scalar-tensor theory with field equations containing only up to 2nd order derivatives of the metric tensor and the scalar field. The scalar field is now allowed to have a number of couplings with different gravitational terms and derivative-like couplings as well. The theory can be further generalised to the beyond-Horndeski theory via the addition of two additional terms in the Lagrangian. In the second part of this review, we will focus on the beyond-Horndeski theory, and demonstrate that, in its context, new solutions describing black holes with an (Anti-)de Sitter-Reissner-

Nordstrom asymptotic behaviour at large distances may be analytically derived. For the derivation of wormhole solutions, we will employ an alternative method, that of applying a disformal transformation to a known solution of the theory. We will show that such an approach can easily lead to new wormhole solutions with a number of particularly attractive properties.

The outline of this review is as follows: in Sect. 2, we will give a brief overview of the type of compact objects arising in the context of GR and of the minimal Einstein-scalar theory. In Sect. 3, we will present the EsGB theory, and search for black holes, wormholes and particle-like solutions in its context. In Sect. 4, we will proceed to consider the beyond-Horndeski theory, and present some recent results on black holes and wormhole solutions. We will present our conclusions and some food-for-thought comments for the future in Sect. 5.

2 Compact Objects in GR and Einstein-Scalar Theory

In this section, we will briefly discuss the main characteristics of three types of compact objects that will be of interest to us, namely black holes, wormholes and particle-like solutions, as these arise in the context of General Relativity and Einstein-scalar theory.

2.1 Black Holes

As is well known, General Relativity admits only three families of black-hole solutions: the Schwarzschild solution [12], which describes a spherically-symmetric, neutral black hole, the Reissner-Nordstrom solution [13], which also describes a spherically-symmetric but charged black hole, and the Kerr(-Newman) solution [14], which describes a rotating, neutral (or, charged) black hole. According to the "no-hair" theorems of GR [15], a BH may be characterized at most by only three parameters—three physical, conserved quantities—namely its mass M, electromagnetic charge Q and angular-momentum J. All three aforementioned GR solutions perfectly comply with this dictation.

The simplest extension of pure General Relativity amounts to the addition of a free, massless scalar field ϕ to the theory. If we assume that this field is a static, spherically-symmetric one, i.e. $\phi = \phi(r)$, its equation of motion simply reads

$$\Box\phi = 0 \;\Rightarrow\; \partial_r[\sqrt{-g}\, g^{rr}\, \partial_r\phi] = 0\,. \tag{1}$$

For an invertible metric tensor with $g \neq 0$, the above leads to the result

$$\phi' \sim g_{rr}\,, \tag{2}$$

which diverges at the horizon. As a result, the simple Einstein-scalar theory does not possess a regular, scalarised black-hole solution.

Even for a self-interacting scalar field with $V(\phi) \neq 0$, Bekenstein's *old scalar "no-hair theorem"* [15] excludes regular black-hole solutions in a general class of minimally-coupled scalar-tensor theories. Starting from the scalar-field equation, multiplying with ϕ, integrating over the exterior spacetime, and performing an integration by parts of the $\Box\phi$ term, we obtain [16]

$$\int_{\mathcal{V}} d^4x \sqrt{-g} \left[\partial_\mu\phi\, \partial^\mu\phi + \phi V'(\phi)\right] + \int_{\partial V} d^3x \sqrt{h}\, \eta^\mu\, \phi \partial_\mu \phi = 0 \,. \tag{3}$$

The boundary term at the end of the above expression vanishes both at the horizon of the black hole, provided that the scalar field remains there finite, and at asymptotic infinity, under the assumption that the self-interacting scalar field falls off sufficiently fast. The first term inside brackets gives $g^{rr}(\partial_r\phi)^2 > 0$ and, thus, the constraint allows for scalarised black-hole solutions only in the case where $\phi V'(\phi) < 0$. However, for e.g. a typical mass term $V(\phi) = m^2\phi^2/2$, this would demand $m^2 < 0$ and would result in a non-physical theory. Generalising this, we may therefore conclude that any theory with a minimally-coupled scalar field and a self-interacting potential satisfying the constraint $\phi V'(\phi) > 0$ does not allow for black-hole solutions with a non-trivial, static scalar hair.

2.2 *Wormholes*

General Relativity admits a second class of compact objects, namely wormholes, which are in fact hidden in the interior of all black-hole solutions that the theory predicts. Unfortunately, these wormhole solutions are not *traversable*.

Taking the Schwarzschild black-hole line-element as a paradigm

$$ds^2 = -\left(1 - \frac{2M}{r}\right) dt^2 + \left(1 - \frac{2M}{r}\right)^{-1} dr^2 + r^2 \left(d\theta^2 + \sin^2\theta\, d\varphi^2\right), \tag{4}$$

it is clear that whereas the exterior region ($r > 2M$) of this background is clearly static, the interior region ($r < 2M$), due to the change of its signature as we cross the horizon, is clearly dynamical. Even the presence of the spacetime singularity at $r = 0$ is not in fact a static feature of the complete Schwarzschild spacetime. A simple but careful analysis [17] reveals that, as the time goes by, a throat appears in the place of the singularity, which connects two asymptotically far-away regions. The radius of the throat expands, reaches its maximum value $r_{max} = 2M$ and then shrinks again and disappears leaving behind the two spacetime singularities of the black and the white hole which together comprise the complete Schwarzschild geometry. Unfortunately, this Einstein-Rosen passage [18–20] opens and closes so quickly that no physical particles, including photons, can pass through.

The Reissner-Nordstrom and Kerr geometries also possess similar internal tunnels. In fact, due to the presence of the internal Cauchy horizons in both of these solutions, the spacetime singularity can always be avoided and the tunnel remains always open. Unfortunately, it is again non-traversable: the internal Cauchy horizons they both possess are unstable, and any small disturbance causes them to collapse and turns them to a spacetime singularity, which unavoidably blocks the passage.

Trying to construct a more general theory than GR, in the context of which a traversable wormhole solution could emerge, we may add as before a scalar field, free of self-interacting. One could also adopt a different perspective [21] on how the spacetime should look like in order to avoid altogether the presence of a horizon or a spacetime singularity, thus enhancing the probability for constructing a traversable wormhole. As we will see in detail in Sect. 4, where we will also follow this different perspective, the addition of a massless scalar field can indeed support a wormhole solution, the well-known Ellis-Bronnikov wormhole [22, 23]. However, the scalar field must be a *ghost* one since its energy density ρ satisfies the constraint $\rho < 0$, and thus violates the energy conditions. The addition of a self-interacting potential also leads to constraints on the energy density and pressure components of the theory which again hint to some exotic form of matter rather than a physical field.

2.3 Particle-Like Solutions

In flat Minkowski spacetime, solutions that are regular and describe different types of distribution of matter are quite common, and are usually termed *solitons*. However, in the context of a pure gravitational theory, such as General Relativity, no such regular solutions emerge.

The same conclusion holds when simple modifications of GR are considered. For instance, if we add again a massless, spherically-symmetric scalar field to the theory, the well-known Fisher/Janis-Newman-Winicour-Wyman solution [24–26] can be found where the metric components and scalar field are given by the expressions

$$|g_{tt}| \sim (r - r_s)^{2s}\,, \quad g_{rr} \sim (r - r_s)^{2(1-s)}\,, \quad \phi \sim D \ln (r - r_s)\,, \tag{5}$$

respectively, with $s = 1/\sqrt{1 + (D/2M)^2}$. In the above, M is the mass of the solution and D a scalar "charge". However, as one may see by calculating the gravitational scalar quantities of the spacetime, this is an irregular solution since the latter diverge at the radius $r_s = M/2s$. Similar results follow if one adds a self-interacting potential for the scalar field.

3 The Einstein-Scalar-GB Theory

According to the brief review of compact objects presented in Sect. 2, one would need to consider a theory beyond pure GR as well as beyond the simple Einstein-scalar theory in order to discover novel solutions describing compact objects which are physically interesting. A more elaborate, generalised theory of gravity could follow by introducing extra fields and/or higher gravitational terms, and could be schematically described by the following action functional

$$S = \int d^4x\,\sqrt{-g}\left[f(R, R_{\mu\nu}, R_{\mu\nu\rho\sigma}, \Phi_i) + \mathcal{L}_X(\Phi_i)\right], \tag{6}$$

where $R_{\mu\nu\rho\sigma}$ is the Riemann tensor, $R_{\mu\nu}$ the Ricci tensor, R the Ricci scalar and Φ_i stands collectively for the different types of fields present in the theory whose properties are described by the Lagrangian $\mathcal{L}_X(\Phi_i)$. In what follows, we will retain, apart from the gravitational field, a single additional, scalar degree of freedom and ignore all other forms of fields. We will nevertheless introduce a higher-derivative term in the form of a quadratic gravitational term. Such a term would naturally go unnoticed in regions of weak gravitational field but could cause significant deviations from GR in the strong field regime.

In particular, we will consider the following generalised, quadratic theory of gravity

$$S = \int d^4x\,\sqrt{-g}\left[\frac{R}{16\pi G} - \frac{1}{2}\,\partial_\mu\phi\,\partial^\mu\phi + f(\phi)\,R^2_{GB}\right], \tag{7}$$

where R^2_{GB} is the so-called Gauss-Bonnet (GB) term

$$R^2_{GB} = R_{\mu\nu\rho\sigma}R^{\mu\nu\rho\sigma} - 4R_{\mu\nu}R^{\mu\nu} + R^2 \tag{8}$$

and $f(\phi)$ is an arbitrary coupling function between the scalar field ϕ and the GB term. The above theory is hardly a new one—in fact, it is common knowledge that it arises as part of the string effective action at low energies, as part of a Lovelock effective theory in four dimensions or as part of an extended scalar-tensor (Horndeski or DHOST) theory. It is a higher-derivative, gravitational theory of gravity yet simpler than one would expect: the particular combination of the gravitational quantities appearing in the definition of the GB term (8) guarantees that the field equations contain only up to 2nd-order derivatives of the metric tensor and the scalar field thus avoiding any Ostrogradski instabilities [27]. It is in the context of this quadratic theory—upgraded to a class of theories due to the general form of the coupling function $f(\phi)$—that we will look for scalarised solutions describing novel black-holes, traversable wormholes and regular particle-like solutions.

3.1 Black-Hole Solutions in Einstein-Scalar-GB Theory

Do we have any reason to believe that the generalised gravitational theory (7) may lead to scalarised black holes, in other words that this theory violates Bekenstein's scalar no-hair theorem? Yes, quite a few indeed. To start with, more than 25 years ago, a novel type of black holes, the so-called *dilatonic* black holes [28, 29], were discovered in the context of the theory (7) with an exponential coupling function, i.e. $f(\phi) = \alpha e^{\phi}$, between the scalar field, or dilaton, and the GB term. Due to the presence of the GB term, the field equations are quite complicated and cannot be solved analytically—therefore, the dilatonic black holes, as well as a large number of variants of this solution were found either in an approximate form or through numerical integration (see [30–47] for an indicative list of works).

Two decades later, another class of scalarised black-hole solutions was found [48, 49] in the context of the shift-symmetric EsGB theory, i.e. for the choice $f(\phi) = \alpha\phi$ of the coupling function. Here, as in the case of the dilatonic black holes, too, α is a coupling constant of the theory. Both the dilatonic and the scalarised shift-symmetric black-hole solutions violated the requirements of the scalar no-hair theorems, both the old [15] and the new versions [50–52] of it, thus paving the way for new black-hole solutions with characteristics not allowed by General Relativity [53, 54].

The obvious question which readily emerges is whether we can find additional classes of black-hole solutions in the context of the theory (7) but for other choices of the coupling function apart from the exponential and the linear one. To investigate this, we reconsidered the Einstein-scalar-GB theory but allowed for a general form of the coupling function $f(\phi)$ [55, 56]. We focused on the simplest possible background, that of a static, spherically-symmetric black hole, and assumed the following form of line-element

$$ds^2 = -e^{A(r)}dt^2 + e^{B(r)}dr^2 + r^2(d\theta^2 + \sin^2\theta\, d\varphi^2)\,, \tag{9}$$

with two unknown metric functions of the radial coordinate. Taking the variation of the action (7) with respect to the scalar field ϕ and the metric tensor $g_{\mu\nu}$, we obtain the scalar field and gravitational field equations

$$\nabla^2\phi + \dot{f}(\phi)R^2_{GB} = 0\,, \qquad R_{\mu\nu} - \frac{1}{2}\,g_{\mu\nu}\,R = T_{\mu\nu}\,, \tag{10}$$

respectively. In the above

$$T_{\mu\nu} = -\frac{1}{4}g_{\mu\nu}(\partial\phi)^2 + \frac{1}{2}\partial_\mu\phi\partial_\nu\phi - \frac{1}{2}(g_{\rho\mu}g_{\lambda\nu} + g_{\lambda\mu}g_{\rho\nu})\eta^{\kappa\lambda\alpha\beta}\tilde{R}^{\rho\gamma}{}_{\alpha\beta}\nabla_\gamma\partial_\kappa f \tag{11}$$

is the energy-momentum tensor of the theory which receives contributions from both the kinetic term of the scalar field and its non-minimal coupling to the quadratic GB term.

The set of field equations (10) involve the two unknown metric functions $A(r)$ and $B(r)$, and the scalar field $\phi(r)$. In fact, the (rr)-component of the gravitational equations takes the form of a 2nd order polynomial for e^B, which therefore may be determined once the solutions for $A(r)$ and $\phi(r)$ are found. The remaining equations reduce indeed to a system of only two independent, ordinary differential equations of second order for the unknown functions $A(r)$ and $\phi(r)$, which is given below schematically

$$A'' = \frac{P}{S}, \qquad \phi'' = \frac{Q}{S}. \tag{12}$$

The quantities (P, Q, S) depend on (r, ϕ, ϕ', A'), and the interested reader may find their exact expressions as well as the remaining technical details of this analysis in [55, 56].

Due to the complexity of the quantities involved, the integration of the system demands numerical integration. However, it may be solved analytically in the small and large regimes of the radial coordinate. The corresponding solutions may be used as boundary solutions for the general solution of the system and, in addition, their form will reveal under which constraints these describe indeed a robust black-hole background.

Starting from the small r-regime, we demand the presence of the most important feature of black-hole geometry, that of a regular horizon. For this, we impose the asymptotic behaviour

$$e^{A(r)} \to 0, \quad e^{-B(r)} \to 0, \quad \phi(r) \to \phi_h\,, \tag{13}$$

as r approaches a certain value r_h. The last demand—the finiteness of the scalar field—encompasses the notion of the regularity of the black-hole horizon. The same must naturally hold for the first and second derivative of the field. However, demanding that, under the behaviour of the metric functions displayed in (13), ϕ'' remains finite at the horizon r_h, the second of the equations in (12) leads to the constraint

$$\phi'_h = \frac{r_h}{4\dot{f}_h}\left(-1 \pm \sqrt{1 - \frac{96\dot{f}_h^2}{r_h^4}}\right). \tag{14}$$

The above formula ensures that, for a selected coupling function and value of the scalar field ϕ_h at the black-hole horizon r_h, choosing this particular value for ϕ'_h leads to a gravitational background that describes the near-horizon geometry of a black hole. No constraint arises on the form of the coupling function $f(\phi)$ itself, which therefore remains arbitrary. The coupling function needs to satisfy an additional constraint which follows from the positivity of the expression under the square root in (14); this may be written as

$$\dot{f}_h^2 < \frac{r_h^4}{96}, \tag{15}$$

and interpreted as a bound on the lower value of the black-hole horizon radius r_h for a given $f(\phi)$ and ϕ_h. It is this characteristic that distinguishes the GB black hole from the Schwarzschild black hole, its analog in the context of GR. Using the above results, the field equations may give the asymptotic forms of the metric functions and scalar field near the horizon, which are found to have the form

$$e^A = a_1(r - r_h) + \dots, \qquad e^{-B} = b_1(r - r_h) + \dots, \tag{16}$$

$$\phi = \phi_h + \phi_h'(r - r_h) + \phi_h''(r - r_h)^2 + \dots, . \tag{17}$$

At the other asymptotic regime, i.e. at large distances from the horizon, we assume, as usually, a power series expansion in $1/r$ for the three unknown functions, namely

$$e^A = 1 + \sum_{n=1}^{\infty} \frac{p_n}{r}, \qquad e^B = 1 + \sum_{n=1}^{\infty} \frac{q_n}{r}, \qquad \phi = \phi_\infty + \sum_{n=1}^{\infty} \frac{d_n}{r}. \tag{18}$$

Substituting these expressions into the field equations, we may determine the arbitrary coefficients (p_n, q_n, d_n). In fact, p_1 and d_1 remain arbitrary and are identified with the black-hole mass M and scalar charge D, respectively. The asymptotic form of the metric and scalar field then take the final form

$$e^A = 1 - \frac{2M}{r} + \frac{MD^2}{12r^3} + \dots, \qquad e^B = 1 + \frac{2M}{r} + \frac{16M^2 - D^2}{4r^2} + \dots, \tag{19}$$

$$\phi = \phi_\infty + \frac{D}{r} + \frac{MD}{r^2} + \frac{32M^2D - D^3}{24r^3} + \frac{12M^3D - 24M^2\dot{f} - MD^3}{6r^4} + \dots. \tag{20}$$

As in the near-horizon analysis, no constraint on $f(\phi)$ arises by demanding a robust black-hole geometry at large distances. A general coupling function f does not interfere with the existence of an asymptotically-flat limit with its exact form making an appearance not earlier than in the 4th order term of the expansion.

In order to construct a complete black-hole solution, we need to find a solution of the field equations (12) which smoothly interpolates between the two asymptotic forms (17) and (20). And this is where the no-hair theorems play a crucial role: if the theory (7) satisfies the requirements of the no-hair theorems, these dictate that no solution that smoothly connects these two asymptotic forms can be found. Let us look briefly into this. We will start from the old no-hair theorem and follow a similar analysis [15, 16]. We will employ the scalar-field equation, multiply by $f(\phi)$ in our case, and integrate over the entire exterior spacetime. Performing an integration by parts, we arrive at the constraint [55–57]

$$\int_{\mathcal{V}} d^4x \sqrt{-g}\, \dot{f}(\phi) \left[\partial_\mu\phi\, \partial^\mu\phi - f(\phi)\, R^2_{GB}\right] - \int_{\partial V} d^3x \sqrt{h}\, \eta^\mu f(\phi) \partial_\mu\phi = 0\,. \tag{21}$$

Due to the dependence of the scalar field solely on the radial coordinate, it holds that $\partial_\mu\phi\,\partial^\mu\phi = g^{rr}(\partial_r\phi)^2 > 0$ and $\eta^\mu\partial_\mu\phi = g^{rr}\partial_r\phi$. The whole boundary term at the end of the above expression vanishes as usual at the black-hole horizon. However, as noted in [57], its value at asymptotic infinity depends on the form of the coupling function. If $f(\phi_\infty) = 0$, then the boundary term vanishes altogether and we need to address only the expression inside the square brackets in the first integral. One may easily see, by using the asymptotic expressions (17) and (20), that the GB term takes on positive values at the two asymptotic regimes [55, 56]—as the exact numerical analysis reveals, in fact it remains positive over the entire exterior regime. Therefore, the emergence of black-hole solutions in this case is allowed only for $f(\phi) > 0$ according to the old no-hair theorem. If, on the other hand, $f(\phi_\infty) \neq 0$, then the boundary term at asymptotic infinity reduces to $f(\phi_\infty)D$. In this case, solutions arise again for $f(\phi) > 0$ but also for a certain interval for negative values of $f(\phi)$ [57].

Let us now turn to Bekenstein's new version of no-hair theorem [50], developed for a theory with a minimally-coupled scalar field, and examine in turn whether the theory (7) evades its own requirements, too. This theorem relies on the particular profile of the $T^r_{\ r}$ component of the energy-momentum tensor. For instance, it demands that, at asymptotic infinity, $T^r_{\ r}$ is positive and decreasing. Indeed, using the asymptotic solution (20), and the exact expression of $T^r_{\ r}$, we find that this is indeed the case. Our result therefore agrees with the one derived in [50] at large distances; this was anticipated since the quadratic GB term is not expected to play any role far away from the black-hole horizon where the curvature of spacetime is small. In the near-horizon regime, the novel no-hair theorem dictates that $T^r_{\ r}$ is negative and increasing. In the context of the Einstein-scalar-GB theory, though, this does not hold any more: indeed, if we use the near-horizon asymptotic solution (17), we find that

$$sign(T^r_r)_h = -sign(\dot{f}_h\phi'_h) = 1 \mp \sqrt{1 - 96\dot{f}^2/r_h^4}\,, \tag{22}$$

where we have used the constraint Eq. (14) for the regularity of the horizon. It is clear that the above expression is always positive-definite, in contrast to the requirement of the novel no-hair theorem. The presence of the GB term near the horizon—where the curvature is strong—changes the profile of $T^r_{\ r}$, and causes the evasion of Bekenstein's new no-hair theorem.

The aforementioned argument was followed also in [28] to prove that the dilatonic theory with $f(\phi) = \alpha e^{\phi}$ evades Bekenstein's theorem, a result which opened the way to find new black-hole solutions, the so-called dilatonic black holes. In the context of the present analysis, it is clear that the exact form of the coupling function is again of no importance. Thus, we set off looking for new black holes in the context of the theory (7) choosing various forms for $f(\phi)$. For each form, we numerically integrated the system of equations (12), by giving the input values (ϕ_h, ϕ'_h). The first quantity was a free parameter constrained only by the condition (15) while the second one was uniquely determined by the regularity constraint (14) of the black-hole horizon. Using this method, every pair of initial values (ϕ_h, ϕ'_h) leads to a

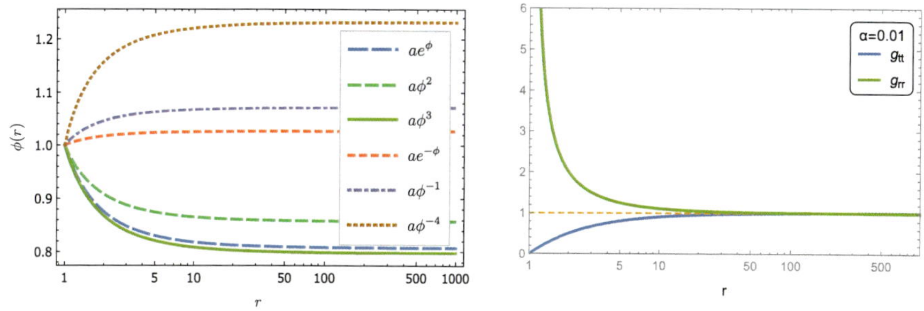

Fig. 1 The profile of the scalar field for a variety of choices for the coupling function $f(\phi)$ (left plot), and the two metric functions (in absolute value, right plot) [55, 56]

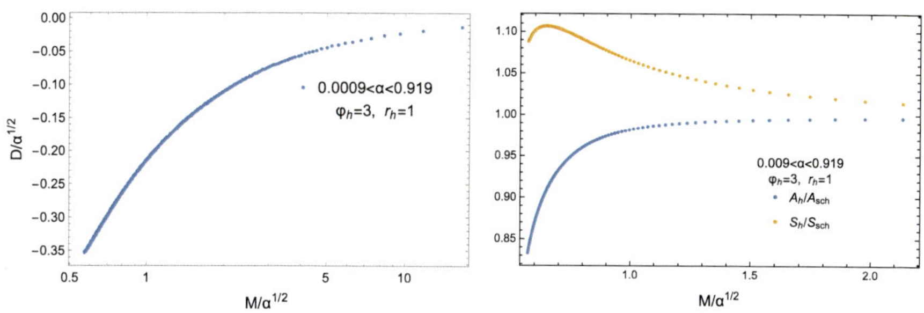

Fig. 2 The scalar charge D (left plot), and the ratios A_h/A_{Sch} and S_h/S_{Sch} (right plot, lower and upper curve respectively) in terms of the mass M, for $f(\phi) = a/\phi$ [55]

regular black-hole solution with a non-trivial scalar hair. In this way, we determined a large number of black-hole solutions with scalar hair for a variety of forms of the coupling function $f(\phi)$: exponential, odd and even power-law, odd and even inverse-power-law. An indicative subset of solutions for the scalar field are depicted in the left plot of Fig. 1 while, on the right plot, we present the profile of the two metric functions (in absolute value) with the characteristic form of an asymptotically-flat black-hole geometry. We should note at this point that our analysis covers both the case of *spontaneous scalarisation* (i.e. the case where the Schwarzschild solution arises as an independent solution) and the case of *natural scalarisation* (where the Schwarzschild solution does not emerge) depending on the particular choice of the coupling function $f(\phi)$.

In Fig. 2, we present some of the properties of the scalarised black holes for the indicative case of $f(\phi) = a/\phi$. The left plot presents the scalar charge D, that characterises the scalar field at infinity. As is clear, this quantity is in fact a function of the black-hole mass, a result that renders the scalar hair secondary. Also, a common characteristic in all cases is that, as the mass of the black hole increases, the scalar charge decreases and eventually vanishes as our black-hole solution matches the Schwarzschild solution. In other words, every massive GB scalarised black hole reduces to the Schwarzschild solution.

The right plot of Fig. 2 depicts two important quantities, the horizon area of the black hole, given by $A_h = 4\pi r_h^2$ and normalised in units of the horizon area of the Schwarzschild solution with the same mass, and the entropy of the black hole normalised again in units of the entropy of the corresponding Schwarzschild solution. Starting from the horizon area, we observe that this ratio is always smaller than unity, which means that all GB black-hole solutions are smaller than their GR analogues. This is due to the effect of the GB term which exerts a positive (outward) pressure and which can be counterbalanced only by "squeezing" further the available energy/mass distribution thus resulting into smaller black holes. We also observe that the area curve stops abruptly at its lower end, thus exhibiting the existence of a lower bound on the horizon area and thus on the mass of the black hole. Beyond this lower value, the black hole ceases to exist—this feature is due to the bound (15) discussed earlier. The entropy of the black-hole solutions may be computed, for an arbitrary form of $f(\phi)$, following various methods [36, 56], and it is found to be

$$S_h = \frac{A_h}{4} + 4\pi f(\phi_h)\,. \tag{23}$$

The profile of the entropy S_h of a GB black hole compared to the one of the Schwarzschild black hole, $S_{Sch} = A_h/4$, depends strongly on the choice for the particular form of the coupling function. For instance, for the choice employed in this case, i.e. the inverse linear form, the entropy ratio comes out to be larger than unity for the entire mass range, a result that renders this particular family of solutions more thermodynamically stable than the corresponding Schwarzschild black hole. For different forms of the coupling function though, the curve of the entropy ratio may lie in whole or in parts below unity thus revealing an instability for the entire or parts of the mass regime. In all cases, as the mass M increases, the entropy ratio always reduces to unity as the Schwarzschild limit is approached.

The theory (7) allows not only asymptotically-flat black-hole solutions but also black holes with an asymptotically Anti-de Sitter behaviour. This follows if we add to the theory a cosmological constant bringing the action functional of the theory to the form [58]

$$S = \frac{1}{16\pi}\int d^4x\sqrt{-g}\left[R - \frac{1}{2}\,\partial_\mu\phi\partial^\mu\phi + f(\phi)R^2_{GB} - 2\Lambda\right]. \tag{24}$$

In this case, the field equations (10) remain unchanged apart from the shift

$$T_{\mu\nu} \rightarrow T_{\mu\nu} - \Lambda\, g_{\mu\nu}\,. \tag{25}$$

Apart from the change in the spacetime background at large distances, we expect that the profile of the scalar field will also be affected. Indeed, imposing as before the regularity of its second derivative ϕ'' at the black-hole horizon, the field equations lead to the modified constraint

$$\phi'_h = \frac{16\Lambda r_h \dot{f}^2\,(\Lambda r_h^2 - 3) + \Lambda r_h^5 - r_h^3 \mp \sqrt{C}}{4\dot{f}[r_h^2 - \Lambda(r_h^4 - \dot{f}^2)]}\,, \tag{26}$$

where all quantities have been evaluated at r_h. The quantity C under the square root stands for the following combination

$$C = 256\Lambda \dot{f}_h^4\left(\Lambda r_h^2 - 6\right) + 32 r_h^2 \dot{f}_h^2\left(2\Lambda r_h^2 - 3\right) + r_h^6 \geq 0\,, \tag{27}$$

and must always be non-negative for ϕ'_h to be real. Under the validity of the constraint (26), the asymptotic form of the solution for the metric function and the scalar field takes the same functional form as the one given in (17).

Assuming the presence of a negative cosmological constant, and expecting the spacetime to assume a form close to that of the Schwarzschild-Anti-de Sitter solution at large distances, we find that the approximate forms for the metric functions in that regime is

$$e^{A(r)} = \left(k - \frac{2M}{r} - \frac{\Lambda_{eff}}{3}r^2 + \frac{q_2}{r^2}\right)\left(1 + \frac{q_1}{r^2}\right)^2, \tag{28}$$

$$e^{-B(r)} = k - \frac{2M}{r} - \frac{\Lambda_{eff}}{3}r^2 + \frac{q_2}{r^2}, \tag{29}$$

where k, M, Λ_{eff} and $q_{1,2}$ are arbitrary constants. Regarding the asymptotic form of the scalar field, this is given by

$$\phi(r) = \phi_\infty + d_1 \ln r + \frac{d_2}{r^2} + \frac{d_3}{r^3} + \dots\,, \tag{30}$$

where again $(\phi_\infty, d_1, d_2, d_3)$ are arbitrary constant coefficients. In principle, the scalar field assumes this form in the case of a linear coupling function, $f(\phi) = a\phi$ [59, 60], however, it may be shown numerically that it holds in the perturbative limit of small GB coupling constant a, for all forms of the coupling function. We notice that the dominant term in the expression of ϕ at large distances has a logarithmic form and not an $1/r$ dependence; as a result, no scalar charge may be attributed to any scalarised solution found. In contrast, the coefficient M in the asymptotic form of the metric functions may be interpreted again as the gravitational mass of the solution.

As in the case of zero cosmological constant, a plethora of solutions emerged from the numerical integration of the set of field equations under the appropriate boundary conditions. Ensuring that the input quantity ϕ'_h is given by the expression (26), black-hole solutions with an asymptotically Anti-de Sitter behaviour emerged for all different choices of the coupling function, namely $f(\phi) = e^{\pm\phi}$, $\phi^{\pm 2n}$, $\phi^{\pm(2n+1)}$, $\ln \phi$, ... [58]. The profiles of both the scalar field and the metric functions are given in the left and right plot, respectively, of Fig. 3.

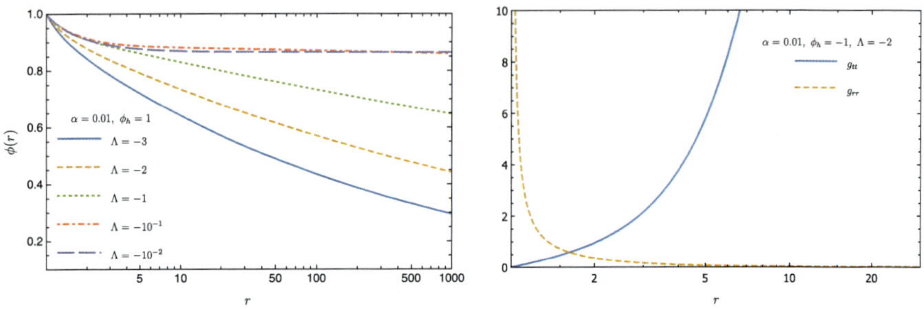

Fig. 3 The solution for the scalar field ϕ (left plot), and for the metric components $|g_{tt}|$ and g_{rr} (right plot) in terms of the radial coordinate r, for $f(\phi) = a\phi^2$ [56]

We may finally promote the negative cosmological constant to a dynamic potential for the scalar field [61] as follows

$$S = \int d^4x \, \sqrt{-g} \left[\frac{R}{16\pi G} - \frac{1}{2}\, \partial_\mu \phi \, \partial^\mu \phi + f(\phi)\, R^2_{GB} - 2\Lambda\, V(\phi) \right]. \tag{31}$$

We will insist again on the case of the negative coupling constant, i.e. $\Lambda < 0$, and consider different choices for the coupling function $f(\phi)$ and the scalar potential $V(\phi)$. Demanding again the regularity of the black-hole horizon and repeating the aforementioned analysis, we may determine a similar constraint on the value of ϕ'_h. This quantity may be used again an an input parameter, together with ϕ_h, for the numerical integration of the field equations, once a specific form for $f(\phi)$ and $V(\phi)$ is chosen. Again, for every possible choice of these two scalar functions, black-hole solutions with a regular horizon and a non-trivial ϕ emerged.

In Fig. 4 (left plot), we present the profile of the scalar field in terms of the radial coordinate for many different (polynomial) choices of the scalar potential, and for fixed form of the coupling function, i.e. $f(\phi) = ae^{\phi}$. We observe that for $\Lambda < 0$, the scalar field oscillates around the zero value where it finally relaxes. As the degree of the polynomial increases, the relaxation time gets longer. In the right plot, we depict the domain of existence of the black-hole solutions under the same choices. Here, we notice a different pattern for the solutions emerging for $V(\phi) = \Lambda\phi^2$ from the one assumed for the other polynomial forms. In the latter cases, the black-hole solutions form branches which are more prominent in the small-mass regime but tend to smooth out and eventually match the horizon value of the corresponding Schwarzschild solution. In the case of a quadratic scalar potential, the solutions present a distinctly different pattern: the solutions form a monotonically decreasing line of existence spanning the whole regime, from large-r_h, low-mass black holes to small-r_h, large-mass black holes. The theory therefore includes *ultra-sparse* black holes, *Schwarzschild-like* black holes and *ultra-compact* black holes. The study of the entropy reveals a similar behaviour in terms of the mass as the horizon radius; as a result, the ultra-sparse black holes are found to be more thermodynamically

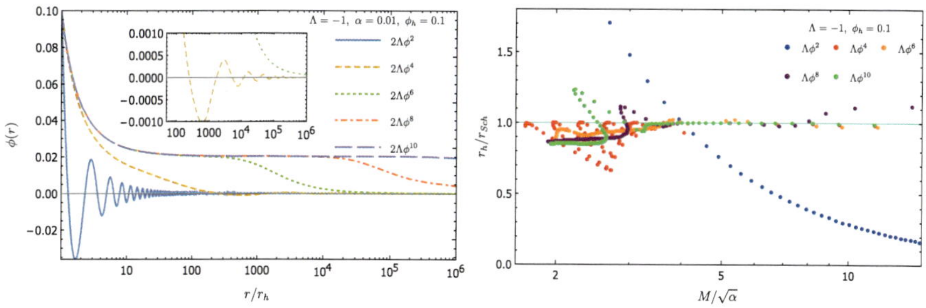

Fig. 4 The profile of the scalar field in terms of the radial coordinate for different (polynomial) choices of the scalar potential (left plot); the lines of existence of the black-hole solutions for the same choices of $V(\phi)$ (right plot) [61]

stable compared to the Schwarzschild-like black holes, and these in turn to be more thermodynamically stable compared to the ultra-compact black holes.

The discussion above summarizes only a small part of the black-hole solutions that have been found in the context of gravitational theories containing the GB or related terms. There is in fact a huge literature in this topic and the interested reader may find more results in [62–151].

3.2 Wormholes in Einstein-Scalar-GB Theory

As we stated in Sect. 2, General Relativity does not accommodate *traversable* wormholes. We also claimed that the Ellis-Bronnikov solution, which emerges in the context of the Einstein-scalar theory, is supported in fact by a ghost scalar field. Is there a particular feature of the wormhole geometry that makes it incompatible with GR or simple extensions of it? Actually, there is.

To demonstrate this, we will employ the Morris-Thorne method for the construction of wormhole solutions [21]. This method also incorporates the alternative approach mentioned in Sect. 2, in which traversable wormhole solutions may emerge only in the absence of a horizon or singularity. According to the analysis of [21], a traversable wormhole may be described by the following line-element

$$ds^2 = -e^{2\Phi(r)}\,dt^2 + \left(1 - \frac{b(r)}{r}\right)^{-1} dr^2 + r^2\,(d\theta^2 + \sin^2\theta\,d\varphi^2)\,. \tag{32}$$

The *red-shift* function Φ must be everywhere finite. The *shape* function b must satisfy $1 - b/r \geq 0$ throughout spacetime, and is allowed to vanish at a single point, i.e. at $r = b = b_0$, where the throat of the wormhole is located. At infinity, Φ and b/r must both vanish to recover the asymptotically flat regime.

To study the geometric structure of the above solution, we construct its embedding diagram. To this end, we equate the line-element of the 2D spacelike equatorial surface, obtained for $t = const.$ and $\theta = \pi/2$, with the one of a 3D Euclidean space:

$$\frac{dr^2}{1 - b/r} + r^2\, d\varphi^2 \equiv dz^2 + d\rho^2 + \rho^2\, d\varphi^2 = \left[\left(\frac{dz}{d\rho}\right)^2 + 1\right] d\rho^2 + \rho^2\, d\varphi^2 . \quad (33)$$

From the above, we obtain immediately that $\rho = r$. We demand that both line-elements describe the same geometry and equate the coefficients of dr^2 in (33). Then, the "fictitious" z coordinate is given by the relation

$$\frac{dz}{dr} = \pm\left(\frac{r}{b} - 1\right)^{-1/2} \Rightarrow z(r) = \pm \int_{b_0}^{r} \left(\frac{r}{b} - 1\right)^{-1/2} dr\,, \quad (34)$$

and describes the lift of the (r, φ)-plane due to the curvature of spacetime. At infinity, as $b/r \to 0$, we obtain $dz/dr \to 0$, and the embedding surface becomes parallel to the (r, φ)-plane exhibiting no lift due to the flatness of this regime. In contrast, near the throat, the slope diverges and this signifies that the embedding surface at this point is vertical. Therefore, $\{\rho(r), z(r)\}$ is a parametric representation of a slice of the embedded $\theta = \pi/2$-plane for a fixed value of the φ coordinate, while the corresponding surface of revolution is the 3-D representation of the wormhole's geometry. The typical embedding diagram of a wormhole is given in Fig. 5.

One could invert the function $z(r)$ to obtain $r(z)$. Then, from Fig. 5, it is clear that $r(z)$ possesses a minimum at the throat and then "flares outwards" as we approach the asymptotically flat region. Therefore, the existence of a throat is mathematically ensured if the following two conditions hold

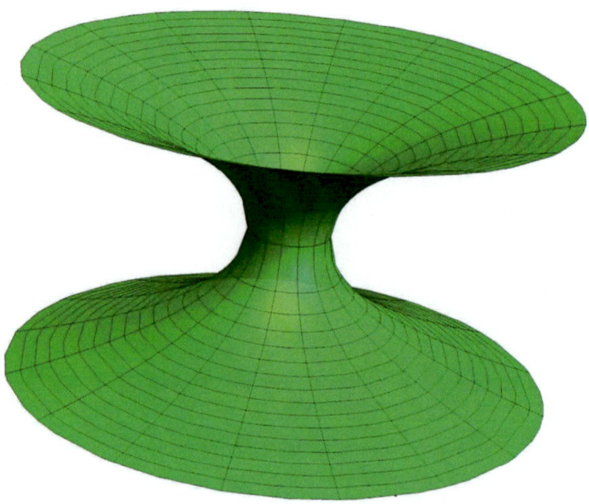

Fig. 5 The embedding diagram of a typical traversable wormhole

$$\left.\frac{dr}{dz}\right|_{r_0} = 0\,, \qquad \left.\frac{d^2r}{dz^2}\right|_{r_0} > 0\,. \tag{35}$$

The first condition is satisfied since, as we showed above, $dz/dr \to 0$ at $r = r_0$. The second condition takes the form

$$\frac{d^2r}{dz^2} = \frac{b - rb'}{2b^2} > 0\,, \tag{36}$$

and is known as the *flaring-out* condition of the wormhole. Therefore, in order to have a throat, the shape function $b(r)$ should satisfy $b - rb' > 0$, which fully justifies its name.

In fact, the flaring-out condition leads to the violation of energy conditions in a wormhole background since it imposes a certain behaviour on the shape function and, through Einstein's equations, on the matter content of the theory. Let us focus on the Null Energy Condition (NEC) which has the form:

$$T_{\mu\nu}n^\mu n^\nu = -g_{tt}\,(T_r^r - T_t^t) = -g_{tt}\,(\rho + p_r) \geq 0, \tag{37}$$

where $T_{\mu\nu}$ is the energy-momentum tensor and n^μ any null vector satisfying $n^\mu n_\mu = 0$. For $g_{tt} < 0$, the above demands that $\rho + p_r \geq 0$. However, through Einstein's equations and using the line-element (32), we obtain

$$8\pi G\,(\rho + p_r) = G_r^r - G_t^t = -\left.\frac{(b - rb')}{r^3}\right|_{r=r_0} < 0\,, \tag{38}$$

where we used the flaring-out condition (36). As a result, the wormhole geometry may be supported by some distribution of matter which violates the energy conditions and instead satisfies the condition $\rho + p_r < 0$. That is why pure GR cannot support a traversable wormhole nor can the Einstein-scalar theory with a physical scalar field do it either.

It becomes therefore clear that a more involved gravitational theory beyond GR is necessary in order to support viable wormhole solutions. Can we find such a physical theory? The Einstein-scalar-GB theory is a theory which violates the energy conditions, as the evasion of no-hair theorems clearly demonstrated. However, this violation is not caused by the introduction of some form of exotic matter but by the direct coupling of a physical scalar field to the quadratic, gravitational GB term. Can this violation then support wormhole solutions?

In [28], where the dilatonic black holes were first discovered, in the context of the EsGB theory with an exponential coupling function, $f(\phi) = a\,e^{\phi}$, another class of solutions were presented with the line-element having the form of a Morris-Thorne solution. By applying a coordinate transformation [152, 153], this was written as

$$ds^2 = -e^{A(\ell)}dt^2 + e^{B(\ell)}d\ell^2 + (\ell^2 + r_0^2)\,(d\theta^2 + \sin^2\theta d\varphi^2)\,, \tag{39}$$

where ℓ is a new spacelike coordinate ranging in the interval $(-\infty, \infty)$. The above line-element describes a wormhole connecting two asymptotically-flat regimes and with a radius throat r_0 at $\ell = 0$. In this new coordinate system, both metric functions and the scalar field assume regular forms in the regime close to the throat, namely

$$e^{A(\ell)} = a_0 + a_1\ell + \dots, \quad e^{B(\ell)} = b_0 + b_1\ell + \dots, \quad \phi(\ell) = \phi_0 + \phi_1\ell + \dots \tag{40}$$

On the other hand, at large distances, we obtain power-law expansions in terms of $(1/r)$

$$e^A \simeq 1 - \frac{2M}{\ell} + \dots, \qquad e^B = 1 + \frac{2M}{\ell} + \dots, \qquad \phi \simeq \phi_\infty + \frac{D}{\ell} + \dots, \tag{41}$$

where M and D are the mass and scalar charge of the wormhole. Note that these expansions are identical to the ones for a black-hole solution with the exception that here M and D are independent quantities.

Employing these asymptotic solutions as boundary conditions, the set of field equations (10) were numerically solved again, this time in quest of wormhole solutions [152, 153]. And such solutions were indeed found, with the spacetime background having the geometry depicted in Fig. 5 and the scalar field assuming the profile presented on the left plot of Fig. 6. However, a true singularity was lurking behind the throat, at the negative ℓ regime. Thus, a traversable, symmetric solution was built by cutting the spacetime at $\ell = 0$ and gluing the positive regular ℓ-regime with its mirror image along $\ell < 0$. At the gluing point $\ell = 0$, cusps were created whose presence could be justified only by introducing a distribution of a (non-exotic!) perfect fluid and a gravitational source term around the throat. This distribution is described by the action functional [152, 153]

$$S_3 = \int d^3x \sqrt{h}\, (\lambda_1 + \lambda_0 e^{\phi} \tilde{R})\,, \tag{42}$$

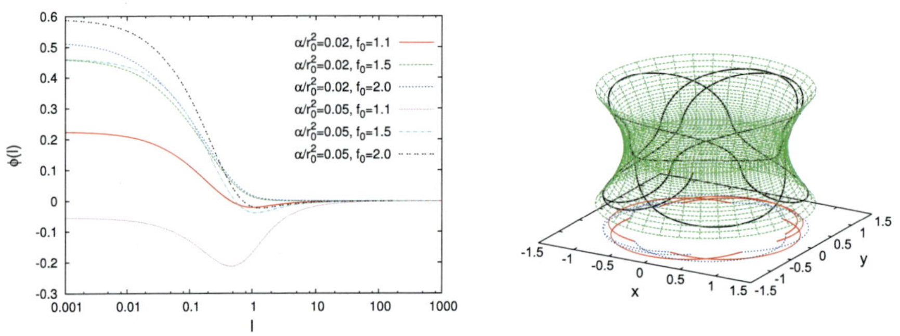

Fig. 6 The solutions for the scalar field for a family of dilatonic wormholes (left plot), and particles trajectories in the background of a dilatonic wormhole (right plot) [152, 153]

where (λ_1, λ_2) are constants and $\tilde{R}$ is the scalar curvature of the 3-dimensional induced spacetime at $\ell = 0$. Under the above construction, a regular wormhole solution is constructed which can be traversed both by null and timelike particles. A set of particle trajectories, which start from one asymptotically-flat regime, traverse the throat, reach the asymptotically-flat regime on the other side of the wormhole and then travel back to the starting point, are presented on the right plot of Fig. 6.

All the above results were derived in the context of the dilatonic Einstein-scalar-GB theory, that is with an exponential coupling function. Can we generalise this analysis and produce similar wormhole solutions in the context of the EsGB theory with alternative forms for $f(\phi)$, as in the case of black holes? In the context of the general theory (7), we looked indeed for viable wormhole solutions following a similar approach as above. Apart from the form of the coupling function, we incorporated also a different form of line-element, namely the following [154]

$$ds^2 = -e^{A(\eta)}dt^2 + e^{\Gamma(\eta)}\left[d\eta^2 + (\eta^2 + \eta_0^2)\,(d\theta^2 + \sin^2\theta d\varphi^2)\right]. \tag{43}$$

This new ansatz modifies the conditions for the existence of a sole minimum in the function $r(z)$ in the embedding diagram, and allows for multiple extremal points and thus for a much richer topology: local maxima correspond to the so-called *equators* whereas local minima give rise to *multiple throats*.

The metric functions and scalar field assume regular forms both near the throat and the far asymptotic regime. The numerical integration yielded a large number of wormhole solutions, for every form of $f(\phi)$ used, with an asymptotically-flat behaviour and with a single throat or with a double throat and an equator. The embedding diagram of a wormhole in the latter case is shown on the left plot of Fig. 7. All solutions were made traversable by employing the same cut and paste technique, where the spacetime with $\ell > 0$ was glued at $\ell = 0$ with its mirror image along $\ell < 0$. As before, a distribution of regular matter around the throat suffices to justify

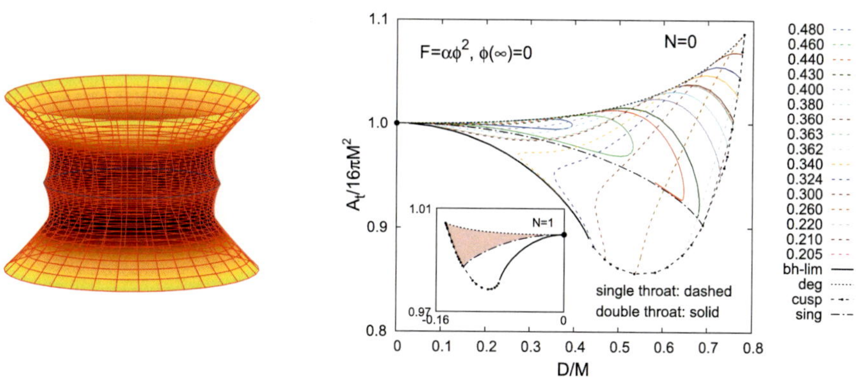

Fig. 7 A wormhole solution with an equator and two throats (left plot), and the domain of existence of wormhole solutions in the case where $f(\phi) = a\phi^2$ (right plot) [154]

the cusps at the gluing point. On the right plot of Fig. 7, we present the domain of existence of wormhole solutions in the indicative case of quadratic coupling function, $f(\phi) = a\phi^2$. All EsGB wormhole solutions are bounded by the corresponding black-hole solutions which form the left boundary of the domain.

For additional works on wormhole solutions arising in the context of generalised gravitational theories involving scalar fields or higher-curvature terms see, for instance, [155–165].

3.3 Particle-Like Solutions in Einstein-Scalar-GB Theory

Let us finally address the last class of compact objects, the gravitational particle-like solutions. As was mentioned, in the context of the minimal Einstein-scalar theory, the corresponding Fisher/Janis-Newman-Winicour-Wyman particle-like solution [24] is irregular. Can we hope to find particle-like solutions which however behave better in the context of the EsGB theory?

With this in mind, we looked for static, spherically-symmetric solutions which describe a regular spacetime, with no singularities or horizons and no throats as well [166, 167]. We chose to work with the following form of line-element

$$ds^2 = -e^{A(r)}dt^2 + e^{\Gamma(r)}\left[dr^2 + r^2\left(d\theta^2 + \sin^2\theta d\varphi^2\right)\right], \tag{44}$$

with two unknown metric functions, $A(r)$ and $\Gamma(r)$, as usual. The above form was inspired by the one employed in our quest for wormhole solutions and resulted, as we saw, in a much richer topology of these solutions around the throat. Here, we will use a similar line-element in the hope that solutions with features not allowed or often observed, when more restrictive ansatzes are used, will now emerge.

The set of field equations of the theory (10) must now be solved for the three unknown functions, the two metric functions and the scalar field $\phi(r)$. If we substitute the above ansatz for the line-element in the field equations, we obtain four, coupled, ordinary differential equations which can further be reduced to a system of only two independent equations for $A(r)$ and $\phi(r)$. The remaining metric function $\Gamma(r)$ can be determined once the solutions for $A(r)$ and $\phi(r)$ are found.

The set of the two independent, second-order differential equations was solved numerically under the proper boundary conditions. We demanded a regular form for both functions over the entire radial regime, an asymptotically-flat behaviour at large distances and a smooth solution close to the origin. At large distances, we obtained indeed an asymptotically-flat behaviour, which was identical to the one for black holes and wormholes, namely

$$e^{A(r)} \simeq 1 - \frac{2M}{r} + \ldots, \qquad \phi \simeq \phi_\infty - \frac{D}{\ell} + \ldots, \tag{45}$$

where M and D are once again the mass and the scalar charge of the solution. The expansion near the origin of the radial coordinate was found to be more involved and to depend on the form of the coupling function $f(\phi)$. For instance, for a quadratic form, $f(\phi) = \alpha\phi^2$, we find the asymptotic solution

$$A(r) = A_0 + A_2 r^2 + A_3 r^3 + \dots , \tag{46}$$

$$\phi = -\frac{c_0}{r} + \phi_0 + \phi_1 r + \phi_2 r^2 + \phi_3 r^3 + \dots . \tag{47}$$

We readily observe that the metric is indeed regular near the origin but the scalar-field expansion features a singular term with a divergent behaviour at small r. In fact, the presence of this term seems to be a generic feature of the expansion of the scalar field near the origin independently of the form of the coupling function. The numerical integration of the system of field equations led to the complete solution which, as expected, interpolated between the two asymptotic forms. The profile of the complete solutions for the metric and scalar field are depicted, for the quadratic coupling function, on the left plot of Fig. 8 [166, 167].

In order to investigate the consequences of the singular term, present in the expansion of the scalar field near the origin, we computed all gravitational scalar invariants as well as the components of the energy-momentum tensor $T_{\mu\nu}$. Despite the singularity in ϕ, all these quantities were found to be finite over the entire radial regime. The exact values of the energy-density and pressure at the origin are given below:

$$\rho(0) = -\frac{3}{32\alpha}, \qquad p(0) = \frac{2}{32\alpha}. \tag{48}$$

We may therefore conclude that the singularity of the scalar field at the origin has no physical consequence, and that the situation resembles more the behaviour of the Coulomb potential which also diverges at the origin of the coordinate system.

The complete profile of the energy-density ρ of an indicative particle-like solution is depicted on the right plot of Fig. 8. We observe that it has a regular, shell-like

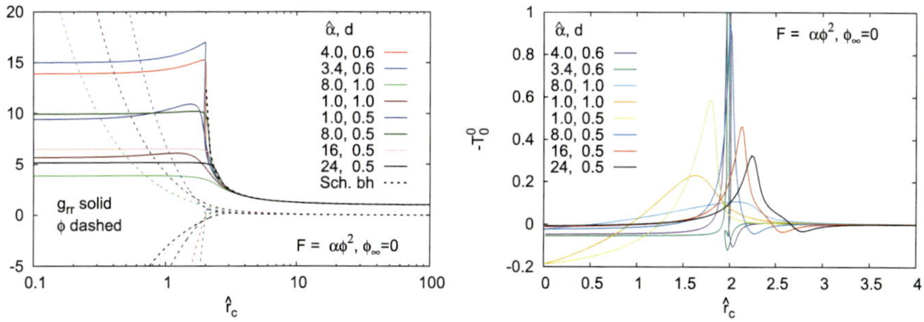

Fig. 8 The solutions for the metric and scalar field (left plot) and the corresponding energy-density (right plot) in the case of the quadratic coupling function [166, 167]

behaviour, and vanishes quickly at a very small distance from the origin. We may therefore conclude that our gravitational particle-like solutions are more *bubble-shaped* than point-like, and clearly qualify as *ultra-compact* objects (see also [168–171]).

The propagation of test particles in the background of the aforementioned gravitational bubble-shaped solutions may lead to interesting observable phenomena. The geodesics for both null and timelike particles may be found from their Lagrangian $\mathcal{L}$ given by the expression

$$2\mathcal{L} = g_{\mu\nu}\dot{x}^\mu\dot{x}^\nu = -e^{f_0}\dot{t}^2 + e^{f_1}\left[\dot{r}^2 + r^2\left(\dot{\theta}^2 + \sin^2\theta\dot{\varphi}^2\right)\right] = -\epsilon\,, \tag{49}$$

where $\epsilon = 0$ and 1 for massless and massive particles, respectively. Employing the two conserved quantities, namely the energy $E = -e^{f_0}\dot{t}$ and the angular-momentum $L = e^{f_1}r^2\dot{\varphi}$ of the particle, and considering motion in the equatorial plane ($\theta = \pi/2$), we obtain the following equation for null particles

$$e^{f_0+f_1}\dot{r}^2 = (E + LV_{\text{eff}})(E - LV_{\text{eff}}), \tag{50}$$

where $V_{\text{eff}} = e^{(f_0-f_1)/2}/r$ is the effective potential for photons. We depict its form in terms of the scaled circumferential coordinate $\hat{R}_c = e^{\Gamma/2}r$ in Fig. 9, for the coupling function $F = \alpha\phi^2$. Whereas, for the less compact particle-like solutions, the effective potential is monotonic, for the more compact solutions, the effective potential V_{eff} is characterised by a pair of extrema. These extrema correspond to circular orbits and thus lead to the presence of *light rings* around these ultra-compact solutions. Bound orbits emerge also for massive particles with radii smaller than or comparable to the radius of the bubble. In fact, it is found that particles may pass over the origin ($r = 0$) or even rest there without encountering any divergence from the singular term in the expression of the scalar field.

We also considered the propagation of a scalar test particle Ψ in this background. Starting from the equation $\Box\Psi = 0$ and expanding the scalar field in spherical harmonics $Y_l^m(\theta, \varphi)$,

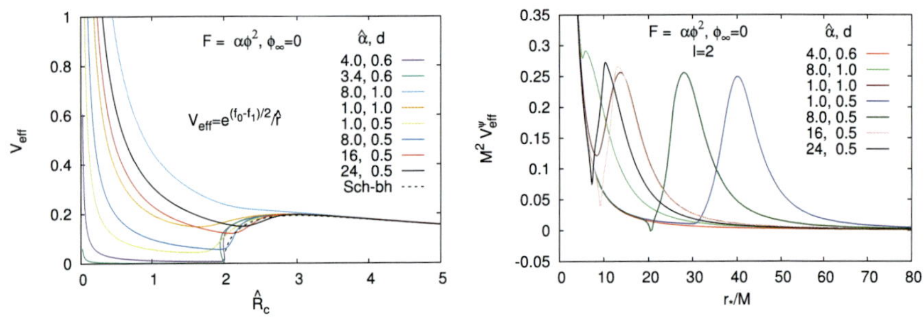

Fig. 9 The effective potential for null particles (left plot) and for test scalar particles (right plot), for a quadratic coupling function [166, 167]

$$\Psi(t, r, \theta, \varphi) = \sum_{l,m} \psi_{l,m}(t, r)\, e^{-f_1/2} Y_l^m(\theta, \varphi)/r \,, \tag{51}$$

we obtain the reduced equation

$$(\partial_t^2 - \partial_{r_*}^2 + V_{\text{eff}}^{\psi})\, \psi_{l,m}(t, r) = 0 \,, \tag{52}$$

where r_* is the tortoise coordinate defined through the relation $dr_* = e^{(f_1 - f_0)/2} dr$, and V_{eff}^{ψ} is the scalar-field effective potential

$$V_{\text{eff}}^{\psi} = e^{f_0 - f_1} \left[\frac{l(l+1)}{r^2} + \frac{2(f_1' + f_0') + r f_1' f_0' + 2r f_1''}{4r} \right] . \tag{53}$$

The form of the scaled effective potential $M^2 V_{\text{eff}}^{\psi}$ versus the scaled tortoise coordinate r_*/M is presented on the right plot of Fig. 9, again for a quadratic coupling function. We observe that, for scalar waves with $l > 0$, the effective potential features an infinite barrier that resides at the origin and a finite local barrier at a larger value of r_*. An incoming scalar-wave with $l > 0$ will be partially transmitted through the finite local barrier and will then undergo a perpetual process of full and partial reflection from the infinite and finite barrier, respectively, thus producing a series of *echoes* in the wave signal.

4 Compact Objects in (Beyond) Horndeski Theory

The Einstein-scalar-GB theory is a special case of a more general tensor-scalar theory of a single scalar degree of freedom with field equations containing only up to 2nd-order derivatives, the *Horndeski theory* [11]. The latter is defined through the Lagrangian

$$S_{\text{H}} = \int d^4x \sqrt{-g}\, (\mathcal{L}_2 + \mathcal{L}_3 + \mathcal{L}_4 + \mathcal{L}_5)$$

with

$$\begin{aligned}
\mathcal{L}_2 =& G_2(X) \,, \qquad \mathcal{L}_3 = -G_3(X)\, \Box\phi, \\
\mathcal{L}_4 =& G_4(X) R + G_{4X} \left[(\Box\phi)^2 - \nabla_\mu \partial_\nu \phi\, \nabla^\mu \partial^\nu \phi \right], \\
\mathcal{L}_5 =& G_5(X) G_{\mu\nu} \nabla^\mu \partial^\nu \phi - \frac{1}{6} G_{5X} \big[(\Box\phi)^3 - 3\, \Box\phi\, \nabla_\mu \partial_\nu \phi\, \nabla^\mu \partial^\nu \phi \\
&+ 2\, \nabla_\mu \partial_\nu \phi\, \nabla^\nu \partial^\rho \phi\, \nabla_\rho \partial^\mu \phi \big].
\end{aligned}$$

The coupling functions G_i depend on the scalar field ϕ only through its kinetic term $X \equiv -\nabla_\mu \phi \nabla^\mu \phi/2$, thus the above theory is the special, *shift-symmetric* Horndeski theory. The symbol G_{iX} above denotes the derivative of the function G_i with respect to X.

Although the theory is more complicated compared to the Einstein-scalar-Gauss-Bonnet theory and the determination of analytic solutions therefore seems unlikely, in fact the presence of all the additional terms and the freedom we have in choosing the exact form of the coupling functions G_i of the theory can help in deriving new analytic solutions. Indeed, whereas this possibility does not exist in the context of the EsGB theory, a wise choice for the form of G_i in Horndeski theory can turn the set of field equations into an integrable set of equations, which can then be solved analytically. For instance, choosing the following coupling functions

$$G_2 = 8\alpha X^2 \,, \quad G_3 = -8\alpha X \,, \quad G_4 = 1 + 4\alpha X \,, \quad G_5 = -4\alpha \ln |X| \,, \tag{54}$$

a static, spherically symmetric black-hole solution of the form

$$ds^2 = -h(r)\, dt^2 + \frac{dr^2}{f(r)} + r^2 d\Omega^2 \,, \tag{55}$$

with $h(r) = f(r)$, was analytically derived in the context of the Horndeski theory with the metric function and scalar field having the following explicit forms [172, 173]

$$h(r) = 1 + \frac{r^2}{2\alpha}\left(1 - \sqrt{1 + \frac{8\alpha M}{r^3}}\right), \qquad \phi' = \frac{\sqrt{h} - 1}{r\sqrt{h}} \,, \tag{56}$$

respectively. The above solution describes an asymptotically-flat black hole with a non-trivial scalar field and a horizon at $r_H = M + \sqrt{M^2 - \alpha}$, where M is the black-hole mass and α the coupling constant of the theory.

The Horndeski theory is certainly not the end of the road when it comes to generalised tensor-scalar theories. One could extend this theory to *beyond Horndeski* by adding the following well-known terms

$$\mathcal{L}_4^{\rm bH} = F_4(X)\varepsilon^{\mu\nu\rho\sigma}\, \varepsilon^{\alpha\beta\gamma}{}_{\sigma}\, \partial_\mu\phi\, \partial_\alpha\phi\, \nabla_\nu\partial_\beta\phi\, \nabla_\rho\partial_\gamma\phi \tag{57}$$

$$\mathcal{L}_5^{\rm bH} = F_5(X)\varepsilon^{\mu\nu\rho\sigma}\, \varepsilon^{\alpha\beta\gamma\delta}\, \partial_\mu\phi\, \partial_\alpha\phi\, \nabla_\nu\partial_\beta\phi\, \nabla_\rho\partial_\gamma\phi\, \nabla_\sigma\partial_\delta\phi \,. \tag{58}$$

Despite the presence of the above terms, the theory remains free of ghosts provided that the coupling functions (G_4, G_5) and (F_4, F_5) of the extended theory satisfy the constraint [174]

$$X G_{5X} F_4 = 3 F_5 \left(G_4 - 2 X G_{4X}\right) . \tag{59}$$

In the context of the beyond Horndeski theory, the complexity of the field equations increases further, and the integrability of the system of equations is easily lost. However, by employing a number of auxiliary functions, we managed to bring the complicated equations of motion to a very simple form, namely [175]

$$
\begin{aligned}
X'\mathcal{A} &= 2\left(\frac{h'}{h} - \frac{f'}{f}\right)\mathcal{B}\,,\\
\frac{h'f}{2h}\mathcal{A} &= G_{2X}r^2 + 2G_{4X} - 2rf\phi' G_{3X} - 2fZ_X\,,\\
2f\frac{h'}{h}\mathcal{B} &= -G_2 r^2 - 2G_4 - 2fZ\,,
\end{aligned}
\tag{60}
$$

where $\mathcal{A}$, $\mathcal{B}$ and Z are functions of $\{G_i, F_i, X\}$ given by the expressions

$$
\begin{aligned}
\mathcal{A} &= 4rZ_X + \phi'\left[r^2 G_{3X} + G_{5X}(1-3f) - 2XfG_{5XX} + 12fX(5F_5 + 2XF_{5X})\right],\\
\mathcal{B} &= rZ - f\phi' X G_{5X} + 12f\phi' X^2 F_5\,,\\
Z &= 2XG_{4X} - G_4 + 4X^2 F_4\,,
\end{aligned}
\tag{61}
$$

and (h, f) are the two metric functions of the static, spherically-symmetric line-element (55).

In the context of the parity-symmetric theory, i.e. with $G_3 = G_5 = F_5 = 0$, the above set of field equations is easily rendered integrable and may lead to a large number of analytic solutions. For instance, for the general class of theories where

$$
G_2 = -2\Lambda - \alpha X + \delta X^m\,, \qquad G_4 = \zeta + \beta X^n\,, \tag{62}
$$

where $(\Lambda, \alpha, \delta, \zeta, \beta)$ are constant parameters of the theory and (m, n) real, integer or rational numbers, and Z chosen to be a constant, a large number of analytical homogeneous solutions with $f(r) = h(r)$ were derived [175]. These solutions were characterized at the small-r regime by zero, one or two horizons, depending on the values of the parameters of the theory, while at large distances they assumed an (Anti-)de Sitter-Reissner-Nordstrom-type of behaviour of the form

$$
h(r) = 1 - \frac{\Lambda_{eff} r^2}{3} - \frac{2M}{r} + \frac{Q^2}{r^2}\,. \tag{63}
$$

In the expression above, M is the ADM mass of the solution, Λ_{eff} the effective cosmological constant and Q a tidal charge. The first parameter is an arbitrary integration constant while the latter two are determined by the coupling parameters of the theory. All solutions are characterized by a non-trivial scalar field, which at large distances behaves as $\phi' \sim 1/h(r)$ while it diverges at the (outer) horizon, when the latter exists; however, X remains everywhere finite as also does the energy-momentum tensor of the theory. By making appropriate, alternative choices for the coupling functions of the theory, non-homogeneous solutions, with $f \neq h$ but with the same asymptotic behaviour, were also found.

In the case of no parity symmetry, when the functions (G_3, G_5, F_5) in principle do not vanish, the set of equations becomes significantly more difficult to solve. Nevertheless, even in this case, our formalism allows for the integration of this set, under convenient choices for the coupling and auxiliary functions of the theory, and new black-hole solutions, which generalise the solution of [172], can be determined, albeit in a non-explicit form over the entire radial regime [175]. Additional black-hole solutions determined in the context of Horndeski or beyond Horndeski theory can be found in [176–178].

In the context of beyond Horndeski theory, wormhole solutions may also be found. However, instead of solving the set of field equations for the desired type of solution, one could follow an alternative technique [179, 180] and apply an appropriate *disformal transformation* [181–183] to a known solution denoted by $(\bar{g}_{\mu\nu}, \bar{\phi})$. A disformal transformation has the general form

$$g_{\mu\nu} = \bar{g}_{\mu\nu} - D(\bar{X})\, \nabla_\mu \phi \, \nabla_\nu \phi \,, \qquad \phi = \bar{\phi}\,, \tag{64}$$

where $D(X)$ is an arbitrary function which characterizes the disformal transformation. According to the above, the components of the metric tensor get transformed in a way that depends on the configuration of the scalar field while the latter remains unchanged. In this way, one could start from a known solution $(\bar{g}_{\mu\nu}, \bar{\phi})$ and obtain a new one $(g_{\mu\nu}, \phi)$. When the "seed solution" $(\bar{g}_{\mu\nu}, \bar{\phi})$ is a solution of Horndeski theory, the disformally transformed solution $(g_{\mu\nu}, \phi)$ is a solution of the *beyond Horndeski* theory [182].

In our analysis [179], we chose to consider as our "seed" solution $(\bar{g}_{\mu\nu}, \bar{\phi})$ the static, spherically-symmetric solution (56). Since $\phi = \phi(r)$, the disformal transformation will result in the relations

$$h = \bar{h}\,, \qquad f = \frac{\bar{f}}{1 + 2D\bar{X}} \equiv \bar{f}\, W(\bar{X})\,, \tag{65}$$

where, for convenience, we have defined a new arbitrary function $W(\bar{X})$. There is clearly an arbitrarily large number of choices for the function $D(\bar{X})$ or $W(\bar{X})$. Its form will be dictated by the desired properties of the new metric tensor. For instance, it is clear that $W(\bar{X})$ should satisfy the following two constraints

$$W(\bar{X}) \geq 0\,, \qquad \lim_{r\to\infty} W(\bar{X}) = 1\,, \tag{66}$$

in order to preserve the signature and asymptotic behaviour of $\bar{g}_{\mu\nu}$. In addition, in order to obtain a spacetime with a throat, we allowed the existence of a root in the expression of $W(\bar{X})$ at a distance $r_0 > r_H$, where r_H is the event horizon of the "seed solution". The presence of this root would result into the vanishing of g^{rr} at this point but not of g_{tt}, which will remain a constant. This is indeed the typical behaviour of the metric tensor close to a wormhole according to the discussion in Sect. 3.2. For example, we could choose [179]

$$W(\bar{X}) = 1 - \frac{r_0}{\lambda}\sqrt{-2\bar{X}} = 1 - \frac{r_0}{\lambda r}\left(1 - \sqrt{\bar{h}}\right), \tag{67}$$

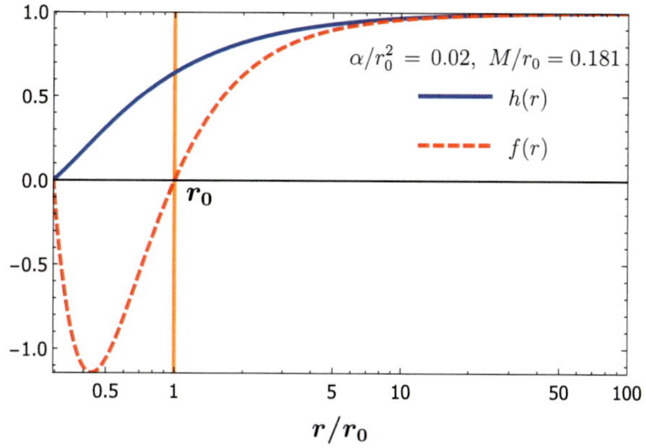

Fig. 10 The two metric functions after the disformal transformation [179]

where λ is a scale parameter of the solution. The root r_0, or the radius of the wormhole, is located at

$$r_0 = \frac{M \pm \sqrt{M^2 - \alpha\lambda^3\,(2-\lambda)^3}}{\lambda(2-\lambda)}\,, \tag{68}$$

and is determined by the mass M of the original black-hole solution, the coupling parameter α and the scalar parameter λ. The profiles of the two metric functions $h(r)$ and $f(r)$ after the disformal transformation are depicted in Fig. 10.

Due to the vanishing of the metric function $f(r)$ at the throat, we focus on the causal part of the spacetime with $r_0 \le r < \infty$. The complete geometry is revealed by setting $r^2 = l^2 + r_0^2$, in which case the line-element reads

$$ds^2 = -H(l)\,dt^2 + \frac{dl^2}{F(l)} + (l^2 + r_0^2)\,d\Omega^2\,, \tag{69}$$

where

$$H(l) = h(r(l))\,, \qquad F(l) = \frac{f(r(l))\,(l^2 + r_0^2)}{l^2}\,. \tag{70}$$

The throat of the wormhole is now located at $l = 0$ and the two asymptotic regimes are reached when $l \to \pm\infty$. The two metric functions $H(l)$ and $F(l)$ are everywhere regular and they both assume constant values at the throat. Their profiles are depicted in Fig. 11. We observe that both metric functions are symmetric under the change $l \to -l$ and that their first derivatives vanish at the throat (all curves in Fig. 11 become horizontal there). As a result, no cusp points appear at the throat and the two parts of the wormhole (the positive and negative l-regimes) are smoothly connected. The wormhole is therefore traversable without the need of any distribution of additional matter, exotic or not.

The spacetime is everywhere regular and smooth, and this is evident in the form of the Kretchmann scalar $\mathcal{K} = R^{\mu\nu\rho\sigma} R_{\mu\nu\rho\sigma}$ presented on the left plot of Fig. 12. We also observe that the non-vanishing curvature is restricted in a narrow region

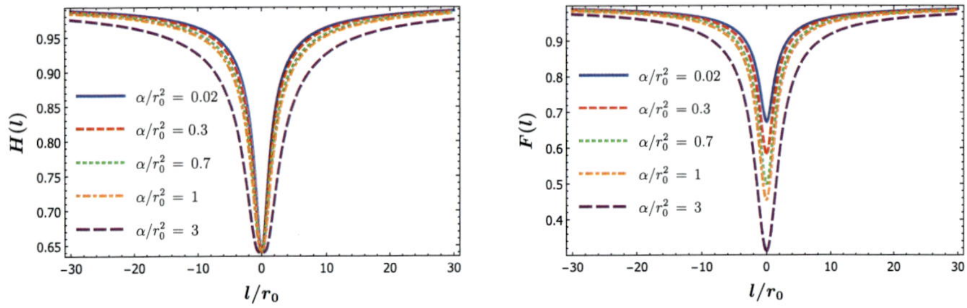

Fig. 11 The two metric functions after the change of variable $r \to l$ [179]

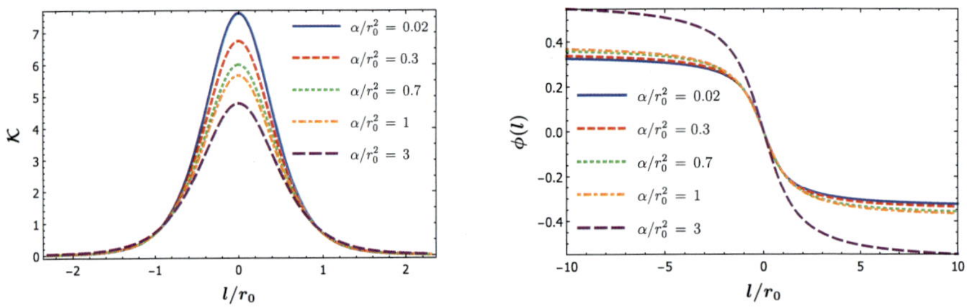

Fig. 12 The Kretchmann scalar $\mathcal{K}$ (left plot), and the profile of the scalar field (right plot) [179]

$-2r_0 \leq l \leq 2r_0$ thus our wormholes may comprise ultra-compact objects. One could construct the embedding diagram of the solution by following a procedure similar to that of Sect. 3.2. Although the details differ, the embedding diagram presents exactly the same form as the one shown in Fig. 5. The profile of the scalar field is depicted on the right plot; this in fact asymmetric under the change $l \to -l$, but is again regular over the entire regime and presents no discontinuities.

One may wonder whether the emergence of these wormhole solutions must again be supported by the violation of the energy conditions. If we focus again on the Null Energy Condition, namely $T_{\mu\nu}n^\mu n^\nu \geq 0$, we arrive at the result

$$8\pi G(T^r_{\ r} - T^t_{\ t}) = G^r_{\ r} - G^t_{\ t} = -\frac{f'(r_0)}{r_0}\,. \tag{71}$$

But one may easily see that the condition $f'(r_0) < 0$ is the flaring-out condition for these wormholes and thus must be always respected. As a result, the NEC is again violated, as one may see also on the left plot of Fig. 13. But this violation is again caused not by the presence of any exotic matter but by the various non-minimal couplings of the scalar field to gravity in this theory. What is also important is that the energy-density ρ of the theory for most of the solutions is everywhere positive which renders our solutions more physically interesting and realistic.

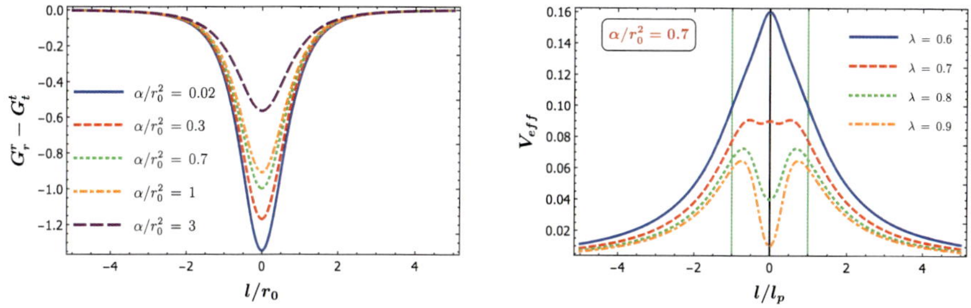

Fig. 13 The violation of the NEC (left plot) and the effective potential of photons propagating in the wormhole background (right plot) [179]

Turning finally to observable phenomena that one could associate with our wormhole solutions, we address again the geodesics of null particles as an indicative example. These are given by the equation

$$W\dot{r}^2 + \frac{h}{r^2} = \frac{E^2}{L^2}\,. \tag{72}$$

Studying the extremal points of the photon's effective potential $V_{eff} = h/r^2$, we find that these are given by the 3rd-order polynomial

$$r^3 - 9M^2 r + 8\alpha M = 0\,. \tag{73}$$

As expected, in the limit $\alpha \to 0$, we obtain two extremal points at $r = 0$ and $r = 3M$, with the first one being stable and the second unstable. As α increases, we find up to 3 extremal points, one at the throat and two more at larger radii. In this case, the light ring around the throat and the more distant one are stable while the intermediate one is unstable; all light rings lie at distances smaller than $3M$, which is the characteristic light-ring radius of the Schwarzschild spacetime. The effective potential for null particles is depicted on the right plot of Fig. 13.

5 Conclusions

In the quest of new black-hole solutions, traversable wormholes and regular particle-like solutions, one is forced to move beyond General Relativity and to consider generalised theories of gravity. Scalar-tensor theories is the simplest extension of GR, however, one must consider non-minimally coupled scalar fields in order for physically-interesting solutions to emerge. The Einstein-scalar-Gauss-Bonnet theory is an indicative example of such a theory that contains a single scalar degree of freedom coupled to the quadratic, gravitational Gauss-Bonnet term, a term which is bound to lead to important modifications from GR in the strong-gravity regime. It was in the context of this theory that we sought for new solutions describing compact objects and being characterized by a non-trivial scalar field.

We demonstrated that the EsGB theory admits a variety of solutions. Scalarised black-hole solutions were discovered first for particular forms of the coupling function between the scalar field and the GB term. In the recent years, it was demonstrated that scalarised, black-hole solutions with a regular horizon and a Minkowski or Anti-de Sitter behaviour at large distances always emerge, independently of the form of the coupling function, provided that appropriate boundary conditions are imposed. These black-hole solutions evade all known forms of the scalar no-hair theorem as the presence of the GB term causes the violation of the theorem's requirements.

Similar analyses for the determination of wormhole solutions soon followed based on the common knowledge that the GB term violates the energy conditions, a requirement for the emergence of a traversable wormhole. Indeed, a plethora of wormhole solutions were found for various forms of the coupling function of the scalar field to the GB term. In order to avoid a spacetime singularity lurking somewhere behind the throat, a cut and paste technique was applied which smoothly connected two regular regimes thus creating a traversable wormhole with a regular scalar field. The study of particle geodesics in this background showed that trajectories starting from one side of the wormhole, crossing the throat, visiting the other side of the wormhole and then returning back to the departure point was a common behaviour.

Gravitational, particle-like solutions were also studied in the context of the EsGB theory causing perhaps no surprise when such solutions readily emerged for various forms of the coupling function. A singularity seemed to plague the expression of the scalar field, however, it was found that this has no physical consequence for the spacetime curvature, energy-momentum tensor or particle trajectories, thus rendering this singularity a harmless, "Coulomb-type" singularity. These regular, particle-like solutions presented a bubble-type distribution of matter with a negative value of energy density at the center, a positive value at the peak of the distribution located very close to the origin and a fast decreasing profile—the latter features render our solutions as ultra-compact objects. The study of null particles in the background of the more compact solutions revealed the existence of a pair of light-rings, an inner stable one and and an outer unstable. The study of a scalar field propagating in the same spacetime led to the result that a series of echoes should characterize the wave signal at infinity, a characteristic observable associated to our particle-like solutions.

The different types of solutions emerging in the context of the EsGB theory are depicted in the domain-of-existence plot of Fig. 14. In there, we may see the line of scalarised black-hole solutions, a one-parameter family of solutions due to the no-hair theorem which acts as a boundary of the wormhole territory. The latter solutions may be characterised by a single throat or by a pair of throats with an equator in between. In the same plot, we also notice different families of particle-like solutions, characterised by an increasing number of nodes of the scalar field as the coupling constant of the theory increases, too. We also observe that some white, uncharted yet, areas still remain in the domain of existence which creates perhaps expectations for future discoveries!

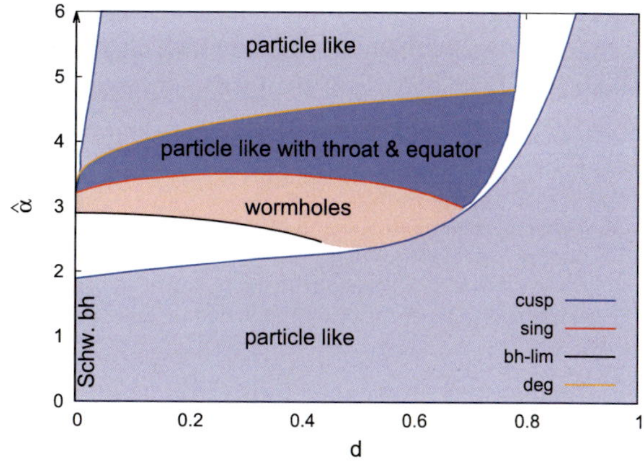

Fig. 14 The different types of solutions emerging in the context of the Einstein-scalar-GB theory [166, 167]

The transition from General Relativity to the Einstein-scalar-Gauss-Bonnet theory resulted in the discovery of a wealth of solutions which, in the former theory, were forbidden. The EsGB theory itself is a subclass of Horndeski theory, which is a much more general scalar-tensor theory characterized by four coupling functions. Although more complicated, the set of field equations can be rendered integrable upon appropriate choices for the four coupling functions. The theory can be further generalised to the beyond-Horndeski theory via the addition of two additional terms in the Lagrangian. In the context of this latter theory, we re-examined the integrability of the set of field equations, formulated these in a particularly convenient form, and investigated the existence of analytical solutions describing black holes. In the context of the parity-symmetric sector of the theory, we analytically determined a large number of black-hole solutions with an (Anti-)de Sitter-Reissner-Nordstrom asymptotic behaviour at large distances. In the case of parity-symmetry breaking, the determination of analytic solutions in a closed form is much more difficult.

Turning to the family of wormhole solutions, we chose to determine these not by analytically solving the field equations but by applying a disformal transformation to a known, "seed", solution of the theory. That was made possible by conveniently choosing the function of the disformal transformation. The resulting wormhole solutions were shown to be generically regular by construction, localised in a very narrow range of the radial coordinate—a feature which renders our solutions as ultra-compact objects—and to be characterised by a non-trivial scalar field. Studying the particle trajectories in the wormhole spacetime, distinct observable signatures were determined such as the emergence of multiple photon rings, some of them stable and all located at radii smaller than the characteristic distance of $3M$ of the Schwarzschild solution.

Due to the larger arbitrariness of the Horndeski and beyond-Horndeski theories, encoded in the forms of the four and six, respectively, undetermined a priori coupling functions, the phase-space of the solutions of the theory has not yet fully been studied. Specific sectors have been studied for particular classes of solutions but the study of the complete theory as well as the charting of the types of solutions that the theory harbors is still lacking. Judging by the variety of solutions that have so far been found and by the wealth of solutions arising in the context of the EsGB theory, which is only a subclass of the theory, we expect a great many discoveries—even surprises—to arise when the complete Horndeski and beyond Horndeski theories are fully investigated.

Acknowledgements I am grateful to my collaborators Georgios Antoniou, Athanasios Bakopoulos, Christos Charmousis, Nicolas Lecoeur, Burkhard Kleihaus, Jutta Kunz and Nikolaos Pappas for our enjoyable and fruitful collaboration. I am also deeply thankful to the organisers of the 11th Aegean Summer School for their kind invitation and for giving me the opportunity to present our results.

References

1. A. Einstein, Sitzungsber. Preuss. Akad. Wiss. Berlin (Math. Phys.) **1915**, 844–847 (1915); Sitzungsber. Preuss. Akad. Wiss. Berlin (Math. Phys.) **1915**, 831–839 (1915); Sitzungsber. Preuss. Akad. Wiss. Berlin (Math. Phys.) **1915**, 778–786 (1915); Annalen Phys. **49**(7), 769–822 (1916)
2. K.A. Bronnikov, Acta Phys. Polon. B **4**, 251–266 (1973)
3. K.S. Stelle, Phys. Rev. D **16**, 953 (1977)
4. T.P. Sotiriou, Lect. Notes Phys. **892**, 3 (2015)
5. E. Berti et al., Class. Quant. Grav. **32**, 243001 (2015)
6. C. Charmousis, Lect. Notes Phys. **769**, 299 (2009); Lect. Notes Phys. **892**, 25–56 (2015)
7. C.G. Callan Jr., I.R. Klebanov, M.J. Perry, Nucl. Phys. B **278**, 78–90 (1986)
8. D.J. Gross, J.H. Sloan, Nucl. Phys. B **291**, 41–89 (1987)
9. R.R. Metsaev, A.A. Tseytlin, Nucl. Phys. B **293**, 385–419 (1987)
10. D. Lovelock, J. Math. Phys. **12**, 498–501 (1971)
11. G.W. Horndeski, Int. J. Theor. Phys. **10**, 363–384 (1974)
12. K. Schwarzschild, Sitzungsber. Preuss. Akad. Wiss. Berlin (Math. Phys.) **1916**, 189–196 (1916); Sitzungsber. Preuss. Akad. Wiss. Berlin (Math. Phys.) **1916**, 424–434 (1916)
13. H. Reissner, Annalen der Physik **50**(9), 106–120 (1916); G. Nordström, Verhandl. Koninkl. Ned. Akad. Wetenschap., Afdel. Natuurk., Amsterdam. **26**, 1201–1208 (1918)
14. R.P. Kerr, Phys. Rev. Lett. **11**, 237–238 (1963)
15. J.D. Bekenstein, Phys. Rev. Lett. **28**, 452 (1972); C. Teitelboim, Lett. Nuovo Cim. **3S2**, 397 (1972)
16. C.A.R. Herdeiro, E. Radu, Int. J. Mod. Phys. D **24**(09), 1542014 (2015)
17. C.W. Misner, K.S. Thorne, J.A. Wheeler, W.H. Freeman (1973). ISBN 978-0-7167-0344-0, 978-0-691-17779-3
18. A. Einstein, N. Rosen, Phys. Rev. **48**, 73–77 (1935)
19. C.W. Misner, J.A. Wheeler, Annals Phys. **2**, 525–603 (1957)
20. M. Visser, *Lorentzian wormholes: From Einstein to Hawking* (American Inst. of Physics, Computational and Mathematical Physics, 1995)
21. M.S. Morris, K.S. Thorne, Am. J. Phys. **56**, 395–412 (1988)
22. H.G. Ellis, Journal of Math. Physics. **14**, 104–118 (1973)
23. K.A. Bronnikov, Acta Physica Polonica. B **4**, 251–266 (1973)
24. I. Fisher, Zh. Eksp, Teor. Fiz. **18**, 636 (1948)
25. A.I. Janis, E.T. Newman, J. Winicour, Phys. Rev. Lett. **20**, 878 (1968)
26. M. Wyman, Phys. Rev. D **24**, 839 (1981)
27. H. Motohashi, T. Suyama, Phys. Rev. D **91**, 085009 (2015)

28. P. Kanti, N.E. Mavromatos, J. Rizos, K. Tamvakis, E. Winstanley, Phys. Rev. D **54**, 5049 (1996)
29. P. Kanti, N.E. Mavromatos, J. Rizos, K. Tamvakis, E. Winstanley, Phys. Rev. D **57**, 6255 (1998)
30. G.W. Gibbons, K.I. Maeda, Nucl. Phys. B **298**, 741 (1988)
31. C.G. Callan Jr., R.C. Myers, M.J. Perry, Nucl. Phys. B **311**, 673 (1989)
32. B.A. Campbell, M.J. Duncan, N. Kaloper, K.A. Olive, Phys. Lett. B **251**, 34 (1990); B.A. Campbell, N. Kaloper, K.A. Olive, Phys. Lett. B **263**, 364 (1991)
33. S. Mignemi, N.R. Stewart, Phys. Rev. D **47**, 5259 (1993)
34. P. Kanti, K. Tamvakis, Phys. Rev. D **52**, 3506 (1995)
35. T. Torii, H. Yajima, K.I. Maeda, Phys. Rev. D **55**, 739 (1997)
36. P. Kanti, K. Tamvakis, Phys. Lett. B **392**, 30 (1997)
37. Z.K. Guo, N. Ohta, T. Torii, Prog. Theor. Phys. **120**, 581 (2008); N. Ohta, T. Torii, Prog. Theor. Phys. **122**, 1477 (2009); K.I. Maeda, N. Ohta, Y. Sasagawa, Phys. Rev. D **80**, 104032 (2009); N. Ohta, T. Torii, Prog. Theor. Phys. **124**, 207 (2010)
38. B. Kleihaus, J. Kunz, E. Radu, Phys. Rev. Lett. **106**, 151104 (2011); B. Kleihaus, J. Kunz, S. Mojica, E. Radu, Phys. Rev. D **93**(4), 044047 (2016)
39. P. Pani, C.F.B. Macedo, L.C.B. Crispino, V. Cardoso, Phys. Rev. D **84**, 087501 (2011); P. Pani, E. Berti, V. Cardoso, J. Read, Phys. Rev. D **84**, 104035 (2011)
40. Y. Bardoux, M.M. Caldarelli, C. Charmousis, JHEP **1205**, 054 (2012)
41. K. Yagi, L.C. Stein, N. Yunes, T. Tanaka, Phys. Rev. D **85**, 064022 (2012). Erratum: [Phys. Rev. D **93**(2), 029902 (2016)]
42. C. Charmousis, T. Kolyvaris, E. Papantonopoulos, M. Tsoukalas, JHEP **1407**, 085 (2014)
43. F. Correa, M. Hassaine, J. Oliva, Phys. Rev. D **89**(12), 124005 (2014)
44. D. Ayzenberg, N. Yunes, Phys. Rev. D **90**, 044066 (2014). Erratum: [Phys. Rev. D **91**(6), 069905 (2015)]
45. J.L. Blazquez-Salcedo, C.F.B. Macedo, V. Cardoso, V. Ferrari, L. Gualtieri, F.S. Khoo, J. Kunz, P. Pani, Phys. Rev. D **94**(10), 104024 (2016)
46. S. Bhattacharya, S. Chakraborty, Phys. Rev. D **95**(4), 044037 (2017); I. Banerjee, S. Chakraborty, S. SenGupta, Phys. Rev. D **96**(8), 084035 (2017)
47. J.L. Blazquez-Salcedo et al., IAU Symp. **324**, 265 (2016)
48. T.P. Sotiriou, S.Y. Zhou, Phys. Rev. Lett. **112**, 251102 (2014)
49. T.P. Sotiriou, S.Y. Zhou, Phys. Rev. D **90**, 124063 (2014); R. Benkel, T.P. Sotiriou, H. Witek, Phys. Rev. D **94**(12), 121503 (2016); Class. Quant. Grav. **34**(6), 064001 (2017)
50. J.D. Bekenstein, Phys. Rev. D **51**(12), R6608 (1995)
51. T.P. Sotiriou, V. Faraoni, Phys. Rev. Lett. **108**, 081103 (2012)
52. L. Hui, A. Nicolis, Phys. Rev. Lett. **110**, 241104 (2013)
53. C.A.R. Herdeiro, E. Radu, Phys. Rev. Lett. **112**, 221101 (2014)
54. E. Babichev, C. Charmousis, JHEP **1408**, 106 (2014)
55. G. Antoniou, A. Bakopoulos, P. Kanti, Phys. Rev. Lett. **120**(13), 131102 (2018)
56. G. Antoniou, A. Bakopoulos, P. Kanti, Phys. Rev. D **97**(8), 084037 (2018)
57. A. Papageorgiou, C. Park, M. Park, Phys. Rev. D **106**(8), 084024 (2022)
58. A. Bakopoulos, G. Antoniou, P. Kanti, Phys. Rev. D **99**(6), 064003 (2019)
59. Y. Brihaye, B. Hartmann, J. Urrestilla, JHEP, **074** (2018). (Y. Brihaye, B, Hartmann, 1806). (arXiv:1810.05108 [gr-qc])
60. Y. Brihaye, B. Hartmann, Phys. Lett. B **772**, 476–482 (2017)
61. A. Bakopoulos, P. Kanti, N. Pappas, Phys. Rev. D **101**(8), 084059 (2020)
62. D.D. Doneva, S.S. Yazadjiev, Phys. Rev. Lett. **120**(13), 131103 (2018)
63. H.O. Silva, J. Sakstein, L. Gualtieri, T.P. Sotiriou, E. Berti, Phys. Rev. Lett. **120**(13), 131104 (2018)
64. D.D. Doneva, S.S. Yazadjiev, JCAP **1804**(04), 011 (2018)
65. H. Motohashi, M. Minamitsuji, Phys. Lett. B **781**, 728 (2018); Phys. Rev. D **98**(8), 084027 (2018)

66. C.A.R. Herdeiro, E. Radu, N. Sanchis-Gual, J.A. Font, Phys. Rev. Lett. **121**(10), 101102 (2018); T. Delsate, C. Herdeiro, E. Radu, Phys. Lett. B **787**, 8 (2018); Y. Brihaye, C. Herdeiro, E. Radu, Phys. Lett. B **788**, 295 (2019)
67. D.D. Doneva, S. Kiorpelidi, P.G. Nedkova, E. Papantonopoulos, S.S. Yazadjiev, Phys. Rev. D **98**(10), 104056 (2018)
68. M. Butler, A.M. Ghezelbash, E. Massaeli, M. Motaharfar, arXiv:1808.03217 [hep-th]
69. B. Danila, T. Harko, F.S.N. Lobo, M.K. Mak, arXiv:1811.02742 [gr-qc]
70. M.M. Stetsko, arXiv:1811.05030 [hep-th]
71. O.J. Tattersall, P.G. Ferreira, M. Lagos, Phys. Rev. D **97**(8), 084005 (2018)
72. S. Mukherjee, S. Chakraborty, Phys. Rev. D **97**(12), 124007 (2018)
73. S. Chakrabarti, Eur. Phys. J. C **78**(4), 296 (2018)
74. E. Berti, K. Yagi, N. Yunes, Gen. Rel. Grav. **50**(4), 46 (2018)
75. Y. Brihaye, B. Hartmann, Class. Quant. Grav. **35**(17), 175008 (2018)
76. K. Prabhu, L.C. Stein, Phys. Rev. D **98**(2), 021503 (2018)
77. Y.S. Myung, D.C. Zou, Phys. Rev. D **98**(2), 024030 (2018). (arXiv:1812.03604 [gr-qc])
78. J.L. Blazquez-Salcedo, D.D. Doneva, J. Kunz, S.S. Yazadjiev, Phys. Rev. D **98**(8), 084011 (2018); J.L. Blazquez-Salcedo, Z. Altaha Motahar, D.D. Doneva, F.S. Khoo, J. Kunz, S. Mojica, K.V. Staykov, S.S. Yazadjiev, arXiv:1810.09432 [gr-qc]
79. R. Benkel, N. Franchini, M. Saravani, T.P. Sotiriou, Phys. Rev. D **98**(6), 064006 (2018)
80. B.H. Lee, W. Lee, D. Ro, Phys. Rev. D **99**(2), 024002 (2019)
81. H. Witek, L. Gualtieri, P. Pani, T.P. Sotiriou, Phys. Rev. D **99**(6), 064035 (2019)
82. H. Motohashi, S. Mukohyama, Phys. Rev. D **99**(4), 044030 (2019)
83. J. Sultana, D. Kazanas, Gen. Rel. Grav. **50**(11), 137 (2018)
84. S. Nojiri, S.D. Odintsov, V.K. Oikonomou, Phys. Rev. D **99**, 044050 (2019)
85. S. Qolibikloo, A. Ghodsi, Eur. Phys. J. C **79**(5), 406 (2019)
86. P.V.P. Cunha, C.A.R. Herdeiro, E. Radu, Phys. Rev. Lett. **123**(1), 011101 (2019)
87. A. Bakopoulos, P. Kanti, N. Pappas, Phys. Rev. D **101**(4), 044026 (2020)
88. M. Minamitsuji, T. Ikeda, Phys. Rev. D 99, 044017. Phys. Rev. D **99**(2019), 104069 (2019)
89. M.M. Stetsko, Phys. Rev. D **99**, 044028 (2019)
90. Y.S. Myung, D.-C. Zou, Phys. Lett. B **790**, 400 (2019)
91. Y. Brihaye, L. Ducobu, Phys. Lett. B **795**, 135 (2019)
92. C.A.R. Herdeiro, E. Radu, Phys. Rev. D **99**, 084039 (2019)
93. T. Kobayashi, Rept. Prog. Phys. **82**, 086901 (2019)
94. H.O. Silva, C.F.B. Macedo, T.P. Sotiriou, L. Gualtieri, J. Sakstein, E. Berti, Phys. Rev. D **99**, 104041 (2019)
95. A. de la Cruz-Dombriz, F.J.M. Torralba, JCAP **1903**, 002 (2019)
96. C.-Y. Wang, Y.-F. Shen, Y. Xie, JCAP **1904**, 022 (2019)
97. P.-A. Cano, A. Ruiperez, JHEP **1905**, 189 (2019)
98. F.M. Ramazanoglu, Phys. Rev. D **99**, 084015 (2019)
99. P.G.S. Fernandes, C.A.R. Herdeiro, A.M. Pombo, E. Radu, N. Sanchis-Gual, Class. Quant. Grav. **36**, 134002 (2019)
100. Y. Brihaye, B. Hartmann, Phys. Lett. B **792**, 244 (2019)
101. M. Saravani, T.P. Sotiriou, Phys. Rev. D **99**, 124004 (2019)
102. C.F.B. Macedo, J. Sakstein, E. Berti, L. Gualtieri, H.O. Silva, T.P. Sotiriou, Phys. Rev. D **99**, 104041 (2019)
103. D.D. Doneva, K.V. Staykov, S.S. Yazadjiev, Phys. Rev. D **99**, 104045 (2019)
104. A. Saffer, H.O. Silva, N. Yunes, Phys. Rev. D **100**, 044030 (2019)
105. T. Anson, E. Babichev, C. Charmousis, S. Ramazanov, JCAP **1906**, 023 (2019)
106. Y.S. Myung, D.-C. Zou, Int. J. Mod. Phys. D **28**, 1950114 (2019)
107. Y. Brihaye, B. Hartmann, JHEP **1909**, 049 (2019)
108. O.J. Tattersall, P.G. Ferreira, Phys. Rev. D **99**, 104082 (2019)
109. N. Andreou, N. Franchini, G. Ventagli, T.P. Sotiriou, Phys. Rev. D **99**, 124022 (2019)
110. Q. Liang, J. Sakstein, M. Trodden, Phys. Rev. D **100**, 063518 (2019)
111. L. Hui, D. Kabat, X. Li, L. Santoni, S.S.C. Wong, JCAP **1906**, 038 (2019)

112. D.Q. Tuan, S.H.Q. Nguyen, Commun. Phys. **29**, 173 (2019)
113. Tuan Do et al., Science **365**, 6454 (2019)
114. P.G.S. Fernandes, C.A.R. Herdeiro, A.M. Pombo, E. Radu, N. Sanchis-Gual, Phys. Rev. D **100**, 084045 (2019)
115. R.A. Konoplya, A. Zhidenko, Phys. Rev. D **100**, 044015 (2019)
116. N. Franchini, T.P. Sotiriou, arXiv:1903.05427 [gr-qc]
117. A. Hees, O. Minazzoli, E. Savalle, Y.V. Stadnik, P. Wolf, B. Roberts, arXiv:1905.08524 [gr-qc]
118. T. Anson, E. Babichev, S. Ramazanov, arXiv:1905.10393 [gr-qc]
119. C. Charmousis, M. Crisostomi, R. Gregory, N. Stergioulas, Phys. Rev. D **100** (2019)(8), 084020
120. M. Khalil, N. Sennett, J. Steinhoff, A. Buonanno, arXiv:1906.08161 [gr-qc]
121. C. de Rham, J. Zhang, arXiv:1907.00699 [hep-th]
122. G. Aguilar-Perez, M. Cruz, S. Lepe, I. Moran-Rivera, arXiv:1907.06168 [gr-qc]
123. R.A. Konoplya, T. Pappas, A. Zhidenko, arXiv:1907.10112 [gr-qc]
124. Y.-X. Gao, D.-J. Liu, arXiv:1908.01346 [gr-qc]
125. T. Ikeda, T. Nakamura, M. Minamitsuji, arXiv:1908.09394 [gr-qc]
126. F.-L. Julie, E. Berti, arXiv:1909.05258 [gr-qc]
127. F.M. Ramazanoglu, K.I. Unluturk, arXiv:1910.02801 [gr-qc]
128. K.V. Aelst, E. Gourgoulhon, P. Grandclement, C. Charmousis, arXiv:1910.08451 [gr-qc]
129. J. Barrientos, F. Cordonier-Tello, C. Corral, F. Izaurieta, P. Medina, E. Rodriguez, O. Valdivia, arXiv:1910.00148 [gr-qc]
130. C. Martinez, R. Troncoso, J. Zanelli, Phys. Rev. D **67**, 024008 (2003)
131. T.J.T. Harper, P.A. Thomas, E. Winstanley, P.M. Young, Phys. Rev. D **70**, 064023 (2004)
132. T. Torii, K. Maeda, M. Narita, Phys. Rev. D **59**, 064027 (1999)
133. E. Winstanley, Found. Phys. **33**, 111 (2003); Class. Quant. Grav. **22**, 2233 (2005)
134. S. Bhattacharya, A. Lahiri, Phys. Rev. Lett. **99**, 201101 (2007)
135. M. Henneaux, C. Martinez, R. Troncoso, J. Zanelli, Phys. Rev. D **70**, 044034 (2004); C. Martinez, R. Troncoso, J. Zanelli, Phys. Rev. D **70**, 084035 (2004); C. Erices, C. Martinez, Phys. Rev. D **97**(2), 024034 (2018)
136. E. Radu, E. Winstanley, Phys. Rev. D **72**, 024017 (2005)
137. A. Anabalon, H. Maeda, Phys. Rev. D **81**, 041501 (2010)
138. D. Hosler, E. Winstanley, Phys. Rev. D **80**, 104010 (2009)
139. C. Charmousis, T. Kolyvaris, E. Papantonopoulos, Class. Quant. Grav. **26**, 175012 (2009); T. Kolyvaris, G. Koutsoumbas, E. Papantonopoulos, G. Siopsis, Gen. Rel. Grav. **43**, 163 (2011)
140. K.I. Maeda, N. Ohta, Y. Sasagawa, Phys. Rev. D **83**, 044051 (2011); Z.K. Guo, N. Ohta, T. Torii, Prog. Theor. Phys. **121**, 253 (2009); N. Ohta, T. Torii, Prog. Theor. Phys. **121**, 959 (2009); N. Ohta, T. Torii, Prog. Theor. Phys. **122**, 1477 (2009)
141. S.G. Saenz, C. Martinez, Phys. Rev. D **85**, 104047 (2012)
142. M.M. Caldarelli, C. Charmousis, M. Hassaine, JHEP **1310**, 015 (2013)
143. P.A. Gonzalez, E. Papantonopoulos, J. Saavedra, Y. Vasquez, JHEP **1312**, 021 (2013)
144. M. Bravo Gaete, M. Hassaine, Phys. Rev. D **88**, 104011 (2013); M. Bravo Gaete, M. Hassaine, JHEP **1311**, 177 (2013)
145. G. Giribet, M. Leoni, J. Oliva, S. Ray, Phys. Rev. D **89**(8), 085040 (2014)
146. J. Ben Achour, H. Liu, arXiv:1811.05369 [gr-qc]
147. J.B. Achour, H. Liu, Phys. Rev. D **99**, 064042 (2019)
148. P. Kanti, A. Bakopoulos, N. Pappas, PoS CORFU2018, 091 (2019)
149. Y. Brihaye, C. Herdeiro, E. Radu, arXiv:1910.05286 [gr-qc]
150. D.D. Doneva, K.V. Staykov, S.S. Yazadjiev, Phys. Rev. D **99**(10), 104045 (2019)
151. Y. Brihaye, C. Herdeiro, E. Radu, Phys. Lett. B **802**, 135269 (2020)
152. P. Kanti, B. Kleihaus, J. Kunz, Phys. Rev. Lett. **107**, 271101 (2011)
153. P. Kanti, B. Kleihaus, J. Kunz, Phys. Rev. D **85**, 044007 (2012)
154. G. Antoniou, A. Bakopoulos, P. Kanti, B. Kleihaus, J. Kunz, Phys. Rev. D **101**(2), 024033 (2020)

155. K.A. Bronnikov, J.C. Fabris, Class. Quant. Grav. **14**, 831–842 (1997)
156. F.S.N. Lobo, Phys. Rev. D **71**, 084011 (2005)
157. S.V. Bolokhov, K.A. Bronnikov, M.V. Skvortsova, Class. Quant. Grav. **29**, 245006 (2012)
158. B. Kleihaus, J. Kunz, Phys. Rev. D **90**, 121503 (2014)
159. M.R. Mehdizadeh, M. Kord Zangeneh, F.S.N. Lobo, Phys. Rev. D **91**(8), 084004 (2015)
160. K.A. Bronnikov, J.C. Fabris, O.F. Piattella, E.C. Santos, Gen. Rel. Grav. **48**(12), 162 (2016)
161. R. Shaikh, S. Kar, Phys. Rev. D **94**(2), 024011 (2016)
162. P. Cañate, J. Sultana, D. Kazanas, Phys. Rev. D **100**(6), 064007 (2019)
163. K.A. Bronnikov, S.V. Bolokhov, M.V. Skvortsova, Int. J. Mod. Phys. D **28**(13), 1941008 (2019)
164. R. Ibadov, B. Kleihaus, J. Kunz, S. Murodov, Phys. Rev. D **102**(6), 064010 (2020)
165. R. Ibadov, B. Kleihaus, J. Kunz, S. Murodov, Symmetry **13**(1), 89 (2021)
166. B. Kleihaus, J. Kunz, P. Kanti, Phys. Lett. B **804**, 135401 (2020)
167. B. Kleihaus, J. Kunz, P. Kanti, Phys. Rev. D **102**, 024070 (2020)
168. V. Cardoso, L.C.B. Crispino, C.F.B. Macedo, H. Okawa, P. Pani, Phys. Rev. D **90**(4), 044069 (2014)
169. J. Keir, Class. Quant. Grav. **33**(13), 135009 (2016)
170. P. Cunha, V.P.E. Berti, C.A.R. Herdeiro, Phys. Rev. Lett. **119**(25), 251102 (2017)
171. Y. Brihaye, B. Hartmann, JHEP **09**, 049 (2019)
172. H. Lu, Y. Pang, Phys. Lett. B **809**, 135717 (2020)
173. R.A. Hennigar, D. Kubizňák, R.B. Mann, C. Pollack, JHEP **07**, 027 (2020)
174. J. Ben Achour, M. Crisostomi, K. Koyama, D. Langlois, K. Noui, G. Tasinato, JHEP **12**, 100 (2016)
175. A. Bakopoulos, C. Charmousis, P. Kanti, N. Lecoeur, JHEP **08**, 055 (2022)
176. E. Babichev, C. Charmousis, A. Lehébel, Class. Quant. Grav. **33**(15), 154002 (2016)
177. E. Babichev, C. Charmousis, A. Lehébel, JCAP **04**, 027 (2017)
178. E. Babichev, C. Charmousis, M. Hassaine, N. Lecoeur, Phys. Rev. D **108**(2), 024019 (2023)
179. A. Bakopoulos, C. Charmousis, P. Kanti, JCAP **05**(05), 022 (2022)
180. N. Chatzifotis, E. Papantonopoulos, C. Vlachos, Phys. Rev. D **105**(6), 064025 (2022)
181. M. Zumalacárregui, J. García-Bellido, Phys. Rev. D **89**, 064046 (2014)
182. M. Crisostomi, M. Hull, K. Koyama, G. Tasinato, JCAP **03**, 038 (2016)
183. J. Ben Achour, M. Crisostomi, K. Koyama, D. Langlois, K. Noui, G. Tasinato, JHEP **12**, 100 (2016)

Dynamics of Scalarization

Daniela D. Doneva and Stoytcho S. Yazadjiev

Abstract Equilibrium solutions describing compact objects such as neutron stars and black holes endowed with a scalar field have been constructed in a number of modified theories of gravity. The mechanisms for developing and sustaining such scalar hair can be different, with the spontaneous scalarization having an important role among them. Studies of the dynamics of such processes is developing at a slower speed due to the high complexity of the problem especially in the nonlinear fully coupled regime. The advance in this topic is crucial, though, for studying a number of astrophysical implications. Here we will discuss some of the developments in this area focusing especially on the scalar-Gauss–Bonnet gravity as a representative example.

1 Introduction

We are currently entering an era of unprecedented accuracy of astrophysical observations, both electromagnetic and gravitational wave ones. They are finally opening up the window towards testing the strong field regime of gravity [1]. Among the main tools in this direction are black hole and neutron star observations that offer the possibility to witness manifestations of physics at its extremes—high spacetime curvature and matter densities, and speeds reaching a large fraction of the speed of light (e.g., in merging binaries). Naturally, a development in the experiments should

D. D. Doneva · S. S. Yazadjiev
Theoretical Astrophysics, Eberhard Karls University of Tübingen, 72076 Tübingen, Germany
e-mail: daniela.doneva@uni-tuebingen.de

D. D. Doneva
INRNE - Bulgarian Academy of Sciences, 1784 Sofia, Bulgaria

S. S. Yazadjiev (✉)
Department of Theoretical Physics, Faculty of Physics, Sofia University, 1164 Sofia, Bulgaria
e-mail: yazad@phys.uni-sofia.bg

Institute of Mathematics and Informatics, Bulgarian Academy of Sciences,
Acad. G. Bonchev St. 8, 1113 Sofia, Bulgaria

© The Author(s), under exclusive license to Springer Nature Switzerland AG 2024
E. Papantonopoulos and N. Mavromatos (eds.), *Compact Objects in the Universe*,
https://doi.org/10.1007/978-3-031-55098-0_3

go hand in hand with profound theoretical studies. With this motivation in mind, the examination of possible deviations from general relativity (GR) in the structure and dynamics of compact objects is advancing rapidly in the last decades.

Even though one can imagine a huge variety of generalizations of Einstein's gravity, still one of the most straightforward and natural possibilities is to consider the existence of a fundamental scalar field that is a mediator of gravitational interaction. We already know that at least one such type of fundamental scalar fields exist, namely the Higgs field, so it is straightforward to assume that other scalar field(s) might be operating on astrophysical scales. An example of such theories of gravity are the Horndeski theories that are the most general extended scalar-tensor theories having field equations of second order [2]. Naturally, a variety of well-known models beyond GR can be considered as a particular subclass of Horndeski theories [3, 4]. This includes the standard scalar-tensor theories, the quantum gravity motivated Gauss–Bonnet theories, etc.

Interestingly, the scalar field often does not have an impact on the black hole structure in a number of modified theories of gravity. This roots in the no-hair theorems according to which black holes are remarkably simple not only in GR but also in a large class of extended scalar tensor theories [5]. Neutron stars on the other hand can evade such constraints much easier because matter acts as a scalar field source. One should note that even though equilibrium properties of black holes in GR and beyond are the same in certain modified theories of gravity, the dynamics might change with respect to Einstein's theory. Up to now, though, this has not been observed in numerical relativity simulations (see e.g., [6]) and even though it is in general possible, it might not give a strong observable effect. Thus, we will restrict ourselves to theories that are able to violate the no-hair theorems and potentially produce large deviations from GR [7].

In that respect, ones of the best examples are the effective field theories that can be viewed as a low-energy limit of quantum gravity. They often involve the addition of curvature invariants coupled to a dynamical scalar field that violates the no-hair theorems. Thus, a scalar field development around black holes can be triggered by the spacetime curvature itself. In the outline of the present paper, we will focus predominantly on the famous Gauss–Bonnet gravity in which the theory of hairy black holes is well developed.

Black hole solutions in scalar-Gauss–Bonnet (sGB) gravity have been studied for the first time few decades ago [8–12], including slowly [11, 13, 14] and rapidly rotating [15–17] solutions. The interest in it was revived a bit later with the discovery that only black holes carry scalar charge in the shift symmetric flavor of the theory (with a linear scalar field coupling) while neutron stars have zero scalar charge thus naturally evading the binary pulsar constraints [18]. Further excitement in the topic was brought by the discovery that black holes and neutron stars in scalar-Gauss–Bonnet theory with a quadratic scalar field coupling can undergo spontaneous scalarization. This is a strong gravity phase transition triggered by a tachyonic instability at large spacetime curvature close to the black hole horizon [19–21]. This mechanism was first observed for neutron stars in the classical scalar-tensor theories [22] and it is the only known dynamical mechanism for endowing black holes (and other compact objects) with scalar hair without altering at all the predictions in the weak field limit (for a review see [23]). Moreover, for a proper choice of the scalar field coupling these hairy black holes are stable [24] and thermodynamically preferred over the bald (GR) one.

Even though static solutions in scalar-Gauss–Bonnet gravity are fairly well understood, constructing numerically the rotating ones is a great challenge [15–17, 25–28]. This is important not only for the purpose of generalizing the existing static models but also because black hole rotation itself can also be a trigger for the scalar field development rather than the pure spacetime curvature [29–32]. Apart from building equilibrium solutions, it is important to study as well the dynamics of spontaneous scalarization. This topic is very involved, though, and due to that reason it is still less studied and understood. In the present work, we focus on the dynamics of scalarization mainly for isolated black holes and neutron stars in scalar-Gauss–Bonnet gravity. We will review some of the most important advances in the field focusing in more detail on certain particular examples. Both approximate as well as fully coupled treatments to the problem will be discussed.

In Section II we will review the basic theory and field equations in scalar-Gauss Bonnet theory. Evolution equations in two particular cases are presented, namely the decoupling limit approximation and the full couples equation including matter in spherical symmetry. Section III is devoted to the dynamics of scalarization for isolated objects covering the cases of scalar field evolution on a Kerr background as well as stellar core collapse. The potential loss of hyperbolicity is also discussed there. We end with Conclusions.

2 Scalar-Gauss–Bonnet Gravity—Field Equations and Time Evolution

The general form of the scalar-Gauss–Bonnet gravity in the presence of matter and neglecting the scalar field potential is given by the following action

$$S = \frac{1}{16\pi}\int d^4x\sqrt{-g}\Big[R - 2\nabla_\mu\varphi\nabla^\mu\varphi + \lambda^2 f(\varphi)\mathcal{R}^2_{GB}\Big] + S_{\rm matter}(g_{\mu\nu}, \Psi), \quad (1)$$

where R is the Ricci scalar with respect to the metric $g_{\mu\nu}$. $\lambda^2 f(\varphi)$ denotes the coupling function of the scalar field φ[1] to the Gauss–Bonnet invariant $\mathcal{R}^2_{GB}$ where λ is the dimensional part having a dimension of *length*, while $f(\varphi)$ is a dimensionless function of φ. The action of the matter is $S_{\rm matter}$ where the matter field are collectively denoted by Ψ. The Gauss–Bonnet invariant is defined in the following way $\mathcal{R}^2_{GB} = R^2 - 4R_{\mu\nu}R^{\mu\nu} + R_{\mu\nu\alpha\beta}R^{\mu\nu\alpha\beta}$ with $R_{\mu\nu\alpha\beta}$ being the curvature tensor.

[1] Note that in the literature different definitions of φ can be found differing most commonly by a factor of 2 or $\sqrt{2}$. Thus, when converting results from one paper to the other one should proceed with a bit of caution.

By varying this action with respect to the metric $g_{\mu\nu}$ and scalar field φ one obtains the following field equations

$$R_{\mu\nu} - \frac{1}{2}Rg_{\mu\nu} + \Gamma_{\mu\nu} = 2\nabla_\mu\varphi\nabla_\nu\varphi - g_{\mu\nu}\nabla_\alpha\varphi\nabla^\alpha\varphi + 8\pi T_{\mu\nu}, \tag{2}$$

$$\nabla_\alpha\nabla^\alpha\varphi = -\frac{\lambda^2}{4}\frac{df(\varphi)}{d\varphi}\mathcal{R}^2_{GB}, \tag{3}$$

where ∇_μ is the covariant derivative with respect to $g_{\mu\nu}$ and $\Gamma_{\mu\nu}$ is defined by

$$\Gamma_{\mu\nu} = -R(\nabla_\mu\Psi_\nu + \nabla_\nu\Psi_\mu) - 4\nabla^\alpha\Psi_\alpha\left(R_{\mu\nu} - \frac{1}{2}Rg_{\mu\nu}\right) + 4R_{\mu\alpha}\nabla^\alpha\Psi_\nu + 4R_{\nu\alpha}\nabla^\alpha\Psi_\mu$$
$$-4g_{\mu\nu}R^{\alpha\beta}\nabla_\alpha\Psi_\beta + 4R^{\beta}{}_{\mu\alpha\nu}\nabla^\alpha\Psi_\beta \tag{4}$$

with

$$\Psi_\mu = \lambda^2\frac{df(\varphi)}{d\varphi}\nabla_\mu\varphi. \tag{5}$$

In the field equations above $T_{\mu\nu}$ is the matter energy-momentum tensor that satisfies the Bianchi identity $\nabla^\mu T_{\mu\nu}$=0.

Performing time evolution of the full system of equations in three spatial dimensions without approximation is a formidable task that became possible only recently. The key problem is that the standard $3 + 1$ formulations giving rise to a well-posed initial value problem in GR turn out to be not suitable in Gauss–Bonnet theory [33]. It was later discovered, though, that a gauge change can cure this problem, at least in the regime of weak coupling of the scalar field to the Gauss–Bonnet invariant [34–36]. An outline of the full well-posed $3 + 1$ formulation in Gauss–Bonnet theory is very lengthy and we refer the interested reader to e.g., [37, 38] for a detailed discussion on the problem. Instead, here we will consider two instructive particular simplified examples that can provide a more pedagogical insight.

The first one will be the formulation of the equations in the so-called decoupling limit approximation where the scalar field dynamics is considered on top of the GR evolution. Thus, the backreaction of the scalar field is neglected. Despite being simplified, such treatment leads to qualitatively and in some cases even quantitatively good picture for the scalar field development. The second will be the evolution of the full coupled system of equations but in the case of spherical symmetry.

2.1 Scalar Field Evolution in Decoupling Limit Approximation

In the current section, we will present a simplified model which is free from heavy technical complications but preserves the leading role of the nonlinearity associated with the coupling function contrary to the perturbative studies [24, 39, 40]. Namely, the spacetime geometry is kept fixed while the dynamics is governed by the scalar field equation. Thus, we isolate the nonlinear scalar field wave equation (3) and consider it on top of a stationary or dynamical GR background. As it turns out, neglecting the scalar field backreaction in such a manner is a good approximation not only qualitatively but also quantitatively for weak scalar fields [41, 42], e.g., in the vicinity of the bifurcation point in the case of spontaneous scalarization.

In order to be more specific, let us consider the particular case of scalar field evolution governed by Eq. (3) on top of a Kerr background geometry. We take the black hole solution in the standard Boyer–Lindquist coordinates

$$\begin{aligned} ds^2 = & -\frac{\Delta - a^2\sin^2\theta}{\Sigma}dt^2 - 2a\sin^2\theta\frac{r^2+a^2-\Delta}{\Sigma}dtd\phi \\ & + \frac{(r^2+a^2)^2 - \Delta a^2\sin^2\theta}{\Sigma}\sin^2\theta d\phi^2 + \frac{\Sigma}{\Delta}dr^2 + \Sigma d\theta^2, \end{aligned} \tag{6}$$

where $\Delta = r^2 - 2Mr + a^2$ and $\Sigma = r^2 + a^2\cos^2\theta$. In that case the Gauss–Bonnet invariant simplifies a lot and is given by

$$\mathcal{R}^2_{GB} = \frac{48M^2}{\Sigma^6}(r^2 - a^2\cos^2\theta)(r^4 - 14a^2r^2\cos^2\theta + a^4\cos^4\theta). \tag{7}$$

Next, we will proceed to present the scalar field equation (3) in the desired formulation. In order to do that let us first introduce a new azimuthal coordinate ϕ_* defined by

$$d\phi_* = d\phi + \frac{a}{\Delta}dr. \tag{8}$$

This allows us to avoid some nonphysical pathologies close to the black hole horizon. A second coordinate change will be to introduce the tortoise coordinate x defined by

$$dx = \frac{r^2+a^2}{\Delta}dr. \tag{9}$$

Thus in the new coordinates (t, x, θ, ϕ_*), and taken on the Kerr background, the scalar field equation (3) takes the following form

$$
\begin{aligned}
&-\left[(r^2+a^2)^2-\Delta a^2\sin^2\theta\right]\partial_t^2\varphi+(r^2+a^2)^2\partial_x^2\varphi+2r\Delta\partial_x\delta\varphi-4Mar\partial_t\partial_{\phi_*}\varphi\\
&+2a(r^2+a^2)\partial_x\partial_{\phi_*}\varphi+\Delta\left[\frac{1}{\sin\theta}\partial_\theta(\sin\theta\partial_\theta\varphi)+\frac{1}{\sin^2\theta}\partial^2_{\phi_*}\varphi\right] \qquad (10)\\
&=-\lambda^2\frac{12M^2\Delta}{\Sigma^5}(r^2-a^2\cos^2\theta)(r^4-14a^2r^2\cos^2\theta+a^4\cos^4\theta)\frac{df(\varphi)}{d\varphi}.
\end{aligned}
$$

As commented above, the scalar field should be weak in order for the decoupling limit approximation to make sense. In that case, and for a stationary background solution, the scalar field equation can be further simplified by assuming that the scalar field is axisymmetric, thus independent of the ϕ_* coordinate.

In order to perform the time evolution, the reduced field equation above has to be supplemented with boundary conditions. The natural ones at the horizon and at infinity is that the scalar field should have the form of an ingoing or outgoing wave, respectively.

2.2 Field Equations in Spherical Symmetry

The second particular example we consider is the case of spherical symmetry. Different formulations of the dimensionally reduced field equations can be found and we will follow more specifically [41].

Since we are working in spherical symmetry the metric can be presented in the following form

$$
ds^2=-e^{2\Phi(r,t)}dt^2+e^{2\Lambda(r,t)}dr^2+r^2(d\theta^2+\sin^2\theta d\phi^2). \qquad (11)
$$

Instead of dealing with the metric function Λ, it is convenient to introduce a new function $m(r,t)$ having the meaning of a local mass enclosed in a sphere with a radius r at the moment t. More specifically

$$
m(r,t)=\frac{r}{2}(1-e^{-2\Lambda}). \qquad (12)
$$

Another widely used simplification is to model the matter as a perfect fluid. In that case the energy-momentum tensor is $T_{\mu\nu}=\rho h u_\mu u_\nu+pg_{\mu\nu}$ with ρ being the rest-mass density and u^μ—the fluid 4-velocity. The specific enthalpy is given by $h=1+e+\frac{p}{\rho}$ where p and e are the fluid pressure and specific internal fluid energy. In spherical symmetry the 4-velocity of the fluid u^μ takes the form

$$
u^\mu=\frac{1}{\sqrt{1-v^2}}(e^{-\Phi},ve^{-\Lambda},0,0) \qquad (13)
$$

where $v=v(t,r)$ is the fluid velocity.

When performing highly dynamical nonlinear simulations, e.g., stellar core collapse or binary merger, the matter usually requires special treatment because of the potential presence of shocks. For that purpose, a standard approach is to use a high-resolution shock-capturing scheme with equations of motion written in flux conservative form using suitable conserved variables. Thus, we can introduce new fluid variable D, S^r and τ that are related to the standard primitive variables ρ, v and p by the following relations

$$D = \frac{\rho e^{\Lambda}}{\sqrt{1-v^2}}, \quad S^r = \frac{\rho h v}{1-v^2}, \quad \tau = \frac{\rho h}{1-v^2} - p - D. \tag{14}$$

Using the field equations, together with the Bianchi identity, one can derive the fluid flux conservative equations

$$\partial_t \mathbf{U} + \frac{1}{r^2}\partial_r \left(r^2 e^{\Phi-\Lambda}\mathbf{f}(\mathbf{U})\right) = \mathbf{s}(\mathbf{U}), \tag{15}$$

where $\mathbf{U}$ is the state vector of the conserved variables, namely $\mathbf{U} = [D, S^r, \tau]$. The flux vector $\mathbf{f}(\mathbf{U})$ and the source vector $\mathbf{s}(\mathbf{U})$ are defined by

$$\mathbf{f}(\mathbf{U}) = [Dv, S^r v + p, S^r - Dv], \tag{16}$$

$$\begin{aligned}\mathbf{s}(\mathbf{U}) = {} & [0, -(\tau + D)e^{\Phi-\Lambda}\Phi' - \frac{2\dot{m}}{r}e^{2\Lambda}S^r - \frac{e^{\Phi+\Lambda}}{r}\left(m' - \frac{m}{r}\right)(S^r v + p) \\ & + \frac{2p}{r}e^{\Phi-\Lambda}, -S^r e^{\Phi-\Lambda}(\Phi' + \Lambda') - \frac{2\dot{m}}{r}e^{2\Lambda}(S^r v + p + \tau + D)],\end{aligned} \tag{17}$$

where the dot and the prime denote differentiation with respect to t and r, respectively.

After dealing with the hydrodynamical representation, we can finally write down the field equations for the metric function Φ and m in spherical symmetry:

$$(1+\mathcal{A})\Phi' + \mathcal{B}\frac{X^2}{r}\dot{m} = X^2\left(4\pi r(s^r v + p) + \frac{m}{r^2}\right) + \frac{(P^2+Q^2)r}{2} + \frac{4m\lambda^2}{r^2}\left(\frac{d^2 f}{d\varphi^2}P^2 + \frac{df}{d\varphi}e^{\Lambda-\Phi}\dot{P}\right), \tag{18}$$

$$(1+\mathcal{A})m' + \mathcal{B}\dot{m} = 4\pi r^2(\tau + D) + \mathcal{A}\frac{m}{r} + \frac{r^2}{2X^2}(P^2+Q^2) + \frac{4m\lambda^2}{rX^2}\left(\frac{d^2 f}{d\varphi^2}Q^2 + \frac{df}{d\varphi}Q'\right), \tag{19}$$

$$\mathcal{C}m' + (1+\mathcal{A})\dot{m} = -4\pi r^2\alpha X^{-1}s^r + \mathcal{C}\frac{m}{r} + e^{\Phi-3\Lambda}r^2 PQ + \frac{4m\lambda^2}{r}e^{\Phi-3\Lambda}\left(\frac{d^2 f}{d\varphi^2}PQ + \frac{df}{d\varphi}P'\right). \tag{20}$$

The functions $\mathcal{A}$, $\mathcal{B}$, and $\mathcal{C}$ are defines as

$$\mathcal{A} = \lambda^2 Q\frac{df}{d\varphi}\frac{2}{r}(1 - 3e^{-2\Lambda}), \quad \mathcal{B} = \frac{4m\lambda^2}{r^2}e^{\Lambda-\Phi}P\frac{df}{d\varphi}, \quad \mathcal{C} = \frac{4m\lambda^2}{r^2}e^{\Phi-\Lambda}P\frac{df}{d\varphi}. \tag{21a}$$

The scalar field equation (3) is of second order and for the numerical integration it is better to transform it into two first order differential equations through the introduction of two new variables P and Q defined as

$$Q = \partial_r \varphi, \qquad P = e^{\Lambda-\Phi}\partial_t \varphi. \tag{22}$$

Thus, the dimensionally reduced Eq. (3) takes the form

$$\begin{aligned}\partial_t P &= e^{\Phi-\Lambda}Q' + e^{\Phi-\Lambda}(\Phi' - \Lambda' + \frac{2}{r})Q - \frac{1}{4}\frac{dV(\varphi)}{d\varphi}e^{\Phi+\Lambda} \\ &+ 2\frac{df(\varphi)}{d\varphi}\frac{\lambda^2}{r^2}e^{\Phi+\Lambda}\left\{(1-e^{-2\Lambda})\left[\frac{1}{r}(\Phi'-\Lambda')e^{-2\Lambda} - 8\pi p + (Q^2 - P^2)e^{-2\Lambda} + \frac{\Gamma_{\theta\theta}}{r^2}\right]\right. \\ &\left. - 2\Phi'\Lambda' e^{-4\Lambda} + 2\dot{\Lambda}^2 e^{-2\Phi-2\Lambda}\right\} \end{aligned} \tag{23}$$

$$\partial_t Q = e^{\Phi-\Lambda}[P' + (\Phi' - \Lambda')P]. \tag{24}$$

As it turns out, solving the above system of equations numerically is more involved compared to standard scalar-tensor theories [42] and we refer the interested reader to [41] for further details.

2.3 Mechanisms of Scalarization

One can distinguish three different types of scalar field development in sGB theory and they depend on the form of the scalar field coupling function $f(\varphi)$. In order to discuss each one of them it will be instructive to expand the function $f(\varphi)$ around $\varphi = 0$

$$f(\varphi) = f_0 + f_1\varphi + f_2\varphi^2 + f_3\varphi^3 + f_4\varphi^4 + \mathcal{O}(\varphi^4), \tag{25}$$

where the coefficients f_i are connected to the i-th derivative of $f(\varphi)$ at $\varphi = 0$.

Since the field equations depend only on the derivatives of $f(\varphi)$, the f_0 can be safely neglected. Having $f_1 \neq 0$ and $f_i = 0$ for $i > 1$ we restore the well known shift symmetric Gauss–Bonnet gravity where $\varphi = 0$ is not a solution of the field equations. Thus, black holes are always endowed with a scalar field [8–10]. The case of $f_1 = 0$ and $f_2 \neq 0$ allows for the so-called spontaneous scalarization [19–21] when $\varphi = 0$ is a solution of the field equations and the theory is equivalent to GR in the weak field regime. A large spacetime curvature around a compact object (e.g., a Kerr black hole or a neutron star) can destabilize the GR solution and ignite the scalar field. In that case, the square of the scalar field effective mass turns negative leading to an exponentially increasing scalar field until it saturated to an equilibrium configuration (see [23] for a review on the topic). Interesting is also the case of $f_1 = f_2 = 0$ and having only higher order term in the expansion. For a sGB theory with such a coupling, $\varphi = 0$ is a solution of the field equations and the effective scalar field mass is zero. Thus the GR compact object solutions, such as Kerr black hole, are always linearly stable. Nevertheless, scalarized black hole and neutron star solutions exist and they can be thermodynamically preferred over the GR one [44, 45]. This is dubbed nonlinear scalarization since only a large nonlinear scalar field perturbation can cause a Kerr black hole (or a neutron star) to reconfigure and end up in a scalarized state.

3 Scalarization of Isolated Compact Objects

Our main focus will be the scalarization of isolated neutron stars and black holes. This can be considered as a model case helping us in understanding the dynamics of more complicated systems, such as binaries. Moreover, the stellar core collapse in scalar-Gauss–Bonnet theory, that will be also reviewed, is the astrophysically relevant mechanism for the birth of scalarized black holes and neutron stars potentially leading to specific observational signatures.

The outline of the section will follow two major directions—scalar field dynamics in the decoupling limit approximation and results for the dynamical evolution of the full coupled system. The former case can give us qualitatively good intuition and in some cases also quantitatively accurate results about the process of scalarization while solving the full system of equations is important for understating how the scalar sector influences the gravitational one. Taking into account the scalar field backreaction allows us to study a plethora of observational signatures, especially gravitational wave ones, but also opens important questions such as the well-possedness of the theory.

3.1 Scalar Field Dynamics in the Decoupling Limit Approximation

The scalar field evolution on top of a Schwarzschild black hole was first considered in the case of shift-symmetric Gauss–Bonnet theory, i.e., when the coupling function $f(\varphi) \sim \varphi$ [46, 47]. Since in this flavour of the Gauss–Bonnet theory, black holes are always endowed with a scalar field, the time evolution of the field equation (10) will drive any scalar field initial data to its equilibrium configuration. Naturally, during this process, strong scalar radiation will be emitted. A similar procedure was applied to the case of spontaneous scalarization, this time including rotating black holes [48], with the main difference that compact objects develop scalar hair only for strong enough spacetime curvature as discussed above. Of course, not only black holes scalarize, neutron stars can develop a nontrivial scalar field as well and as a matter of fact this process is easier to be triggered in the presence of matter. Spontaneous scalarization of neutron stars in the decoupling limit approximation was considered in [45] while the shift symmetric case was addressed in [49].

Let us discuss in more detail the case of spontaneous scalarization of a Kerr black hole [48] as a representative example for the scalar field dynamics in scalar-Gauss–Bonnet theory. As commented above, in this case, we should employ a coupling function being quadratic in the scalar field in the leading order expansion. Pure quadratic term, i.e., $f(\varphi) = \varphi^2$, leads always to unstable hairy black holes, though, and the scalar field grows exponentially indefinitely [24]. In order to cure this instability one can consider a coupling function of the form

$$f(\varphi) = \frac{\epsilon}{2\beta}\left(1 - e^{-\beta\varphi^2}\right), \tag{26}$$

that has the same leading order expansion as $f(\varphi) = \varphi^2$ but allows for the existence of stable scalarized solutions. In the expression above β is a parameter and $\epsilon = \pm 1$. In the static case only $\epsilon = +1$ can lead to black hole scalarization while for $\epsilon = -1$ we can have spin-induced scalarization where the source for the scalar field is the black hole rotating itself [27–29, 31].[2] Below we will discuss separately scalar field growth for static and rotating black holes. All simulations are performed for $\beta = 12$ that leads to hairy black hole solutions with a relatively weak scalar field where the decoupling limit is a good approximation.

3.1.1 Nontoratating Black Holes

Let us consider the evolution of a small scalar field perturbation on top of a Schwarzschild black hole within scalar-Gauss–Bonnet theory employing the coupling function (26) with $\epsilon > 0$ which allows for spontaneous scalarization of nonrotating black holes. Furthermore, we choose the ratio M/λ, where M is the black hole mass and λ is the dimensional part of the scalar coupling function, in such a way that the Schwarzschild solution is unstable and thus prone to spontaneous scalarization. This instability manifests itself as an exponential growth of the scalar field starting from an arbitrarily small perturbation. Of course, the scalar field can not increase indefinitely like in the linearized case [31] and its growth is eventually halted when the amplitude becomes large enough so that the nonlinearities in the scalar field equation (10) start being important.

In order to demonstrate the process, snapshots of the scalar field radial profile at different times are presented in Fig. 1 with the x-axis being the normalized tortoise coordinate x/λ. The time evolution of the scalar field at the horizon, denoted by φ_H, is plotted as well in Fig. 2. In both figures two Schwarzschild black holes are used as backgrounds—one of them being relatively close to the bifurcation point with mass

[2] Note that neutron stars scalarize in the static case for both signs of ϵ due to the presence of matter.

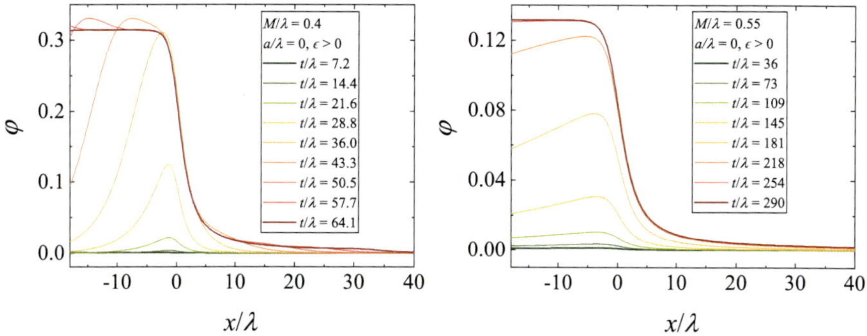

Fig. 1 Time evolution of the scalar field radial profile for two black holes with $M/\lambda = 0.4$ *(left panels)* and $M/\lambda = 0.55$ *(right panels)*. The x-axis represents the normalized tortoise coordinate x/λ. The coupling function is Eq. (26), where $\beta = 12$. Different colors of the lines depict the scalar field profile at different snapshots in time. The solid dark red line represents the final state. Adapted from [50]

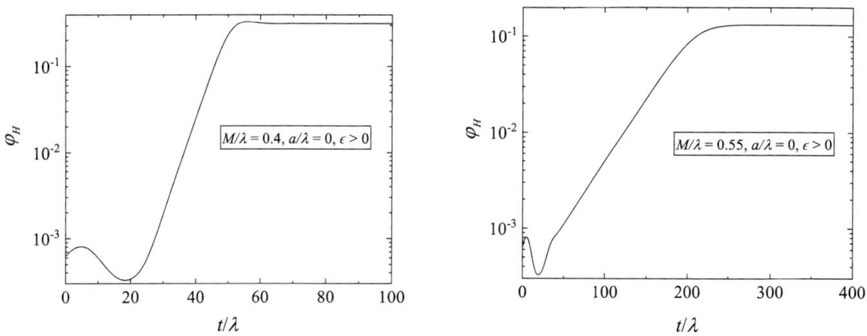

Fig. 2 The scalar field at the horizon as a function of time, for the same models presented in Fig. 1. Adapted from [50]

d$M/\lambda = 0.55$ (right panels)[3] while the other one is away from the bifurcation having $M/\lambda = 0.4$ (left panels).

If the background black hole solution was stable, the small initial scalar field perturbation that is imposed as initial data would decay exponentially in time and the energy of the scalar field will be radiated away through scalar gravitational waves. As evident from the figures, in case of instabilities the time evolution is qualitatively different—after a phase of exponential growth the scalar field saturates to an equilibrium. The numerical investigations show that indeed the growth time of the instability and the final scalar field profile after saturation do not dependent on the initial data. In addition, this growth is faster for black holes with smaller masses, i.e., further away from the bifurcation point. Thus, the time for reaching the end state decreases for smaller black hole masses.

[3] The exact location of the bifurcation point is at $(M/\lambda)_{\rm crit} = 0.587$.

3.1.2 Rotating Black Holes

Here we will address the more complicated case of scalar field development on top of a Kerr black hole governed by Eq. (10). In order to demonstrate that the qualitative picture remains the same for a wide variety of couplings being quadratic in the scalar field in leading order expansion, in this subsection we will discuss a different form of $f(\varphi)$, namely

$$f(\varphi)_{\text{CaseII}} = \frac{\epsilon}{\beta^2}\left(1 - \frac{1}{\cosh(\beta\varphi)}\right). \tag{27}$$

In addition, we will discuss the cases of both $\epsilon > 0$ and $\epsilon < 0$: While the former one is an extension of the results presented in the previous subsection to the case of a Kerr black hole, the second one is particularly interesting since it allows for the phenomenon of spin-induced scalarization as discussed above.

Snapshots of the scalar field profile at different times, for both signs of $\epsilon > 0$ and $\epsilon < 0$, are depicted in Figs. 3 and 4 respectively. In contrast to the static case, the snapshots are two-dimensional surfaces dependent on the tortoise coordinate x/λ and on θ. For each sign of ϵ, two Kerr black holes are taken as a background solution. When $\epsilon > 0$, i.e., static black holes can scalarize, the model with $M/\lambda = 0.56, a/\lambda = 0.3$ represents a moderately rotating black hole being able to scalarize also in the $a \to 0$ limit. The case of $M/\lambda = 0.59, a/\lambda = 0.5$ on the other hand allows for spontaneous scalarization of a Kerr black hole even though a Schwarzschild solution with such M/λ is linearly stable and thus bald. For spin-induced scalarization, we always need a large black hole spin in order to trigger the scalar field, namely $a/M > 0.5$. Thus,

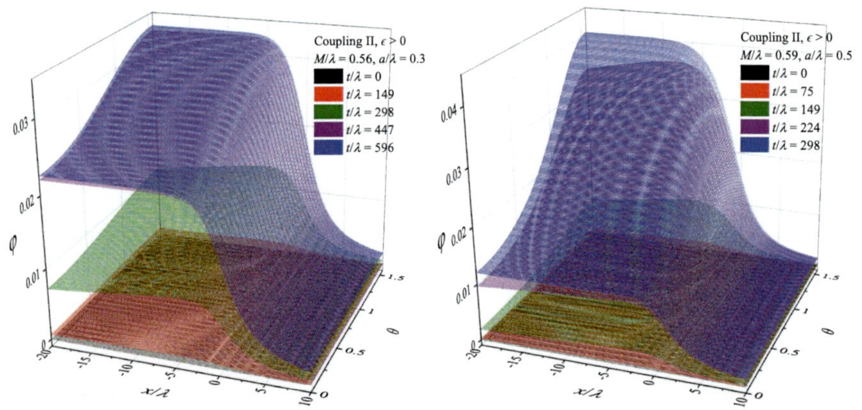

Fig. 3 Snapshots of the two-dimensional scalar field profile taken in a $x/\lambda - \theta$ plane, where x/λ is the tortoise coordinate. The coupling function employed is (27) with $\beta = 12$ and $\epsilon > 0$. Two background Kerr black holes are considered: A moderately rotating model with $M/\lambda = 0.56$, $a/\lambda = 0.3$ *(left panel)* and a more rapidly rotating black hole with $M/\lambda = 0.59$, $a/\lambda = 0.5$ *(right panel Credit* [19]*)*

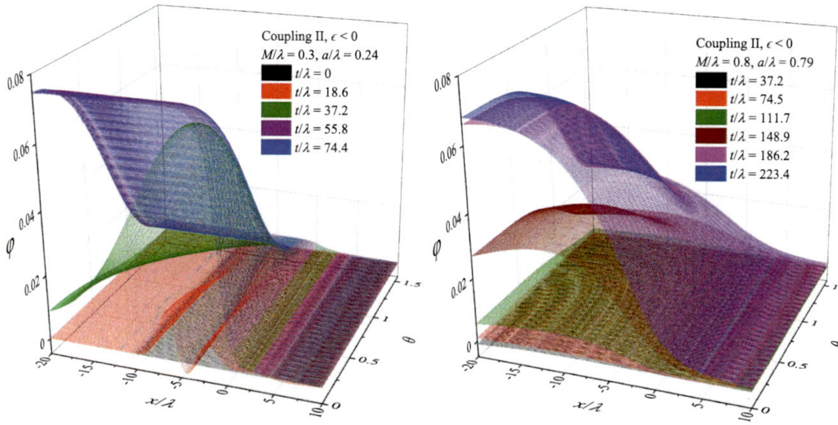

Fig. 4 Similar to Fig. 3 but for the case of spin-induced scalarization with $\epsilon < 0$. The coupling function employed is (27) with $\beta = 12$. The background models are a moderately rotating Kerr black hole with $M/\lambda = 0.3$, $a/\lambda = 0.24$ *(left panel)* and a model close to the extremal limit with $M/\lambda = 0.8$, $a/\lambda = 0.79$ *(right panel)*

the two background models are take as $M/\lambda = 0.3$, $a/\lambda = 0.24$ and $M/\lambda = 0.8$, $a/\lambda = 0.79$, where the second one is very close to the extremal limit.

Similar to the static case, a small scalar field perturbation grows rapidly until it settles to an equilibrium. The major difference in both cases is that while the equilibrium scalar field that is reached has a maximum at the equator, $\theta = \pi/2$, for the standard scalarization with $\epsilon > 0$, this maximum for the spin-induced scalarization with $\epsilon < 0$ happens at $\theta = 0$. This is connected with the structure, and more precisely the sign, of the Gauss–Bonnet invariant close to the black hole horizon.

3.2 *Scalar Field Evolution on the Background of a Neutron Star*

The evolution of a neutron star in scalar-Gauss–Bonnet theory is a much less studied topic. The case of shift symmetric Gauss–Bonnet theory with $f(\varphi) = \varphi$ and more specifically the collapse of a spherically symmetric neutron star towards a black hole within this theory, adopting the decoupling limit approximation, was considered in [51]. The evolution of the scalar field is presented in Fig. 5 where the initial data is a GR neutron star without a scalar field that collapses to a Schwarzschild black hole in a few tens of milliseconds. Since in the shift symmetric Gauss–Bonnet theory, a neutron star will be always endowed with scalar hair, the evolution quickly drives these initial data to a quasi-equilibrium scalar field configuration around the neutron star. What is plotted in the figure is the subsequent evolution when the star collapses

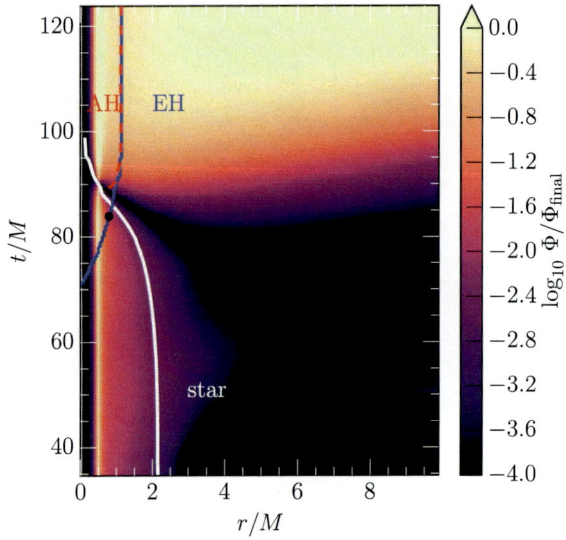

Fig. 5 Simulated space-time diagram of a collapsing neutron star in shift symmetric sGB gravity showing the value of the scalar field Φ relative to its final configuration $\Phi_{\rm final}$. The stellar surface is depicted with a white line, the event horizon with blue, and the apparent horizon in red. The black dot on the blue line denoted the time of peak scalar flux emitted during the collapse process. Credit [51]

and a black hole is formed. Since the scalar field is sourced by the spacetime curvature, it grows much stronger as the black hole forms.

Similar characteristics of the scalar field evolution were also observed in the case of a collapse of a normal stellar core towards a neutron star or a black hole this time in a scalar-Gauss–Bonnet theory admitting spontaneous scalarization [41]. We will discuss this topic further in the next subsections because the simulations in [41] where actually performed both in the decoupling limit approximation and in the full theory.

The scalar field evolution when backreaction is neglected can also be used to gain intuition about (in)stability of neutron star (or black hole) solutions. This is particularly valuable especially in cases when the solution structure is more complicated and more than one scalarized branches exist. For such theories, the scalar field time evolution can only drive it (independent of the initial data) towards the stable scalarized configuration(s). It is often the case that even if multiple scalarized branches are present only one of them is stable (see e.g., [52]). Thus typically only one of the scalarized solutions can be reproduced in such dynamical simulations. This approach was successfully applied for determining (in)stability of spontaneously scalarized neutron stars in [45] borrowing the methodology developed previously for black holes [44]. Of course, this is a rigorous proof only for instability while a potentially stable branch can be destabilized if one considers the coupled evolution. Nevertheless, as the results in [52] indicate, often the intuition based on such approximate approaches is accurate enough.

3.3 *Beyond Decoupling Limit Approximation*

3.3.1 Evolution of Isolated Black Holes and Loss of Hyperbolicity

With the advance of numerical relativity simulations in theories beyond general relativity it was quickly realized that well-posedness is often an issues and loss of hyperbolicity can be observed [33]. Taking as an example the sGB theory, the reason is that the scalar field enters nontrivially in the generalized Einstein equation (2) that effectively changes the principle symbol of the system of PDEs. This problem is much easier to tackle in spherical symmetry and that is why dynamics of isolated static black holes was considered first. This includes the shift-symmetric sGB theory [42, 53], the Einstein-dilaton-Gauss–Bonnet gravity with an exponential coupling [54], and the case of spontaneous scalarization [55]. The general causes for hyperbolicity loss were examined in detail in [49] for a spherically symmetric system.

Advance in numerically solving the full $3 + 1$ equations, without assuming spacetime symmetry, was possible only after the theoretical discovery that at least in the weak coupling regime hyperbolicity can be regained by introducing the so-called Modified Generalized Harmonic Coordinate (mGHC) formulation [34–36]. That led to the development of numerical codes being able to handle black hole evolution in three spacial dimension [38, 56–58]. As expected, though, once we deviate enough from the weak coupling regime where the results for the well-posedness of the mGHC formulation are no longer guaranteed, loss of hyperbolicity is observed as well. This effectively limits the maximum scalar field a black hole can sustain and the results from the nonlinear simulations indicate that the energy contribution associated with the scalar field should be less than a few percent of the total back hole mass.

Interestingly, similar observations were made also for theories admitting the standard spontaneous scalarization and the spin-induced scalarization [57, 59]. The maximum scalar field a black hole can sustain in the former case is presented in Fig. 6 and as one can see its total energy contribution can be only a few percent of the black hole mass [57]. One can claim that potentially a gauge change can cure this problem. The results in [36] and the simulations in [59] indicate, though, that the loss of hyperbolicity is dominated by physical modes and is thus unavoidable. A potential way out of this problem is to introduce an additional coupling between the Ricci scalar and the scalar field [60]. Even though results in spherical symmetry are promising [61] still analysis in the general $3D$ case are needed. Introducing a fixing equation might also be used to cure the loss of hyperbolicity [61] even though see [62] for possible important shortcomings of this approach.

One should always remember, though, that hyperbolicity loss appears only for a larger coupling strength where the effective field theory treatment of the sGB gravity is no longer justified. This was explicitly demonstrated also in the numerical simulations in [59]. Namely, for models that develop an elliptic region the weak coupling condition is larger than unity. Therefore, one can not neglect higher-order correction in the action in addition to the Gauss–Bonnet invariant. Thus, the results indicate that as long as we treat sGB gravity as an effective field theory within the regime of its validity, well-posedness is guaranteed.

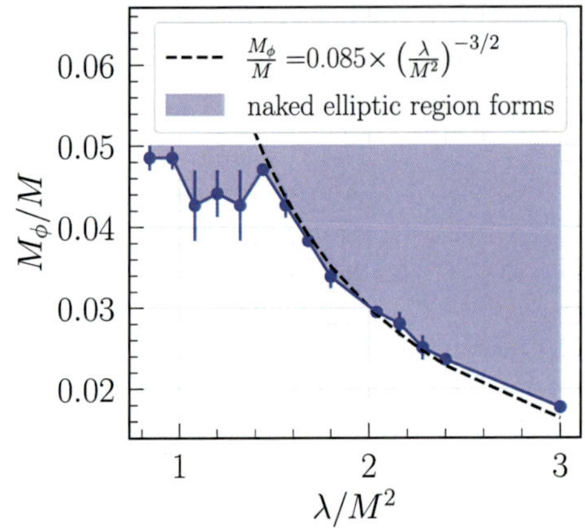

Fig. 6 The scalar field energy scaled to the total black holes mass M_ϕ/M as a function of the coupling parameters λ/M^2 is plotted for the threshold models on the border of developing an elliptic region outside the black holes. Shaded blue indicates the region to the left of the line where the system losses hyperbolicity. The scalar field coupling to the Gauss–Bonnet invariant is taken of the form $f(\varphi) = \frac{\lambda}{2}\phi^2 + \frac{\sigma}{4}\phi^4$. Note that the definitions of φ and the coupling constants deviate slightly from the ones presented in Sect. 2 and we refer the reader to [57] for further detail. Credit [57]

3.3.2 Formation of Isolated Scalarized Black Holes and Neutron Stars Through Core-Collapse

The actual astrophysical process leading to the formation of an isolated neutron star or a black hole is the core collapse of a stellar core. Simulating this process is very involved not only because of the difficulty of performing general relativistic hydrodynamic simulations but also because the microphysics should be taken into account accurately. Due to the uncertainties and difficulties in pure GR, core collapse simulations are performed rarely in modified gravity. Nevertheless, it was demonstrated that in certain cases the effect of scalar field dynamics, as the collapse proceeds, can lead to strong observable gravitational wave signal [63–65]. Here we will review briefly the results for stellar core collapse in scalar-Gauss–Bonnet gravity since, as we will see below, it provides richer phenomenology even compared to the standard scalar-tensor theories due to the possibility to scalarize not only a neutron star but also a black hole.

In a core collapse simulation the time evolution is heavily dependent on the microphysics taken into account but the end state, at least in GR, is pretty simple—either a neutron star or a black hole forms. If we allow for GR modifications the picture can be much richer. The reason is that depending on the theory parameters and the progenitor properties, a scalar field can develop or disappear at different phases of the progenitor evolution. The possible outcomes, in case of a nonrotating progenitor and a coupling function allowing for spontaneous scalarization, e.g., Eq. (26), are summarized in Fig. 7. The scenarios are divided into two major classes depend-

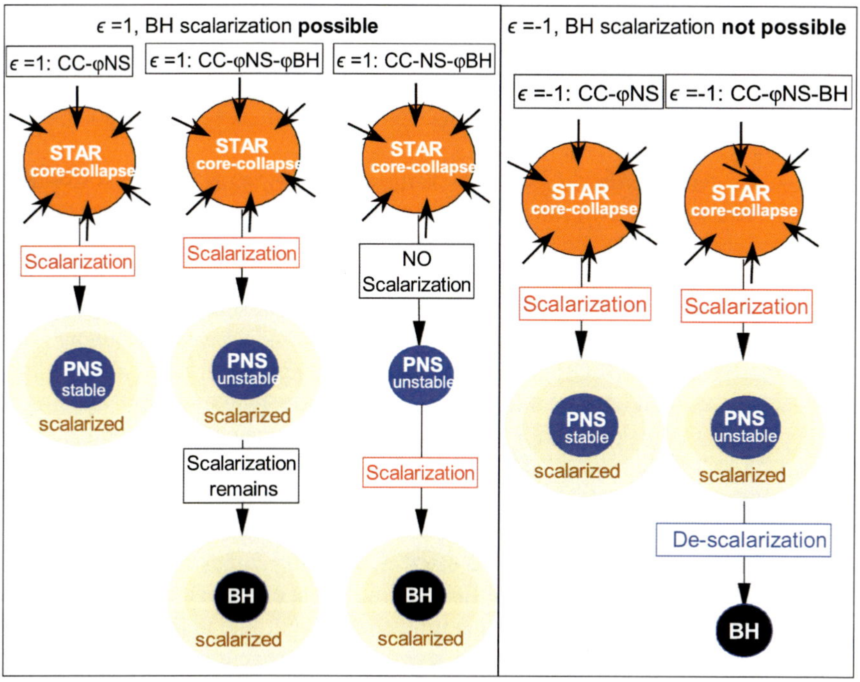

Fig. 7 Possible outcomes of stellar core collapse in sGB gravity in the case of a nonrotating progenitor and coupling function allowing for spontaneous scalarization, e.g., (26). Credit [41]

ing on the sign of the ϵ parameter in $f(\varphi)$. As discussed, for $\epsilon = 1$, both neutron stars and black holes can scalarize while for $\epsilon = -1$ scalarization is not possible for static black holes while neutron stars can still develop nontrivial scalar field [20, 66]. For a theory admitting spontaneous scalarization, where the GR solutions with zero scalar field are also solutions of the more general Gauss–Bonnet field equations, the scalar field can be triggered only when the spacetime curvature becomes large enough. This happens as the progenitor compactness increases during the collapse and can be observed either during the formation of the protoneutron star (e.g., the scenario "$\epsilon = 1$: CC-φNS" or "$\epsilon = -1$: CC-φNS" in Fig. 7) or later on if it collapses to a black hole ("$\epsilon = 1$: CC-φNS-φBH" and "$\epsilon = 1$: CC-NS-φBH" in Fig. 7). Interestingly, for $\epsilon = -1$, scalarization can appear only temporarily at the protoneutron star stage that is an effect similar to the one observed in standard scalar-tensor theories [43] (the scenario "$\epsilon = -1$: CC-φNS-BH" in Fig. 7).

Let us note that if a relatively fast rotating black hole is formed after the core collapse, it might be possible to develop spin-induced scalarization and thus have other core collapse scenarios for $\epsilon = -1$. Producing a rapidly rotating protoneutron star after a core collapse, though, seems to be a rare event [67, 68].

Solving the full coupled system of field equations in the presence of matter (presented in Sect. 2.2), even in spherical symmetry, is very involved. Even more, the coupling between the scalar field and the Gauss–Bonnet invariant, that influences significantly the generalized Einstein equations, requires a change of the original numerical approaches in GR and standard scalar-tensor theories. Thus a modification in the spirit of black hole evolution in Gauss–Bonnet theory can be implemented [42] (see [41] for further details).

Below, we will shortly review some of the most interesting scenarios listed in Fig. 7. For that purpose two progenitors from the Woosley & Heger's catalog [70] will be used. They are denoted as $z12$ and $z40$ having masses $12M_{\odot}$ and $40M_{\odot}$ before the collapse and primordial metallicity. The first one, at least in GR, leads to a collapse to a stable protoneutron star. The second one forms a meta-stable protoneutron star that subsequently collapses to a black hole due to follow-up accretion of matter after bounce. In the considered sGB gravity this does not change because the scalar field is kept relatively weak, not being able to influence dramatically the collapse dynamics.

The combination of progenitor masses and values of the coupling parameters λ and β are chosen such that the scalar field develops either at the neutron or the black hole stage of the collapse. In general, the critical central energy density ρ_{bifurc}, above which scalarization is possible, is controlled by λ, while β is responsible for the strength of the scalar field [66]: For a fixed λ, typically larger β leads to weaker scalar fields. Details on the microphysics treatment, the employed equation of state and the numerical approach to solve the problem can be found in [41, 43, 71].

Core collapse to a scalarized neutron star.—For the progenitor $z12$, a PNS forms as an outcome of the collapse. Spontaneous scalarization develop during the evolution for both $\epsilon = 1$ and $\epsilon = -1$ if β and λ are properly chosen. Here we will present only the case with $\epsilon = 1$. The second case $\epsilon = -1$ has a similar dynamical evolution and the only qualitative difference is in the final equilibrium scalar field profile that is dictated up to a large extend by the properties of the Gauss–Bonnet invariant $\mathcal{R}^2_{GB}$.

Figure 8 is an example of the core collapse scenario "$\epsilon = 1$: CC-φNS" in Fig. 7. In that case, the lighter progenitor z12 collapses to a scalarized protoneutron star. The upper panel of the figure shows the snapshots of the scalar field and fluid velocity evolution while the lower panel depicts the central value of the scalar field and the scalar change $(r\varphi)|_{r\to\infty}$ as a function of time. Concentrating on the lower panel, one clearly sees how the scalar field is ignited and later on it reaches some quasi-equilibrium value. This coincides with the moment when the protoneutron star compactness is enough so that it can act as a scalar field source. The spontaneous appearance of a nonzero scalar charge at the associated retarded time indicates that in sGB gravity the scalar wave propagates at the speed of light.

Collapse to a black hole.—If the progenitor is even more massive, e.g., z40, the star will undergo a second phase of a collapse to a black hole due to follow-up accretion. In that case for $\epsilon = 1$ the scalar field will grow further since the curvature, that is the scalar field source, increases. The evolution in this case is presented in Fig. 9. For $\epsilon = -1$, though, static black holes can not sustain scalar hair. Thus, the temporarily developed scalar field during the unstable proto-neutron star stage will be completely lost once the black hole is formed. The evolution is very similar to the one presented in Fig. 9 with the main difference that at the point of black hole

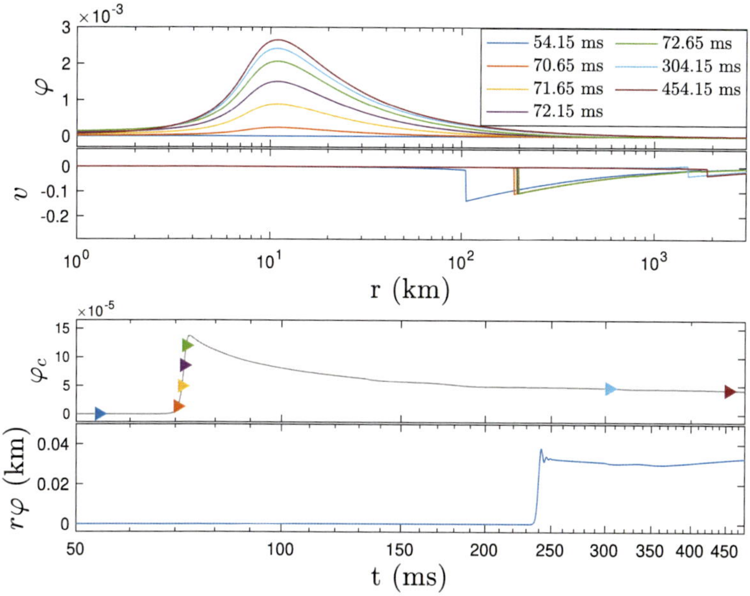

Fig. 8 Collapse of the z12 progenitor to a scalarized neutron star following the path "$\epsilon = 1$: CC-φNS" in Fig. 7. The coupling parameters are set to $\epsilon = 1$, $\lambda = 100$, and $\beta = 14\,000$. *(upper panel)* Temporal snapshots of scalar field φ and fluid velocity v. *(lower panel)* The evolution of the central value of the scalar field φ_c and the scalar charge $r\varphi$, taken far away from the star. Markers in the bottom panels indicate the time of snapshots having the same colors in the corresponding upper panels. Credit [41]

formation the scalar field outside the black hole apparent horizon quickly drops to zero.

In the described processes of (de)scalarization a significant amount of scalar gravitational radiation is emitted, reaching several orders of magnitude above the typical energy emitted through tensorial gravitational waves during core collapse [72, 73]. Thus, the scalar waves offer a much more efficient channel of energy loss. Unfortunately, in the standard sGB gravity, this radiation is not detectable since it does not induce directly tonsorial gravitational wave emission, e.g., no breathing modes are present in contrast to the core collapse in scalar-tensor theories [63]. Of course, in a more general form of sGB gravity such breathing modes can be present [74] that would allow us to set constraints on the corresponding theory from future gravitational wave observations.

Since the simulations we have discussed involve solving the full coupled system of field equations, naturally the problem of hyperbolicity loss, that we discussed above, appears as well. As a results, the scalar field should be limited to very small value in order to preserve hyperbolicity. Since the scalar field strength decreases with increasing β, one has to work with very high values of this parameter in order to be able to perform the evolution. For example, in some of the simulations above β

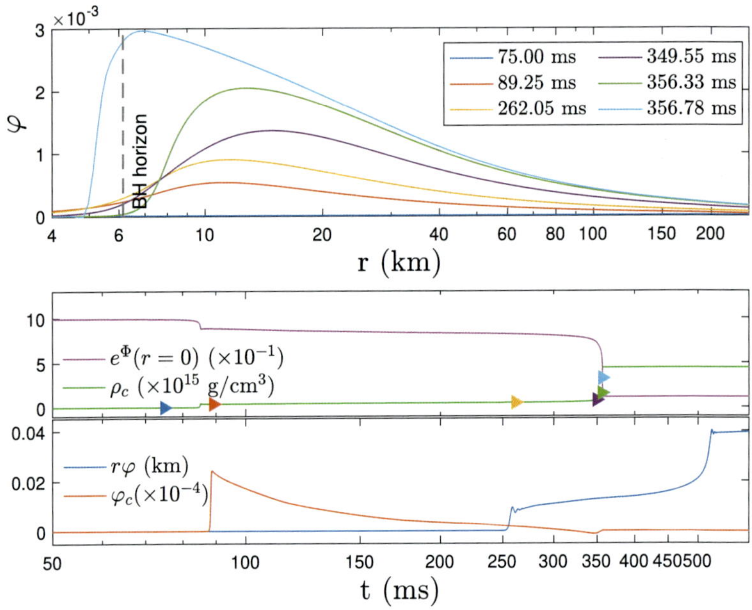

Fig. 9 The core collapse of the z40 progenitor realizing the scenario "$\epsilon = 1$: CC-φNS-φBH" in Fig. 7, namely a collapse of the more massive z40 progenitor to a scalarized black hole through a scalarized protoneuton star. The values of the coupling parameters are chosen to be $\epsilon = 1$, $\lambda = 120$, $\beta = 900\,000$. In the lower panel the evolution of the metric function e^{Φ} and the central energy density ρ_c is also plotted showing the formation of a black hole at late times. Credit [41]

reaches a few hundreds of thousands. Thus, for the presented models, the total energy of the scalar field reaches no more that $\lesssim 1\%$ of the compact object mass. This is in agreement with the binary neutron star merger simulations in the shift symmetric sGB gravity [75] as well as the black hole nonlinear evolution results presented in Fig. 6 [57].

4 Conclusion

In this work, we reviewed some aspects of the very dynamics of the scalarization of isolated black holes and neutron stars within the scalar-Gauss–Bonnet gravity. We tried to keep the exposition as concise as possible and present only those essential facts and results which form our physical understanding of the scalarization dynamics evading sophisticated mathematical constructions. First, we discussed the scalarization dynamics of static and rotating black holes in the decoupling limit approximation. Although simplified, that approximate approach is free from heavy technicalities and creates good insight into the physics of the scalarization process.

We also discussed the formation of scalarized spherically symmetric black holes and neutron stars in the scalar-Gauss–Bonnet gravity through a stellar core collapse in its full non-linearity. Recent important advances were also briefly reviewed, especially the loss of hyperbolicity in the strong field regime of the Gauss–Bonnet gravity. While the field equations of the Gauss–Bonnet gravity are well-possed in the weak field regime, in the strong field regime those field equations can lose their hyperbolic character rendering in this way the Cauchy problem ill-defined. It has to be emphasized, however, that the mentioned problems with the hyperbolicity loss appear only for larger couplings where the effective field theory treatment of the sGB gravity is no longer justified. At first glance, it may seem that the loss of hyperbolicity is a serious drawback of the theory, however, as we mentioned in the main text, adding an additional coupling of the scalar field to the scalar Ricci curvature mitigates the loss of hyperbolicity at least in spherical symmetry. This demonstrates that including specific additional interactions may be the right way of recovering the well-posedness in sGB and other effective field theories.

An important future direction of investigation is to study the influence of additional couplings on the hyperbolicity in the general case without assuming spherical symmetry. Such studies, although very challenging from a numerical point of view, have already been started by several groups and we hope that we will have a clear understanding of the problem in the near future. It is important to also understand whether approximate approaches recovering to a certain extent some of the higher energy degrees are viable in general realistic cases such as binary evolution.

Acknowledgements This study is in part financed by the European Union- NextGenerationEU, through the National Recovery and Resilience Plan of the Republic of Bulgaria, project No BG-RRP-2.004-0008-C01. DD acknowledges financial support via an Emmy Noether Research Group funded by the German Research Foundation (DFG) under grant no. DO 1771/1-1.

References

1. E. Berti et al., Class. Quant. Grav. **32**, 243001 (2015). https://doi.org/10.1088/0264-9381/32/24/243001. arXiv:1501.07274 [gr-qc]
2. G.W. Horndeski, Int. J. Theor. Phys. **10**, 363 (1974). https://doi.org/10.1007/BF01807638
3. T. Kobayashi, M. Yamaguchi, J. Yokoyama, Prog. Theor. Phys. **126**, 511 (2011). https://doi.org/10.1143/PTP.126.511. arXiv:1105.5723 [hep-th]
4. T. Kobayashi, Rept. Prog. Phys. **82**, 086901 (2019). https://doi.org/10.1088/1361-6633/ab2429. arXiv:1901.07183 [gr-qc]
5. M. Heusler, *Black Hole Uniqueness Theorems*, Cambridge Lecture Notes in Physics (Cambridge University Press, 1996). https://doi.org/10.1017/CBO9780511661396
6. J. Healy, T. Bode, R. Haas, E. Pazos, P. Laguna, D. Shoemaker, N. Yunes, Class. Quant. Grav. **29**, 232002 (2012). https://doi.org/10.1088/0264-9381/29/23/232002. arXiv:1112.3928 [gr-qc]
7. C.A.R. Herdeiro, E. Radu, Int. J. Mod. Phys. D **24**, 1542014 (2015). https://doi.org/10.1142/S0218271815420146. arXiv:1504.08209 [gr-qc]
8. S. Mignemi, N.R. Stewart, Phys. Rev. D **47**, 5259 (1993). https://doi.org/10.1103/PhysRevD.47.5259. arXiv:hep-th/9212146

9. P. Kanti, N.E. Mavromatos, J. Rizos, K. Tamvakis, E. Winstanley, Phys. Rev. D **54**, 5049 (1996). https://doi.org/10.1103/PhysRevD.54.5049. arXiv:hep-th/9511071
10. T. Torii, H. Yajima, K.-I. Maeda, Phys. Rev. D **55**, 739 (1997). https://doi.org/10.1103/PhysRevD.55.739. arXiv:gr-qc/9606034
11. P. Pani, V. Cardoso, Phys. Rev. D **79**, 084031 (2009). https://doi.org/10.1103/PhysRevD.79.084031. arXiv:0902.1569 [gr-qc]
12. T.P. Sotiriou, S.-Y. Zhou, Phys. Rev. Lett. **112**, 251102 (2014). https://doi.org/10.1103/PhysRevLett.112.251102. arXiv:1312.3622 [gr-qc]
13. D. Ayzenberg, N. Yunes, Phys. Rev. D **90**, 044066 (2014), note [Erratum: Phys.Rev.D 91, 069905 (2015)]. https://doi.org/10.1103/PhysRevD.90.044066. arXiv:1405.2133 [gr-qc]
14. A. Maselli, P. Pani, L. Gualtieri, V. Ferrari, Phys. Rev. D **92**, 083014 (2015). https://doi.org/10.1103/PhysRevD.92.083014. arXiv:1507.00680 [gr-qc]
15. B. Kleihaus, J. Kunz, E. Radu, Phys. Rev. Lett. **106**, 151104 (2011). https://doi.org/10.1103/PhysRevLett.106.151104. arXiv:1101.2868 [gr-qc]
16. B. Kleihaus, J. Kunz, S. Mojica, Phys. Rev. D **90**, 061501 (2014). https://doi.org/10.1103/PhysRevD.90.061501. arXiv:1407.6884 [gr-qc]
17. B. Kleihaus, J. Kunz, S. Mojica, M. Zagermann, Phys. Rev. D **93**, 064077 (2016). https://doi.org/10.1103/PhysRevD.93.064077. arXiv:1601.05583 [gr-qc]
18. K. Yagi, L.C. Stein, N. Yunes, Phys. Rev. D **93**, 024010 (2016). https://doi.org/10.1103/PhysRevD.93.024010. arXiv:1510.02152 [gr-qc]
19. D.D. Doneva, S.S. Yazadjiev, Phys. Rev. Lett. **120**, 131103 (2018). https://doi.org/10.1103/PhysRevLett.120.131103. arXiv:1711.01187 [gr-qc]
20. H.O. Silva, J. Sakstein, L. Gualtieri, T.P. Sotiriou, E. Berti, Phys. Rev. Lett. **120**, 131104 (2018). https://doi.org/10.1103/PhysRevLett.120.131104. arXiv:1711.02080 [gr-qc]
21. G. Antoniou, A. Bakopoulos, P. Kanti, Phys. Rev. Lett. **120**, 131102 (2018). https://doi.org/10.1103/PhysRevLett.120.131102. arXiv:1711.03390 [hep-th]
22. T. Damour, G. Esposito-Farèse, Phys. Rev. Lett. **70**, 2220 (1993). https://doi.org/10.1103/PhysRevLett.70.2220
23. D.D. Doneva, F.M. Ramazanoğlu, H.O. Silva, T.P. Sotiriou, S.S. Yazadjiev (2022). arXiv:2211.01766 [gr-qc]
24. J.L. Blázquez-Salcedo, D.D. Doneva, J. Kunz, S.S. Yazadjiev, Phys. Rev. D **98**, 084011 (2018). https://doi.org/10.1103/PhysRevD.98.084011. arXiv:1805.05755 [gr-qc]
25. P.V.P. Cunha, C.A.R. Herdeiro, E. Radu, Phys. Rev. Lett. **123**, 011101 (2019). https://doi.org/10.1103/PhysRevLett.123.011101. arXiv:1904.09997 [gr-qc]
26. L.G. Collodel, B. Kleihaus, J. Kunz, E. Berti, Class. Quant. Grav. **37**, 075018 (2020). https://doi.org/10.1088/1361-6382/ab74f9. arXiv:1912.05382 [gr-qc]
27. C.A.R. Herdeiro, E. Radu, H.O. Silva, T.P. Sotiriou, N. Yunes, Phys. Rev. Lett. **126**, 011103 (2021). https://doi.org/10.1103/PhysRevLett.126.011103. arXiv:2009.03904 [gr-qc]
28. E. Berti, L.G. Collodel, B. Kleihaus, J. Kunz, Phys. Rev. Lett. **126**, 011104 (2021). https://doi.org/10.1103/PhysRevLett.126.011104. arXiv:2009.03905 [gr-qc]
29. A. Dima, E. Barausse, N. Franchini, T.P. Sotiriou, Phys. Rev. Lett. **125**, 231101 (2020). https://doi.org/10.1103/PhysRevLett.125.231101. arXiv:2006.03095 [gr-qc]
30. S. Hod, Phys. Rev. D **102**, 084060 (2020). https://doi.org/10.1103/PhysRevD.102.084060. arXiv:2006.09399 [gr-qc]
31. D.D. Doneva, L.G. Collodel, C.J. Krüger, S.S. Yazadjiev, Phys. Rev. D **102**, 104027 (2020). https://doi.org/10.1103/PhysRevD.102.104027. arXiv:2008.07391 [gr-qc]
32. D.D. Doneva, L.G. Collodel, C.J. Krüger, S.S. Yazadjiev, Eur. Phys. J. C **80**, 1205 (2020). https://doi.org/10.1140/epjc/s10052-020-08765-3. arXiv:2009.03774 [gr-qc]
33. L. Bernard, L. Lehner, R. Luna, Phys. Rev. D **100**, 024011 (2019). https://doi.org/10.1103/PhysRevD.100.024011. arXiv:1904.12866 [gr-qc]
34. A.D. Kovács, H.S. Reall, Phys. Rev. Lett. **124**, 221101 (2020). https://doi.org/10.1103/PhysRevLett.124.221101. arXiv:2003.04327 [gr-qc]
35. A.D. Kovács, H.S. Reall, Phys. Rev. D **101**, 124003 (2020). https://doi.org/10.1103/PhysRevD.101.124003. arXiv:2003.08398 [gr-qc]

36. H.S. Reall, Phys. Rev. D **103**, 084027 (2021). https://doi.org/10.1103/PhysRevD.103.084027. arXiv:2101.11623 [gr-qc]
37. H. Witek, L. Gualtieri, P. Pani, Phys. Rev. D **101**, 124055 (2020). https://doi.org/10.1103/PhysRevD.101.124055. arXiv:2004.00009 [gr-qc]
38. L. Aresté Saló, K. Clough, P. Figueras (2023). arXiv:2306.14966 [gr-qc]
39. J.L. Blázquez-Salcedo, D.D. Doneva, S. Kahlen, J. Kunz, P. Nedkova, S.S. Yazadjiev, Phys. Rev. D **101**, 104006 (2020). https://doi.org/10.1103/PhysRevD.101.104006. arXiv:2003.02862 [gr-qc]
40. J.L. Blázquez-Salcedo, D.D. Doneva, S. Kahlen, J. Kunz, P. Nedkova, S.S. Yazadjiev, Phys. Rev. D **102**, 024086 (2020). https://doi.org/10.1103/PhysRevD.102.024086. arXiv:2006.06006 [gr-qc]
41. H.-J. Kuan, D.D. Doneva, S.S. Yazadjiev, Phys. Rev. Lett. **127**, 161103 (2021). https://doi.org/10.1103/PhysRevLett.127.161103. arXiv:2103.11999 [gr-qc]
42. J.L. Ripley, F. Pretorius, Phys. Rev. D **101**, 044015 (2020). https://doi.org/10.1103/PhysRevD.101.044015. arXiv:1911.11027 [gr-qc]
43. D. Gerosa, U. Sperhake, C.D. Ott, Class. Quant. Grav. **33**, 135002 (2016). https://doi.org/10.1088/0264-9381/33/13/135002. arXiv:1602.06952 [gr-qc]
44. D.D. Doneva, S.S. Yazadjiev, Phys. Rev. D **105**, L041502 (2022). https://doi.org/10.1103/PhysRevD.105.L041502. arXiv:2107.01738 [gr-qc]
45. D.D. Doneva, C.J. Krüger, K.V. Staykov, P.Y. Yordanov (2023). arXiv:2306.16988 [gr-qc]
46. R. Benkel, T.P. Sotiriou, H. Witek, Phys. Rev. D **94**, 121503 (2016). https://doi.org/10.1103/PhysRevD.94.121503. arXiv:1612.08184 [gr-qc]
47. R. Benkel, T.P. Sotiriou, H. Witek, Class. Quant. Grav. **34**, 064001 (2017). https://doi.org/10.1088/1361-6382/aa5ce7. arXiv:1610.09168 [gr-qc]
48. D.D. Doneva, S. Kiorpelidi, P.G. Nedkova, E. Papantonopoulos, S.S. Yazadjiev, Phys. Rev. D **98**, 104056 (2018). https://doi.org/10.1103/PhysRevD.98.104056. arXiv:1809.00844 [gr-qc]
49. A.H.K. R, J.L. Ripley, N. Yunes, Phys. Rev. D **107**, 044044 (2023). https://doi.org/10.1103/PhysRevD.107.044044. arXiv:2211.08477 [gr-qc]
50. D.D. Doneva, S.S. Yazadjiev, Phys. Rev. D **103**, 064024 (2021). https://doi.org/10.1103/PhysRevD.103.064024. arXiv:2101.03514 [gr-qc]
51. A. Hegade KR, E.R. Most, J. Noronha, H. Witek, N. Yunes, Phys. Rev. D **105**, 064041 (2022). https://doi.org/10.1103/PhysRevD.105.064041. arXiv:2201.05178 [gr-qc]
52. J.L. Blázquez-Salcedo, D.D. Doneva, J. Kunz, S.S. Yazadjiev, Phys. Rev. D **105**, 124005 (2022). https://doi.org/10.1103/PhysRevD.105.124005. arXiv:2203.00709 [gr-qc]
53. J.L. Ripley, F. Pretorius, Phys. Rev. D **99**, 084014 (2019). https://doi.org/10.1103/PhysRevD.99.084014. arXiv:1902.01468 [gr-qc]
54. F. Corelli, M. De Amicis, T. Ikeda, P. Pani, Phys. Rev. D **107**, 044061 (2023). https://doi.org/10.1103/PhysRevD.107.044061. arXiv:2205.13007 [gr-qc]
55. J.L. Ripley, F. Pretorius, Class. Quant. Grav. **37**, 155003 (2020). https://doi.org/10.1088/1361-6382/ab9bbb. arXiv:2005.05417 [gr-qc]
56. W.E. East, J.L. Ripley, Phys. Rev. D **103**, 044040 (2021). https://doi.org/10.1103/PhysRevD.103.044040. arXiv:2011.03547 [gr-qc]
57. W.E. East, J.L. Ripley, Phys. Rev. Lett. **127**, 101102 (2021). https://doi.org/10.1103/PhysRevLett.127.101102. arXiv:2105.08571 [gr-qc]
58. L. Aresté Saló, K. Clough, P. Figueras, Phys. Rev. Lett. **129**, 261104 (2022). https://doi.org/10.1103/PhysRevLett.129.261104. arXiv:2208.14470 [gr-qc]
59. D.D. Doneva, L. Aresté Saló, K. Clough, P. Figueras, S.S. Yazadjiev (2023). arXiv:2307.06474 [gr-qc]
60. G. Ventagli, G. Antoniou, A. Lehébel, T.P. Sotiriou, Phys. Rev. D **104**, 124078 (2021). https://doi.org/10.1103/PhysRevD.104.124078. arXiv:2111.03644 [gr-qc]
61. N. Franchini, M. Bezares, E. Barausse, L. Lehner, Fixing the dynamical evolution in scalar-Gauss-Bonnet gravity (2022). arXiv:2206.00014 [gr-qc]
62. A. Coates, F.M. Ramazanoğlu (2023). arXiv:2307.07743 [gr-qc]

63. U. Sperhake, C.J. Moore, R. Rosca, M. Agathos, D. Gerosa, C.D. Ott, Phys. Rev. Lett. **119**, 201103 (2017). https://doi.org/10.1103/PhysRevLett.119.201103. arXiv:1708.03651 [gr-qc]
64. R. Rosca-Mead, C.J. Moore, M. Agathos, U. Sperhake, Class. Quant. Grav. **36**, 134003 (2019). https://doi.org/10.1088/1361-6382/ab256f. arXiv:1903.09704 [gr-qc]
65. R. Rosca-Mead, U. Sperhake, C.J. Moore, M. Agathos, D. Gerosa, C.D. Ott, Phys. Rev. D **102**, 044010 (2020). https://doi.org/10.1103/PhysRevD.102.044010. arXiv:2005.09728 [gr-qc]
66. D.D. Doneva, S.S. Yazadjiev, JCAP **04**, 011. https://doi.org/10.1088/1475-7516/2018/04/011. arXiv:1712.03715 [gr-qc]
67. S.B. Popov, R. Turolla, Astrophys. Space Sci. **341**, 457 (2012). https://doi.org/10.1007/s10509-012-1100-z. arXiv:1204.0632 [astro-ph.HE]
68. A. Noutsos, D. Schnitzeler, E. Keane, M. Kramer, S. Johnston, Mon. Not. Roy. Astron. Soc. **430**, 2281 (2013). https://doi.org/10.1093/mnras/stt047. arXiv:1301.1265 [astro-ph.GA]
69. T.E. Riley et al., Astrophys. J. Lett. **887**, L21 (2019). https://doi.org/10.3847/2041-8213/ab481c. arXiv:1912.05702 [astro-ph.HE]
70. S.E. Woosley, A. Heger, Phys. Rept. **442**, 269 (2007). https://doi.org/10.1016/j.physrep.2007.02.009. arXiv:astro-ph/0702176
71. E. O'Connor, C.D. Ott, Class. Quant. Grav. **27**, 114103 (2010). https://doi.org/10.1088/0264-9381/27/11/114103. arXiv:0912.2393 [astro-ph.HE]
72. D. Radice, V. Morozova, A. Burrows, D. Vartanyan, H. Nagakura, Astrophys. J. Lett. **876**, L9 (2019). https://doi.org/10.3847/2041-8213/ab191a. arXiv:1812.07703 [astro-ph.HE]
73. E. Abdikamalov, G. Pagliaroli, D. Radice (2020). arXiv:2010.04356 [astro-ph.SR]
74. T. Evstafyeva, M. Agathos, J.L. Ripley, Phys. Rev. D **107**, 124010 (2023). https://doi.org/10.1103/PhysRevD.107.124010. arXiv:2212.11359 [gr-qc]
75. W.E. East, F. Pretorius, Phys. Rev. D **106**, 104055 (2022). https://doi.org/10.1103/PhysRevD.106.104055. arXiv:2208.09488 [gr-qc]

EFT Approach to Black Hole Scalarization and Its Compatibility with Cosmic Evolution

Cristián Erices, Simón Riquelme, and Nicolás Zalaquett

Abstract We address the issue of black hole scalarization and its compatibility with cosmic inflation and big bang cosmology from an effective field theory (EFT) point of view. In practice, using a well-defined and healthy toy model which (in part) has been broadly considered in the literature, we consider how higher-order theories of gravity, up to cubic operators in Riemann curvature, fit within this context. Interestingly enough, we find that already at this minimal level, there is a non-trivial interplay between the Wilson coefficients which are otherwise completely independent, constraining the parameter space where scalarization may actually occur. Conclusively, we claim that the EFT does exhibit black hole scalarization, remaining compatible with the inflationary paradigm, and admitting General Relativity as a cosmological attractor.

1 Introduction

There is an overwhelming amount of evidence that establishes General Relativity (GR) as the most successful theory describing the gravitational interaction in the so-called "weak-field" regime. However, the recent gravitational-wave observations [1–3] announce the emergence of a new era in astronomy, which will provide the possibility of testing gravity in the "strong-field" regime. This mostly unexplored

C. Erices (✉)
Universidad Central de Chile, Vicerrectoría Académica, Toesca 1783, Santiago, Chile
e-mail: cristian.erices@ucentral.cl

Universidad Católica del Maule, Av. San Miguel 3605, Talca, Chile

S. Riquelme
Grupo de Cosmología y Astrofísica Teórica, Departamento de Física, FCFM, Universidad de Chile, Blanco Encalada 2008, Santiago, Chile
e-mail: sriquelm@ing.uchile.cl

N. Zalaquett
Physics Ludique Research, Las Urbinas 87, Santiago, Chile
e-mail: nzalaquett@ludique.cl

© The Author(s), under exclusive license to Springer Nature Switzerland AG 2024
E. Papantonopoulos and N. Mavromatos (eds.), *Compact Objects in the Universe*,
https://doi.org/10.1007/978-3-031-55098-0_4

scenario has motivated new physical grounds where modifications of GR may take place. In this context, there exist theories that exhibit a phenomenon dubbed "spontaneous scalarization". This process is indeed a distinctive manifestation of gravitational interactions in the strong-field regime. The earliest of such theories is the Damour and Esposito-Farèse (DEF) model [4, 5], where scalarization occurs in neutron stars. Neutron stars are able to acquire a non-trivial structure since the theory exhibits non-perturbative deviations from GR only in the strong-field regime. However, spontaneous scalarization was later discovered to happen in other compact objects, such as black holes, in the context of higher-order curvature theories [6, 7].

In the strong-field regime of gravity, the effects of higher-order curvature operators become more significant, leading (generically) to the appearance of ghost degrees of freedom [8]. As it is well-known, the Gauss-Bonnet invariant is a purely topological term in four dimensions, but it may become dynamical and "Ostrogradsky ghosts-free" when a coupling to a scalar field degree of freedom is considered. This feature is the essence of Einstein Scalar-Gauss-Bonnet (ESGB) gravity, which has become a focus of great interest during the last few years. Several hairy black hole solutions and compact objects have been studied in the four-dimensional ESGB theories [9–14]. The phenomenon of spontaneous scalarization requires that the theory admits the GR solution with a trivial scalar field *and* the scalarized hairy black hole solutions (or even interior ones like, e.g., neutron and/or boson stars) [15, 16]. Indeed, when the mass of the Schwarzschild black hole is below a certain value, the former becomes unstable in regions of strong curvature [6], while the latter emerges as a physically favorable configuration.

Since then, a plethora of coupling functions [17–19], as well as several attempts to generalize the spontaneous scalarization framework, have been considered to get a better understanding of the scalarization process. In [20], scalarization is studied within a theory where the scalar field is coupled to the Ricci scalar, while [21] proceeds in analogy but under the presence of the Chern-Simons invariant. Scalarized black hole solutions and compact objects in asymptotically flat spacetime of ESGB theories were also investigated in [22–24]. Moreover, the spontaneous scalarization of black holes extended to AdS/dS asymptotics were studied in [25–28]. Interestingly enough, asymptotically AdS black holes allow for connecting scalarization with holographic phase transitions in the dual theory [29, 30]. Furthermore the authors in [31] discuss the spontaneous scalarization in $f(R)$ gravity in the presence of a minimally coupled scalar field with a self-interacting potential; while recently, within ESGB theory, spin-induced black hole spontaneous scalarization, which is the outcome of linear tachyonic instability triggered by rapid rotation, has been explored in [32–37].

Another interesting class of modified gravity theories based on higher-order curvature invariants was constructed in [38] by using cubic contractions of the Riemann tensor. This theory, dubbed "Einsteinian Cubic Gravity" (ECG), possesses basic healthiness properties such as: (i) having a spectrum identical to that of Einstein gravity, i.e., the metric perturbation *on a maximally symmetric background* propagates only a transverse massless graviton; (ii) it is neither topological nor trivial in four dimensions, and (iii) it is defined such that it is independent of the number of dimensions. It is well-known that, in general, such terms contribute with fourth-order derivatives of the metric in the field equations. However, as it was shown in [39–41], the original form of the theory is sufficient to admit spherically symmetric black hole solutions with a second-order differential equation for the metric function. Moreover, an extra cubic contribution, which is trivial for a spherically symmetric black hole ansatz at the level of the field equations, allows one to additionally accommodate a Friedmann-Lemaître-Robertson-Walker (FLRW) solution with second-order field equations for the scale factor, leading to a "purely geometric" inflationary period [42]. Black hole solutions were also obtained in [43], alongside the examination of a well-defined cosmology for modified cubic gravity and higher-order terms such as the quartic, quintic and even an infinite tower of higher-curvature corrections to the Einstein-Hilbert action [44, 45]. All these features make this theory physically interesting, and hence, worth of further investigation.

So far curvature-induced scalarization has been explored in scalar-tensor theories where the scalar degree of freedom is excited within the extreme curvature regime through invariants made up of the Riemann tensor. However, up to now as far as the authors have noticed, the theories have not followed a "strict" Effective Field Theory (EFT) approach, which has led to hasty conclusions about scalarization of cosmological spacetimes. While it is true that in Reference [46] the author does acknowledge the necessity of treating ESGB as an EFT, we claim such a model is quantum-mechanically incomplete since it arbitrarily sets the Wilson coefficient β of the marginal operator $\varphi^2 R$ to zero, which is doomed to be loop-generated, as there is no symmetry argument to forbid it. Consequently, the aim of this paper is three-fold. First, we want to go a step further by addressing a scalar-tensor EFT that exhibits curvature-induced scalarization, triggered by a set of suitable invariants made up of Riemann tensor, up to cubic order. Second, it is of interest then to investigate within this framework, how the new operators modify a previously claimed catastrophic instability triggered by quantum fluctuations during the inflationary stage in ESGB theory [47]. Third, we explore the Big Bang Cosmology (BBC) of the model, and check that GR is indeed a late-time cosmological attractor as experiments seem to demand [48].

The article is organized as follows: Sect. 2 introduces our model and discusses how it achieves black hole scalarization; Sect. 3 considers the theory of perturbations around a FLRW background and discusses the (in-)stability of such a spacetime geometry; Sect. 4 studies whether (or not) the model has GR as a cosmological attractor; Finally, Sect. 5 discusses our findings, concludes, and ellaborates on future venues of exploration. Appendices A, B, and C are very brief overviews of the EFT approach, unstable mode existence condition, and scalar pertubation theory in dS space, respectively.

2 The Model and Black Hole Scalarization

We want to consider a theory that exhibits scalarization triggered by higher-order curvature operators (other than the Gauss-Bonnet invariant, which is the one usually considered in the literature) and also satisfies some basic criteria, in order to have a well-posed and healthy gravitational system. Consequently, we construct a model that possesses a spectrum identical to that of Einstein gravity *on a maximally symmetric background*, and is neither topological nor trivial in four spacetime dimensions.

The purely geometric theory was constructed in [38]. Essentially, they introduced a new linearization method for higher-order theories of gravity on maximally symmetric spacetimes. To cubic order, such a procedure allows us to identify the (naive) physical degrees of freedom that are propagated by the metric perturbation through a set of three second order equations of motion. Besides the usual massless graviton, one finds a massive spin-2 fluctuation and a massive scalar mode fluctuation. Crucially, at least one of them is doomed to be a ghost, so we actually need to get rid of them both for the theory to become sensible. Imposing then an "Einsteinian spectrum" condition restricts the coupling parameters of the EFT delivering the so-called Einsteinian Cubic Gravity theory [38]. The generalized version of the above construction to higher dimensions and higher order curvature operators, known as "Generalized Quasi-Topological Gravity" (GQTG), was studied in [41].

We start by recalling the cubic operator $\mathcal{P}$ in ECG theory, which reads

$$\mathcal{P} = 12R_{\mu\ \nu}^{\ \rho\ \sigma}R_{\rho\ \sigma}^{\ \gamma\ \delta}R_{\gamma\ \delta}^{\ \mu\ \nu} + R_{\mu\nu}^{\ \ \rho\sigma}R_{\rho\sigma}^{\ \ \gamma\delta}R_{\gamma\delta}^{\ \ \mu\nu} - 12R_{\mu\nu\rho\sigma}R^{\mu\rho}R^{\nu\sigma} + 8R^{\nu}_{\ \mu}R^{\mu}_{\ \rho}R^{\rho}_{\ \nu}, \tag{2.1}$$

while the operator $\mathcal{C}$, found in GQTG [41], is given by the combination

$$\mathcal{C} = R_{\mu\nu\rho\sigma}R^{\mu\nu\rho}_{\ \ \ \ \delta}R^{\sigma\delta} - \frac{1}{4}R_{\mu\nu\rho\sigma}R^{\mu\nu\rho\sigma}R - 2R_{\mu\nu\rho\sigma}R^{\mu\rho}R^{\nu\sigma} + \frac{1}{2}R_{\mu\nu}R^{\mu\nu}R. \tag{2.2}$$

The latter leads to null contributions in the equations of motion (EOM) when evaluated on a static, spherically symmetric ansatz, a feature that has led some authors to arbitrarily neglect it altogether. However, it has been proven that it is the exact combination $\mathcal{P} - 8\mathcal{C}$, the one that leads to cosmologies with a well-posed initial value problem [42]. Therefore this is the combination of third-order curvature invariants we shall consider in this paper.

In order to explore the phenomenon of scalarization, we must include a scalar field without spoiling the conditions already mentioned. Therefore, we will then consider the dynamical system determined by the action

$$S[g_{\mu\nu}, \varphi] = \int d^4x\sqrt{-g}\left[\frac{M_{\rm Pl}^2}{2}R + \frac{\alpha}{M_{\rm Pl}^2}(\mathcal{P} - 8\,\mathcal{C}) - \frac{1}{2}g^{\mu\nu}\nabla_\mu\varphi\nabla_\nu\varphi + f\left(\varphi/M_{\rm Pl}\right)\mathcal{I} + \cdots\right], \tag{2.3}$$

where $f(\varphi/M_{\rm Pl})$ is a dimensionless "coupling function" between a canonically-normalized scalar field φ and a set of curvature invariants given by

$$\mathcal{I} = -\beta\, M_{\rm Pl}^2 R + \gamma\, \mathcal{G} - \frac{\lambda}{M_{\rm Pl}^2}\left(\mathcal{P} - 8\mathcal{C}\right), \tag{2.4}$$

where $\mathcal{G}$ stands for the well-known Gauss-Bonnet operator $\mathcal{G} \equiv R_{\mu\nu\alpha\beta}R^{\mu\nu\alpha\beta} - 4R_{\mu\nu}R^{\mu\nu} + R^2$. Note that $\mathcal{G}$, being a topological invariant, does not satisfy property (ii) unless it is coupled to the scalar field, so this explains why it is only considered within $\mathcal{I}$. Moreover, α, β, γ, and λ are *dimensionless* coupling constants, which are expected to be $\mathcal{O}(1)$ numbers from an EFT point of view (see Appendix A). Hereafter, for simplicity, we will refer to this theory as "Scalar-Einsteinian Cubic Gravity" (SECG).

It should be emphasized that here we are assuming a $\varphi \to -\varphi$ (discrete) symmetry as well as a $\varphi \to \varphi +$ constant (shift) symmetry of the scalar Lagrangian, where the latter is only spoiled by gravitational interactions as given by $f(\varphi/M_{\rm Pl})\,\mathcal{I}$ and higher-order operators represented by the ellipsis in (2.3). Note that in this paper we will *not* set the (reduced) Planck scale $M_{\rm Pl} = 2.4 \times 10^{18}$ GeV to unity as it is usually done in the literature since we want to keep track of it to easily emphasize its role of being the ultimate EFT cut-off of any gravitational system.

The EOM that stem from extremizing the action $S[g_{\mu\nu}, \varphi] = \int d^4x\sqrt{-g}\,\mathcal{L}$ read

$$R^{\alpha\beta\rho}{}_{\mu}P_{\nu\rho\alpha\beta} + 2\nabla^{\alpha}\nabla^{\beta}P_{\alpha\mu\nu\beta} + \frac{1}{2}\nabla_{\mu}\varphi\nabla_{\nu}\varphi + \frac{1}{2}g_{\mu\nu}\mathcal{L} = 0, \tag{2.5}$$

$$\Box\varphi + f_{,\varphi}\left(\varphi/M_{\rm Pl}\right)\mathcal{I} = 0, \tag{2.6}$$

where ∇_{μ} is the covariant derivative compatible with the spacetime metric $g_{\mu\nu}$, $\Box\varphi \equiv \frac{1}{\sqrt{-g}}\partial_{\mu}\left(\sqrt{-g}\,g^{\mu\nu}\partial_{\nu}\varphi\right)$, $f_{,\varphi} \equiv \frac{df}{d\varphi}$, and $P_{\alpha\beta\mu\nu}$ is defined as[1]

$$\begin{aligned} P_{\alpha\beta\mu\nu} &\equiv \frac{\partial\mathcal{L}}{\partial R^{\alpha\beta\mu\nu}} \\ &= \frac{1}{2}M_{\rm Pl}^2,\, g_{\alpha[\mu}g_{\nu]\beta} + \frac{\alpha}{M_{\rm Pl}^2}S_{\alpha\beta\mu\nu} + f\left(\varphi/M_{\rm Pl}\right)\left(-\frac{\beta}{2}M_{\rm Pl}^2,\, g_{\alpha[\mu}g_{\nu]\beta} + \gamma\, Q_{\alpha\beta\mu\nu} - \frac{\lambda}{M_{\rm Pl}^2}S_{\alpha\beta\mu\nu}\right), \end{aligned} \tag{2.7}$$

where

$$Q_{\alpha\beta\mu\nu} \equiv 4g_{\beta[\mu}R_{\nu]\alpha} - 4g_{\alpha[\mu}R_{\nu]\beta} + 2g_{\alpha[\mu}g_{\nu]\beta}R + 2R_{\alpha\beta\mu\nu}, \tag{2.8}$$

$$\begin{aligned} S_{\alpha\beta\mu\nu} \equiv\ & 4R_{\alpha[\mu}R_{\nu]\beta} + 8R_{\alpha\beta\rho[\mu}R_{\nu]}{}^{\rho} + 3R_{\alpha\beta}{}^{\rho\sigma}R_{\mu\nu\rho\sigma} + 4\left(2R^{\rho}{}_{[\alpha}R_{\beta]\rho\mu\nu} + 9R_{\beta\rho\sigma[\nu}R_{\mu]}{}^{\sigma\rho}{}_{\alpha} + RR_{\alpha\beta\mu\nu}\right) \\ & + 2g_{\alpha[\mu}\left(6R_{\nu]\rho}R_{\beta}{}^{\rho} - 2R_{\nu]\beta}R + g_{\nu]\rho}\left(R_{\rho\sigma\gamma\delta}R^{\rho\sigma\gamma\delta} - 2R_{\rho\sigma}R^{\rho\sigma}\right) + 2R^{\rho\sigma}R_{\nu]\sigma\beta\rho} - 2R_{\nu]\rho\sigma\gamma}R_{\beta}{}^{\rho\sigma\gamma}\right) \\ & + 4g_{\beta[\mu}\left(R_{\nu]\alpha}R - 3R_{\nu]\rho}R_{\alpha}{}^{\rho} - R_{\nu]\sigma\alpha\rho}R^{\rho\sigma} + R_{\nu]\rho\sigma\gamma}R_{\alpha}{}^{\rho\sigma\gamma}\right). \end{aligned} \tag{2.9}$$

[1] We use a normalized antisymmetrization convention $A_{[\mu\nu]} = \frac{1}{2}(A_{\mu\nu} - A_{\nu\mu})$.

The EOM for the scalar field fluctuation $\delta\varphi \equiv \varphi - \varphi_0$ is given by

$$\left[\Box + f_{,\varphi\varphi}(\varphi_0/M_{\rm Pl})\,\mathcal{I}\right]\delta\varphi = 0, \tag{2.10}$$

where φ_0 is the scalar field background, while the d'Alembertian operator and $\mathcal{I}$ are computed in a fixed background. To prove that this theory admits black hole scalarization, we start by noting that the Schwarzschild black hole solution is also a trivial solution of the scalar-tensor cubic theory. This may be achieved by a suitable coupling function satisfying both $f_{,\varphi}(0) = 0$ and $f_{,\varphi\varphi}(0) > 0$. The first condition ensures $\varphi_0 = 0$ is a solution of the theory while the second condition has been understood to be necessary for the emergence of a tachyonic instability in the scalar-Gauss-Bonnet model. It is not difficult, though cumbersome, to strictly prove that the linearized Einstein field equations are the same as in ECG, and therefore property (i) is satisfied, provided the aforementioned conditions are fulfilled. Taking for simplicity[2] $f(x) = \frac{1}{2}x^2 + \ldots$, so that $f_x(0) = 0$ and $f_{,xx}(0) = 1 > 0$, we may read off from (2.10) an effective mass squared $m^2_{\rm eff}$ of the form

$$m^2_{\rm eff} = -f_{,\varphi\varphi}\left(\varphi_0/M_{\rm Pl}\right)\mathcal{I} = \beta R - \frac{\gamma}{M^2_{\rm Pl}}\mathcal{G} + \frac{\lambda}{M^4_{\rm Pl}}\left(\mathcal{P} - 8\mathcal{C}\right). \tag{2.11}$$

It so happens that a black hole is a solution with $R = 0$ and $\mathcal{G} > 0$, so we clearly see that γ must be positive ($\gamma > 0$) in order to have a tachyonic instability $\left(m^2_{\rm eff} < 0\right)$, as the presence of the cubic term should and will be taken to be immaterial because it is further suppressed by the Planck scale for "natural" values of λ. Therefore, in this article we shall always assume

$$\gamma > 0, \tag{2.12}$$

as we are interested in scalarized black hole solutions. Let us now consider perturbations on a fixed Schwarzschild background. The symmetry of such a spacetime allows for a decomposition of the perturbation using separation of variables, meaning

$$\delta\varphi = \frac{u(r)}{r}e^{-i\omega t}\,Y_{lm}\left(\theta, \phi\right), \tag{2.13}$$

where $Y_{lm}\left(\theta, \phi\right)$ are the usual spherical harmonics. After substitution of (2.13) into the EOM (2.10), and using tortoise coordinates defined through $dr_* = dr\left(1 - \frac{r_g}{r}\right)^{-1}$, with $r_g \equiv \mathcal{M}/4\pi M^2_{\rm Pl}$ standing for the Schwarzschild radius of the black hole of mass $\mathcal{M}$, we obtain a "Schrödinger-like" equation of the form

---

[2] Indeed, after imposing the discrete $\varphi \to -\varphi$ symmetry, we may think of $f(\varphi/M_{\rm Pl})$ as an even function which is Taylor expanded, so that the first non-trivial term is of $\mathcal{O}(\varphi^2/M^2_{\rm Pl})$. Moreover (as will be argued below) it will always be the case that $\varphi \ll M_{\rm Pl}$, so the additional higher-order terms, which do not contribute to the scalar field fluctuation mass, are further suppressed. In this sense, there is no loss of generality in assuming the simple form $f(x) = \frac{1}{2}x^2$, as we do in this manuscript.

$$\frac{d^2u}{dr_*^2} + \omega^2 u = V_{\rm eff}(r)u, \tag{2.14}$$

where the effective potential $V_{\rm eff}$ is defined as

$$V_{\rm eff}(r) = \left(1 - \frac{r_g}{r}\right)\left(\frac{l(l+1)}{r^2} + \frac{r_g}{r^3} - \frac{\gamma}{M_{\rm Pl}^2}\frac{12\,r_g^2}{r^6} + \frac{\lambda}{M_{\rm Pl}^4}\frac{84\,r_g^3}{r^9}\right). \tag{2.15}$$

There exists a sufficient condition for the existence of an unstable mode which is given by (see Appendix B)

$$\int_{-\infty}^{\infty} dr_*\, V_{\rm eff}(r_*) = \int_{r_g}^{\infty} dr\, \frac{V_{\rm eff}(r)}{\left(1 - \frac{r_g}{r}\right)} < 0. \tag{2.16}$$

For this condition to hold, spherically symmetric perturbations ($l = m = 0$) in a Schwarzschild background require

$$5\,(r_g M_{\rm Pl})^4 - 24\,(r_g M_{\rm Pl})^2\,\gamma + 105\,\lambda < 0. \tag{2.17}$$

This implies that

$$\left(\frac{\mathcal{M}}{M_{\rm Pl}}\right)^2 \in \left[q_-, q_+\right] \qquad \text{if} \quad 0 < \lambda \leq \frac{48}{175}\gamma^2, \tag{2.18}$$

$$\left(\frac{\mathcal{M}}{M_{\rm Pl}}\right)^2 \in \left[0, q_+\right] \qquad \text{if} \quad \lambda \leq 0, \tag{2.19}$$

where we have defined

$$q_\pm \equiv \frac{16\,\pi^2}{5}\left(12\,\gamma \pm \sqrt{144\,\gamma^2 - 525\,\lambda}\right). \tag{2.20}$$

Interestingly enough, the Schwarzschild background is unstable for a specific range of masses given by the above bounds. Moreover, we see that scalarization may only occur when

$$\lambda \leq \frac{48}{175}\gamma^2, \tag{2.21}$$

which is a non-trivial constraint between the couplings. This is still compatible with γ and λ both being $\mathcal{O}(1)$ numbers. However, the sign of λ is clearly not fixed by this condition. Note that when $\lambda = 0$, Eqs. (2.17) and (2.19) imply

$$\mathcal{M}^2 < \frac{384\,\pi^2}{5}\,\gamma\, M_{\rm Pl}^2. \tag{2.22}$$

A back-of-the-envelope calculation shows then that within our naive quadratic theory (endowed with $M_{\rm Pl}$ as the *only* relevant scale), the maximum mass of Schwarzschild black holes that may be scalarized is of order 10^{-37} solar masses. This fact certainly precludes any possibility of such a version of SECG theory to be compared with observations. In the next section, we will modify the theory in order to solve this and some other "problems" that we will find along the way. Additionally, it is crucial to stress that if we want GR solutions to be admissible in the model we have been considering, we need to check that $\varphi = \varphi^{(0)} = 0$ is the asymptotic value that φ needs to take for unscalarized configurations. Let us now consider how scalarization may occur within a cosmological setting.

3 Perturbations on a FLRW Background

In a FLRW background with metric

$$ds^2 = g_{\mu\nu} dx^\mu dx^\nu = -dt^2 + a(t)^2 \delta_{ij}\, dx^i dx^j, \quad \text{and} \quad H \equiv \frac{\dot{a}}{a}, \tag{3.1}$$

it so happens that the EOM for the fluctuation reads

$$\delta\ddot{\varphi} + 3H\delta\dot{\varphi} - \frac{\nabla^2 \delta\varphi}{a^2} + m_{\rm eff}^2\, \delta\varphi = 0, \tag{3.2}$$

which, by Fourier expanding $\delta\varphi \sim \int d\omega\, d^3\boldsymbol{k}\, \delta\varphi(\omega, \boldsymbol{k})\, e^{-i(\omega t - \boldsymbol{k}\cdot\boldsymbol{x})}$, implies that

$$\omega^2 = \frac{k^2}{a^2} + m_{\rm eff}^2, \tag{3.3}$$

when neglecting the slow change of ω on the time scales shorter than H^{-1}. Moreover, for a FLRW spacetime

$$R = 6\left(2H^2 + \dot{H}\right), \quad \mathcal{G} = 24H^2\left(H^2 + \dot{H}\right), \quad \mathcal{P} - 8\mathcal{C} = -48H^4\left(2H^2 + 3\dot{H}\right), \tag{3.4}$$

so that the effective mass squared becomes

$$m_{\rm eff}^2 = 6\beta\left(2H^2 + \dot{H}\right) - \frac{24\,\gamma}{M_{\rm Pl}^2} H^2\left(H^2 + \dot{H}\right) - \frac{48\,\lambda}{M_{\rm Pl}^4} H^4\left(2H^2 + 3\dot{H}\right). \tag{3.5}$$

Let us entertain for a second the case $\beta = \lambda = 0$. The effective mass would then read

$$m_{\rm eff}^2 = -\frac{24\,\gamma}{M_{\rm Pl}^2} H^2\left(H^2 + \dot{H}\right) = -\frac{24\,\gamma}{M_{\rm Pl}^2} H^2 \frac{\ddot{a}}{a}, \tag{3.6}$$

so that, as long as $\gamma > 0$,

$$m_{\text{eff}}^2 < 0 \Longleftrightarrow \ddot{a} > 0, \tag{3.7}$$

implying the existence of a tachyonic instability during any (quasi)-de Sitter (dS) phase of our universe, as it is emphasized in [47] ($\gamma = 1$ for them). However, we see *no reason*, on EFT grounds, to arbitrarily set $\beta = \lambda = 0$. Moreover, as previously noted, we expect that all β, γ, and λ are $\mathcal{O}(1)$ numbers, and yet, it is still the case that the dimensionless ratio $(H/M_{\text{Pl}})^2$ does generate a clear hierarchy among the contributions to the effective mass in (3.5). Thus, the appropriate thing to do is to consider the effective mass, which by using the standard definition $\epsilon \equiv -\dot{H}/H^2$ we now write

$$m_{\text{eff}}^2 = 12\left[\beta\left(1 - \frac{\epsilon}{2}\right) - 2\,\gamma\,(1-\epsilon)\,\zeta - 8\,\lambda\left(1 - \frac{3}{2}\,\epsilon\right)\zeta^2\right]H^2, \tag{3.8}$$

as an expansion in the small ratio

$$\zeta \equiv \left(\frac{H}{M_{\text{Pl}}}\right)^2. \tag{3.9}$$

Say we keep only the zeroth-order term in such an expansion in ζ so that

$$m_{\text{eff}}^2 = 12\beta\left(1 - \frac{\epsilon}{2}\right)H^2. \tag{3.10}$$

We see that since during inflation $\epsilon \ll 1$, the overall sign of m_{eff}^2 is simply the sign of β itself. If we now consider the leading non-trivial contribution in the expansion in ζ, the effective mass squared may in principle be negative as long as

$$\beta\left(1 - \frac{\epsilon}{2}\right) < 2\,\gamma\,(1-\epsilon)\,\zeta. \tag{3.11}$$

For inflation we may safely take $H \sim 10^{13}$ GeV, so we may assume that $\zeta \sim 10^{-11}$ (strictly positive), implying that

$$\frac{\beta}{\gamma} \lesssim \mathcal{O}(10^{-11}). \tag{3.12}$$

We see that if $\beta > 0$ this condition implies a big hierarchy between β and γ, which is clearly not expected nor justified from an EFT perspective. Instead, if $\beta < 0$, the condition may be naturally satisfied. Again then, the possibility of developing a tachyonic instability during inflation is ultimately related to the sign of β. In other words, it is still the case that

$$m_{\text{eff}}^2 < 0 \Longleftrightarrow \beta < 0. \tag{3.13}$$

If we assume (for the sake of simplicity) $\beta > 0$, $\gamma = 0$, $\lambda > 0$, requiring the cubic operator contribution to render a negative effective mass squared would imply that

$$\frac{\beta}{\lambda} \lesssim \mathcal{O}(10^{-22}), \tag{3.14}$$

which of course worsens the prospects for the naive idea that higher-order operators may play some non-trivial role on the phenomenon of scalarization during inflation. At the risk of being a little pedantic, we repeat; the sign of β univocally determines the fate of scalarization in this weakly broken shift-symmetric theory.

For a dS solution $\epsilon = 0$ so

$$m_{\text{eff}}^2 = \left(12\beta - 24\gamma\zeta - 96\lambda\zeta^2\right) H^2. \tag{3.15}$$

Let us come back for a (second) second to the $\beta = \lambda = 0$ case. Doing so allows us to estimate the corresponding would-be γ constant in the model discussed by the authors of [47]. Taking, as they do,[3] $\bar{\lambda} \sim 10^{19}\,\text{GeV}^{-1}$, this would imply a colossal value for $\gamma = \frac{1}{4} M_{\text{Pl}}^2 \bar{\lambda}^2$, namely

$$\gamma \sim 10^{74}. \tag{3.16}$$

Sweeping this issue under the rug for the time being, one may go ahead and estimate the ratio of the instability time t_{inst} to the age of the universe $t_0 \sim 1/H_0$ when taking $H_0 \simeq 10^{-43}$ GeV,

$$\frac{t_{\text{inst}}}{t_0} \sim \frac{H_0}{m_{\text{eff}}} \sim \frac{1}{2\sqrt{6\gamma}} \frac{M_{\text{Pl}}}{H_0} \sim 10^{23}, \tag{3.17}$$

and we learn that the instability is not noticeable during current dark energy domination.

Instead, during inflation, the estimation delivers

$$\frac{t_{\text{inst}}}{t_{\text{inf}}} \sim \frac{1}{N} \frac{H_{\text{inf}}}{m_{\text{eff}}} \sim \frac{1}{2\sqrt{6\gamma} N} \frac{M_{\text{Pl}}}{H_{\text{inf}}} \sim 10^{-34}, \tag{3.18}$$

where $N \sim 10^2$ is the required number of e-folds to overcome the classical shortcomings of BBC. Equation (3.18) informs us of a very short instability time, implying a catastrophic instability of the theory during inflation, unless the initial values of the field are finely tuned [47]. Nevertheless, it should be common wisdom that any perturbative scheme around an *unstable* ground state is doomed to exhibit "runaway" behavior. The quantum analysis of such finely tuned initial conditions system was

[3] Here, $\bar{\lambda}$ stands for λ in [47] and has nothing to do with our own λ introduced in this paper.

indeed made in [47], with foreseeable results. However, before moving forward to the analisys of the more sensible theory where $\beta \neq 0$, we should address the issue of getting such a big order of magnitude for the estimation of γ in (3.16).

Say we instead had chosen from the onset a coupling function of the same functional form $f\,(\varphi/M_{\rm Pl})$ but with $M_{\rm Pl} \to M$, where M stands for some *other* additional scale that we may fix later on. Now the effective mass reads

$$m_{\rm eff}^2 = -f_{,\varphi\varphi}\,(\varphi_0/M)\,\mathcal{I} = \beta\frac{M_{\rm Pl}^2}{M^2}R - \frac{\gamma}{M^2}\mathcal{G} + \frac{\lambda}{M^2M_{\rm Pl}^2}\,(\mathcal{P}-8\mathcal{C})\,, \tag{3.19}$$

or

$$\begin{aligned} m_{\rm eff}^2 &= 6\beta\frac{M_{\rm Pl}^2}{M^2}\left(2H^2+\dot{H}\right) - \frac{24\,\gamma}{M^2}H^2\left(H^2+\dot{H}\right) - \frac{48\,\lambda}{M^2M_{\rm Pl}^2}H^4\left(2H^2+3\dot{H}\right) \\ &= 12\left[\beta\left(1-\frac{\epsilon}{2}\right)\frac{M_{\rm Pl}^2}{M^2} - 2\,\gamma\,(1-\epsilon)\,\frac{H^2}{M^2} - 8\,\lambda\left(1-\frac{3}{2}\,\epsilon\right)\frac{H^4}{M^2M_{\rm Pl}^2}\right]H^2. \end{aligned} \tag{3.20}$$

Let us then consider one last time the $\beta = \lambda = 0$ case. For scalarization to be relevant for astrophysical black holes we need to pick $M = L^{-1}$, with L being the characteristic length-scale of the compact object which is usually taken to be $L \sim 10$ km., thus arriving at

$$M = 1.98 \times 10^{-20}\,\text{GeV} \xrightarrow{\gamma=\frac{1}{4}M^2\bar{\lambda}^2} \gamma \sim 10^{-2}, \tag{3.21}$$

which is a much more sensible number than the one estimated in (3.16). There is still a pending issue though. Staring at (3.19) it is clear that the coefficient accompanying the Ricci scalar, $\tilde{\beta} \equiv \beta\,M_{\rm Pl}^2/M^2$, cannot be an order one number unless β is tuned to be extremely small because the ratio $M_{\rm Pl}^2/M^2 \sim 10^{76}$. This implies we would be dealing with a field of super-Planckian mass, which is clearly off the EFT description and makes no sense. The way out of this conundrum is to acknowledge the fact that the mass scale should not only be modified within the coupling function f but within the curvature invariant $\mathcal{I}$ as well, so we should really take

$$\mathcal{I} = -\beta\,M^2R + \gamma\,\mathcal{G} - \frac{\lambda}{M^2}\,(\mathcal{P}-8\mathcal{C})\,, \tag{3.22}$$

which implies that the effective mass is finally given by the expression in (3.20) after carrying out the replacement $M_{\rm Pl} \to M$, so that

$$m_{\rm eff}^2 = 12\left[\beta\left(1-\frac{\epsilon}{2}\right) - 2\,\gamma\,(1-\epsilon)\,\chi - 8\,\lambda\left(1-\frac{3}{2}\,\epsilon\right)\chi^2\right]H^2, \quad \text{with} \quad \chi \equiv \left(\frac{H}{M}\right)^2. \tag{3.23}$$

It is quite interesting that the introduction of the new energy scale M in the problem, besides the "kinematical" ($M_{\rm Pl}$) and the "dynamical" (H) ones, "naturalizes" an otherwise finely tuned EFT. Moreover, such an energy scale corresponds to nothing but the actual physical extension of the object to become scalarized. We will have more to say about this later.

With the emergence of the scale M, a Schwarzschild black hole of mass $\mathcal{M}$ may be now scalarized provided

$$\left(\frac{\mathcal{M}}{M}\right)^2 \in [q_-, q_+] \qquad \text{if} \;\; 0 < \lambda \leq \frac{48}{175}\gamma^2, \tag{3.24}$$

$$\left(\frac{\mathcal{M}}{M}\right)^2 \in [0, q_+] \qquad \text{if} \;\; \lambda \leq 0, \tag{3.25}$$

with $q_\pm \equiv \frac{16\pi^2}{5}\left(12\,\gamma \pm \sqrt{144\,\gamma^2 - 525\,\lambda}\right)$ as before. We see now that when $\lambda = 0$,

$$\mathcal{M}^2 < \frac{384\,\pi^2}{5}\gamma\left(\frac{M_{\rm Pl}^4}{M^2}\right), \tag{3.26}$$

which is the bound found in the ESGB theory, up to normalization [6]. The difference between the inequalities in (2.22) and (3.26) is crucial, since the former would imply that only black holes with extremely small masses may be scalarized, and these are definitely not observable. The introduction of the mass scale M given by (3.21) makes it possible for black holes of up to 180 solar masses to become scalarized. Hereafter, we will refer to SECG as the theory with this new scale incorporated.

In Reference [6], the authors consider a positive coupling constant of the Gauss-Bonnet operator, a condition that we also demand in (2.12) in order for scalarization of Schwarzschild black holes to become attainable. However, and unlike ESGB, the instability of Schwarzschid black holes is not only triggered once the curvature at the horizon reaches a minimum value within our model, as it must also *not* exceed an upper bound. Such bounds are determined by the coupling constant λ associated with the cubic operators in the action. This means that black holes with extremely large curvature are not prone to become scalarized. This last assertion may be quantified by computing the Kretschmann scalar at the horizon $\mathcal{K}_H$, so that the above relations imply that scalarization will occur only when

$$\frac{\mathcal{K}_H}{M_{\rm Pl}^4} \in \left[\frac{3072\,\pi^4}{q_+^2}, \frac{3072\,\pi^4}{q_-^2}\right] \qquad \text{if} \;\; 0 < \lambda \leq \frac{48}{175}\gamma^2, \tag{3.27}$$

$$\frac{\mathcal{K}_H}{M_{\rm Pl}^4} \in \left[0, \frac{3072\,\pi^4}{q_-^2}\right] \qquad \text{if} \;\; \lambda \leq 0. \tag{3.28}$$

Finally, despite the fact that the estimation of the instability times in (3.17) and (3.18) remain the same, the conclusion that present-day acceleration is safe from the instability, but inflation may not be, is precipitous. Indeed, as we have been firmly trying to emphasize, there are no compelling reasons for having a vanishing β value. Let us consider the leading term in (3.20), meaning

$$m_{\text{eff}}^2 = 12\beta H^2, \tag{3.29}$$

where we have taken the dS approximation, $\epsilon = 0$. We then find that

$$\frac{t_{\text{inst}}}{t_0} \sim \frac{H_0}{m_{\text{eff}}} \sim \frac{1}{\sqrt{12\,|\beta|}} \sim \frac{0.29}{\sqrt{|\beta|}}, \tag{3.30}$$

which is marginally different than one, unless β is tuned to *not* be an order one number. Moreover, in the case of inflation, with $N \sim 10^2$, the relevant ratio reads

$$\frac{t_{\text{inst}}}{t_{\text{inf}}} \sim \frac{1}{N}\frac{H_{\text{inf}}}{m_{\text{eff}}} \sim \frac{1}{\sqrt{12\,|\beta| N}} \sim \frac{2.9 \times 10^{-3}}{\sqrt{|\beta|}}. \tag{3.31}$$

The estimation in (3.30) seems to be the end of the story for us, as present-day cosmology would be largely unstable under scalarization within our EFT. Understanding that there is complete loss of generality in arbitrarily setting $\beta = 0$, and the expected hierarchy in the set of curvature operators, has led to the conclusion that in order to have good stable cosmological evolution we need to take β **strictly positive**. In other words, there is *no* chance of having a tachyonic instability during *any* (quasi)-dS stage of the cosmic evolution of our universe, and because of this, we will only consider the case

$$\beta > 0 \tag{3.32}$$

in the rest of this paper.

Before we may consider the model within BBC, it is worth mentioning why we do *not* expect that the standard reheating processes [49] should be influenced in any appreciable way by the presence of the scalar field φ. Say we have a canonically normalized inflaton field denoted by Φ, which by the end of inflation begins to oscillate at the bottom of its potential $V(\Phi)$. If Φ had no couplings with the rest of the world, its energy density would naively scale like pressureless matter, $\rho_\Phi \propto a^{-3}$, and the universe would end up empty. The necessary coupling to Standard Model (SM) degrees of freedom yields, on the other hand, a continuity equation of the form $\dot{\rho}_\Phi + 3H\rho_\Phi = -\Gamma_\Phi^{\text{SM}}\rho_\Phi$, where Γ_Φ^{SM} is the inflaton decay rate to SM particles. In the presence of φ one then should expect an equation of the form $\dot{\rho}_\Phi + 3H\rho_\Phi = -\left(\Gamma_\Phi^{\text{SM}} + \Gamma_\Phi^{\varphi}\right)\rho_\Phi$, where Γ_Φ^{φ} is the inflaton decay rate to φ particles. At this stage it should be clear that all we need, for reheating to remain intact, is that $\Gamma_\Phi^{\text{SM}} \ll \Gamma_\Phi^{\varphi}$. The problem lies in that, on EFT grounds, there should be a marginal operator of the form $\mathcal{L} \ni \mu\,\varphi^2\Phi^2$ where μ is a dimensionless coupling constant, which in principle implies sizeable $\Phi \to \varphi\varphi$ and $\Phi\Phi \to \varphi\varphi$ decays. However, such an operator is highly suppressed through the enforcement of φ shift symmetry, which is only broken through gravitational couplings. The latter implies that $\mu \lll 1$, so that it is technically natural to expect that $\Gamma_\Phi^{\text{SM}} \ll \Gamma_\Phi^{\varphi}$. Let us now check if our SECG model has the nice and desirable property of having GR as a late-time cosmic attractor.

4 General Relativity as a Cosmic Attractor

The scalar EOM (2.6) in a FLRW background is given by

$$\ddot{\varphi} + 3H\dot{\varphi} + m_{\text{eff}}^2\,\varphi = 0, \tag{4.1}$$

where

$$m_{\text{eff}}^2 = \beta R - \frac{\gamma}{M^2}\mathcal{G} + \frac{\lambda}{M^4}\left(\mathcal{P} - 8\mathcal{C}\right), \tag{4.2}$$

and depends on the cosmological background. As usual, to study the evolution of the scale factor we concentrate in the "time-time" component of the modified Einstein equations

$$M_{\text{Pl}}^2 G_{tt} = \rho_{\text{eff}} + \rho_a, \tag{4.3}$$

where ρ_a denotes the energy densities of the several BBC components of the cosmic fluid, while ρ_{eff} denotes an effective energy density associated with the presence of the cubic operator and the scalar field, and it is given by

$$\rho_{\text{eff}} \equiv \rho_{\mathcal{PC}} + \rho_{\varphi}, \tag{4.4}$$

$$\rho_{\mathcal{PC}} = -\frac{48\,\alpha}{M_{\text{Pl}}^2}H^6, \tag{4.5}$$

$$\rho_{\varphi} = \frac{1}{2}\dot{\varphi}^2 + 6\left(\beta - 4\,\gamma\,\chi - 24\,\lambda\,\chi^2\right)H\varphi\dot{\varphi} + 3\left(\beta + 8\,\lambda\,\chi^2\right)H^2\varphi^2. \tag{4.6}$$

We shall demand usual cosmic evolution, meaning

$$\rho_a \approx 3M_{\text{Pl}}^2 H^2, \tag{4.7}$$

which directly implies that

$$|\rho_{\mathcal{PC}}| \ll \rho_a, \tag{4.8}$$

given the expectation (due to EFT reasoning) that $\alpha \ll (M_{\text{Pl}}/H)^4$. In fact, notice that even during inflation, which is the cosmic stage when H attains its highest possible value, $(M_{\text{Pl}}/H)^4 \sim 10^{21} \gg \alpha$. In other words, the cubic energy density is indeed negligible when compared with BCC cosmic fluid densities. Moreover, as we do not want φ to play any role in late-time cosmology, we shall assume, for the time being, that

$$\rho_{\varphi} \ll \rho_a, \tag{4.9}$$

though, we acknowledge, it will be mandatory to check if such an assumption is dynamically consistent. Using the Friedmann Eq. (4.7) and the continuity equation

$$\dot{\rho}_a + 3H\rho_a\,(1 + w_a) = 0, \quad \text{where} \quad w_a \equiv \frac{p_a}{\rho_a}, \tag{4.10}$$

we may put the curvature invariants in the useful form

$$R = 6\left(2H^2 + \dot{H}\right) = \frac{\rho_a}{M_{\text{Pl}}^2}\,(1 - 3\,w_a)\,, \tag{4.11}$$

$$\mathcal{G} = 24H^2\left(H^2 + \dot{H}\right) = -\frac{4}{3}\left(\frac{\rho_a}{M_{\text{Pl}}^2}\right)^2 (1 + 3\,w_a)\,, \tag{4.12}$$

$$\mathcal{P} - 8\mathcal{C} = -48H^4\left(2H^2 + 3\dot{H}\right) = \frac{8}{3}\left(\frac{\rho_a}{M_{\text{Pl}}^2}\right)^3 \left(\frac{5}{3} + 3\,w_a\right). \tag{4.13}$$

We are interested in the behavior of the scalar field during different cosmological stages. In Table 1 below we summarize the signs of the Ricci scalar R, the Gauss-Bonnet operator $\mathcal{G}$, and the cubic operators $\mathcal{P} - 8\mathcal{C}$.

These signs, together with the signs of the dimensionless coupling constants β, γ, and λ, control the sign of the effective mass. The fact that the aforementioned operators have different mass dimension implies that they will scale differently with time. The cubic operator is then naively expected to be dominant during early times. Considering the expression for the effective mass in (3.23), we see that during radiation domination (RD), $\epsilon = 2$, and

$$m_{\text{eff,R}}^2 = 24\,\chi\,(\gamma + 8\,\lambda\,\chi)\,H^2. \tag{4.14}$$

Staring at (4.14) we observe that during RD the dynamics of the scalar field is determined by the higher-order curvature operators, as it is not sensitive to β, and we will explicitly see such behavior once we numerically solve the scalar field equation. Moreover during matter domination (MD), $\epsilon = \frac{3}{2}$, and one finds that

$$m_{\text{eff,M}}^2 = 3\left(\beta + 4\,\gamma\,\chi + 40\,\lambda\,\chi^2\right)H^2. \tag{4.15}$$

Table 1 Signs of the Ricci scalar, the Gauss-Bonnet invariant, and the cubic invariants during different cosmological eras

	Radiation	Matter	Dark energy
$\mathcal{P} - 8\mathcal{C}$	> 0	> 0	< 0
$\mathcal{G}$	< 0	< 0	> 0
R	0	> 0	> 0

Throughout MD it so happens that $H \ll M$, so we may confirm that, within that era, the contributions from the Gauss-Bonnet and cubic operators are largely subdominant, as expected from dimensional analysis. Finally during dark energy domination (DED), $\epsilon = 0$, so

$$m^2_{\text{eff,DE}} = 12\left(\beta - 2\,\gamma\,\chi - 8\,\lambda\,\chi^2\right) H^2, \tag{4.16}$$

and since during this stage $H \lll M$, it is clear that the "destiny of the spacetime itself" relies on the sign of β. Luckily for us, having taken the positive option, our current cosmology is safe.

Actually, it has been pointed out in [48], that the strict $\gamma = \lambda = 0$ limit is intimately related with a linearized version of the seminal DEF model for scalarization [5], which has been proven, for $\beta > 0$, to have GR as a cosmological attractor [4].[4] As non-linearities are not important for this analysis, it is expected that once the $\mathcal{G}$ and $\mathcal{P} - 8\mathcal{C}$ operators become negligible, the cosmic evolution will match that of the DEF model, meaning the scalar field will be naturally driven to $\varphi_0 = 0$. As the transition from MD to DED does not qualitatively change the dynamics of φ and the influence of higher-order operators becomes more and more marginal, GR with $\varphi_0 = 0$ is expected to be an attractor.

Let us verify all the above assertions through a quantitative study of the scalar field dynamics by expressing the scalar EOM in terms of the redshift z instead of cosmic time t, so that the scalar field equation now reads

$$\varphi''_{(a)}(z) + f_a(z)\varphi'_{(a)}(z) + q_a(z)\varphi_{(a)}(z) = 0, \tag{4.18}$$

where

$$f_a(z) \equiv \frac{H'(z)}{H(z)} - \frac{2}{z+1}, \tag{4.19}$$

$$q_a(z) \equiv \frac{3}{(1+z)^2}\left[\beta\,(1 - 3\,\omega_a) + 4\,\gamma\,\chi(z)\,(1 + 3\,\omega_a) + 8\,\lambda\,\chi(z)^2\,(5 + 9\,\omega_a)\right], \tag{4.20}$$

[4] The exact relation implies that $\beta = \beta_{\text{DEF}}/8$, where β_{DEF} is a DEF parameter [50]. Using the latter, we may try to constrain the magnitude and sign of β by analyzing how the DEF model fits within the Parametrized Post-Newtonian (PPN) formalism [51]. The relation between β and the PPN parameter β_{PPN} is schematically given by

$$\beta_{\text{PPN}} - 1 = 8\,\beta\,\mathfrak{C}, \tag{4.17}$$

with $\mathfrak{C} > 0$ some constant which tends to zero in the GR limit [52, 53]. As reported in [51], current planetary data has constrained β_{PPN} to values pretty close to unity. Indeed, according to data from the Messenger spacecraft that orbited Mercury, $\beta_{\text{PPN}} - 1 = (-4.1 \pm 7.8) \times 10^{-5}$. On the other hand, data stemming from the perihelion advance of Mars obtained by the Mars Reconnaissance Orbiter imposes a slightly weaker bound given by $\beta_{\text{PPN}} - 1 = (0.4 \pm 2.4) \times 10^{-4}$. In short, we see that such bounds allow, through the use of (4.17), both positive and negative values for our β parameter.

the primes denote differentiation with respect to z, $\varphi'(z) \equiv \frac{d\varphi(z)}{dz}$, and $\chi(z) \equiv (H(z)/M)^2$. We note that the form of the Hubble friction and "mass" terms $f_a(z)$ and $q_a(z)$ depend on the effective energy density that drives the cosmic evolution. Explicitly, we have that during

$$\text{RD:} \quad f_{\rm r}(z) = 0, \qquad q_{\rm r}(z) = \frac{24}{(1+z)^2}\,\chi(z)\left[\gamma + 8\,\lambda\,\chi(z)\right], \qquad \omega_{\rm r} = \frac{1}{3}, \tag{4.21}$$

$$\text{MD:} \quad f_{\rm m}(z) = -\frac{1}{2\,(1+z)}, \qquad q_{\rm m}(z) = \frac{3}{(1+z)^2}\left[\beta + 4\,\gamma\,\chi(z) + 40\,\lambda\,\chi(z)^2\right], \qquad \omega_{\rm m} = 0, \tag{4.22}$$

$$\text{DED:} \quad f_{\rm m}(z) = -\frac{2}{1+z}, \qquad q_{\rm de}(z) = \frac{12}{(1+z)^2}\left[\beta - 2\,\gamma\,\chi(z) - 8\,\lambda\,\chi(z)^2\right], \qquad \omega_{\rm de} = -1. \tag{4.23}$$

The numerical analysis is set to start at $z_i = 10^{10}$, right before big bang nucleosynthesis (BBN) epoch. A natural initial value (see the discussion below (4.26)) for the dimensionless field[5] $\tilde{\varphi} \equiv \varphi/\sqrt{2}M_{\rm Pl}$ is given by

$$\tilde{\varphi}_i = \frac{\varphi_i}{\sqrt{2}M_{\rm Pl}} \simeq \frac{H}{\sqrt{2}M_{\rm Pl}} \ll 1, \tag{4.24}$$

as φ is basically coupled with the thermal bath. Moreover, it is also natural to expect, as will be argued below, that $\dot{\tilde{\varphi}}_i \simeq H(z_i)\tilde{\varphi}_i$, which implies that

$$|\tilde{\varphi}_i'| \simeq \left|\frac{\tilde{\varphi}_i}{z_i}\right| \ll 1. \tag{4.25}$$

Considering that the scalar field to cosmic fluid energy densities ratio $\rho_\varphi(z)/\rho_a(z)$, with $\rho_\varphi(z)$ and $\rho_a(z)$ as given by the "z-version" of (4.6) and (4.7), respectively, turns out to be

$$\begin{aligned}\frac{\rho_\varphi(z)}{\rho_a(z)} &= \frac{1}{3}\,(1+z)^2\,\tilde{\varphi}'(z)^2 - 4\,(1+z)\left[\beta - 4\,\gamma\,\chi(z) - 24\,\lambda\,\chi(z)^2\right]\tilde{\varphi}(z)\tilde{\varphi}'(z)\\ &+ 2\left[\beta + 8\,\lambda\,\chi(z)^2\right]\tilde{\varphi}(z)^2,\end{aligned} \tag{4.26}$$

one may foresee that the initial conditions (4.24) and (4.25) are indeed compatible with the previous assumption that $\rho_\varphi(z) \ll \rho_a(z)$, so the scalar field remains cosmologically subdominant, as required by observations. Actually, the fact of the matter is that a field value $\tilde{\varphi} \sim 1$ constitutes a Planckian regime which is not compatible with an EFT description because such a scenario most probably falls into the so-called "swampland" of quantum gravity (see [54] and references therein); all in all, $\tilde{\varphi}_i \ll 1$ should *not* be considered fine tuning but the natural choice for the problem at hand.

[5] Incidentally, such a field redefinition recovers the quadratic Lagrangian considered in [48].

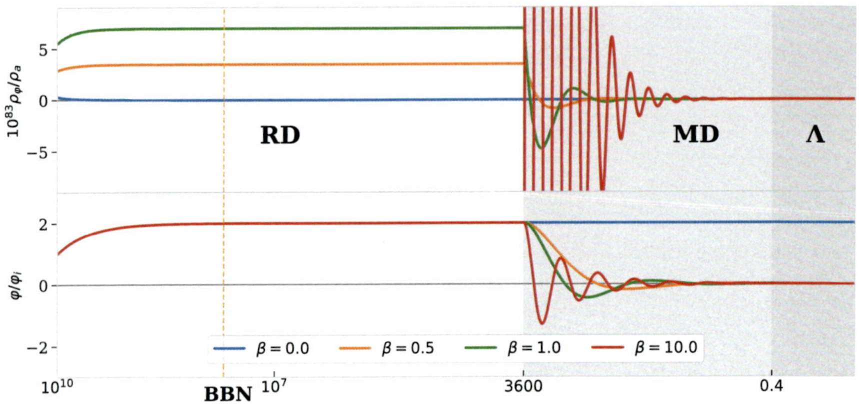

Fig. 1 *Top panel*: Effective energy density ρ_φ relative to the energy density of the cosmic fluid ρ_a. *Bottom panel*: Scalar field value relative to its initial value fixed at $z_i = 10^{10}$. The values of the coupling constants are taken to be $\gamma = 1$ and $\lambda = 48/175$, so they saturate the inequality in (2.21)

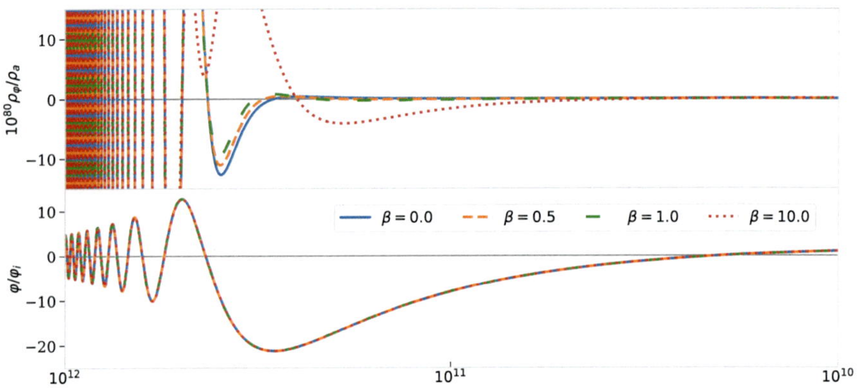

Fig. 2 Relative effective density and scalar field value for high redshift. The values of the coupling constants are the same as those of Fig. 1

In Fig. 1 below we show the evolution of both the dimensionless scalar field φ/φ_i and the dimensionless ratio ρ_φ/ρ_a for $z < z_i$ for different values of β and fixed values of γ and λ.

As previously noted, the contributions from higher-order curvature terms do become relevant during the very early stage of the universe. We may explicitly observe this, for high redshift $z > z_i$, in Fig. 2.

We confirm that the solution is *strongly* consistent with our initial assumption, as actually $\rho_\varphi(z) \lll \rho_a(z)$ for the whole range of numerical integration which goes from $z = 0$ to $z = 10^{12}$. Moreover, φ is basically a constant throughout most of that range with the exception of cosmological "phase transitions" redshifts and very early and late times.

During early times, or high redshift, $m_{\rm eff}^2$ dominates over Hubble friction within the scalar field Eq. (4.1). However, as we "move" forward in time, $m_{\rm eff}^2$ decays much faster than the Hubble friction which rapidly takes over, so it is expected that the scalar field freezes to a constant way before entering MD era. To confirm that this is indeed the case, we may estimate the time at which the potential and the Hubble friction become comparable, since afterwards it should only take a couple of Hubble times for $\dot{\varphi}$ to effectively vanish. Explicitly, using (4.14), the condition $m_{\rm eff,R} \lesssim H$ is equivalent to $\sqrt{24\,\gamma\,\chi + 192\,\lambda\,\chi^2} \lesssim 1$, which by taking $\gamma = 1$, $\lambda = 48/175$ and $L = 10\,$km., along with the fact that during RD $H(z) = H_0\,(1+z)^2$ with $H_0 \simeq 10^{-43}$ GeV, comes to $z \approx 1.97 \times 10^{11}$. For even higher redshift values the relative scalar field and the relative energy density oscillate with ever increasing frequency as can be seen in Fig. 2. We also explicity note that within this regime the relative scalar field is completely insensitive to the value of β as the Ricci scalar identically vanishes, while the relative density does marginally depend on such a constant even though all the curves, for high enough z, eventually converge.

On the other hand, by the time the MD era begins, the Ricci scalar stops being trivial, and in fact it entirely determines the relative scalar and energy density evolution because the higher-order operators become irrelevant considering that $\chi \sim 10^{-36} \lll 1$ when $z = 3600$. Looking at Fig. 1 we observe that the strict $\beta = 0$ case is actually problematic because the relative field stays frozen at its RD value and this goes against the premise of having an unscalarized $\varphi_0 = 0$ asymptotic solution, which is actually mandatory, in view of weak field constraints. To the contrary, the same figure also shows that for any $\beta \neq 0$ scenario, the desired behavior of an asymptotically vanishing scalar field is attained. Moreover, as mentioned in [48], φ oscillations may in principle affect the formation of large scale structures in the universe through the redefinition of (an effective) Newton's constant $G_{\rm eff} \equiv G/\left(1 - 2\,\beta\,\tilde{\varphi}^2\right)$ with $\tilde{\varphi} \equiv \varphi/\sqrt{2}M_{\rm Pl}$ as before. However, it is clear that $\tilde{\varphi}^2 \lll 1$ and, since $\beta \sim \mathcal{O}(1)$ number, $G_{\rm eff} \approx G$ for all purposes, so GR truly becomes a trustworthy cosmological attractor. As it was expected, the scalar field profile in SECG exhibits a manifest deviation from its quadratic counterpart [48] for very high cosmological redshifts. This is depicted in Fig. 3.

In Fig. 3 each color stands for the profiles obtained within both the quadratic theory [48] (continuous curves) and SECG (dashed curves) for different values of the astrophysical length $L = M^{-1}$. We observe that the scalar field equation in SECG only gets a small correction (as it should) with respect to the Gauss-Bonnet quadratic theory during radiation domination up to a specific redshift value where both terms become "competitive". From (4.21) we see that this roughly happens when $\gamma \approx 8\,\lambda\,\chi$ which implies, for the $L = 10$ km. case, a redshift value $z \approx 3.7 \times 10^{11}$, which lies before BBN epoch. Such a redshift value marks the "breakdown" of the EFT expansion (EFTBD), in the sense that perturbativity is lost and we should *not* trust the naive model anymore. In other words, in reality, any behavior of the system beyond the EFTBD point should not be taken seriously as it does not represent sensible

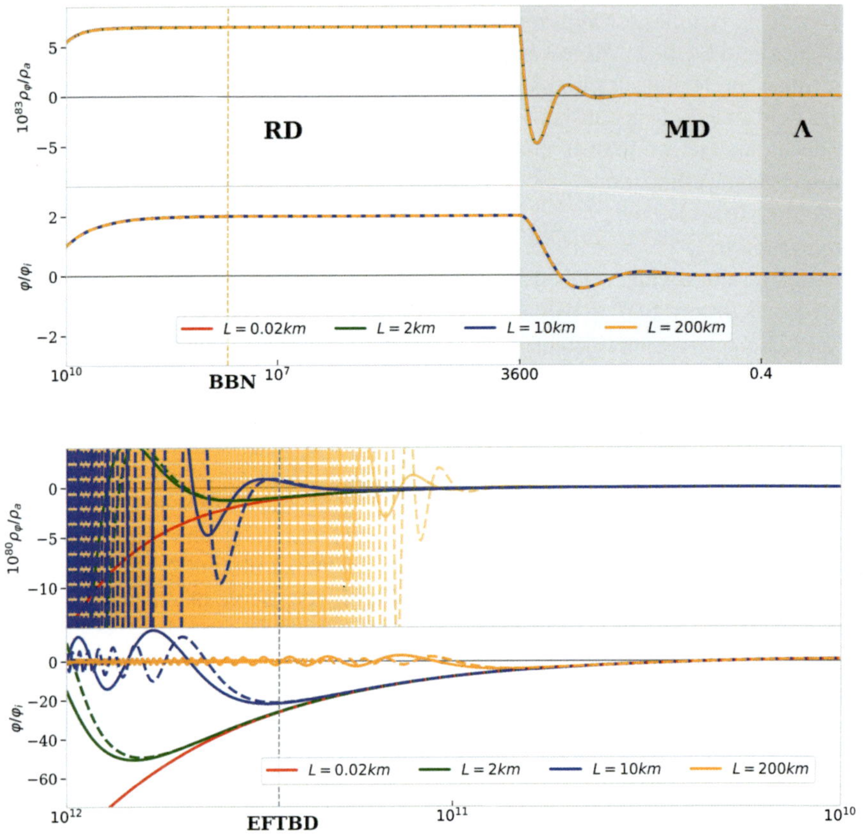

Fig. 3 Relative effective density and scalar field. The upper panel shows their value for $z < z_i$ while the lower one does so for $z > z_i$. Each color represents a theory with a specific value of the "astrophysical length" L. The continuous and dashed curves represent the profile stemming from the quadratic theory [48] and SECG, respectively. The values of the coupling constants are taken to be $\beta = 1$, $\gamma = 1$, and $\lambda = 48/175$

perturbative physics because the system becomes strongly-coupled.[6] Effectively, by the time the cubic contribution becomes relevant, new non-trivial physics in the form of an ultraviolet (UV) completion, like inflation, necessarily becomes a much better description of the primordial cosmic plot. Interestingly, by the EFTBD time the size of the universe is about four kilometers. It can also be visually checked that in general, given that the redshift corresponding to the EFTBD point scales as $z \propto L^{-1/2}$, the larger the astrophysical length L, the smaller the redshift z where the

[6] As it is well-known, one possibility of rendering the theory perturbative again is that new degrees of freedom enter the game as it is further discussed in Appendix A.

continuous and dashed curves formally "separate". This implies that the bigger the scale M, the further back in time the EFT is appropriate to describe the phenomenon of scalarization within a cosmological setting.

5 Discussion and Conclusions

In this paper we have followed the principles of EFT in order to address the issue of black hole scalarization and its compatibility with standard cosmic evolution in higher-order theories of gravity, using what we have dubbed "Scalar-Einstenian Cubic Gravity" as a well-motivated, healthy toy model. Within the process, we have learned several lessons concerning the naive approach one may have to such a seemingly straightforward model-building exercise. For example, it became obvious by using "naturalness" arguments, the complete necessity of introducing a "foreign" new scale in the problem, which is actually associated with the physical extension of the would-be scalarized compact object. Interestingly enough, such a procedure also improves the phenomenological prospects for scalarization in the sense that, without a new scale, the mass of the black holes that may become scalarized would be of order 10^{-37} solar masses, so they clearly lie way out of any observational scope. Moreover, we have confirmed the fact that the relative signs of the dimensionless parameters of the model play a crucial role within the process of spacetime scalarization as they actually determine whether or not scalarized solutions may emerge from dynamics.

In the context of tackling the phenomenon of scalarization through the construction of a model in a systematic manner we conclude that the claim for the existence of a catastrophic instability during inflation [47], although classically true, is not really justified from an EFT perspective. This is because there is *no* symmetry reason for discarding the marginal operator $\varphi^2 R$, which shall be quantum-mechanically generated anyway, and will actually lead the dynamics whenever the theory is weakly-coupled. Moreover, picking the right sign for such an operator in the theory is enough to render inflation safe from the tachyonic instability while still scalarizing compact objects through the irrelevant operator $\varphi^2\mathcal{G}$, hence *scalarization remains compatible with the inflationary paradigm.*

Next we addressed the issue of the compatibility of scalarization and BBC evolution. To do so, we followed [48] and integrated the scalar field equation to find that, indeed under very sensible assumptions for the initial conditions of the system, *the theory admits GR as a cosmological attractor.* The departure from GR does become significant for high redshift $\left(z \sim 10^{11}\right)$, way before BBN, where the relative scalar field and energy density deviate from its quasi-constant behavior. The departure, however, between SECG theory and the quadratic one, occurs around the same high redshift and this roughly defines the breakdown of the weakly-coupled EFT description, implying the necessity of a UV completion. The understanding of the interplay between the different scales in the game is further elucidated in Appendix A.

Finally, let us note that it is quite remarkable that by using a criterion based on the existence of an unstable mode for a scalar field fluctuation living on a Schwarzschild background we may get important information about the appropriate EFT that captures the phenomenon of black hole scalarization. In short, such an unstable mode signals the instability of the underlying geometry as the curvature surpasses a specific threshold, and its tendency to consequently develop "scalar hair", i.e, a non-zero scalar profile. By imposing a sensible reality condition for the would-be scalarized black hole mass we were able to find a non-trivial constraint between otherwise completely independent "Wilson coefficients" of the EFT expansion. It would be nice to find, through the application of some other physical requirements, more relations between the EFT coefficients so that we would be able to further constrain the parameter space of the model.

For example, if we abandon the goal of scalarizing astrophysical objects by entertaining the idea of dealing with more fundamental "primordial" black holes, we may consider the free theory of scalar fluctuations on a (quasi)-dS inflationary spacetime, whose analytic solutions are well-known. As detailed in Appendix C, the relevant quantity for the scalar fluctuation dynamics in a dS background is the "index" $\nu^2 \equiv \frac{9}{4} - \frac{m_{\rm eff}^2}{H^2}$, which actually needs *not* to be necessarily positive. The whole "cosmological collider physics" programme of Arkani-Hamed and Maldacena [55], which emphasizes the fact that inflation pertains the highest energy *observable* natural process, is actually quite simple for scalar fields; particles of mass $m \sim H$ during inflation may leave observable characteristic imprints, so they become "fossils" of a fundamental high-energy physics era, depending on the actual value of the ν-index. Then, as our scalar "particle" has a zeroth-order ν-index given by $\nu^2 \simeq \frac{9}{4} - 12\,\beta$, the imprint of such a particle would be linked to the value of the EFT parameter β, which we only demand to be positive (so BBC does *not* get scalarized) but whose magnitude is otherwise undetermined. This opens up the possibility of eventually measuring the actual value of the β parameter. More kinematical and dynamical conditions among EFT parameters are waiting to be explored and it is our aim to do so in the near future.

Let us close by stating that it would be great to derive that β ultimately *needs* to be positive in order for the EFT to be compatible with a quantum gravity UV-completion, using the so-called "positivity constraints" programme [56]. Moreover, we want to understand under what circumstances the shift symmetry is only spoiled by gravitational interactions. We are already working in a realistic particle physics model that truly embeds scalarization within the framework of spontaneous symmetry breaking, and we shall report our findings elsewhere [57].

Acknowledgements CE is supported by PAI Grant N° 77190046, by FONDECYT Grant N° 11221063 and Universidad Central de Chile through Proyecto Interno CIP2020039. CE would like to thank the Department of Physics at the National Technical University of Athens (Greece) and Centro de Estudios Científicos at Valdivia (Chile) for warm hospitality. SR acknowledges support from Fondecyt Postdoctorado project N° 3190554. CE, SR, and NZ thank an anonymous referee who helped improving the original manuscript for our potential readers.

Appendix A: EFT's in a Nutshell

Say the final theory of some physical system may be written in terms of two fields ϕ and σ, meaning the UV complete theory has a Lagrangian of the form

$$\mathcal{L}[\phi, \sigma] = \mathcal{L}[\phi] + \mathcal{L}[\sigma] + \mathcal{L}_{\text{int}}[\phi, \sigma]. \tag{A.1}$$

We will assume that ϕ is a *light* field while σ is a *heavy* field, meaning $m_\phi < \Lambda$ and $M_\sigma > \Lambda$, where m_ϕ and M_σ are the masses of ϕ and σ, respectively, and Λ is the cut-off of the EFT. To derive the EFT, encapsulated in the effective action S_{eff}, we use the so-called *Wilsonian approach* formally defined through the path integral

$$e^{i S_{\text{eff}}[\phi]} = \int \mathscr{D}\sigma \, e^{i S[\phi, \sigma]}. \tag{A.2}$$

As it is well-known, at tree-level, performing such an integral is equivalent to using the field equations to eliminate the heavy field. The complete path integral includes, however, loop corrections involving the heavy fields, implying that the effective action is, in principle, non-local. However, at low energies, meaning $E \ll \Lambda \leq M_\sigma$, one may get an approximately local EFT action with a systematic expansion of the form

$$\mathcal{L}_{\text{eff}}[\phi] = \mathcal{L}[\phi] + \sum_i c_i(g) \, \frac{\mathcal{O}_i[\phi]}{(M_\sigma)^{\Delta_i - 4}}, \tag{A.3}$$

where the $c_i = c_i(g)$ are dimensionless constants which are expected to be of order one. It is important to emphasize that the meaning of c_i being order one numbers is somehow *not* universal. In this paper we declare that any number $c_i \sim \left\{10^{-2}, 10^2\right\}$ should be considered an order one number (basically, not super small, not super big). Anyhow, these numbers depend on the couplings g of the UV theory, while the $\mathcal{O}_i$ are local operators of mass dimension Δ_i. The procedure described above will generate all operators compatible with the symmetries of the UV theory, implying that the absence of any of such terms, or an anomalously small corresponding coefficient, should be understood as fine-tuning (an important exception to this statement is the case of a so-called *accidental symmetry*, which however, is *not* relevant for our discussion, so we do not consider it).

We may take the point of view that GR and any low-energy field description encapsulate the relevant degrees of freedom for describing the physics way below the Planck scale M_{Pl}. Therefore, for a higher-order EFT of gravity such as the one we deal with in this paper, it is natural to take M_{Pl} as the cut-off and so we do. The introduction of the scale M however, does become necessary in order to "naturalize" otherwise finely-tuned coefficients of the EFT, as emphasized elsewhere within the article. In principle, one may think that the operators containing such a scale are the remains after integrating out dynamical degrees of freedom of mass M. This implies that it is natural to expect the hierarchy $m_{\rm eff}^2 \ll M^2 \ll M_{\rm Pl}^2$. However, the perturbative condition $H^2 \ll M^2$ is not fundamental, but a restriction of the weakly-coupled EFT description.

Appendix B: Unstability Condition on the Effective Potential

The condition on the effective potential for the existence of at least one unstable mode

$$\int_{-\infty}^{\infty} dr_* \, V_{\rm eff}(r_*) < 0, \tag{B.1}$$

is important and deserves to be elucidated for the interested reader. To this purpose we closely follow Reference [58], where the authors prove the above condition for the existence of at least one bound state in a non-relativistic one-dimensional quantum mechanical system. Let us consider the time-independent "Schrödinger-like" equation for the radial fluctuation u,

$$\omega^2 u = -\frac{d^2 u}{dr_*^2} + V_{\rm eff}(r) u. \tag{B.2}$$

Recalling that the time dependence of the scalar field fluctuation (2.13) goes like $\sim e^{-i\omega t}$, it is clear that

$$\omega^2 < 0 \Longleftrightarrow \text{exponentially growing solution} \Longleftrightarrow \text{unstable mode.} \tag{B.3}$$

Moreover, for a localized perturbation, it is certainly reasonable to demand that $u(r_*)$ is a continuous and bounded function with boundary conditions given by $u \to 0$ and $du/dr_* \to 0$ as $r_* \to \infty$, which incidentally also implies that $V_{\rm eff}(r) \to 0$ in this limit, as can be seen from (2.15). Let us now consider a normalized Gaussian as a trial function for the radial perturbation u, namely,

$$u_\varepsilon(r) = \left(\frac{\varepsilon}{\pi}\right)^{1/4} e^{-\varepsilon r^2/2}, \quad \varepsilon > 0. \tag{B.4}$$

Multiplying (B.2) by u_ε and integrating yields

$$\omega^2(\varepsilon) = -u_\varepsilon \frac{du_\varepsilon}{dr_*}\bigg|_{-\infty}^{\infty} + \int_{-\infty}^{\infty} dr_* \left[\left(\frac{du_\varepsilon}{dr_*}\right)^2 + V_{\rm eff}(r_*) u_\varepsilon^2\right] = \frac{\varepsilon}{2} + \sqrt{\frac{\varepsilon}{\pi}} \int_{-\infty}^{\infty} dr_* V_{\rm eff}(r_*) e^{-\varepsilon r_*^2}, \tag{B.5}$$

where we have enforced the aforementioned boundary conditions. We may write Eq. (B.5) as

$$\frac{\omega^2(\varepsilon)}{\sqrt{\varepsilon}} = \frac{\sqrt{\varepsilon}}{2} + \mathfrak{I}(\varepsilon), \tag{B.6}$$

where we define

$$\mathfrak{I}(\varepsilon) \equiv \frac{1}{\sqrt{\pi}} \int_{-\infty}^{\infty} dr_* V_{\text{eff}}(r_*) e^{-\varepsilon r_*^2} , \tag{B.7}$$

which is a continuous function of ε. Note that in the limit $\varepsilon \to 0, \omega^2(\varepsilon)/\sqrt{\varepsilon} \to \mathfrak{I}(0)$. Consequently, if $\mathfrak{I}(0) < 0$, then $\exists\, \delta > 0$ such that $\mathcal{I}(\varepsilon) < 0$ for $\varepsilon \in [0, \delta)$. Hence, for a sufficiently small value of ε, $\omega^2(\varepsilon)/\sqrt{\varepsilon} < 0$ implying $\omega^2 < 0$. We may conclude then that a sufficient (though not necessary) condition on V_{eff} for the existence of at least one unstable mode is given by (B.1).

Appendix C: Scalar Perturbation in a de Sitter Background

Let us consider dS spacetime in conformal coordinates $dt = a(\eta)d\eta$ with line element

$$ds^2 = \frac{-d\eta^2 + d\boldsymbol{x}^2}{H^2\eta^2}. \tag{C.1}$$

In such coordinates the Klein-Gordon equation for the scalar fluctuation, $\left(\Box - m_{\text{eff}}^2\right)\delta\varphi = 0$, reads

$$\delta\varphi_{\boldsymbol{k}}'' - \frac{2}{\eta}\delta\varphi_{\boldsymbol{k}}' + k^2\delta\varphi_{\boldsymbol{k}} + \frac{m_{\text{eff}}^2}{H^2\eta^2}\delta\varphi_{\boldsymbol{k}} = 0, \tag{C.2}$$

where $m_{\text{eff}}^2 \equiv 12(\beta - 2\gamma\chi - 8\lambda\chi^2)H^2$, $\chi \equiv (H/M)^2$, and $\delta\varphi_{\boldsymbol{k}}' \equiv \frac{d\delta\varphi_{\boldsymbol{k}}}{d\eta}$ within this appendix. Defining the field $\psi \equiv a\,\delta\varphi$, the above equation may be put in a "Mukhanov-Sasaki" form, meaning

$$\psi_{\boldsymbol{k}}'' + \left(k^2 - \frac{\nu^2 - \frac{1}{4}}{\eta^2}\right)\psi_{\boldsymbol{k}} = 0, \tag{C.3}$$

where $\nu \equiv \sqrt{\frac{9}{4} - \frac{m_{\text{eff}}^2}{H^2}}$. The solution to (C.3) is well-known. It is given by

$$\psi_{\boldsymbol{k}} = -ie^{i\left(\nu+\frac{1}{2}\right)\frac{\pi}{2}}\frac{\sqrt{\pi}}{2}H(-\eta)^{3/2}H_{\nu}^{(1)}(-k\eta) \qquad \text{if} \quad \frac{m_{\text{eff}}^2}{H^2} \leq \frac{9}{4}, \tag{C.4}$$

$$\psi_{\boldsymbol{k}} = -ie^{-\frac{\pi}{2}\tilde{\nu}+i\frac{\pi}{4}}\frac{\sqrt{\pi}}{2}H(-\eta)^{3/2}H_{i\tilde{\nu}}^{(1)}(-k\eta) \qquad \text{if} \quad \frac{m_{\text{eff}}^2}{H^2} > \frac{9}{4}, \tag{C.5}$$

where $\tilde{\nu} \equiv \sqrt{\frac{m_{\text{eff}}^2}{H^2} - \frac{9}{4}}$, and usual Bunch-Davies initial conditions have been imposed. It is useful to consider the behavior of the mode functions after horizon exit, $k\eta \to 0$. One finds that for $m_{\text{eff}}^2/H^2 \leq 9/4$,

$$\psi_{\boldsymbol{k}} \to -e^{i\left(\nu+\frac{1}{2}\right)\frac{\pi}{2}} \frac{2^{\nu-1}}{\sqrt{\pi}} \Gamma(\nu) \frac{H}{k^{\nu}} (-\eta)^{-\nu+\frac{3}{2}}\,, \qquad 0 < \nu \leq \frac{3}{2},$$
$$\psi_{\boldsymbol{k}} \to e^{i\frac{\pi}{4}} \frac{1}{\sqrt{\pi}} H\,(-\eta)^{3/2} \ln(-k\eta)\,, \qquad \nu = 0, \tag{C.6}$$

and for $m_{\text{eff}}^2/H^2 > 9/4$,

$$\psi_{\boldsymbol{k}} \to -ie^{-\frac{\pi}{2}\tilde{\nu}+i\frac{\pi}{4}} \frac{\sqrt{\pi}}{2} H\,(-\eta)^{3/2} \left[\frac{1}{\Gamma(i\tilde{\nu}+1)} \left(\frac{-k\eta}{2}\right)^{i\tilde{\nu}} - i\frac{\Gamma(i\tilde{\nu})}{\pi} \left(\frac{-k\eta}{2}\right)^{-i\tilde{\nu}} \right]. \tag{C.7}$$

Recalling that during inflation $H \simeq 10^{13}$ GeV, it is crucial to impose a perturbative hierarchy $H^2 \ll M^2 \ll M_{\text{Pl}}^2$, so that $\chi \ll 1$, and it is then safe to approximate $m_{\text{eff}}^2 \simeq 12\beta H^2$. The latter approximation implies that $\nu^2 \simeq \frac{9}{4} - 12\beta$, so that (C.4) and (C.5) are appropriate whenever $\beta \lesssim \frac{3}{16}$ and $\beta > \frac{3}{16}$, respectively. Notice that by taking, for definiteness, $M \in \left[10^{14}, 10^{17}\right]$ GeV, implies we are dealing with a would-be scalarized object of physical length $L \in \left[1.23 \times 10^5, 123\right] \ell_{\text{Pl}}$, with $\ell_{\text{Pl}} = 1.61 \times 10^{-35}$ m standing for the famous Planck length.

We see that in the $k\eta \to 0$ limit, (C.6) contains a decay factor $(-\eta)^{-\nu+3/2}$, while (C.7) contains the decay factor $(-\eta)^{3/2}$ and an oscillation factor of the form $\eta^{\pm i\tilde{\nu}}$. Both decay factors indicate the well-known fact that massive field perturbations eventually die off. In the context of inflation, information about the mass spectrum is encoded in the power of the momentum ratio in the squeezed-limit bispectrum (see [59] for more details). In short, the squeezed-limit of the bispectrum may be considered as a "particle detector" for massive particles. Importantly, it is possible to show that for large masses, meaning $m_{\text{eff}} > \mathcal{O}(H)$, the signal acquires a "Boltzmann suppression" $e^{-\pi\tilde{\nu}}$ since too massive particles cannot be thermally produced during inflation, so this factor limits the mass range for such a particle detector. Therefore one needs to focus on particles whose masses are not too much heavier than H. This is exactly our case at hand, because $m_{\text{eff}} \simeq \sqrt{12\beta}H$, with $\beta \sim \mathcal{O}(1)$ number, as expected from an EFT perspective.

References

1. LIGO Scientific, Virgo collaboration, B.P. Abbott et al., Observation of gravitational waves from a binary black hole merger. Phys. Rev. Lett. **116**, 061102 (2016). https://doi.org/10.1103/PhysRevLett.116.061102. [arXiv:1602.03837]
2. LIGO Scientific, Virgo collaboration, B.P. Abbott et al., GW170817: observation of gravitational waves from a binary neutron star inspiral. Phys. Rev. Lett. **119**, 161101 (2017). https://doi.org/10.1103/PhysRevLett.119.161101. [arXiv:1710.05832]
3. LIGO Scientific, Virgo, Fermi-GBM, INTEGRAL collaboration, B.P. Abbott et al., Gravitational waves and gamma-rays from a binary neutron star merger: GW170817 and GRB 170817A. Astrophys. J. Lett. **848**, L13 (2017). https://doi.org/10.3847/2041-8213/aa920c. [arXiv:1710.05834]
4. T. Damour, K. Nordtvedt, General relativity as a cosmological attractor of tensor-scalar theories. Phys. Rev. Lett. **70**, 2217 (1993). https://doi.org/10.1103/PhysRevLett.70.2217
5. T. Damour, G. Esposito-Farèse, Nonperturbative strong-field effects in tensor-scalar theories of gravitation. Phys. Rev. Lett. **70**, 2220 (1993). https://doi.org/10.1103/PhysRevLett.70.2220
6. D.D. Doneva, S.S. Yazadjiev, New Gauss-Bonnet Black Holes with Curvature-Induced Scalarization in Extended Scalar-Tensor Theories. Phys. Rev. Lett. **120**, 131103 (2018). https://doi.org/10.1103/PhysRevLett.120.131103. [arXiv:1711.01187]
7. H.O. Silva, J. Sakstein, L. Gualtieri, T.P. Sotiriou, E. Berti, Spontaneous scalarization of black holes and compact stars from a gauss-bonnet coupling. Phys. Rev. Lett. **120**, 131104 (2018). https://doi.org/10.1103/PhysRevLett.120.131104
8. K.S. Stelle, Renormalization of higher derivative quantum gravity. Phys. Rev. D **16**, 953 (1977). https://doi.org/10.1103/PhysRevD.16.953
9. S. Mignemi, N.R. Stewart, Charged black holes in effective string theory. Phys. Rev. D **47**, 5259 (1993). https://doi.org/10.1103/PhysRevD.47.5259. [arXiv:hep-th/9212146]
10. P. Kanti, N.E. Mavromatos, J. Rizos, K. Tamvakis, E. Winstanley, Dilatonic black holes in higher curvature string gravity. Phys. Rev. D **54**, 5049 (1996). https://doi.org/10.1103/PhysRevD.54.5049. [arXiv:hep-th/9511071]
11. T. Torii, H. Yajima, K.-I. Maeda, Dilatonic black holes with Gauss-Bonnet term. Phys. Rev. D **55**, 739 (1997). https://doi.org/10.1103/PhysRevD.55.739. [arXiv:gr-qc/9606034]
12. D. Ayzenberg, N. Yunes, Slowly-rotating black holes in Einstein-Dilaton-Gauss-Bonnet Gravity: Quadratic order in spin solutions. Phys. Rev. D **90**, 044066 (2014). https://doi.org/10.1103/PhysRevD.90.044066. [arXiv:1405.2133]
13. B. Kleihaus, J. Kunz, E. Radu, Rotating black holes in dilatonic Einstein-Gauss-Bonnet theory. Phys. Rev. Lett. **106**, 151104 (2011). https://doi.org/10.1103/PhysRevLett.106.151104. [arXiv:1101.2868]
14. D.D. Doneva, S.S. Yazadjiev, Neutron star solutions with curvature induced scalarization in the extended Gauss-Bonnet scalar-tensor theories. JCAP **04**, 011 (2018). https://doi.org/10.1088/1475-7516/2018/04/011. [arXiv:1712.03715]
15. G. Antoniou, A. Bakopoulos, P. Kanti, Evasion of no-hair theorems and novel black-hole solutions in gauss-bonnet theories. Phys. Rev. Lett. **120**, (2018). https://doi.org/10.1103/physrevlett.120.131102
16. G. Antoniou, A. Bakopoulos, P. Kanti, Black-hole solutions with scalar hair in Einstein-Scalar-Gauss-Bonnet theories. Phys. Rev. D **97**, (2018). https://doi.org/10.1103/physrevd.97.084037
17. D.D. Doneva, S. Kiorpelidi, P.G. Nedkova, E. Papantonopoulos, S.S. Yazadjiev, Charged gauss-bonnet black holes with curvature induced scalarization in the extended scalar-tensor theories. Phys. Rev. D **98**, (2018). https://doi.org/10.1103/physrevd.98.104056
18. C.A. Herdeiro, E. Radu, N. Sanchis-Gual, J.A. Font, Spontaneous scalarization of charged black holes. Phys. Rev. Lett. **121**, (2018). https://doi.org/10.1103/physrevlett.121.101102
19. P.G.S. Fernandes, C.A.R. Herdeiro, A.M. Pombo, E. Radu, N. Sanchis-Gual, Spontaneous scalarisation of charged black holes: coupling dependence and dynamical features. Class. Quantum Gravity. **36**, 134002 (2019). https://doi.org/10.1088/1361-6382/ab23a1

20. C.A. Herdeiro, E. Radu, Black hole scalarization from the breakdown of scale invariance. Phys. Rev. D **99**, (2019). https://doi.org/10.1103/physrevd.99.084039
21. Y. Brihaye, C. Herdeiro, E. Radu, The scalarised schwarzschild-nut spacetime. Phys. Lett. B **788**, 295 (2019). https://doi.org/10.1016/j.physletb.2018.11.022
22. D.D. Doneva, K.V. Staykov, S.S. Yazadjiev, R.Z. Zheleva, Multiscalar gauss-bonnet gravity: Hairy black holes and scalarization. Phys. Rev. D **102**, (2020). https://doi.org/10.1103/physrevd.102.064042
23. D. Astefanesei, C. Herdeiro, J. Oliveira, E. Radu, Higher dimensional black hole scalarization. J. High Energy Phys. **2020**, (2020). https://doi.org/10.1007/jhep09(2020)186
24. P. Cañate, S.E.P. Bergliaffa, Novel exact magnetic black hole solution in four-dimensional extended scalar-tensor-gauss-bonnet theory. Phys. Rev. D **102**, (2020). https://doi.org/10.1103/physrevd.102.104038
25. A. Bakopoulos, G. Antoniou, P. Kanti, Novel black-hole solutions in einstein-scalar-gauss-bonnet theories with a cosmological constant. Phys. Rev. D **99**, (2019). https://doi.org/10.1103/physrevd.99.064003
26. Y. Brihaye, C. Herdeiro, E. Radu, Black hole spontaneous scalarisation with a positive cosmological constant. Phys. Lett. B **802**, 135269 (2020). https://doi.org/10.1016/j.physletb.2020.135269
27. A. Bakopoulos, P. Kanti, N. Pappas, Large and ultracompact gauss-bonnet black holes with a self-interacting scalar field. Phys. Rev. D **101**, (2020). https://doi.org/10.1103/physrevd.101.084059
28. K. Lin, S. Zhang, C. Zhang, X. Zhao, B. Wang, A. Wang, No static regular black holes in einstein-complex-scalar-gauss-bonnet gravity. Phys. Rev. D **102**, (2020). https://doi.org/10.1103/physrevd.102.024034
29. H. Guo, S. Kiorpelidi, X.-M. Kuang, E. Papantonopoulos, B. Wang, J.-P. Wu, Spontaneous holographic scalarization of black holes in einstein-scalar-gauss-bonnet theories. Phys. Rev. D **102**, (2020). https://doi.org/10.1103/physrevd.102.084029
30. Y. Brihaye, B. Hartmann, N.P. Aprile, J. Urrestilla, Scalarization of asymptotically anti–de sitter black holes with applications to holographic phase transitions. Phys. Rev. D **101**, (2020). https://doi.org/10.1103/physrevd.101.124016
31. Z.-Y. Tang, B. Wang, T. Karakasis, E. Papantonopoulos, Curvature scalarization of black holes in f(R) gravity. Phys. Rev. D **104**, 064017 (2021). https://doi.org/10.1103/PhysRevD.104.064017. [arXiv:2008.13318]
32. L.G. Collodel, B. Kleihaus, J. Kunz, E. Berti, Spinning and excited black holes in einstein-scalar-gauss–bonnet theory. Class. Quantum Gravity. **37**, 075018 (2020). https://doi.org/10.1088/1361-6382/ab74f9
33. A. Dima, E. Barausse, N. Franchini, T.P. Sotiriou, Spin-induced black hole spontaneous scalarization. Phys. Rev. Lett. **125**, (2020). https://doi.org/10.1103/physrevlett.125.231101
34. D.D. Doneva, L.G. Collodel, C.J. Krüger, S.S. Yazadjiev, Spin-induced scalarization of kerr black holes with a massive scalar field. Eur. Phys. J. C **80** (2020). https://doi.org/10.1140/epjc/s10052-020-08765-3
35. C.A. Herdeiro, E. Radu, H.O. Silva, T.P. Sotiriou, N. Yunes, Spin-induced scalarized black holes. Phys. Rev. Lett. **126**, (2021). https://doi.org/10.1103/physrevlett.126.011103
36. E. Berti, L.G. Collodel, B.Kleihaus, J. Kunz, Spin-induced black hole scalarization in einstein-scalar-gauss-bonnet theory. Phys. Rev. Lett. **126**, (2021). https://doi.org/10.1103/physrevlett.126.011104
37. S.-J. Zhang, B. Wang, A. Wang, J.F. Saavedra, Object picture of scalar field perturbation on kerr black hole in scalar-einstein-gauss-bonnet theory. Phys. Rev. D **102**, (2020). https://doi.org/10.1103/physrevd.102.124056
38. P. Bueno, P.A. Cano, Einsteinian cubic gravity. Phys. Rev. **D94**, 104005 (2016). https://doi.org/10.1103/PhysRevD.94.104005. [arXiv:1607.06463]
39. P. Bueno, P.A. Cano, Four-dimensional black holes in einsteinian cubic gravity. Phys. Rev. D **94**, 124051 (2016). https://doi.org/10.1103/PhysRevD.94.124051

40. R.A. Hennigar, R.B. Mann, Black holes in einsteinian cubic gravity. Phys. Rev. D **95**, 064055 (2017). https://doi.org/10.1103/PhysRevD.95.064055
41. R.A. Hennigar, D. Kubizňák, R.B. Mann, Generalized quasitopological gravity. Phys. Rev. D **95**, 104042 (2017). https://doi.org/10.1103/PhysRevD.95.104042
42. G. Arciniega, J.D. Edelstein, L.G. Jaime, Towards geometric inflation: the cubic case. Phys. Lett. B **802**, 135272 (2020). https://doi.org/10.1016/j.physletb.2020.135272
43. A. Cisterna, N. Grandi, J. Oliva, On four-dimensional einsteinian gravity, quasitopological gravity, cosmology and black holes. Phys. Lett. B **805**, 135435 (2020). https://doi.org/10.1016/j.physletb.2020.135435
44. G. Arciniega, P. Bueno, P.A. Cano, J.D. Edelstein, R.A. Hennigar, L.G. Jaime, Geometric inflation. Phys. Lett. B **802**, 135242 (2020). https://doi.org/10.1016/j.physletb.2020.135242
45. C. Erices, E. Papantonopoulos, E.N. Saridakis, Cosmology in cubic and $f(P)$ gravity. Phys. Rev. D **99**, 123527 (2019). https://doi.org/10.1103/PhysRevD.99.123527. [arXiv:1903.11128]
46. M. Herrero-Valea, The shape of scalar Gauss-Bonnet gravity. JHEP **03**, 075 (2022). https://doi.org/10.1007/JHEP03(2022)075. [arXiv:2106.08344]
47. T. Anson, E. Babichev, C. Charmousis, S. Ramazanov, Cosmological instability of scalar-Gauss-Bonnet theories exhibiting scalarization. JCAP **06**, 023 (2019). https://doi.org/10.1088/1475-7516/2019/06/023. [arXiv:1903.02399]
48. G. Antoniou, L. Bordin, T.P. Sotiriou, Compact object scalarization with general relativity as a cosmic attractor. Phys. Rev. D **103**, 024012 (2021). https://doi.org/10.1103/PhysRevD.103.024012. [arXiv:2004.14985]
49. L. Kofman, A.D. Linde, A.A. Starobinsky, Reheating after inflation. Phys. Rev. Lett. **73**, 3195 (1994). https://doi.org/10.1103/PhysRevLett.73.3195. [arXiv:hep-th/9405187]
50. N. Andreou, N. Franchini, G. Ventagli, T.P. Sotiriou, Spontaneous scalarization in generalised scalar-tensor theory. Phys. Rev. D **99**, 124022 (2019). https://doi.org/10.1103/PhysRevD.99.124022. [arXiv:1904.06365]
51. C.M. Will, The confrontation between general relativity and experiment. Living Rev. Rel. **17**, 4 (2014). https://doi.org/10.12942/lrr-2014-4. [arXiv:1403.7377]
52. T. Damour, G. Esposito-Farese, Tensor multiscalar theories of gravitation. Class. Quantum Gravity. **9**, 2093 (1992). https://doi.org/10.1088/0264-9381/9/9/015
53. T. Anson, E. Babichev, S. Ramazanov, Reconciling spontaneous scalarization with cosmology. Phys. Rev. D **100**, 104051 (2019). https://doi.org/10.1103/PhysRevD.100.104051. [arXiv:1905.10393]
54. E. Palti, The swampland: introduction and review. Fortschr. Phys. **67**, 1900037 (2019). https://doi.org/10.1002/prop.201900037. [arXiv:1903.06239]
55. N. Arkani-Hamed, J.Maldacena, Cosmological collider physics. arXiv:1503.08043
56. A. Adams, N. Arkani-Hamed, S. Dubovsky, A. Nicolis, R. Rattazzi, Causality, analyticity and an IR obstruction to UV completion. JHEP **10**, 014 (2006). https://doi.org/10.1088/1126-6708/2006/10/014. [arXiv:hep-th/0602178]
57. C. Erices, S. Riquelme, In preparation
58. W.F. Buell, B.A. Shadwick, Potentials and bound states. Am. J. Phys. **63**, 256 (1995). https://doi.org/10.1119/1.17935
59. P.D. Meerburg, M. Münchmeyer, J.B. Muñoz, X. Chen, Prospects for cosmological collider physics. JCAP **03**, 050 (2017). https://doi.org/10.1088/1475-7516/2017/03/050. [arXiv:1610.06559]

Conformal Renormalization and Energy Functionals in Anti-De Sitter Gravity

Rodrigo Olea

Abstract Within a holographic framework, we explore the physical consequences of embedding Einstein-AdS gravity in Conformal Gravity in four and six dimensions. In the bulk, the procedure is equivalent to Holographic Renormalization, as the Einstein-AdS action appears augmented by the correct boundary counterterms. In codimension-2 surfaces, 4D Conformal Gravity induces a conformal invariant which, for given conditions on the ambient space and the surface itself, reproduces different functionals: Renormalized Area, Willmore energy and Reduced Hawking Mass.

1 Introduction

One of the key ingredients of the correspondence anti-de Sitter/Conformal Field Theory (AdS/CFT) [1–3] is the existence of a conformal structure at the boundary of asymptotically AdS (AAdS) spacetimes. The boundary field theory, therefore, lives on a curved background metric $g_{(0)ij}$, which is a representative of a conformal class of metrics. Because this tensor is fixed at the conformal boundary, it acts as a source for the holographic reconstruction of the bulk spacetime.

The dynamics of AdS gravity in $d+1$ dimensions is governed by the Einstein–Hilbert action

$$I_{\mathrm{EH}} = \frac{1}{16\pi G} \int\limits_M d^{d+1}x \sqrt{-g}\,(R - 2\Lambda)\,, \tag{1.1}$$

where the cosmological constant Λ is negative and expressed as $\Lambda = -d(d-1)/2\ell^2$ in terms of the AdS radius ℓ, which sets a distance scale in the theory.

Solutions of the theory are geometries which satisfy the Einstein equation

$$G_{\mu\nu} = R_{\mu\nu} - \frac{1}{2}\, g_{\mu\nu}\, R + \Lambda\, g_{\mu\nu}\,. \tag{1.2}$$

R. Olea (✉)
Departamento de Ciencias Físicas, Facultad de Ciencias Exactas, Universidad Andres Bello, Sazié 2212, Piso 7, Santiago, Chile
e-mail: rodrigo.olea@unab.cl

© The Author(s), under exclusive license to Springer Nature Switzerland AG 2024

E. Papantonopoulos and N. Mavromatos (eds.), *Compact Objects in the Universe*,
https://doi.org/10.1007/978-3-031-55098-0_5

and which has the same asymptotic behavior as the corresponding vacuum state, that is, global AdS space.

General properties of AAdS spacetimes were analyzed in detail by Fefferman and Graham in Ref. [4]. As a defining feature, these spaces posses a pole of order two when approaching the conformal boundary. That is the reason why it is useful to describe them in the coordinate frame

$$ds^2 = \frac{\ell^2}{z^2}dz^2 + \frac{1}{z^2}\bar{g}_{ij}(x,z)\,dx^i dx^j\,, \tag{1.3}$$

where the metric $\bar{g}_{ij}(x,z)$ is regular at the boundary $z \to 0$, and gets away from it into the bulk as a power series in the radial coordinate

$$\bar{g}_{ij}(x,z) = g_{(0)ij} + z^2 g_{(2)ij} + \cdots\,, \tag{1.4}$$

where logarithmic terms appear in odd (bulk) dimensions.

The computation of holographic correlators in AdS gravity relies on the asymptotic expansion of the canonical momentum, defined for the radial evolution of the theory, which can be obtained from the modified gravitational action

$$I_{\mathrm{Dir}} = I_{\mathrm{EH}} - \frac{1}{8\pi G}\int\limits_{\partial M} d^d x\sqrt{-h}\,K\,. \tag{1.5}$$

This action results from the addition of the Gibbons–Hawking term, which guarantees a well-posed variational principle for a Dirichlet boundary condition for the metric. In this case, the boundary term is expressed in terms of a normal (spatial) foliation of the spacetime, that is,

$$ds^2 = N^2(z)dz^2 + h_{ij}(z,x)\,dx^i dx^j\,, \tag{1.6}$$

which, in turn, defines the extrinsic curvature as

$$K_{ij} = \frac{1}{2N}\,\partial_z h_{ij}\,. \tag{1.7}$$

Then, the trace of the extrinsic curvature is taken with respect to the the induced metric, as it is a proper tensor of the boundary.

The variation of this new action takes the form

$$\delta I_{\mathrm{Dir}} = \frac{1}{16\pi G}\int\limits_{M} d^{d+1}x G^{\mu\nu}\delta g_{\mu\nu} + \int\limits_{\partial M} d^d x\frac{1}{2}\sqrt{-h}\,\pi^{ij}\delta h_{ij}\,, \tag{1.8}$$

where π^{ij} is the canonical momentum. The fact the momentum can be read off from the surface term in the above variation is a consequence of having eliminated the *acceleration* (second normal derivative of the metric) in the original Lagrangian, by the addition of the Gibbons–Hawking term. This means that the momentum in the definition

$$\pi^{ij} = \frac{2}{\sqrt{-h}} \frac{\delta I_{\text{Dir}}}{\delta h_{ij}} = \frac{1}{8\pi G}\left(K^{ij} - h^{ij} K\right), \tag{1.9}$$

coincides with the one obtained from the variation of a first-order Lagrangian with respect to the normal derivative of the boundary metric, $\partial_z h_{ij}$.

In the Gauss-normal frame, the component of the Einstein equation $G_{iz} = 0$ implies the conservation of the momentum π^{ij} along the boundary directions. An appropriate contraction with an asymptotic Killing vector would allow to construct a conserved current $J^i = \pi^{ij}\xi_j$ which, in turn, leads to a conserved quantity. However, this charge turns out to be divergent for AAdS spacetimes. Indeed, this is closely related to the fact that a well-defined action principle for the metric h_{ij} does not produce a finite variation of the action.

A similar issue is rendered manifest by the FG expansion of the canonical momentum. At holographic order in z, this tensor contains physical information of the gravity theory (e.g., conformal anomaly) which makes possible the connection to a boundary field theory [5, 6]. A proper extraction of such information requires the cancellation of the divergent terms in the infrared sector by the use of a renormalization scheme. In AAdS gravity, this can be achieved by the addition of local counterterms, i.e., covariant functionals of the boundary metric, such that they cancel divergences at the conformal boundary, and they still preserve a Dirichlet boundary condition for the metric h_{ij}. These counterterms are required on top of the Gibbons–Hawking term

$$I_{\text{HR}} = I_{EH} - \frac{1}{8\pi G}\int\limits_{\partial M} d^d x\,\sqrt{-h}\,K + \int\limits_{\partial M} d^d x\, L_{ct}(h, \mathcal{R}, D\mathcal{R})$$

and they depend on the boundary curvature $\mathcal{R}_{ijk\ell}$ and covariant derivative D_i defined for the induced metric [7, 8]. Their form is worked out from the near-boundary resolution of the Einstein equations, in a procedure known as Holographic Renormalization. The first terms in the series are explicitly given by

$$8\pi G\, L_{ct} = \frac{d-1}{\ell}\sqrt{-h} + \frac{\ell\sqrt{-h}}{2(d-2)}\mathcal{R} + \frac{\ell^3\sqrt{-h}}{2(d-2)^2(d-4)}\left(\mathcal{R}^{ij}\mathcal{R}_{ij} - \frac{d}{4(d-1)}\mathcal{R}^2\right) + \cdots \tag{1.10}$$

In doing so, a renormalized AdS action is attained, giving rise to a regulated quasilocal energy-momentum tensor

$$T^{ij} = \pi^{ij} + \frac{2}{\sqrt{-h}}\frac{\delta L_{ct}}{\delta h_{ij}}. \tag{1.11}$$

Once a renormalized Brown–York stress tensor has been constructed, the holographic stress tensor $\langle T^{ij}\rangle$, associated to the holographic source $g_{(0)ij}$, can be read off from the limit in the rescaled tensor

$$\langle T^{ij}\rangle = \lim_{z\to 0}\frac{1}{z^{d-2}}T^{ij}\,, \tag{1.12}$$

or, equivalently,

$$\delta I_{\mathrm{HR}} = \frac{1}{2}\int\limits_{\partial M}\sqrt{-g_{(0)}}\langle T^{ij}\rangle\,\delta g_{(0)ij}\,. \tag{1.13}$$

Holographic Renormalization is a well defined, systematic method to render the action finite. However, it is technically involved to carry out in higher dimensions or higher-curvature gravity. That point is part of the motivation to try a more geometric approach to the problem of renormalization of AdS gravity, as discussed below.

2 Kounterterms in AdS Gravity

For Einstein-AdS spaces the gravity action is proportional to the volume of the spacetime, which is divergent as a consequence of the asymptotic behavior. In the FG frame, infinities in the spacetime volume appear as integer powers in the radial coordinate, but also a logarithmic, anomalous term in even boundary dimensions.

In particular, the definition of thermodynamic quantities out of this bulk functional, for AdS black holes, faces the problem of divergences in the asymptotic region. In order to show this point, one may consider the simple example of the Euclidean version of Schwarzschild-AdS geometries

$$ds^2 = f^2(r)d\tau^2 + \frac{dr^2}{f^2(r)} + r^2 d\Omega^2_{d-1}\,, \tag{2.1}$$

where $\tau = it$ is the Euclidean time and the metric function is given by

$$f^2(r) = 1 - \frac{2\omega GM}{r^{d-2}} + \frac{r^2}{\ell^2}\,. \tag{2.2}$$

Here, ω is a numerical factor that depends on the dimension.

The evaluation of the Euclidean action I^E gives rise to the thermodynamic relation for static black holes, in terms of the Gibbs free energy $G = TI^E$,

$$TI^E_{bulk} = \frac{(d-2)}{(d-1)}M - TS + \lim_{r\to\infty}\frac{V(S^{d-1})}{8\pi G}\frac{r^d}{\ell^2}\,, \tag{2.3}$$

where T is the Hawking temperature, S is the black hole entropy and M is the Hamiltonian mass.

Thus, it is quite evident that the bulk action fails to reproduce the standard formula

$$G = U - TS\,, \tag{2.4}$$

where the internal energy is the conserved quantity associated to invariance under time translations. The unusual factor in the mass in Eq. (2.3) can be corrected by the addition of the Gibbons–Hawking term. However, the Euclidean action is still divergent. Different background-dependent methods, which considers the subtraction of the action evaluated on global AdS space, can be implemented. Even though this procedure leads to a proper thermodynamics in a number of cases, it cannot account for thermodynamic properties of gravitational objects with a nontrivial horizon/boundary topology. Indeed, as a book-keeping counter-example one notices that cannot match the asymptotic geometry of gravitational instantons (Taub-NUT/Bolt AdS) to the one of global AdS.

In the context of AdS/CFT correspondence [1–3], the removal of infrared divergences in the Euclidean action is possible by the addition of a suitable series of counterterms, as discussed above [7, 8]. Counterterms prescribed by the asymptotic resolution of the Einstein equation realizes a renormalization scheme which is background independent. A direct consequence of this fact is the appearance of a vacuum energy for asymptotically AdS black holes, i.e., the internal energy of the thermodynamic system is shifted by a constant $U = M + E_0$ [8]

$$E_0 = (-1)^n \frac{(2n-1)!!^2}{(2n)!} \frac{\mathrm{Vol}(S^{2n-1})}{8\pi G} \ell^{2n-2}\,, \tag{2.5}$$

in $D = 2n + 1$ dimensions, for a spherical transversal section.

From a purely gravitational viewpoint, this vacuum energy does not change the dynamic processes governing the physics of black holes. Nevertheless, within the framework of gauge/gravity duality, the existence of E_0 in the gravity side of the correspondence is interpreted as the Casimir energy of the boundary CFT on a given geometry. In that respect, the vacuum/Casimir energy provides a criterion to select the field theory at the conformal boundary dual to AAdS spaces [7].

However, from a practical point of view, there is no general prescription for the counterterm series in all dimensions. Indeed, as one increases the dimension of the spacetime, more terms are needed for the regularity of the action and its variation. At the same time, there is no criterion to apply, such that the new terms could be guessed in a way.

The previous argument serves as a motivation to explore a different approach to renormalization of AdS gravity, such that holographic correlators can also be computed in an alternative method. Such idea is supported by a few arguments, based on defining features of asymptotically AdS spacetimes.

Counterterm method was originally formulated in terms of a gravity action with Dirichlet boundary conditions for the metric $\delta h_{ij}|_{\partial M} = 0$. However, it was later pointed out that this condition is ill defined at the conformal boundary. As a matter of fact, as it becomes manifest in the FG frame, the expansion of the metric is

$$h_{ij} = \frac{g_{(0)ij}}{z^2} + \cdots , \tag{2.6}$$

what makes both the metric and its variation divergent as $z \to 0$. However, as the holographic description of AAdS spacetimes considers the the metric $g_{(0)ij}$ at the conformal boundary as a source, it is convenient to rethink of the variational problem within that picture. With a holographic boundary condition instead, the infrared divergences induced by the asymptotic behavior of the metric are removed by local counterterms. The addition of counterterms, therefore, is not only required by the renormalization of the variation of the action, but also connected to the variational principle for the metric $g_{(0)ij}$ at the conformal boundary [9].

On the other hand, the pole of order two in z at the conformal boundary is also reflected in the extrinsic curvature. Indeed, as K_{ij} is defined as the normal derivative of the boundary metric, for standard asymptotic AdS behavior, they match at their leading order [10]

$$K_{ij} = \frac{1}{\ell} \frac{g_{(0)ij}}{z^2} + \cdots . \tag{2.7}$$

The introduction of a proper length scale in AdS gravity makes clear that this accident is not longer valid in flat gravity, such that holography in that case is of a different sort.

The simple facts pointed out above give the chance to consider the addition of extrinsic counterterms to the gravitational action

$$I_{\text{ren}} = I_{\text{EH}} + c_d \int\limits_{\partial M} d^d x \sqrt{-h}\, B_d(f(h), K) , \tag{2.8}$$

where c_d is a given coupling. This makes sense as long as the holographic description of the gravity theory is not spoilt. Thus, on top of the fact that the variation of the total action is finite, it has to be expressed in terms of the variation of the holographic source

$$\delta I_{\text{ren}} = \frac{1}{2} \int\limits_{\partial M} d^d x \sqrt{-g_{(0)}}\, \tau^{ij}\, \delta g_{(0)ij} . \tag{2.9}$$

It is worth sounding a note of caution because, at this early stage of the formulation of this alternative renormalization scheme, one cannot guarantee that the new renormalized action would be able to reproduce the same holographic correlators as in the standard procedure.

Another piece of evidence in favor of the existence of extrinsic counterterms comes from a simple notion of energy in AAdS spacetimes, which is known as Conformal Mass [11–13]. The formula by Ashtekar–Magnon–Das considers a conserved charge with a linear dependence on the electric part of the Weyl tensor $\mathcal{E}^i_{\ j}$, which is obtained as the double contraction with a normal vector to the boundary n_μ, i.e.,

$$\mathcal{E}^i_j = W^{i\mu}{}_{j\nu} n_\mu n^\nu \,. \tag{2.10}$$

The charge is suitably expressed as a surface integral

$$Q_{\text{AMD}}[\xi] = \frac{1}{8\pi G} \frac{\ell}{d-2} \int\limits_{\Sigma} d^{d-1}x \sqrt{\sigma}\, u_i \, \mathcal{E}^i_j \, \xi^j \,, \tag{2.11}$$

where Σ is a surface which appears as the intersection between a constant-time slice and a $r = const$ cylinder. The geometry of this codimension-2 region is given by the metric σ_{mn}. The above definition also involves a timelike unit vector u_i and a set of boundary Killing vectors $\{\xi^i\}$.

Physical properties as mass and angular momentum of asymptotically AdS black holes can be readily obtained from the AMD expression. Therefore, it is natural to think that the electric part of the Weyl tensor differs from the quasilocal stress tensor by subleading terms which produce the Weyl anomaly. Indeed, while the tensor $\mathcal{E}^i_j$ is traceless, the trace of the stress tensor is nonvanishing at the holographic order

$$T^i_i = z^d \mathcal{A} \,, \tag{2.12}$$

in even-dimensional boundaries. As a consequence, it is reasonable to expect that, on general grounds, T^i_j will always differ from $\mathcal{E}^i_j$ by a holographic term $\Delta^i_{\ j}$, whose trace gives rise to the conformal anomaly. This idea has a concrete realization in the case of five spacetime dimensions [12], such that

$$T^i_j = \mathcal{E}^i_j + \Delta^i_{\ j} \,, \quad \Delta^i_i = z^4 \mathcal{A} \,. \tag{2.13}$$

In other words, the electric part of the Weyl tensor is, to certain extent, equivalent to the quasilocal stress tensor. This fact has conceptual implications, which are discussed in detail below.

One may start by noticing that the general expression for the Weyl tensor

$$W^{\alpha\beta}_{\mu\nu} = R^{\alpha\beta}_{\mu\nu} - S^\alpha_\mu \delta^\beta_\nu + S^\beta_\mu \delta^\alpha_\nu - S^\beta_\nu \delta^\alpha_\mu + S^\alpha_\nu \delta^\beta_\mu \,, \tag{2.14}$$

in terms of the Schouten tensor

$$S^\alpha_\mu = \frac{1}{d-1} \left(R^\alpha_\mu - \frac{1}{2d} \delta^\alpha_\mu R \right) \,, \tag{2.15}$$

takes a much simpler form in Einstein gravity ($R_{\mu\nu} = -d/\ell^2 g_{\mu\nu}$)

$$W^{\alpha\beta}_{(E)\mu\nu} = R^{\alpha\beta}_{\mu\nu} + \frac{1}{\ell^2}\,\delta^{\alpha\beta}_{\mu\nu}\,. \tag{2.16}$$

Then, the electric part of the Weyl tensor can be conveniently rewritten as

$$\mathcal{E}^i_j = -\left(R^{ik}{}_{jk} + \frac{d-1}{\ell^2}\,\delta^i_j\right), \tag{2.17}$$

by virtue of the tracelessness property of W, i.e.,

$$\mathcal{E}^i_j = W^{iz}{}_{jz} = -W^{ik}{}_{jk}\,. \tag{2.18}$$

Somehow unexpectedly, the stress tensor can be re-expressed in such way that contains nonlinear terms in the extrinsic curvature

$$T^i_j = -\left(\mathcal{R}^i_j - K^i_j K + K^i_k K^k_j + \frac{3}{\ell^2}\delta^i_j\right) + \Delta^i_j\,, \tag{2.19}$$

where the Gauss–Codazzi relation for the boundary components of the Riemann tensor has been used. This may sound contradicting because the quasilocal stress tensor (1.11) contains only linear terms in K_{ij} which, in turn, come from the canonical momentum (1.9). The rest of the contributions are purely intrinsic as they are derived from the counterterm series. The only possibility to reconcile the expressions for the stress tensor (1.11) and (2.19) is to consider the asymptotic expansion of the fields involved in AdS gravity. In doing so, one should be able to truncate the formula (2.19) to the linear in K_{ij} without loss of information on the conserved quantities of AAdS solutions.

Then, if one believes that nonlinear terms in the extrinsic curvature in the stress tensor has an origin in boundary terms which are nonlinear in K_{ij}, one has to depart from a renormalization scheme based on a Dirichlet boundary condition in h_{ij}. Such argument provides support to a renormalization scheme of AdS gravity based on the addition of extrinsic counterterms (Kounterterms).

In even-dimensional AdS gravity ($d+1=2n$), the Kounterterms take a compact form, as a consequence of the use of a parametric integral [14]

$$\begin{aligned} B_{2n-1} = 2n\int_0^1 & dt\,\delta^{j_1\cdots j_{2n-1}}_{i_1\cdots i_{2n-1}}\,K^{i_1}_{j_1}\left(\frac{1}{2}\mathcal{R}^{i_2 i_3}_{j_2 j_3} - t^2 K^{i_2}_{j_2}K^{i_3}_{j_3}\right)\times\cdots \\ & \cdots\times\left(\frac{1}{2}\mathcal{R}^{i_{2n-2}\,i_{2n-1}}_{j_{2n-2}\,j_{2n-1}} - t^2 K^{i_{2n-2}}_{j_{2n-2}}K^{i_{2n-1}}_{j_{2n-1}}\right), \end{aligned} \tag{2.20}$$

whose coupling is chosen such that it guarantees the removal of the highest-order divergences in the action

$$c_{2n-1} = -\frac{(-\ell^2)^{n-1}}{16\pi G n(2n-2)!}\,. \tag{2.21}$$

On the other hand, the Kounterterms which renormalize the gravity action in odd spacetime dimensions ($d+1 = 2n+1$) adopt a slightly different form, in terms of a double-parametric integral [15]

$$B_{2n} = 2n\sqrt{-h}\int_0^1 dt \int_0^t ds \quad \delta^{j_1\cdots j_{2n}}_{i_1\cdots i_{2n}} K^{i_1}_{j_1}\delta^{i_2}_{j_2}\left(\frac{1}{2}\mathcal{R}^{i_3 i_4}_{j_3 j_4} - t^2 K^{i_3}_{j_3}K^{i_4}_{j_4} + \frac{s^2}{\ell^2}\delta^{i_3}_{j_3}\delta^{i_4}_{j_4}\right)\times\cdots$$
$$\cdots\times\left(\frac{1}{2}\mathcal{R}^{i_{2n-1} i_{2n}}_{j_{2n-1} j_{2n}} - t^2 K^{i_{2n-1}}_{j_{2n-1}}K^{i_{2n}}_{j_{2n}} + \frac{s^2}{\ell^2}\delta^{i_{2n-1}}_{j_{2n-1}}\delta^{i_{2n}}_{j_{2n}}\right)\,. \tag{2.22}$$

The coupling of this boundary structure is singled out once again by the vanishing of the variation of the total action at leading order, that is,

$$c_{2n} = \frac{(-\ell^2)^{n-1}}{2^{2n+2}\pi G n(n-1)!^2}\,. \tag{2.23}$$

In even bulk dimensions, the Kounterterms are proportional to the nth Chern form, what is a Chern–Simons density for the Lorentz group. The parametric integral comes from the representation of the homotopy operator, which allows to find the explicit form of the Chern–Simons form L^{CS}_{2n-1} in $2n-1$ dimensions, from the defining relation [16]

$$\mathrm{tr}(F^n) = dL^{CS}_{2n-1}(A)\,. \tag{2.24}$$

The left-hand side in the above formula is a topological term for a Lie group, in terms of an invariant tensor defined for the corresponding generators and the field strength $F = dA + A\wedge A$, where A is a Lie algebra-valued connection 1-form. In doing so, the Chern–Simons density whose exterior derivative is the topological term takes the form

$$L^{CS}_{2n+1}(A) = n\int_0^1 dt\,\mathrm{tr}\left[AF_t^n\right]\,, \tag{2.25}$$

where F_t is the curvature associated to a connection $A_t = tA$, trivially interpolating between 0 and A. After writing previous relations for the particular case of the Lorentz group and taking global issues (topology) into consideration, one ends up with the Euler theorem

$$\int_{M_{2n}} \mathcal{E}_{2n} = (4\pi)^n n!\chi(M_{2n}) + \int_{\partial M_{2n}} B_{2n-1}\,. \tag{2.26}$$

This implies that, by virtue of the Euler theorem, the Kounterterms can be traded off by the Euler term in the bulk,

$$\mathcal{E}_{2n} = \frac{1}{2^n}\delta^{\mu_1\ldots\mu_{2n}}_{\nu_1\ldots\nu_{2n}} R^{\nu_1\nu_2}_{\mu_1\mu_2}\ldots R^{\nu_{2n-1}\nu_{2n}}_{\mu_{2n-1}\mu_{2n}}\,. \tag{2.27}$$

up to the Euler characteristic of the manifold.[1]

In sharp contrast, in odd bulk dimensions, the expression for the Kounterterms (2.22) is not a standard structure in the mathematical literature. While this object was originally constructed on a rather case by case basis, it possesses remarkable properties of interest. Even though one cannot provide a cleaner derivation of it, one may suggest that the presence of a double parametric integral resembles the construction of Transgression Forms, which are gauge-invariant extensions of Chern–Simons densities

$$L^{TF}_{2n+1}(A,\bar{A}) = (n+1)\int\limits_0^1 dt\,\mathrm{tr}\left[\left(A-\bar{A}\right)F_t^n\right]. \tag{2.28}$$

Transgression Forms accommodate another connection $\bar{A}$, such that the interpolating gauge field adopts the expression $A_t = tA + (1-t)\bar{A}$. Such density can be suitably rewritten as the difference between two Chern–Simons forms (associated to each connection) and a contact term at the boundary

$$L^{TF}_{2n+1}(A,\bar{A}) = L^{CS}_{2n+1}(A) - L^{CS}_{2n+1}(\bar{A}) + d\beta_{2n}(A,\bar{A})\,. \tag{2.29}$$

The contact term $\beta_{2n}(A,\bar{A})$ is an even-form, which couples in a nontrivial manner both connection fields and is given by the formula

$$\beta_{2n}(A,\bar{A}) = \int\limits_0^1 dt \int\limits_0^t ds\,\mathrm{tr}\left[A_t\left(A-\bar{A}\right)F_{st}^{n-1}\right], \tag{2.30}$$

where the field strength F_{st} is the complicated interpolation $F_{st} = sF_t + s(s-1)A_t^2$.

What is remarkable of having an expression for the extrinsic counterterms in all spacetime dimensions is that one can perform generic computations of the Euclidean action for any d. In point of fact, one may prove that the evaluation of the Euclidean version of the Kounterterms for Schwarzschild-AdS black holes produces

$$T\,c_d\int\limits_{\partial M} B_d = \frac{M}{(D-2)} + E_0 - \lim_{r\to\infty}\frac{V(S^{D-2})}{8\pi G}\frac{r^d}{\ell^2}, \tag{2.31}$$

that is, the missing fraction in the mass in Eq. (2.3) and exactly the same divergent term, with the opposite sign. Thus, Kounterterms reproduce the correct black hole thermodynamics in any dimension in a background-independent fashion.

---

[1] Conventions and properties of the Kronecker symbol of order p are given in "Appendix 1: Useful Identities".

3 Counterterms Versus Kounterterms: A Comparison

It is worthwhile to notice that, in the expression for B_{2n-1}, there is no linear term in the extrinsic curvature. This is, in a way, a reflection of a different guiding principle to construct the counterterm series. A similar situation occurs for B_{2n}, where there is indeed a term linear in K, but that it does not carry the same factor as the Gibbons–Hawking term. Therefore, it is inevitable to have variations of the extrinsic curvature in the surface term when the total action is varied. However, the extrinsic counterterms can be compared directly with the standard ones, by adding and substracting the Gibbons–Hawking term. In what follows, the procedure is illustrated in the simple example of four spacetime dimensions.

In that case, the AdS gravity action, renormalized by the addition of Kounterterms

$$I_{\text{ren}} = I_{\text{EH}} + \frac{\ell^2}{16\pi G}\int\limits_{\partial M} d^3x\,\sqrt{-h}\,\delta^{i_1 i_2 i_3}_{j_1 j_2 j_3}\,K^{j_1}_{i_1}\left(\frac{1}{2}\,\mathcal{R}^{j_2 j_3}_{i_2 i_3}(h) - \frac{1}{3}\,K^{j_2}_{i_2}K^{j_3}_{i_3}\right), \tag{3.1}$$

what can be rendered in a more standard form by adding zero in a clever way

$$I_{\text{ren}} = I_{EH} - \frac{1}{8\pi G}\int\limits_{\partial M} d^3x\sqrt{-h}\,K + \int\limits_{\partial M} d^3x\,L_{ct}\,, \tag{3.2}$$

where the second contribution at the boundary is written as

$$L_{\text{ct}} = \frac{\ell^2}{16\pi G}\sqrt{-h}\delta^{i_1 i_2 i_3}_{j_1 j_2 j_3}\,K^{j_1}_{i_1}\left(\frac{1}{2}\,\mathcal{R}^{j_2 j_3}_{i_2 i_3}(h) - \frac{1}{3}\,K^{j_2}_{i_2}K^{j_3}_{i_3} + \frac{1}{\ell^2}\,\delta^{j_2}_{i_2}\delta^{j_3}_{i_3}\right). \tag{3.3}$$

The use of the notation L_{ct} suggests that this expression has something to do with the counterterm series, though it is still far from obvious what is the actual relation. It is therefore necessary to convert extrinsic terms into intrinsic ones, what can be made by considering the Fefferman–Graham expansion in the extrinsic curvature

$$K^i_j = \frac{1}{\ell}\delta^i_j - \ell S^i_j(h) + \mathcal{O}(\mathcal{R}^2)\,. \tag{3.4}$$

Here, $S^i_j(h)$ is the Schouten tensor of the boundary metric

$$S^i_j(h) = \frac{1}{d-2}(\mathcal{R}^i_j(h) - \frac{1}{2(d-1)}\delta^i_j\mathcal{R}(h))\,. \tag{3.5}$$

The asymptotic form of the extrinsic curvature can be expressed as a power-series expansion. When plugged in Eq. (3.3), the strategy is collecting terms at different orders in the holographic coordinate z from the

$$L_{ct} = \frac{\ell^2}{16\pi G}\frac{\sqrt{-g}}{z^3}\,\delta^{i_1i_2i_3}_{j_1j_2j_3}\left(\frac{\delta^{j_1}_{i_1}}{\ell} - \ell S^{i_1}_{j_1}\right)\times$$
$$\times\left(\frac{1}{2}\mathcal{R}^{j_2j_3}_{i_2i_3}(h) - \frac{1}{3}\left(\frac{\delta^{j_2}_{i_2}}{\ell} - \ell S^{i_2}_{j_2}\right)\left(\frac{\delta^{j_3}_{i_3}}{\ell} - \ell S^{i_3}_{j_3}\right) + \frac{1}{\ell^2}\delta^{j_2}_{i_2}\delta^{j_3}_{i_3}\right) + \cdots$$

In doing so, the suitable expansion of Kounterterms turn into the standard series of intrinsic counterterms [17]

$$L_{\text{ct}} = \frac{1}{8\pi G}\frac{\sqrt{-g}}{z^3}\left(\frac{2}{\ell} + \frac{\ell}{2}z^2\mathcal{R}(g)\right) + \mathcal{O}(z)\,, \tag{3.6}$$
$$= \frac{1}{8\pi G}\sqrt{-h}\left(\frac{2}{\ell} + \frac{\ell}{2}\mathcal{R}(h)\right). \tag{3.7}$$

The approach in higher even dimensions considers a procedure analog to the four-dimensional case. The corresponding Einstein-AdS action plus Kounterterms looks like

$$I_{\text{ren}} = I_{\text{EH}} + c_{2n-1}\int\limits_{\partial M} d^dx\,\sqrt{-h}\,B_{2n-1}\,. \tag{3.8}$$

The proper addition and subsequent substraction of the Gibbons–Hawking term at the boundary produces

$$I_{\text{ren}} = I_{\text{Dir}} + \int\limits_{\partial M} d^{2n-1}x\,L_{\text{ct}}\,, \tag{3.9}$$

where the last term is written as

$$L_{\text{ct}} = \sqrt{-h}\left(c_{2n-1}B_{2n-1} + \frac{1}{8\pi G}K\right). \tag{3.10}$$

The above expression can be recast into the equivalent form

$$L_{\text{ct}} = \frac{(-\ell^2)^n}{8\pi G(2n-2)!}\sqrt{-h}\,\delta^{i_1\cdots i_{2n-1}}_{j_1\cdots j_{2n-1}}\int\limits_0^1 dt\,K^{j_1}_{i_1}\left[\left(\frac{1}{2}\mathcal{R}^{j_2j_3}_{i_2i_3} - t^2K^{j_2}_{i_2}K^{j_3}_{i_3}\right)\times\cdots\right.$$
$$\left.\cdots\times\left(\frac{1}{2}\mathcal{R}^{j_{2n-2}j_{2n-1}}_{i_{2n-2}i_{2n-1}} - t^2K^{j_{2n-2}}_{i_{2n-2}}K^{j_{2n-1}}_{i_{2n-1}}\right) + \frac{(-1)^n}{\ell^{2n-2}}\delta^{j_2}_{i_2}\cdots\delta^{j_{2n-1}}_{i_{2n-1}}\right], \tag{3.11}$$

such that one can expand the extrinsic curvature as in Eq. (3.4). The expansion of this polynomial in the extrinsic curvature and further collection of terms of different orders in the boundary curvature leads to the series

$$L_{\mathrm{ct}} = \frac{\sqrt{-h}}{8\pi G}\left[\frac{(2n-2)}{\ell} + \frac{\ell}{2(2n-3)}\mathcal{R}+ \right.$$
$$\left. + \frac{\ell^3}{2(2n-3)^2(2n-5)}\left(2\mathcal{R}^{ij}\mathcal{R}_{ij} - \frac{(2n+1)}{4(2n-2)}\mathcal{R}^2 - \frac{(2n-3)}{4}\mathcal{R}^{ijkl}\mathcal{R}_{ijkl}\right) + \cdots\right], \quad (3.12)$$

which contains a term Rie^2 that is rather unusual. As a matter of fact, the asymptotic resolution of the Einstein equation can never fix the Riemann tensor at the boundary, such that quadratic contributions in this object are not present in the standard counterterms. The only way to reconcile the current expansion with holographic methods is to trade off the Rie^2 term by the other curvature-squared invariants

$$\mathcal{R}^{ijkl}\mathcal{R}_{ijkl} = \mathcal{W}^{ijkl}\mathcal{W}_{ijkl} + \frac{4}{(2n-3)}\left(\mathcal{R}^{ij}\mathcal{R}_{ij} - \frac{1}{2(2n-2)}\mathcal{R}^2\right), \quad (3.13)$$

where $\mathcal{W}_{ijkl} = \mathcal{W}_{ijkl}(h)$ is the Weyl tensor of the boundary metric. The resulting expression matches the standard counterterms series for even bulk dimensional AdS gravity

$$L_{\mathrm{ct}} = \frac{\sqrt{-h}}{8\pi G}\left[\frac{(2n-2)}{\ell} + \frac{\ell}{2(2n-3)}\mathcal{R} + \frac{\ell^3}{2(2n-3)^2(2n-5)}\left(\mathcal{R}^{ij}\mathcal{R}_{ij} - \frac{(2n-1)}{4(2n-2)}\mathcal{R}^2\right)\right.$$
$$\left. - \frac{\ell^3}{8(2n-3)(2n-5)}\mathcal{W}^{ijkl}\mathcal{W}_{ijkl} + \cdots\right]. \quad (3.14)$$

So, faced with the question whether Kounterterms can provide the same renormalization as the one prescribed by holographic techniques, the honest answer is that there is a mismatch which, in a way, relies on conformal properties of the boundary. This mismatch is a series of higher curvature/derivative terms, whose first nonvanishing contribution is of the $Weyl^2$ type [18]

$$I_{\mathrm{ren}} = I_{\mathrm{HR}} - \frac{\ell^3}{64\pi G(2n-3)(2n-5)}\int\limits_{\partial M}\sqrt{-h}\,\mathcal{W}^{ijkl}\mathcal{W}_{ijkl} + \cdots. \quad (3.15)$$

A similar result is obtained for odd spacetime dimensions, as well.

The last term in Eq. (3.15) is zero for a large class of AAdS spaces, what explains the fact the addition of Kounterterms can correctly account for thermodynamics of static topological black holes, what was discussed above. However, for a high enough dimension, the extra terms are divergent in the holographic coordinate z and thus, potentially risky for the evaluation of the action and its variation. In fact, that is the case of gravitational instantons, whose boundary geometry leaves them out of the category of asymptotically conformally flat solutions in Einstein gravity. One could simply adopt the alternative of patching up the gravitational theory by considering additional counterterms which cancel the mismatching terms. Instead, there seems to be a more fundamental form to address the problem: embedding Einstein gravity in Conformal Gravity.

4 Beyond Kounterterms: Conformal Renormalization

In the previous section, it was pointed out the fact that the agreement between the addition of Kounterterms and Holographic Renormalization was possible for AAdS spaces with conformally flat boundary. In that case, a formula for the adequate counterterms can be readily worked out in any dimension [18].

In four spacetime dimensions, one can derive further properties of the AdS action, properly renormalized by the use of Kounterterms. Indeed, employing the Euler–Gauss–Bonnet theorem

$$\int\limits_{\partial M} d^3x\,\sqrt{-h}\,B_3 = \int\limits_{M} d^4x\,GB - 32\pi^2\chi(M)\,, \tag{4.1}$$

where GB stands for the Gauss–Bonnet term

$$GB = \sqrt{-g}\left(R_{\mu\nu\sigma\lambda}R^{\mu\nu\sigma\lambda} - 4R_{\mu\nu}R^{\mu\nu} + R^2\right)\,, \tag{4.2}$$

one is able to write down boundary, extrinsic counterterms as a topological term in the bulk. The resulting action takes the form which was originally used to derive the fineteness of the conserved quantities in AdS gravity [19] by the Noether–Wald method [20]

$$I_{\text{ren}} = \frac{1}{16\pi G}\int\limits_{M} d^4x\sqrt{-g}\left[(R-2\Lambda) + \frac{\ell^2}{4}(R_{\mu\nu\sigma\lambda}R^{\mu\nu\sigma\lambda} - 4R_{\mu\nu}R^{\mu\nu} + R^2)\right] - \frac{\pi\ell^2}{2G}\chi(M)\,, \tag{4.3}$$

up to a term proportional to the Euler characteristic of the manifold. It is quite remarkable the fact that the renormalized action matches the expression by MacDowell and Mansouri in Ref. [21]

$$I_{\text{ren}} = \frac{\ell^2}{256\pi G}\int\limits_{M} d^4x\sqrt{-g}\,\delta^{\mu_1\cdots\mu_4}_{\nu_1\cdots\nu_4}\left(R^{\nu_1\nu_2}_{\mu_1\mu_2} + \frac{\delta^{\nu_1\nu_2}_{\mu_1\mu_2}}{\ell^2}\right)\left(R^{\nu_3\nu_4}_{\mu_3\mu_4} + \frac{\delta^{\nu_3\nu_4}_{\mu_3\mu_4}}{\ell^2}\right) - \frac{\pi\ell^2}{2G}\chi(M)\,, \tag{4.4}$$

which, in turn, for Einstein spaces can be cast into the form

$$I_{\text{ren}} = \frac{\ell^2}{64\pi G}\int\limits_{M} d^4x\,\sqrt{-g}\,W_{(E)\mu\nu\sigma\lambda}W^{\mu\nu\sigma\lambda}_{(E)} - \frac{\pi\ell^2}{2G}\chi(M)\,. \tag{4.5}$$

The latter fact suggests a connection to Conformal Gravity, which can be realized within a holographic context. At this point, the concept of Conformal Renormalization is introduced as a scheme which supersedes the addition of Kounterterms by a proper embedding of Einstein theory in Conformal Gravity. In four dimensions, this is justified by the fact that Einstein spaces form a well-defined sector of this higher-derivative gravity. The perspective that this framework provides the renormalization of the action is inspired by the fact that the action of Conformal Gravity is finite for general AAdS conditions [22]. In this line of reasoning, one may also expect that the renormalization property would be inherited by the Einstein sector.

Furthermore, in Ref. [23], it is proposed a concrete, holographic mechanism which allows to pass from a generic solution of CG to one of Einstein gravity. In fact, at the level of the asymptotic expansion for the metric $\bar{g}_{ij}(x, z)$ in Eq. (1.4) features a mode linear in z

$$\bar{g}_{ij}(x, \rho) = g_{(0)ij}(x) + z g_{(1)ij}(x) + z^2 g_{(2)ij}(x) + \cdots . \tag{4.6}$$

Therefore, the Fefferman–Graham expansion for AAdS spaces in CG contains a new holographic source, as a consequence of field equations which are of higher derivatives in the metric tensor. By switching off this new mode, one goes back to the standard holographic description of Einstein spaces. This can be made consistent by taking Neumann boundary conditions on the metric at the conformal boundary.

The action of Conformal Gravity considers an arbitrary coupling and a Weyl-squared Lagrangian, i.e.,

$$I_{CG} = \alpha \int\limits_{M} d^4x \sqrt{-g}\, W_{\mu\nu\sigma\lambda} W^{\mu\nu\sigma\lambda} , \tag{4.7}$$

as it is the integral of the only conformal invariant in four spacetime dimensions.

The corresponding equations of motion for the theory are given in terms of the Bach tensor

$$B_{\mu\nu} = \nabla^{\lambda} C_{\mu\nu\lambda} + S^{\lambda\sigma} W_{\lambda\mu\sigma\nu} = 0 , \tag{4.8}$$

which, in turn, is defined in terms of the Cotton tensor

$$C^{\mu}_{\ \nu\lambda} = \nabla_{\nu} S^{\mu}_{\lambda} - \nabla_{\lambda} S^{\mu}_{\nu} , \tag{4.9}$$

which is a third-derivative object. It is clear that the Einstein condition, expressed as $S^{\mu}_{\nu} = -\frac{1}{2\ell^2}\delta^{\mu}_{\nu}$ produces a vanishing Cotton $C_{\mu\nu\lambda} = 0$ and, therefore, any Einstein space satisfies $B_{\mu\nu} = 0$. The holographic boundary conditions which allows to obtain Einstein gravity from CG can be rewritten in terms of the traceless Ricci tensor

$$H^{\mu}_{\nu} = R^{\mu}_{\nu} - \frac{1}{D} R \delta^{\mu}_{\nu} = 0 , \tag{4.10}$$

which is a defining feature of any Einstein space. The vanishing of this tensor $\partial_z \bar{g}_{ij} = g_{(1)ij} = 0$, which is the Neumann condition discussed in Ref. [23], but also the cancellation of the stress tensor associated to the new holographic source $g_{(1)ij}$. Due to the fact the field equations are of higher order in the normal derivative, the coefficient $g_{(2)ij}$ appears as free data in CG. The kinematic approach developed in Ref. [24], based on asymptotic symmetries which preserve the FG form of the metric, implies that this mode can be chosen as $g_{(2)ij} = -\ell^2 S_{(0)ij}$, as in Einstein gravity. All of the above implies that, when the CG action is evaluated on Einstein spaces, it recovers the MacDowell–Mansouri form of AdS gravity (4.4) for $\alpha = \ell^2/(64\pi G)$. Therefore, the action includes the boundary counterterms prescribed by Holographic Renormalization, that is,

$$I_{\text{CG}}\left[E\right] = I_{\text{HR}} \,. \tag{4.11}$$

In six spacetime dimensions, the addition of Kounterterms at the boundary can be traded off by a bulk topological term, by virtue of the Euler theorem

$$\int\limits_{M} d^6x\sqrt{-g}\,\mathcal{E}_6 = 3!\,(4\pi)^3\,\chi\left(M\right) + \int\limits_{\partial M} d^5x\sqrt{-h}\,B_5\,, \tag{4.12}$$

where the Euler term $\mathcal{E}_6$ is defined as

$$\mathcal{E}_6 = \frac{1}{2^n}\delta^{\mu_1\ldots\mu_6}_{\nu_1\ldots\nu_6}R^{\nu_1\nu_2}_{\mu_1\mu_2}\cdots R^{\nu_5\nu_6}_{\mu_5\mu_6}\,. \tag{4.13}$$

Thus, the equivalent form of the AdS action considers its renormalization by the inclusion of the Euler topological term

$$I_{\text{ren}} = \frac{1}{16\pi G}\int\limits_{M} d^6x\sqrt{-g}\left(R + \frac{20}{\ell^2} - \frac{\ell^4}{72}\mathcal{E}_6\right), \tag{4.14}$$

up to the Euler characteristic, for a suitable coupling fixed such that the action vanishes for global AdS space. As pointed out in the previous section, there is a mismatch between this notion of renormalized action and the one obtained in the standard AdS/CFT framework.

The first piece of evidence which hints at a connection between renormalized AdS action (with topological terms) and Conformal Gravity in six dimensions, is the fact the former can be appropriately cast as a polynomial of the Weyl tensor for Einstein spaces [25, 26]

$$I_{\text{ren}} = \frac{\ell^4}{16\pi G}\int\limits_{M} d^6x\sqrt{-g}\,P_6\left(W_{(\text{E})}\right), \tag{4.15}$$

where $P_6\left(W_{(\text{E})}\right)$ considers quadratic and cubic contractions of $W_{(E)}$, i.e.,

$$P_6(W_{(\text{E})}) = \frac{1}{2\,(4!)\,\ell^2}\delta^{\nu_1\ldots\nu_4}_{\mu_1\ldots\mu_4}W^{\mu_1\mu_2}_{(E)\nu_1\nu_2}W^{\mu_3\mu_4}_{(E)\nu_3\nu_4} - \frac{1}{(4!)^2}\delta^{\nu_1\ldots\nu_6}_{\mu_1\ldots\mu_6}W^{\mu_1\mu_2}_{(E)\nu_1\nu_2}W^{\mu_3\mu_4}_{(E)\nu_3\nu_4}W^{\mu_5\mu_6}_{(E)\nu_5\nu_6}\,. \tag{4.16}$$

On the other hand, Conformal Gravity in six dimensions is expressed as a linear combination of the basis of conformal invariants

$$I_1 = W_{\alpha\beta\mu\nu} W^{\alpha\sigma\lambda\nu} W_{\sigma}{}^{\beta\mu}{}_{\lambda} , \tag{4.17}$$

$$I_2 = W_{\mu\nu\alpha\beta} W^{\alpha\beta\sigma\lambda} W_{\sigma\lambda}{}^{\mu\nu} , \tag{4.18}$$

$$I_3 = W_{\mu\rho\sigma\lambda} \left(\delta^{\mu}_{\nu} \Box + 4R^{\mu}_{\nu} - \frac{6}{5} R \delta^{\mu}_{\nu} \right) W^{\nu\rho\sigma\lambda} + \nabla_{\mu} V^{\mu} , \tag{4.19}$$

with

$$\begin{aligned} V_{\mu} = {} & 4R_{\mu}{}^{\lambda\rho\sigma} \nabla^{\nu} R_{\nu\lambda\rho\sigma} + 3R^{\nu\lambda\rho\sigma} \nabla_{\mu} R_{\nu\lambda\rho\sigma} - 5R^{\nu\lambda} \nabla_{\mu} R_{\nu\lambda} \\ & + \frac{1}{2} R \nabla_{\mu} R - R^{\nu}_{\mu} \nabla_{\nu} R + 2R^{\nu\lambda} \nabla_{\nu} R_{\lambda\mu} , \end{aligned} \tag{4.20}$$

where each one is a six-derivative object.

An adequate sum of these three invariants, in terms of a single coupling α, and the extensive use of Bianchi identity, led to Lu, Pang and Pope to a particular CG action [27]

$$I_{\text{LPP}} = \int\limits_{M} d^6x \sqrt{-g}\, \mathcal{L}_{\text{LPP}} + \int\limits_{\partial M} d^5x \sqrt{-h}\, n_{\mu} J^{\mu} , \tag{4.21}$$

where the Lagrangian density and the corresponding surface term are given by

$$\mathcal{L}_{\text{LPP}} = \alpha \left(\frac{1}{4!} \delta^{\nu_1 \ldots \nu_6}_{\mu_1 \ldots \mu_6} W^{\mu_1\mu_2}_{\nu_1\nu_2} W^{\mu_3\mu_4}_{\nu_3\nu_4} W^{\mu_5\mu_6}_{\nu_5\nu_6} + \frac{1}{2} \delta^{\nu_1 \ldots \nu_5}_{\mu_1 \ldots \mu_5} W^{\mu_1\mu_2}_{\nu_1\nu_2} W^{\mu_3\mu_4}_{\nu_3\nu_4} S^{\mu_5}_{\nu_5} + 8C^{\mu\nu\lambda} C_{\mu\nu\lambda} \right) , \tag{4.22}$$

$$J^{\mu} = \alpha \left(8W^{\mu\kappa\lambda\nu} C_{\kappa\lambda\nu} - W^{\kappa\lambda}_{\nu\sigma} \nabla^{\mu} W^{\nu\sigma}_{\kappa\lambda} \right) . \tag{4.23}$$

In that reference, the authors managed to prove that the theory admits Schwarzschild-AdS black hole as a solution of the field equations. Later, it was shown that the whole class of Einstein spaces defines a proper sector of the space of solutions of LPP Conformal Gravity [28]. This is possible by expressing the field equations in terms of the Weyl, Cotton and Schouten tensors.

As a hand-waving argument, one would be able to justify the existence of such CG action by the conformal completion of the Einstein action supplemented with a topological term. This conformal covariantization procedure, loosely defined yet, is schematically shown in the following table. This idea might shed some light on a generalization to higher even dimensions of conformal gravity actions which contain an Einstein sector.

Conformal Covariantization		
Einstein	Conformal	CI's
$\delta^{\nu_1\cdots\nu_6}_{\mu_1\cdots\mu_6} W^{\mu_1\mu_2}_{(E)\nu_1\nu_2} W^{\mu_3\mu_4}_{(E)\nu_3\nu_4} W^{\mu_5\mu_6}_{(E)\nu_5\nu_6}$	$\delta^{\nu_1\cdots\nu_6}_{\mu_1\cdots\mu_6} W^{\mu_1\mu_2}_{\nu_1\nu_2} W^{\mu_3\mu_4}_{\nu_3\nu_4} W^{\mu_5\mu_6}_{\nu_5\nu_6}$	$32(2I_1+I_2)$
$-\frac{1}{\ell^2}\delta^{\nu_1\cdots\nu_4}_{\mu_1\cdots\mu_4} W^{\mu_1\mu_2}_{(E)\nu_1\nu_2} W^{\mu_3\mu_4}_{(E)\nu_3\nu_4}$	$\delta^{\nu_1\cdots\nu_5}_{\mu_1\cdots\mu_5} W^{\mu_1\mu_2}_{\nu_1\nu_2} W^{\mu_3\mu_4}_{\nu_3\nu_4} S^{\mu_5}_{\nu_5} +$ $16C^{\mu\nu\lambda}C_{\mu\nu\lambda} + \nabla^{\mu} J_{\mu}$	$4I_1 - I_2 -$ I_3

It is important to stress that the WWS term in Eq. (4.22) appears supplemented by a $Cotton^2$ contribution and a surface term, as dictated by conformal invariance. Then, taking the Einstein condition in LPP Conformal Gravity at the level of the action produces not only the polynomial in the Weyl tensor $W_{(E)}$, but also a boundary term

$$I_{\mathrm{LPP}}[E] = -4!\alpha \int\limits_M d^6x\sqrt{-g}\, P_6\left(W_{(E)}\right) - \alpha \int\limits_{\partial M} d^5x\sqrt{-h}\, n^{\mu}\left(W^{\kappa\lambda}_{(E)\nu\sigma}\nabla_{\mu} W^{\nu\sigma}_{(E)\kappa\lambda}\right). \tag{4.24}$$

The new boundary term, once a FG expansion is performed, brings in an additional counterterm which is quadratic in the Weyl tensor of the boundary metric

$$I_{\mathrm{LPP}}\,[E] = \frac{1}{16\pi G}\int\limits_M d^6x\sqrt{-g}\left(R + \frac{20}{\ell^2} - \frac{\ell^4}{72}\mathcal{E}_6\right) + \frac{\ell^3}{192\pi G}\int\limits_{\partial M} d^5x\sqrt{-h}\,\mathcal{W}^{ijkl}(h)\mathcal{W}_{ijkl}(h)\,, \tag{4.25}$$

as a consequence of a power-counting argument. Here, in order to make connection to the Einstein sector of the theory, the coupling is singled out as $\alpha_{\mathrm{E}} = -\frac{\ell^4}{384\pi G}$.

In other words, the conformal completion of the Einstein gravity which leads to LPP Conformal Gravity, naturally accommodates an extra boundary term. Then, when back in Einstein gravity, the problem of the mismatch between Kounterterms and counterterms in Eq. (3.15) is solved, as the renormalized action agrees with the one given by Holographic Renormalization

$$I_{\mathrm{LPP}}\,[E] = I_{\mathrm{HR}}\,. \tag{4.26}$$

5 Energy Functionals from Conformal Gravity in Four Dimensions

The proposal for the holographic computation of entanglement entropy (EE) [29] shows the relevance of finding suitable codimension-2 energy functionals within the framework of gauge/gravity duality. Such formula was proven in Ref. [30], where the authors developed a method to construct EE functionals dual to a generic gravity theory. This is essentially based on the holographic application of the replica trick which produces conical singularities in the bulk. In particular, the evaluation of the Einstein–Hilbert action in such manifold

$$\int_{M^{(\alpha)}} d^4x\sqrt{g}R^{(\alpha)} = \int_{M} d^4x\sqrt{g}R + 4\pi\left(1-\alpha\right)\mathcal{A}\left[\Sigma\right], \tag{5.1}$$

induces the area functional $\mathcal{A}\left[\Sigma\right] = \int\limits_{\Sigma} d^2y\sqrt{\gamma}$ on a codimension-2 surface Σ with metric γ_{ab}, where the conical deficit is given by $2\pi(1-\alpha)$.

Since, in the context of gauge/gravity duality, one is interested on surfaces which appear anchored to the conformal boundary, the area functional will be divergent. It is reasonable to think that the renormalization of area in codimension-2 is induced by the renormalization of the bulk gravity action. However, the use of standard counterterms and further projection in codimension-2 does not enlighten the discussion on general properties of renormalized area [31].

In what follows, a different strategy is adopted, as the implications of Conformal Renormalization for four-dimensional AdS gravity are explored in the definition of energy functionals for surfaces. The starting point is the Euclidean version of the action for CG

$$I_{\mathrm{CG}} = \frac{\ell^2}{64\pi G}\int_{M} d^4x\sqrt{g}\, W_{\mu\nu}^{\sigma\lambda}W_{\sigma\lambda}^{\mu\nu} - \frac{\pi\ell^2}{2G}\chi\left(M\right), \tag{5.2}$$

which properly reduces to the action (4.4) when evaluated on Einstein spaces.

The contribution in codimension-2 of CG is rendered explicit when evaluated on the orbifold $M^{(\alpha)}$, which contains a conical defect. The equivalent form of the CG action, in terms of quadratic curvature terms, is given by

$$I_{\mathrm{CG}} = \frac{\ell^2}{64\pi G}\int_{M} d^4x\sqrt{g}\left(R_{\mu\nu\sigma\lambda}R^{\mu\nu\sigma\lambda} - 2R_{\mu\nu}R^{\mu\nu} + \frac{1}{3}R^2\right) - \frac{\pi\ell^2}{2G}\chi\left(M\right), \tag{5.3}$$

where one can use the expressions for the Euler characteristic and the quadratic terms in the curvature given in Ref. [32]. Thus, the CG action in this manifold

$$I_{CG}^{(\alpha)} = \frac{\ell^2}{64\pi G_N}\int_{M^{(\alpha)}} d^4x\sqrt{g}\left|W^{(\alpha)}\right|^2 - \frac{\pi\ell^2}{2G_N}\chi\left(M^{(\alpha)}\right), \tag{5.4}$$

produces

$$I_{\mathrm{CG}}^{(\oslash)} = I_{\mathrm{CG}} + \frac{(1-\alpha)}{4G_N}L_{\Sigma} + \mathcal{O}\left((1-\alpha)^2\right), \tag{5.5}$$

where the first order term in codimension-2 is

$$L_{\Sigma} = \frac{\ell^2}{2}\int_{\Sigma} d^2y\sqrt{\gamma}K_{\Sigma} - 2\pi\ell^2\chi\left(\Sigma\right). \tag{5.6}$$

The object K_Σ is a 2D conformal covariant term defined on the surface Σ as [33]

$$K_\Sigma = W^{ab}_{ab} - P^{(A)}_{ab} P^{ab}_{(A)}, \tag{5.7}$$

where $P^{(A)}_{ab} = k^{(A)}_{ab} - \frac{1}{2}\gamma_{ab}k^{(A)}$ is the traceless part of the extrinsic curvature in codimension-2 (the label (A) denotes the normal directions to the surface).

Renormalized Area. For ambient spaces which are of Einstein type, the Weyl tensor takes the form (2.16), such that its double subtrace can be expressed using Gauss-Codazzi relation as

$$W^{ab}_{ab} = \mathscr{R} - k^{(A)}k_{(A)} + k^{(A)}_{ab}k^{ab}_{(A)}, \tag{5.8}$$

where $\mathscr{R}$ is the Ricci scalar of the metric γ_{ab}. On the other hand, a defining feature of minimal surfaces Σ_{min} is that the trace of any of the extrinsic curvatures vanishes, i.e., $k^{(A)} = 0$. For those conditions, the functional (5.6) turns into the Alexakis–Mazzeo formula [34]

$$L_{\Sigma_{\text{min}}}[E] = \mathcal{A}_{\text{ren}} = \frac{\ell^2}{4}\int\limits_{\Sigma_{\text{min}}} d^2y\sqrt{\gamma}\,\delta^{ab}_{cd}\left(\mathscr{R}^{cd}_{ab} + \frac{1}{\ell^2}\delta^{cd}_{ab}\right) - 2\pi\ell^2\chi\left[\Sigma_{\text{min}}\right]. \tag{5.9}$$

(Conformal) Willmore Energy. The fact that the functional L_Σ is a conformal invariant makes it particularly convenient to obtain a more general version of Willmore energy, defined for compact surfaces. Because one is dealing with a general Weyl tensor in Eq. (5.6), its double subtrace is given by

$$W^{ab}_{ab} = R^{ab}_{ab} - 2S^a_a. \tag{5.10}$$

Extensive use of Gauss–Codazzi relations leads to a functional rewritten as

$$L_\Sigma = \frac{\ell^2}{2}\int\limits_\Sigma d^2y\sqrt{\gamma}\left[\mathscr{R} - 2\left(H^{(A)}\right)^2 - 2S^a_a\right] - 2\pi\ell^2\chi\left(\Sigma\right), \tag{5.11}$$

where $H^{(A)} = \frac{1}{2}k^{(A)}$ is the mean curvature of Σ.

For a compact surface Σ_{comp}, the integral of the Ricci scalar exactly cancels out the Euler characteristic, such that the general formula reduces to

$$L_{\Sigma_{\text{comp}}} = -\ell^2\int\limits_{\Sigma_{\text{comp}}} d^2y\sqrt{\gamma}\left[\left(H^{(A)}\right)^2 + S^a_a\right]. \tag{5.12}$$

For a surface embedded in an arbitrary submanifold M_3, defined as a constant-time slice of an AAdS space in the bulk, the above expression corresponds to the Conformal Willmore Energy [35]. In that situation, one can well drop the label (A), as the extrinsic curvature computed for a normal pointing along the time direction always vanishes.

In the case the ambient space is global AdS, the standard Willmore Energy functional $\mathcal{W}(\Sigma)$ is recovered when embedded in $\mathbb{R}^3$. This is achieved by mapping the metric into an unphysical conformal frame $\hat{g}$ which represents the Euclidean flat space $\mathbb{R}^3$. Thus, the new Schouten tensor is identically zero, what leads to [36, 37]

$$\begin{aligned} L_{\Sigma_{\text{comp}}}\left[\mathbb{R}^3\right] &= -\ell^2 \int\limits_{\Sigma_{\text{comp}}} d^2y\sqrt{\hat{\gamma}}\hat{H}^2 , \\ &= -\ell^2 \mathcal{W}(\Sigma) . \end{aligned} \tag{5.13}$$

Reduced Hawking Mass. Non-extremal surfaces are not anchored orthogonally to the conformal boundary of AAdS spaces. As a matter of fact, the contact points with the boundary may induce divergences in the evaluation of the formula (5.9). On the other hand, the derivation of L_Σ does not make use of any constraint on the surface Σ, such that it may lead to a generalization of renormalized area for non-minimal surfaces.

Indeed, for arbitrary surfaces immersed in Einstein ambient spaces, the funtional L_Σ can be cast in the form

$$L_\Sigma[E] = \frac{\ell^2}{4} I_H(\Sigma) - 2\pi\ell^2\chi(\Sigma) , \tag{5.14}$$

where

$$I_H[\Sigma] = 2\int\limits_\Sigma d^2y\sqrt{\gamma}\left[\mathscr{R} + \frac{2}{\ell^2} - \frac{1}{2}\left(k^{(A)}\right)^2\right] , \tag{5.15}$$

is the Reduced Hawking Mass [38]. An interesting, equivalent way to express the above functional is

$$L_\Sigma[E] = \mathcal{A}_{\text{ren}}[\Sigma] - \frac{\ell^2}{4}\int\limits_\Sigma d^2y\sqrt{\gamma}\left(k^{(A)}\right)^2 , \tag{5.16}$$

what makes manifest that $L_\Sigma[E]$ is proportional to the renormalized area when Σ is minimal.

6 Conclusions

This work hints at a rather intriguing connection between renormalization of functionals (defined in the bulk or on a surface) and conformal symmetry. In Sect. 4, a proper embedding of Einstein gravity in Conformal Gravity leads to an AdS gravity action augmented with the correct counterterms. This procedure has been dubbed in recent literature as Conformal Renormalization. In Sect. 5, it is shown that conformal invariance in the bulk is inherited by codimension-2 functionals, obtained by evaluating Conformal Gravity action on a manifold with conical singularities. In doing so, and for different types of surfaces and ambient spaces, one is able to reproduce the notions of Renormalized Area, Willmore Energy and Reduced Hawking Mass.

Acknowledgements I am indebted to my collaborators Giorgos Anastasiou, Ignacio J. Araya and Cristóbal Corral for paving the road to the notion of Conformal Renormalization. I would also like to thank Laura Andrianopoli, Nicolas Boulanger, Riccardo D'Auria, Danilo Díaz, Olivera Miskovic, Lucrezia Ravera, Marika Taylor and Mario Trigiante for interesting discussions. This work was funded by ANID Grants N° 1190533 *Black holes and asymptotic symmetries* and N° 1230492 *Holographic description of anti-de Sitter gravity*, and ANID/ACT210100 Anillo Grant *Holography and its applications to High Energy Physics, Quantum Gravity and Condensed Matter Systems*.

Appendix 1: Useful Identities

Totally anti-symmetric Kronecker delta of rank p**.** Throughout the text, it is often used the totally anti-symmetric Kronecker delta of order p,

$$\delta^{\mu_1\mu_2\cdots\mu_p}_{\nu_1\nu_2\cdots\nu_p} = \det[\delta^{\mu_1}_{\nu_1}\cdots\delta^{\mu_p}_{\nu_p}]\,. \tag{6.1}$$

When considering the contraction of k indices, it follows the relation

$$\delta^{\mu_1\cdots\mu_k\cdots\mu_p}_{\nu_1\cdots\nu_k\cdots\nu_p}\delta^{\nu_1}_{\mu_1}\cdots\delta^{\nu_k}_{\mu_k} = \frac{(N-p+k)!}{(N-p)!}\,\delta^{\mu_{k+1}\cdots\mu_p}_{\nu_{k+1}\cdots\nu_p}\,, \tag{6.2}$$

where $k \leq p$ and N is the range of the indices.

References

1. J.M. Maldacena, The Large N limit of superconformal field theories and supergravity. Adv. Theor. Math. Phys. **2**, 231–252 (1998). arXiv:hep-th/9711200
2. S.S. Gubser, I.R. Klebanov, A.M. Polyakov, A Semiclassical limit of the gauge/string correspondence. Nucl. Phys. B **636**, 99 (2002). arXiv:hep-th/0204051
3. E. Witten, Anti-de Sitter space and holography. Adv. Theor. Math. Phys. **2**, 253 (1998). arXiv:hep-th/9802150
4. C. Fefferman, C.R. Graham, *The Mathematical Heritage of Élie Cartan* (Lyon, 1984);(Asterisque, 1985) 95
5. M. Henningson, K. Skenderis, The Holographic Weyl anomaly. JHEP **07**, 023 (1998). arXiv:hep-th/9806087
6. K. Skenderis, Lecture notes on holographic renormalization. Class. Quant. Grav. **19**, 5849 (2002). arXiv:hep-th/0209067

7. V. Balasubramanian, P. Kraus, A stress tensor for anti-de sitter gravity. Commun. Math. Phys. **208**, 413 (1999). arXiv:hep-th/9902121
8. R. Emparan, C.V. Johnson, R.C. Myers, Surface terms as counterterms in the AdS / CFT correspondence. Phys. Rev. D **60**, 104001 (1999). arXiv:hep-th/9903238
9. I. Papadimitriou, K. Skenderis, Thermodynamics of asymptotically locally AdS spacetimes. JHEP **08**, 004 (2005). arXiv:hep-th/0505190
10. E. Witten, A note on boundary conditions in Euclidean gravity. Rev. Math. Phys. **33**, 2140004 (2021)
11. A. Ashtekar, A. Magnon, Asymptotically anti-de Sitter space-times. Class. Quant. Grav. **1**, L39 (1984)
12. A. Ashtekar, S. Das, Asymptotically anti-de sitter space-times: conserved quantities. Class. Quant. Grav. **17**, L17 (2000). arXiv:hep-th/9911230
13. D.P. Jatkar, G. Kofinas, O. Miskovic, R. Olea, Conformal Mass in AdS gravity. Phys. Rev. D **89**, 124010 (2014). arXiv:1404.1411 [hep-th]
14. R. Olea, Mass, angular momentum and thermodynamics in four-dimensional Kerr-AdS black holes. JHEP **06**, 023 (2005). arXiv:hep-th/0504233
15. R. Olea, Regularization of odd-dimensional AdS gravity: Kounterterms. JHEP **04**, 073 (2007). arXiv:hep-th/0610230
16. T. Eguchi, P.B. Gilkey, A.J. Hanson, Gravitation, gauge theories and differential geometry. Phys. Rept. **66**, 213 (1980)
17. O. Miskovic, R. Olea, Topological regularization and self-duality in four-dimensional anti-de Sitter gravity. Phys. Rev. D **79**, 124020 (2009). arXiv:0902.2082 [hep-th]
18. G. Anastasiou, O. Miskovic, R. Olea, I. Papadimitriou, Counterterms, Kounterterms, and the variational problem in AdS gravity. JHEP **08**, 061 (2020). arXiv:2003.06425 [hep-th]
19. R. Aros, M. Contreras, R. Olea, R. Troncoso, J. Zanelli, Conserved charges for gravity with locally AdS asymptotics. Phys. Rev. Lett. **84**, 1647–1650 (2000). arXiv:gr-qc/9909015
20. V. Iyer, R.M. Wald, Some properties of Noether charge and a proposal for dynamical black hole entropy. Phys. Rev. D **50**, 846–864 (1994). arXiv:gr-qc/9403028
21. S.W. MacDowell, F. Mansouri, Unified geometric theory of gravity and supergravity. Phys. Rev. Lett. **38**, 739 (1977) [erratum: Phys. Rev. Lett. **38**, 1376 (1977)]
22. D. Grumiller, M. Irakleidou, I. Lovrekovic, R. McNees, Conformal gravity holography in four dimensions. Phys. Rev. Lett. **112**, 111102 (2014). arXiv:1310.0819 [hep-th]
23. J. Maldacena, Einstein gravity from conformal gravity. arXiv:1105.5632 [hep-th]
24. C. Imbimbo, A. Schwimmer, S. Theisen, S. Yankielowicz, Diffeomorphisms and holographic anomalies. Class. Quant. Grav. **17**, 1129–1138 (2000). arXiv:hep-th/9910267
25. O. Miskovic, R. Olea, M. Tsoukalas, Renormalized AdS action and Critical Gravity. JHEP **08**, 108 (2014). arXiv:1404.5993 [hep-th]
26. G. Anastasiou, I.J. Araya, C. Arias, R. Olea, Einstein-AdS action, renormalized volume/area and holographic Rényi entropies. JHEP **08**, 136 (2018). arXiv:1806.10708 [hep-th]
27. H. Lü, Y. Pang, C.N. Pope, Black holes in six-dimensional conformal gravity. Phys. Rev. D **87**, 104013 (2013). arXiv:1301.7083 [hep-th]
28. G. Anastasiou, I.J. Araya, R. Olea, Einstein gravity from conformal gravity in 6D. JHEP **01**, 134 (2021). arXiv:2010.15146 [hep-th]
29. S. Ryu, T. Takayanagi, Holographic derivation of entanglement entropy from AdS/CFT. Phys. Rev. Lett. **96**, 181602 (2006). arXiv:hep-th/0603001
30. A. Lewkowycz, J. Maldacena, Generalized gravitational entropy. JHEP **08**, 090 (2013). arXiv:1304.4926 [hep-th]
31. M. Taylor, W. Woodhead, Renormalized entanglement entropy. JHEP **08**, 165 (2016). arXiv:1604.06808 [hep-th]
32. D.V. Fursaev, A. Patrushev, S.N. Solodukhin, Distributional geometry of squashed cones. Phys. Rev. D **88**, 044054 (2013). arXiv:1306.4000 [hep-th]
33. S.N. Solodukhin, Entanglement entropy, conformal invariance and extrinsic geometry. Phys. Lett. B **665**, 305–309 (2008). arXiv:0802.3117 [hep-th]

34. S. Alexakis, R. Mazzeo, Renormalized area and properly embedded minimal surfaces in hyperbolic 3-manifolds. Commun. Math. Phys. **297**, 621–651 (2010)
35. A. Mondino, H.T. Nguyen, Global conformal invariants of submanifolds. Annales de l Institut Fourier **68**(6), 2663–2695 (2018)
36. P. Fonda, D. Seminara, E. Tonni, On shape dependence of holographic entanglement entropy in AdS_4/CFT_3. JHEP **12**, 037 (2015). arXiv:1510.03664 [hep-th]
37. G. Anastasiou, J. Moreno, R. Olea, D. Rivera-Betancour, Shape dependence of renormalized holographic entanglement entropy. JHEP **09**, 173 (2020). arXiv:2002.06111 [hep-th]
38. S. Fischetti, T. Wiseman, A bound on holographic entanglement entropy from inverse mean curvature flow. Class. Quant. Grav. **34**, 125005 (2017). arXiv:1612.04373 [hep-th]

Black Hole Spectroscopy

Black-Hole Spectroscopy: Quasinormal Modes, Ringdown Stability and the Pseudospectrum

Kyriakos Destounis and Francisco Duque

Abstract Black-hole spectroscopy is a powerful tool to probe the Kerr nature of astrophysical compact objects and their environment. The observation of multiple ringdown modes in gravitational waveforms could soon lead to high-precision gravitational-wave spectroscopy, thus it is critical to understand if the quasinormal mode spectrum itself is affected by astrophysical environments, quantum corrections, and other generic modifications. In this chapter, we will review the black-hole spectroscopy program and its challenges regarding quasinormal mode detection, the overtone status and the recent evidence that supports the existence of nonlinearities in the spectrum of black holes. We will then discuss a newly introduced non-modal tool in black-hole physics, namely the pseudospectrum; a mathematical notion that can shed light on the spectral stability of quasinormal modes, and discuss its novel applications in black holes and exotic horizonless compact objects. We will show that quasinormal modes generically suffer from spectral instabilities, explore how such phenomena can further affect black-hole spectroscopy, and discuss potential ringdown imprints and waveform stability issues in current and future gravitational-wave detectors.

K. Destounis (✉)
Dipartimento di Fisica, Sapienza Universitá di Roma, Piazzale Aldo Moro 5, 00185 Roma, Italy
e-mail: kyriakos.destounis@uniroma1.it

INFN, Sezione di Roma, Piazzale Aldo Moro 2, 00185 Roma, Italy

F. Duque
CENTRA, Departamento de Física, Instituto Superior Técnico–IST, Universidade de Lisboa, Avenida Rovisco Pais 1, 1049 Lisboa, Portugal

Max Planck Institute for Gravitational Physics (Albert Einstein Institute), Am Mühlenberg 1, 14476 Potsdam, Germany

© The Author(s), under exclusive license to Springer Nature Switzerland AG 2024
E. Papantonopoulos and N. Mavromatos (eds.), *Compact Objects in the Universe*,
https://doi.org/10.1007/978-3-031-55098-0_6

1 Introduction

The observation of gravitational waves (GWs) from compact binary sources has become a primary avenue of scientific exploration. GW detections have become routine, leading to a novel era of GW astrophysics [1]. Systematic multimessenger observations [2] from the Laser Interferometer Gravitational-Wave Observatory (LIGO), Virgo and KAGRA [3, 4], as well as future ground- and space-based interferometers, such as the Laser Interferometer Space Antenna (LISA) [5, 6], Taiji [7] and TianQin [8, 9], will unequivocally strengthen our understanding of black holes (BHs) and strong-field gravity [6, 10]. GWs carry fundamental information regarding the remnant's externally observable quantities, their surrounding environment [11–16] and their potential deviations from general relativity (GR) [17, 18].

According to uniqueness theorems in GR [19], the merger of two isolated BHs eventually leads to a stationary compact object described by at most three parameters: mass, charge (which is astrophysically expected to be negligible) and angular momentum. Due to the dissipative nature of GWs, binaries typically undergo three distinct stages, before forming a final, stable configuration, namely the inspiral, merger and ringdown. We are now able to accurately simulate binaries and extract fundamental information regarding the initial parameters of their constituents, such as their respective masses and spin, as well as the externally-observable properties of the remnant. Such simulations are usually quite cumbersome and demand the use of powerful numerical tools [20]. Nevertheless, for the inspiral and ringdown stages we can take advantage of semi-analytic models which agree surprisingly well with numerical relativity simulations. During the inspiral stage of extreme-mass-ratio systems [21–27], as well as during the ringdown stage, BH perturbation theory provides the most appropriate schemes to treat the evolution of these scenarios.

In particular, by simply analyzing the ringdown signal of perturbed BHs and compact objects, through an infinite (discrete) set of vibrational spectra called quasinormal modes (QNMs) [28–30], one can extract paramount information regarding the nature of the final remnants; a task called BH spectroscopy [31]. So far, there are various claims that besides the fundamental QNM, several overtones have been detected from gravitational waveforms, that are critical for parameter estimation and validity tests of GR [32–36]. Nevertheless, we are not yet entirely certain if the BH spectroscopy program is suitable for non-vacuum objects residing in astrophysical environments [11–16], quantum BHs with horizon-scale corrections [37], exotic compact objects [38–40] or novel configurations that incorporate modifications of gravity [41–48]; ingredients that are widely present in the Universe and can affect the QNM spectrum significantly [49–58]. Recent disputes regarding the detection of overtones [59, 60], as well as the presence of nonlinear QNMs [61–65] in numerical simulations can pose significant limitations in GW data analysis and BH spectroscopy. To exacerbate on the aforesaid issues, very recent developments on BH QNM instabilities [51–54, 56, 66], when various external perturbations to the vacuum solutions are taken into account, may also affect overtones and even the fundamental mode [55, 57, 67].

More specifically, consider that a perturbation is added to the BH effective potential. If the new perturbed QNMs of the dirty BH have disproportionately migrated to the complex plane with respect to the scale of the perturbation introduced to the effective BH potential, then the BH QNMs exhibit a *spectral instability*. Spectral instabilities are crucial in what regards the QNM overtone status and can be quantified by employing the mathematical notion of the pseudospectrum [68]. It is the aim of this chapter to discuss contemporary non-modal tools of perturbation theory, describe how to employ those novel schemes in order to shed more light on the nature of ringdown physics, and to further discuss the underlying spectral features of perturbed compact objects that may limit the goals of BH spectroscopy and ringdown waveform modeling.

2 Black-Hole Perturbation Theory and Quasinormal Modes

When binaries merge, they form a single object which undergoes a characteristic vibrational stage until it relaxes to a final stationary state. This state is approached via the ringdown phase. Due to its simplicity, it can be analyzed with perturbative schemes, by assuming that the background spacetime is slightly perturbed by very small fluctuations. In this limit, the Einstein field equations reduce to a master wave equation with an effective potential [69–72] describing the radiative degrees of freedom of the gravitational field. The resulting equation can then be integrated in the time domain via scattering of waves off the BH effective potential [73] or in the frequency domain to reveal the frequency content of the ringdown signal [74, 75]. In both cases, the dynamics of the signal is sufficiently-described by a superposition of exponentially damped sinusoids. The oscillation frequency and decay timescale of the ringdown is associated with the real and imaginary parts of the QNM frequencies, which contain information regarding the underlying geometry of the remnant [28–30]. The QNM spectrum of BHs in GR has been extensively analyzed with various analytic and numerical methods [76], though the spectral content of ringdown signals generated by compact binaries is less understood. Even so, there are important indications that several BH oscillation harmonics, including the fundamental mode and higher overtones, as well as various multipolar components, are crucial to fully understand the GW ringdown [31, 32].

In what follows we adopt a static and spherically-symmetric background of the form

$$ds^2 = -f(r)dt^2 + f(r)^{-1}dr^2 + r^2\left(d\theta^2 + \sin^2\theta d\varphi^2\right), \tag{1}$$

where $f(r)$ is a lapse function that describes the spacetime background through its externally-observable parameters. The propagation of linear perturbations Ψ on Eq. (1) is described by the wave-like equation

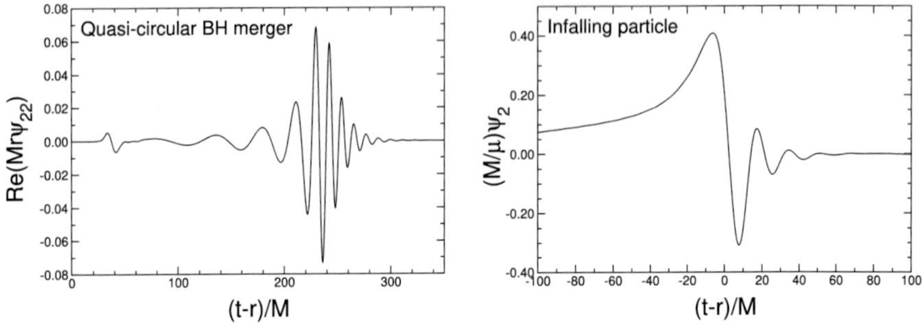

Fig. 1 Left: The waveform produces by a quasi-circular BH binary merger, where M is the total mass of the binary and Ψ_{22} is $\ell = m = 2$ multipolar component of the Weyl scalar Ψ_4. Right: The ringdown signal produced by an infalling particle with μ its mass onto a Schwarzschild BH with mass M, where Ψ_2 is the Zerilli function for $\ell = 2$. In both cases, r is the waveform extraction radius. Notice the similarities in the ringdown stage on both cases demonstrated. The figure is an adaptation from [29]

$$\left(\frac{\partial^2}{\partial t^2} - \frac{\partial^2}{\partial r_*^2} + V\right)\Psi(t,r) = 0, \quad V = f(r)\left(\frac{\ell(\ell+1)}{r^2} + (1-s^2)\frac{f'(r)}{r}\right), \quad (2)$$

where $dr/dr_* = f(r)$ is the tortoise coordinate, ℓ is the angular number that appears from the decomposition of the perturbation onto spherical harmonics [29], $s = 0, \pm 1, \pm 2$ correspond to scalar, electromagnetic and axial gravitational perturbations, respectively, and prime denotes radial differentiation. Equation (2) can be evolved in the time domain [77] in order to obtain the corresponding ringdown signal of perturbations on the fixed background (1). Figure 1 represents such a scenario where the background is a Schwarzschild BH (with $f(r) = 1 - 2M/r$, where M is the BH mass), together with a quasi-circular BH merger that produces a similar ringdown. On the other hand, the time-domain perturbation evolution Eq. (2) can be Fourier-transformed onto a time-independent Schrödinger-like equation in the frequency domain

$$\frac{\partial^2\Psi(r)}{\partial r_*^2} + \left(\omega^2 - V\right)\Psi(r) = 0. \quad (3)$$

To obtain the QNM spectrum [28–30], we apply appropriate boundary conditions for asymptotically-flat BHs, such that at the event horizon $r = r_h$ we only accept purely ingoing waves while at infinity we impose purely outgoing waves, i.e.

$$\Psi \to \begin{cases} e^{-i\omega r_*}, & r_* \to -\infty \quad (r \to r_h) \\ e^{i\omega r_*}, & r_* \to \infty \quad (r \to \infty) \end{cases} \quad (4)$$

A typical illustration of the QNM spectrum of Schwarzschild BHs is shown in Fig. 2.

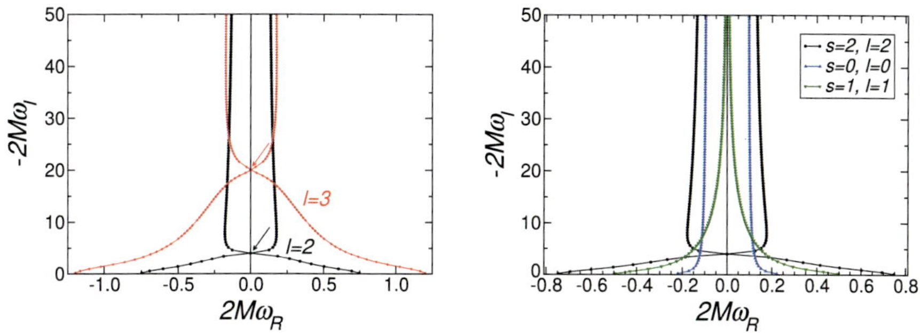

Fig. 2 Left: Real ω_R and imaginary ω_I parts of axial gravitational QNMs of a Schwarzschild BH with mass M and varying angular number ℓ. Right: Real ω_R and imaginary ω_I parts of scalar ($s = \ell = 0$), electromagnetic ($s = \ell = 1$) and axial gravitational ($s = \ell = 2$) QNMs of a Schwarzschild BH with mass M. The figure is taken from [29]

2.1 *Quasinormal Mode Detection Status*

The status of the fundamental mode detection is well-established since the first GW detection, GW150914 [78], though it is insufficient to recover accurately the mass and spin of the final remnant and perform validity tests of GR among other tests regarding BH environments, quantum effects and the existence of exotic compact objects that can mimick BHs in particular scenarios. The status of QNM overtones and if in fact we have managed to detect them in the loudest event to date, is currently under scrutiny, even though there was a strong certainty on their detection [32, 33, 35, 60].

In fact, the overtone status has been recently challenged regarding its robustness. In Ref. [59] it was pointed out that the uncertainty in the measured start time cannot be ignored, leading to a significant decrease in the evidence for overtones (see Ref. [79] for earlier investigations). Additionally, GW150914-like injections in neighboring segments of the real detector noise at high sampling-rate have shown that noise can induce artificial evidence for overtones, indicating how overtone detection is quite sensitive against data-analysis choices. Such specific choices were the source of apparent initial disagreement [59, 60, 80]. When the time uncertainty is included, and identical settings employed, all analyses consistently point to a detection significance slightly below 2σ, as independently obtained through both time-domain [60, 80] and frequency-domain [81, 82] methods. Furthermore, nonlinearities have been recently found at the early stage of BH ringdowns [61, 62, 83–86]. Such effects are expected to strongly impact the above analyses and their interpretation, due to the fact that the fitting begins arbitrarily close to the peak of the merger, which is highly dynamical and sensitive to time offsets and initial transients [87].

The above issues created turmoil in the BH spectroscopy program, though at least the nonlinearity investigations now agree. Yet, the detection status of overtones is still under discussion, but with time the program will resolve this issue as well. Nevertheless, this chapter will argue how spectral instabilities may add another thorn in ringdown physics and how this might affect BH spectroscopy. This is particularly important for the overtone discussion, since they are the ones mostly affected from spectral instabilities.

3 Early Evidence of a Spectral Instability

The first evidence of a spectral instability in BH physics dates back to the 90 s, when Nollert attempted to solve the non-completeness issue of QNMs [49]. Due to the backscattering of perturbations at infinity, which lead to a late-time polynomial tail after the exponential ringdown [77] and a branch cut in the complex plane of the retarded Green function of the wave Eq. [88, 89] (this does not occur on non-asymtotically flat spacetimes), one cannot built a complete set of QNMs that describe the ringdown signal as a whole. Nollert's experiment consisted of a large number of step barriers in order to approximate and ultimately replace the continuous effective potential of gravitational perturbations in Schwarzschild BHs. His task was successful; the QNM spectrum of the step barrier potential could form a complete set though with a very strong drawback. The new *perturbed spectrum* was entirely different from that of the *unperturbed spectrum* (with continuous potential) of the BH [49]. Nevertheless, he later found that regardless that the perturbed QNMs were significantly different, the ringdown signal still behaved as if the unperturbed fundamental mode was present, and only at late times the new perturbed modes would appear [50].

In Fig. 3, the methodology of approximating the potential with steps is demonstrated as well as the respective spectra that depend on the number of steps used. As already discussed, this was an unexpected result at the time, since the more steps were used, the more the perturbed QNMs were branching out and showing similarities with the w-modes of neutron stars [90]. The reason behind the spectra changing so significantly is obvious. When utilizing steps to approximate a potential, you cut out the asymptotic behavior where the first step appears. In all cases and no matter the number of steps used, besides infinity, the branch cut in the Green function

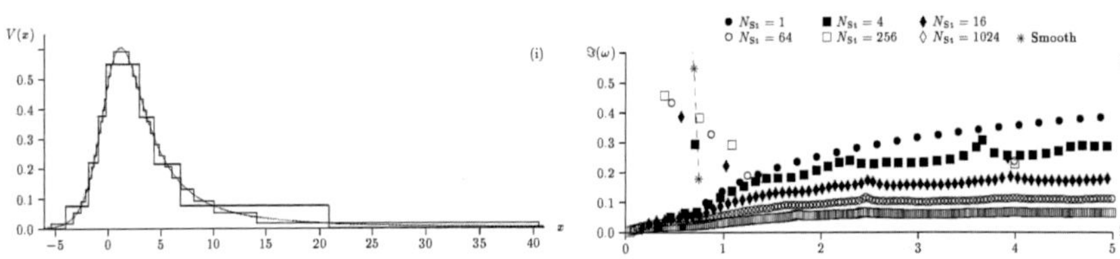

Fig. 3 Left: The continuous BH potential from (2), with $\ell = s = 2$, shown with a black line, and the various choices of step barriers to approximate it. Right: The corresponding spectrum resulting from the approximated potential, where N_{St} is the number of steps in the potential. For reference, the modes shown with asterisks correspond to the continuous potential and their trend is shown with a dashed line. The figure has been adapted from [49]

is disregarded from the problem, which is the sole reason why the polynomial tail appears. One can then build a complete base of QNMs that describes the ringdown signal, since the branch cut is absent, but since the asymptotic behavior is altered abruptly at some distance x in Fig. 3, or equivalently at some r_* in Eq. (3), it is natural that the spectrum will change, since the boundary conditions are intrinsically altered. This work paved the way to further understand the connection between QNMs of BHs and their role in ringdown signals and eventually led to the realization that BH spectroscopy may suffer from deep mathematical issues; in this case, spectral instabilities. For a closer look into this, see also [91].

4 The Need for Non-modal Tools: The Pseudospectrum

To amply exploit the abilities of BH spectroscopy, one must better understand the relative excitation of each QNM [50] and the possible instability of the spectrum itself under small perturbations of the scattering potential [49], as demonstrated in the previous section. Because astrophysical BHs are not isolated, but rather are permeated by highly dynamic environments, this last question is of paramount importance. For example, if the BH's surrounding is *perturbed* in a small way by an accretion disk or matter shell, then the QNM spectrum may surprisingly change by a unequally large amount [11, 13, 49, 55]. This migratory behavior of QNMs may have important implications for interpreting GW observations.

For selfadjoint operators, the spectral theorem underlies the notion of normal modes of vibration (which provide an orthonormal basis) and guarantees the stability of the eigenvalues under fluctuations of the eigenvalue operator. In other words, a small-scale perturbation to the operator leads to a new spectrum that migrates in the complex plane within a region of radius comparable to the scale of the perturbation, desgnating *spectral stability*. In stark contrast, the lack of a spectral theorem in the non-selfadjoint case entails, in general, the loss of completeness in the set of eigenfunctions as well as their orthogonality. Thus, the eigenvalues may show strong sensitivity to small-scale environmental perturbations of the operator, as Nollert first found [49]. In these cases, the perturbed spectrum migrate to an extent that is orders of magnitude larger than the perturbation's scale, designating a *spectral instability*. Operators that describe BH oscillations and QNMs fall into the second category.

4.1 *The Mathematical Notion of the Pseudospectrum*

It turns out that there is a powerful mathematical notion called the pseudospectrum [92], which is able to capture the extent to which systems controlled by non-selfadjoint operators exhibit spectral instabilities. The pseudospectrum is a continuous superset of the spectrum and can be formally defined through the *resolvent operator* which portrays a 'topographical map' of spectral migration under generic perturbations of order ϵ to the underlying eigenvalue operator. Formally, the pseudospectrum is defined as follows. Given $\epsilon > 0$, the ϵ-pseudospectrum $\sigma^\epsilon(A)$ of an operator A can be characterized as

$$\sigma^\epsilon(A) = \{z \in \mathbb{C}, \exists\, \delta A \in M_n(\mathbb{C}), ||\delta A|| < \epsilon : z \in \sigma(A + \delta A)\}, \tag{5}$$

$$\sigma^\epsilon(A) = \{z \in \mathbb{C} : ||R_A(z)|| = ||(z\mathbb{I} - A)^{-1}|| > 1/\epsilon\}, \tag{6}$$

with z a complex number, δA a complex operator, $R_A(z)$ the resolvent operator of A at the point z, $\mathbb{I}$ the identity operator and $\sigma(A)$ the spectrum of A. When $\epsilon \to 0$, then $\sigma^0(A) = \sigma(A)$.

Selfadjoint (normal) operators satisfy $\left[A, A^\dagger\right] = 0$, therefore the spectral theorem of normal operators ensures that eigenfunctions form an orthonormal basis and that the perturbed eigenvalues of $\sigma(A + \delta A)$, in accord to Eq. (5), are spectrally stable, i.e. the perturbed spectra will reside in concentric circles of radius ϵ. Spectral stability can also be understood in a more robust way by Eq. (6). Assume a point in the complex plane z. Asking if it is an eigenvalue is not a robust question (numerically) since the equation $\det|z\mathbb{I} - A| = 0$ has to be satisfied exactly, which needs fine-tuning or extremely accurate predictions. Nevertheless, selfadjoint operators abide with an eigenvalue analysis since it gives good predictions regarding stability. On the contrary, non-selfadjoint operators (non-normal) lack a spectral theorem and lead to weak control of eigenfunction completeness and stability. Therefore, a more well-posed question to ask is that of Eq. (6), i.e. how close to an eigenvalue a point of the complex plane is through the resolvent operator. Hence, the pseudospectrum, defined either by (5) or (6), can answer questions regarding eigenvalues, their spectral stability and associated eigenfunction completeness. Figure 4 portrays a schematic representation of hypothetical normal and non-normal operators spectra and ϵ-pseudospectra. The differences between the structure of these operators is quite obvious. The normal operator forms concentric pseudospectra contours around the eigenvalues, while the non-normal operator contours extend further from the eigenvalues and do not form concentric circles with specified radii. The three-dimensional Fig. 5 for the non-normal case demonstrates the topographic structure of the pseudospectrum through the resolvent's norm in the complex z-plane.

As demonstrated, strong non-normality, occuring in quite generic settings, leads to a severely uncontrolled spectrum under perturbations of the governing operator, thus by using the pseudospectrum in BH physics we can formally capture the potential sensitivity of the QNMs to small perturbations at the level of the non-perturbed operator that described BH QNMs, without the need of systematically introducing perturbations to the operator, as suggested by Eq. (5). As a side note, we need to emphasize that the definition of the pseudospectrum, and thus any statement on spectral stability, depends delicately on the choice of the norm used, which directly connects with a scalar product. The scalar product fixes the norm that quantifies the extent of large and small perturbations $||\delta A||$. Physically, one of the logical choices for a norm is the one that captures the system's energy, since it conventionally encodes the size of the physical perturbation with respect to the physics of the problem. In what

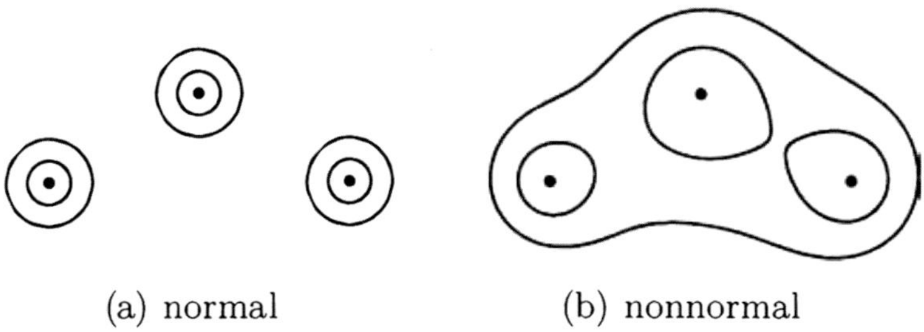

Fig. 4 The geometry of pseudospectra of a self-adjoint (normal) **a** and a non-selfadjoint (non-normal) **b** operator. In each plot, the contours represent the boundary of $\sigma^{\epsilon}(A)$ for two values of ϵ, while the eigenvalues are shown with black dots. The figure is adapted from [95]

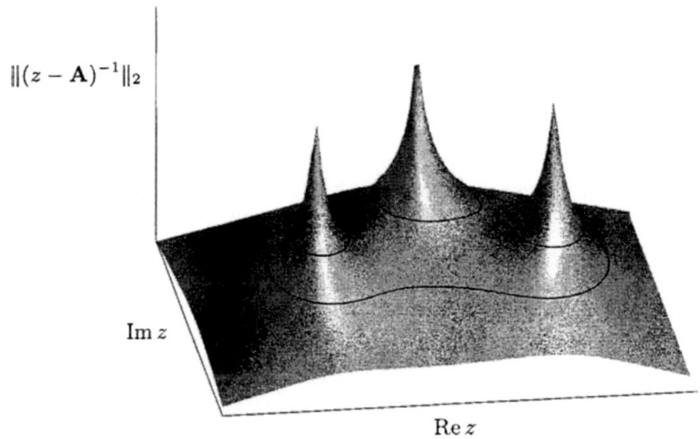

Fig. 5 The resolvent norm as a function of z in a specified region of the complex plane for case **b** in Fig. 4. The contours shown in black match those in Fig. 4, (**b**). The peaks correspond to eigenvalues of the assumed operator, where the norm of the resolvent diverges. The figure is adapted from [95]

follows, the so-called *energy norm*, or energy scalar product, which is systematically discussed in [93] and introduced decades ago in [94], will be used. We point the reader to [93] and in the Appendices of [51] for a complete discussion of the energy norm as well as how it can be incorporated in the resolvent operator and eventual pseudospectrum computation.

Naturally, there is a variety of non-modal tools, such as the *eigenvalue condition number*, the *spectral, numerical, pseudospectral abscissa* and the *Kreiss constant*, among others, that help to further understand the transient early-time behavior of the matrix evolution operator and the existence of pseudoresonances [68]. In this chapter, we will stay clear from these concepts and focus on the *pseudospectrum of compact objects*, though we encourage the reader to take a close look at [53, 68] for further details.

4.2 Hydrodynamics and the Pseudospectrum

Mathematical tools that do not depend on the calculation of eigenvalues are not new in the literature of science. The non-modal notion of the pseudospectrum has been used for decades in issues such as quantum mechanics, where the introduction of non-selfadjoint operators in PT-symmetric quantum mechanics entails that the associated spectrum is insufficient to draw full, quantum-mechanically relevant conclusions [96]. Another application, which is also extremely pertinent to spherically-symmetric BH perturbations, is on one-dimensional wave equations with dissipative boundary conditions which suffer from similar limitations in spectral predictions [94].

Probably one of the most interesting application of the pseudospectrum and its potential to predict spectral instabilities and even nonlinear phenomena at the linear operator level occurred in the field of hydrodynamics. The linearized Navier-Stokes equations are typically used to theoretically understand if a laminar (stable) fluid flow will become turbulent (unstable) for a choice of parameters that characterize the initial fluid flow, such as its speed. Eigenvalue analysis predicted that there is a critical (or no) parameter of the fluid flow beyond which the instability occurs [97]. To the contrary, in the laboratory, transition to turbulence was observed for much smaller fluid parameters than those predicted by modal analyses [98, 99]. These anomalies between theory and experiment of subcritical transition to turbulence and their initial explanation have been recognized for decades as bad attributes of the linearization around the laminar flow.

Later, it was understood that the failure of the eigenvalue analysis may be attributed to a misguided search for unstable eigenvalues of the linearized system. The pseudospectrum gave the answer to the aforementioned anomalies in a seminal work of Trefethen et al. [95]. The analysis performed in hydrodynamics and the Navier-Stokes equations [95], and later the books written regarding the pseudospectrum of various physical systems [68, 92], revealed that the pseudospectra of the linearized problem implied that small perturbations to the smooth flow may be amplified by factors of order 10^5 by a linear mechanism, even though all the eignevalues of the system decay monotonically. This occurs due to the non-orthogonality of the eingenfunctions and the non-normality of the underlying operator of the hydrodynamic system. The most notable results from [95] are demonstrated in Fig. 6, where the spectrum is shown with in the dark gray area. It is evident that the subcritical transition to turbulence is encoded in the pseudospectrum of the linearized Navier-Stokes equations, in total agreement with experiments, due to the protrusion of the pseudospectral contours in the upper half of the complex plane, where unstable modes reside. This is an indication that the tiniest perturbation to the governing operator of the linear system can destabilize the evolution and lead to turbulent flow, even though all eigenvalues lie on the lower half of the complex plane, which designates modal stability. For further indications regarding spectral instabilities and the need for non-modal tools, we refer the reader to [68, 95], though in what follows, we will focus solely on the pseudospectrum.

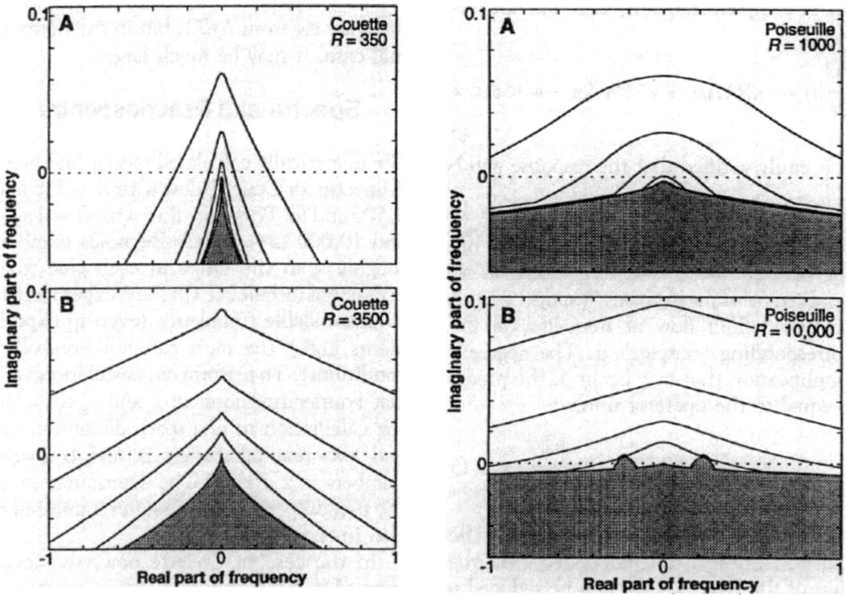

Fig. 6 Left: Spectrum (gray area) and ϵ-pseudospectra (solid curves) in the complex plane of the linearized Navier-Stokes evolution operator for Couette hydrodynamic flow with specific Reynolds number R that is proportional to the speed of the flow. The pseudospectral contour lines from outer to inner correspond to $\epsilon = 10^{-2}$, $10^{-2.5}$, 10^{-3}, $10^{-3.5}$. The horizontal dashed lines designate the real axis, above which unstable modes lie. The pseudospectrum extends significantly on the upper half of the complex plane, in which by convention, unstable modes reside. Right: Same as Left but for Poiseuille hydrodynamic flow. For $R > 5772$, two bumps of the eigenvalue area extend into the upper half of the complex plane. The figure is adapted from [95]

4.3 Black Holes and Their Pseudospectrum in a Nutshell

The pseudospectrum was recently introduced in the context of BH physics to address BH QNM instabilities. The seminal analysis in [51] indicates an overall spectral instability in Schwarzschild BH QNMs. Similar spectral instability have been found in subextremal and extremal Reissner-Nordström (RN) BH QNMs [54], exotic horizonless objects [56], as well as BH with de Sitter [100] and anti-de Sitter asymptotics [101]. The mathematical tools introduced in [51] have the potential to be used in order to address many interesting problems in gravitational physics, such as testing GR with GWs. Such a test would require the ability to extract various QNMs from the ringdown of BH mergers. By detecting the fundamental mode, we can determine the mass and spin of the final BH. Nevertheless, if the overtones could be accurately extracted, they would provide a test of GR, since overtones differ a lot in different theories of gravity [102–107]. It turns out that the predicted instability in these overtones may limit the abilities of BH spectroscopy [52] though further investigations are needed to assess to what extent a BH spectral instability at the ringdown signal can be detected with LIGO and LISA. Since the pseudospectrum is hardwired to predict transient phenomena and may assess up to which extent a BH QNM instability can limit the abilities of BH spectroscopy [52], it is of high importance to further explore its applicability in the vast zoo of BH physics. In what follows, we present a plethora of recent findings that regard spectral stability of BH and exotic compact object (ECO) QNMs, their corresponding pseudospectra, a more modern analysis of spectral instability detection in ringdown signals in line with [50], as well as future directions to be explored.

4.4 A Pseudospectrum Calculator Primer

In order to numerically calculate pseudospectra of operators and matrices, and in particular those that describe the QNM problem of asymptotically flat BHs, it has been found in [108–110] that an appropriate coordinate transformation is necessary in order to build a matrix, that describes the QNM problem, and later discretize it. This transformation is known as the *hyperboloidal scheme* to QNMs which extends the domain of $r_* \in [-\infty, \infty]$ and introduce in Eq. (2) the dimensionless quantities

$$\bar{t} = t/\lambda, \quad \bar{x} = r_*/\lambda, \quad \tilde{V}_\ell = \lambda^2 V_\ell, \tag{7}$$

where λ is an appropriate length scale to be defined according to the spacetime in study. We consider coordinates (τ, x) that implement the hyperboloidal foliation [51, 108]

$$\bar{t} = \tau - h(x), \quad \bar{x} = g(x), \tag{8}$$

where the height function $h(x)$ implements the hyperboloidal slicing by bending the Cauchy initial data to horizon and infinity penetrating data [111, 112] (see Fig. 7 for a schematic representation) and the function $g(x)$ introduces a spatial compactification from $\bar{x} \in [-\infty, \infty]$ to $x \in [a, b]$. The functions $h(x)$, $g(x)$ depend on the spacetime under study and are not fixed for all compact objects (see e.g. [51, 56, 113]).

Through Eqs. (7) and (8), Eq. (2) becomes [51, 54, 108, 109]

$$-\ddot{\phi} + L_1\phi + L_2\dot{\phi} = 0, \tag{9}$$

where an overdot denotes differentiation with respect to τ, and the differential operators are given by [51]

$$L_1 = \frac{1}{w(x)}\big[\partial_x\left(p(x)\partial_x\right) - q_\ell(x)\big], \quad L_2 = \frac{1}{w(x)}\big[2\gamma(x)\partial_x + \partial_x\gamma(x)\big]. \tag{10}$$

The height and compactification functions $h(x)$ and $g(x)$ enter the above operators via

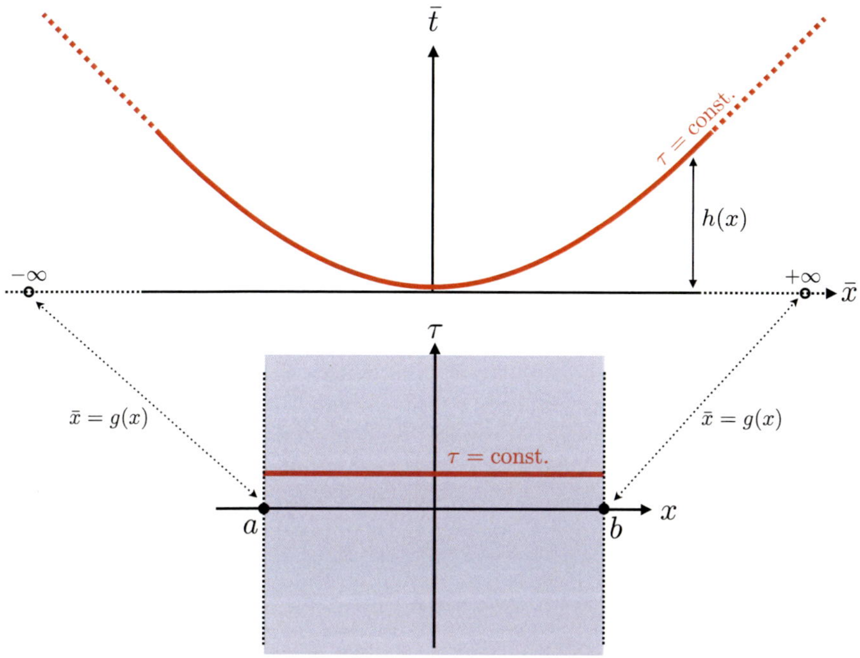

Fig. 7 Schematic representation of the implementation of the hyperboloidal scheme, together with the spatial compactification. The height function $h(x)$ bends the Cauchy data upwards to penetrate the horizon and null infinity, while the function $g(x)$ compactifies the domain from $\bar{x} \in [-\infty, \infty]$ to $x \in [a, b]$. The figure is taken from [51]

$$w(x) = \frac{g'^2 - h'^2}{|g'|}, \quad p(x) = |g'|^{-1}, \quad \gamma(x) = \frac{h'}{|g'|}, \quad q_\ell(x) = \lambda^2 |g'| \, V. \tag{11}$$

Equation (9) can be re-written as a matrix evolution problem through a first-order reduction in time by introducing $\psi = \dot{\phi}$. Equation (9) becomes

$$\dot{u} = i L u, \quad L = \frac{1}{i} \begin{pmatrix} 0 & 1 \\ L_1 & L_2 \end{pmatrix}, \quad u = \begin{pmatrix} \phi \\ \psi \end{pmatrix}, \tag{12}$$

which has the eigenfunction solution

$$u(\tau, x) = e^{iL\tau} u(0, x), \tag{13}$$

in terms of the evolution operator $e^{iL\tau}$. By further performing a harmonic decomposition $u(\tau, x) \sim u(x) e^{i\omega\tau}$ in Eq. (12) we arrive at the eigenvalue equation

$$L \, u_{n\ell} = \omega_{n\ell} \, u_{n\ell}, \tag{14}$$

where $\omega_{n\ell}$ is an infinite set of eigenvalues of the operator L, with $n \geq 0$ labelling the overtone number. The calculation of QNMs through the hyperboloidal framework ultimately reduces to the eigenvalue problem of the operator L, which, in turn, contains information concerning the boundary conditions of the spacetime metric at the geometric level. It is important to note that the blocks L_1 and L_2 multiply ϕ and $\dot{\phi}$, respectively. Hence, we can intuitively understand that L_1 contains terms associated with the effective potential (it includes $q_\ell \sim V_\ell$) and L_2 incorporates the boundary conditions at the event horizon and infinity (dissipative effects) due to its proportionality with the height function that applies the hyperboloidal slicing. This will prove important later on and serve as a sanity check for using the energy norm [51, 93].

To bring the aforementioned system in a numerical format, we discretize the compactified domain of the operator $L, x \in [0, 1]$, with $N + 1$ Chebyshev grid points, while to discretize the differential operators we apply the Chebyshev differentiation matrices [68, 92]. Once L is discrete, the BH QNMs follow from the eigenvalues of the matrix. For pseudospectra, one implements the Chebyshev-discretized version of the energy norm to calculate the adjoint $L^\dagger$, which is exhaustively detailed in [93], the Appendices of [51] and in [54]. The ϵ-pseudospectrum $\sigma^\epsilon(L)$ in the energy norm is given by [51]

$$\sigma^\epsilon(L) = \{\omega \in \mathbb{C} : s^{\min}(\omega\mathbb{I} - L) < \epsilon\}, \tag{15}$$

where $s^{\min}$ is the minimum of the generalized singular value decomposition, which incorporates the adjoint in the energy scalar product

$$s^{\min}(\mathcal{M}) = \min\{\sqrt{\omega} : \omega \in \sigma(\mathcal{M}^\dagger \mathcal{M})\}, \qquad \mathcal{M} = \omega\mathbb{I} - L. \tag{16}$$

5 Pseudospectrum of Black Holes

In the following section, we present the most interesting results obtained in the seminal work of Ref. [51] when calculating the pseudospectrum of a toy model BH, which includes a Pöschl-Teller potential and the pseudospectrum of Schwarzschild BH QNMs, as well as the pseudospectrum and its rich phenomenology in RN BH QNMs [54].

5.1 Pöschl-Teller Potential: A Toy Model

In this toy model, the Schwarzschild effective potential for QNMs in Eq. (2) is replaced by the Pöschl-Teller function

$$V(\bar{x}) = V_0/\cosh^2(\bar{x}), \quad \bar{x} \in (-\infty, \infty), \tag{17}$$

which is typically used in QNM studies [114, 115], since for an appropriate choice of V_0, one can reproduce the fundamental axial gravitational QNM of a Schwarzschild BH with good accuracy. Moreover, the hyperboloidal scheme has already been tested in this setup [116]. A key attribute of the Pöschl-Teller potential is that it decays exponentially on both boundaries, thus does not include a branch cut in the complex plane of the Green's function, and linear field fluctuations decay exponentially at infinity.

5.1.1 Pseudospectrum: Spectral Stability and Instability

By applying the methods outlined in Sect. 4.4, we can bring the wave equation, with V_ℓ as in Eq. (17), in a matrix evolution problem form, that allows us to find the spectrum and pseudospectrum (see [51] for a detailed description of the steps involved). It has been noted that the block matrix L_2 contains the dissipative boundary conditions needed for QNMs. With the Pöschl-Teller potential, one can perform the following validity check outlined in [51]: by setting $L_2 = 0$ and keeping L_1 as is, one obtains a selfadjoint operator, with eigenvalues that are normal modes. Such an operator is relevant by itself, since it corresponds to the azimuthal mode $m = 0$ of a wave propagating on a sphere with a constant unit potential, which is a conservative, rather than a dissipative, system. The eigenfunctions are then the Legendre polynomials $\phi_n(x) = P_n(x)$, with real eigenvalues $\omega_n^\pm = \pm\sqrt{1 + \ell(\ell+1)}$. This provides a robust test case for the norm chosen to calculate pseudospectra and we should expect the existence of a spectral theorem and a flat pseudospectrum with concentric circles forming around the normal modes of the system.

Figure 8 demonstrates exactly how a selfadjoint operator's pseudospectrum should look like. The absence of dissipative effects, such as the boundary conditions for QNMs, force the pseudospectrum to form concentric circles around the eigenvalues of the conservative system, where the circles have radii equal to the order of perturbation in the energy norm, i.e. ϵ, as better shown in the bottom panel of Fig. 8. The rest of the complex plane (shown in the upper panel) appears flat, i.e. having the same dark blue color everywhere and can only be accessed by the perturbed normal modes with very large-scale perturbations. Therefore, Fig. 8 represents how spectral stability manifests in conservative systems and provides a reassuring test of the whole numerical scheme introduced in [51] regarding the use of the energy norm.

Of course, our interest towards BH QNMs and their spectral stability requires that we apply the proper ingoing boundary conditions at the (would-be) event horizon and outgoing at infinity, as the hyperboloidal approach already implements, meaning that $L_2 \neq 0$. On the left panel of Fig. 9, the spectral instability of the eigenvalues is obvious with pseudospectral contours extending logarithmically to asymptotic regimes that are disproportionally larger than those assumed in the operator L, through the resolvent norm. Therefore, as a rule of thumb, the opening of the pseudospectral lines designates spectral instability, which is justified here due to the non-normality of the respective operator. Take for example the second contour line from the top on the left panel of Fig. 9. This roughly corresponds to a perturbation in L of order $\sim 10^{-25}$.

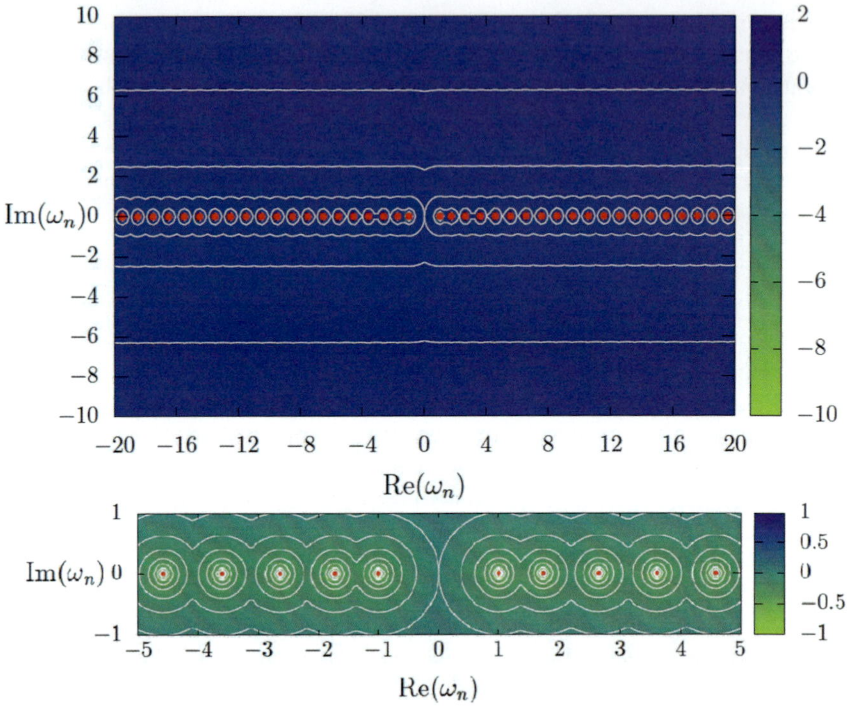

Fig. 8 Pseudospectrum of a normal operator ($L_2 = 0$) with a Pöschl-Teller potential with number of grid points $N = 100$. The white lines correspond to different ϵ pseudospectral contour lines which become more negative close to the eigenvalues (red dots). The colorbar on the right is in logarithmic scale meaning that a value of, e.g. -4 corresponds to $\log[\sigma^{-4}(L)]$ or simply a perturbation in L with energy norm 10^{-4}. The figures have been adapted from [51]

Nevertheless, from the 10th mode and above, the eigenvalues have the freedom to migrate at any point from the specific contour and above boundlessly and disproportionately with respect to the scale of the perturbation. The contours discussed are therefore the bound of eigenvalue migration below which the remaining eigenvalues remain stable under the particular perturbation of order ϵ. Even so, a larger norm perturbation can further destabilize more overtones, and even the fundamental mode as the pseudospectrum entails. Notice that we only show a part of the complex plane; the contour lines extend far and wide and behave smoothly as long as the number of interpolation grid points used is large enough.

An important aspect of the results shown in the left panel of Fig. 9 is that the assumed perturbation is not only constrained to the part of the operator that contains the effective potential, but rather affects L in its entirety. This result, although non-trivial and interesting for other systems, is not straightforwardly connected to perturbations of the effective potential of a BH due to environmental effects, quantum corrections, and other effects. To understand if the pseudospectrum is a meaningful tool for BH physics, the authors in Ref. [51] perturbed only the Pöschl-Teller

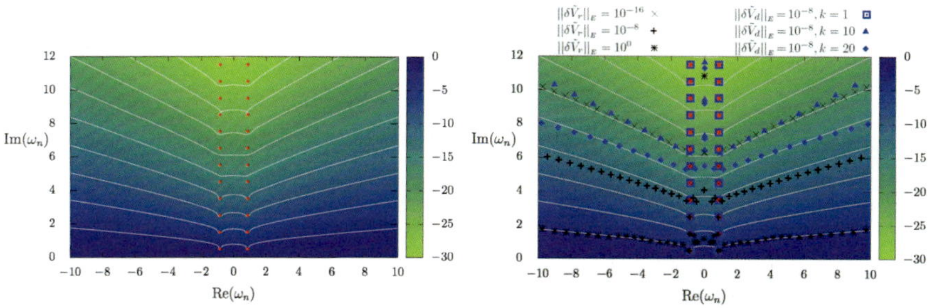

Fig. 9 Left: Pseudospectrum of the wave Eq. (2) with a Pöschl-Teller potential and $N = 100$ grid points. The white lines correspond to different ϵ pseudospectral contour lines, which become more negative close to the eigenvalues (red dots). The colorbar on the right is in logarithmic scale, meaning that a value of, e.g. -15 corresponds to $\log[\sigma^{-15}(L)]$ or simply a perturbation in L with energy norm 10^{-15}. Right: The pseudospectrum from Left with over-plotted perturbed QNMs resulting from random δV_r and deterministic δV_d perturbations to the potential. The perturbations are normalized such that in the energy norm they acquire the specified normalized value shown above the right panel. The figures have been adapted from [51]

potential part of L_2, by adding two different kinds of energy norm perturbations: (i) random perturbations to the potential, designated as $||\delta V_r||_E$ and (ii) deterministic perturbations designated as $||\delta V_d||_E \sim \cos(2\pi k x)$, where k is the frequency of the perturbation. Both types of perturbations are normalized such that their absolute value in the energy norm is a dimensionless number. By merely choosing a random perturbation of order 10^{-16}, as shown in the right panel of Fig. 9, we observe that from the 7th mode and on, a QNM migration, that practically lies on top of the respective pseudospectral contour lines, occurs. Increasing the order of the perturbation destabilizes even lower overtones. Since random additions of numbers on the effective potential do not necessarily have a physical meaning, the authors in [51] also included deterministic perturbations by adding a sinusoid with some scale and frequency. This is a more physical scenario of a perturbing agent such as a quantum effect. The results in this case are also fascinating: by fixing the order of the perturbation to 10^{-8} and choosing a small fluctuation frequency k, there is no observed migration, and the system may falsely seem to exhibit spectral stability. Nevertheless, the increment of k has the same effect with random perturbations, where larger frequencies induce destabilization of lower overtones. Significantly, no matter the order of random and the order/frequency combinations of deterministic perturbations, the fundamental mode remains stable to these fluctuations, even though the pseudospectrum entails that even the fundamental mode can migrate for appropriate ϵ. This observation practically designates the difference between perturbing only the potential of the system or perturbing its whole operator. Nonetheless, the fact that a tiny addition to the effective potential can lead to the destabilization of overtones is fundamental for performing validity tests of GR and spectral instabilities, if present in BHs, need to be taken seriously for the BH spectroscopy program to be successful.

5.2 Schwarzschild Black Holes

Schwarzschild BHs are exact static and spherically-symmetric solutions of the vacuum Einstein field equations. The line element describing them has the form of Eq. (1) with

$$f(r) = 1 - \frac{2M}{r}, \tag{18}$$

where M is gravitational mass of the object. They possess an event horizon at $r = 2M$, which satisfies $f(2M) = 0$.

Scalar, electromagnetic and axial gravitational perturbations are described by Eq. (2). The axial gravitational effective potential of a perturbed Schwarzschild BH, known as the Regge-Wheeler potential is isospectral (same QNM spectrum) with the polar (Zerilli) sector, and has the form

$$V = \left(1 - \frac{2M}{r}\right)\left(\frac{\ell(\ell+1)}{r^2} - \frac{6M}{r^3}\right). \tag{19}$$

The potential in Eq. (19) decays exponentially at the event horizon but in contrast to the Pöschl-Teller function, it decays polynomially at infinity, thus it includes a branch cut in the complex plane of the Green's function. In this case, the counterpart of the power-law cutoff in the hyperboloidal formulation used in [51] gives rise to a continuous part in the spectrum, besides the typical convergent QNMs, that is not converging as N increases. These continuous eigenvalues are not QNMs and can be unabiguously identified, though their presence must be considered when computing the pseudospectra. Thus, the benchmark case of the Pöschl-Teller potential will prove crucial for realistic scenarios.

5.2.1 Pseudospectrum of Schwarzschild Black Holes and Perturbed Quasinormal Modes

Again, by applying the methods outlined in Sect. 4.4, we can bring the wave equation, with V as in Eq. (19), into a matrix evolution form, that allows us to find the QNMs and pseudospectra of Schwarzschild BH QNMs (see [51] for a detailed description of the steps and functions involved). Besides the pseudospectral analysis, the steps of adding sinusoid perturbations to the BH potential is repeated to demonstrate that the pseudospectra contour lines correspond to the destabilization boundaries of perturbed QNMs above which they can migrate in the complex plane disproportionally. Figure 10 conforms with the above discussion, i.e. the pseudospectral contour lines for different level sets ϵ extend to the edges of the studied range of the complex plane in a logarithmic manner, as in the Pöschl-Teller case. A significant difference here is the appearance of eigenvalues along the imaginary axis. These modes are non-convergent as N is increased, they are continuous and they are associated to the branch-cut in the complex plane of solutions of the Green's function [88] that

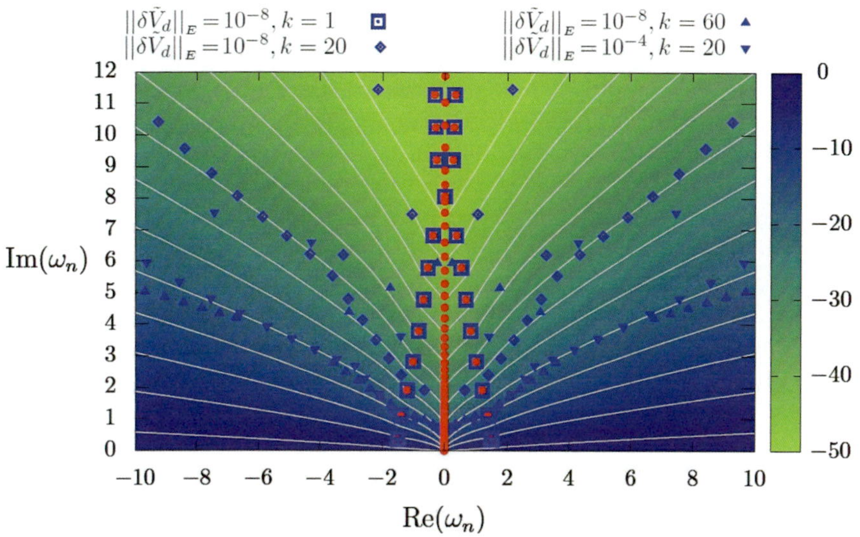

Fig. 10 Pseudospectrum of Schwarzschild BH QNMs $N = 200$ grid points. The white lines correspond to different ϵ pseudospectral contour lines which become more negative close to the eigenvalues (red dots). The colorbar on the right in is logarithmic scale meaning that a value of, e.g. -30 corresponds to $\log[\sigma^{-30}(L)]$ or simply a perturbation in L with energy norm 10^{-30}. On top of the pseudospectrum, perturbed QNMs resulting from deterministic δV_d perturbations to the BH potential are plotted with blue markers that vary according to the oscillation frequency k of the perturbation. They are normalized such that in the energy norm they acquire the specified normalized value shown above the figure. The figure has been adapted from [51]

leads to the inverse-polynomial decay of perturbations at infinity [77, 117–119]. Even though they do not belong in the QNM spectrum of Schwarzschild BHs, they significantly affect the pseudospectrum close to the imaginary axis by bending it further downwards. This is in contrast to the Pöschl-Teller case where the potential decays exponentially on both sides and the branch-cut is absent, thus a smoother pseudospectrum is obtained even at the imaginary axis. These non-convergent modes pose an open issue regarding their existence in asymptotically flat BHs and may be connected to the hyperboloidal approach, which explicitly includes null infinity into the grid. Nevertheless, they do not taint the underlying migration of perturbed QNMs since the opening of the contour lines is a clear-cut imprint of spectral instability.

It is important to note that besides the random perturbations that were utilized in Pöschl-Teller potentials, deterministic perturbations for the Schwarzschild case are more physically relevant for two reasons. Firstly, the results are qualitatively the same as those arising from random perturbations. Secondly, deterministic sinusoidal perturbations to the effective potential may have an astrophysical relevance to explore the effects of perturbations (assessment of "long range/low frequency" versus "small scale/high frequency" perturbations), as well as those arising from quantum gravity phenomenology ("small scale/high frequency" effective fluctuations).

In Fig. 10, besides the pseudospectrum, four cases of different combinations of perturbation norm and frequency are over-plotted, shown with different blue markers. The perturbed spectra confirm once more that perturbations to the potential almost coincide with the pseudospectra contour lines following the behavior found in [49] (shown in Fig. 3). Yet, there is a crucial difference. The fundamental mode is never destabilized, no matter the scale or frequency of the perturbation, in contrast to what the pseudospectrum suggests for sufficiently large ϵ. The same holds for random perturbations. On the other hand, overtones are prone to migrate on the branches of maximal migration as the perturbation frequency increases. This provides a definitive connection of perturbed QNMs, resulting by adding perturbations to the Schwarzschild potential, and pseudospectra branches, which led the authors in [51] to conclude that perturbed QNMs track pseudospectra contour lines, thus "christening" them "Nollert-Price" branches.

5.3 Reissner-Nordström Black Holes

RN BHs are exact solutions of the Einstein-Maxwell equations that include, besides the BH mass M, an electric charge Q. In this case, the lapse function takes the form

$$f(r) = 1 - \frac{2M}{r} + \frac{Q^2}{r^2}. \tag{20}$$

RN BHs possess two horizons, an event horizon at $r = r_+$ and an inner (Cauchy) horizon r_-, beyond which the deterministic nature of the field equations is lost [120, 121]. Their exact forms are

$$r_\pm = M \pm \sqrt{M^2 - Q^2}. \tag{21}$$

RN BHs allow us to perform further tests of geometrical aspects of pseudospectra universality, which could not be performed with Schwarzschild BHs. Due to the existence of two horizons, there exist two choices of spacetime slicings for the hyperboloidal approach [109], namely the *areal radius fixing gauge*, where we fix $r = r_+/x$[1] and the *Cauchy horizon fixing gauge*, where we fix $x_- \to \infty$, that corresponds to the location of the Cauchy horizon at the compactified coordinate. The hyperboloidal approach can be applied in both slicings, and tests have shown that different gauges yield the same results, supporting the geometrical nature of the pseudospectrum [54]. Furthermore, RN BHs possess an extremal limit when $Q = M$, in which $r_+ = r_-$. Approaching this limit gives rise to a new family of near-extremal, purely imaginary and long-lived modes as $Q \to M$ [121–127]. At exact extremality, RN BHs become marginally stable [128, 129] and allow for the formation of local

[1] We remind that $x \in [0, 1]$ is the appropriately compactified radial coordinate through a choice of $g(x)$.

horizon hair [130]. Finally, the gravitational-led QNMs with angular number ℓ are isospectral with electromagnetic-led QNMs with angular number $\ell - 1$ when the BH is extremal [131–134].

Here, we focus on scalar and polar gravitoelectric perturbations, described by the potential [54, 135–137]

$$V = \frac{f(r)}{r^2}\left[\ell(\ell+1) + \frac{r_+}{r}\left(\mu - \kappa^2\nu\frac{r_+}{r}\right)\right], \tag{22}$$

with

$$\nu = 3n_p - 1, \quad \kappa = Q/r_+ = \frac{Q/M}{1+\sqrt{1-(Q/M)^2}} \quad \mu = n_p(1+\kappa^2) - (1-n_p)\mathfrak{m}_\pm,$$

$$\mathfrak{m}_\pm = \frac{1+\kappa^2}{4}\left[1 \pm 3\sqrt{1+\frac{4\kappa^2 A}{(1+\kappa^2)^2}}\right], \quad A = \frac{4}{9}(\ell+2)(\ell-1).$$

In the above parameterization of equations, scalar perturbations are recovered for $n_p = 1$, whereas electromagnetic-led ($\mathfrak{m}_-$) and gravitational-led ($\mathfrak{m}_+$) perturbations (which reduce to electromagnetic and gravitational perturbations of Schwarzschild BHs in the uncharged limit) are recovered for $n_p = -1$.

5.3.1 Pseudospectrum of Reissner-Nordström Black Holes

By using the areal radius fixing gauge, we have calculated the spectra and pseudospectra of RN BHs, for various type of perturbations with a constant grid resolution of $N = 200$ Chebyshev points. Due to the asymptotic flatness of spacetime, similar non-convergent eigenvalues appears in the imaginary axis which from now on we designate with blue dots for clarity. For more details regarding the functions and gauges involved for the hyperboloidal approach, as well as the numerical calculation of the pseudospectrum in the energy norm, we refer the reader to [54].

In Fig. 11 we show the electromagnetic-led and gravitational-led spectra and pseudospectra of a RN BH with charge $Q/M = 0.5$. The pseudospectrum is qualitatively similar to those obtained for Schwarzschild BHs (see Fig. 10), with pseudospectral contour lines designating a spectral instability of QNMs and a logarithmically-extending pattern away from QNMs. In fact, according to [52], the QNM-free regions are asymptotically bounded from below by logarithmic curves of the form

$$\mathrm{Im}(\omega) \sim C_1 + C_2 \log[\mathrm{Re}(\omega) + C_3]. \tag{23}$$

Remarkably, appropriate choices of the constants C_1, C_2 and C_3 can lead to a fit that holds also at intermediate values of $\mathrm{Re}(\omega)$ and even close to the QNMs; a region one can hardly consider as asymptotic.

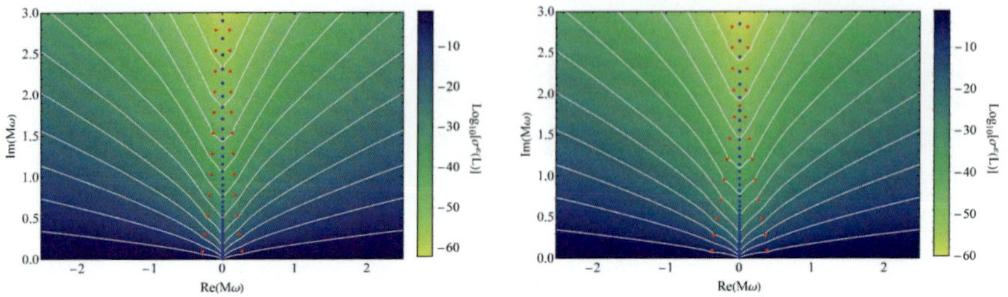

Fig. 11 Pseudospectra of a BH with charge $Q/M = 0.5$. Left: $\ell = 1$ electromagnetic-led QNMs and ϵ-pseudospectra boundaries. Right: Same for $\ell = 2$ gravitational-led QNMs. The contour levels $\log_{10}(\epsilon)$ range from -55 (top level) to -5 (bottom level) with steps of 5. The blue dots designate branch-cut (non-convergent) modes while the red dots designate QNMs. The figure has been taken from [54]

5.3.2 The Extremal Limit

By examining the spectra and pseudospectra of extremal RN for scalar perturbations we find that their overall behavior is the same as the one found for gravitoelectric pseudospectra.

The pseudospectrum in extremality possesses logarithmic asymptotics and exhibits a zero-frequency mode ($M\omega = 0$), associated with the Aretakis instability [128, 129]. In general, modes that reside close to the real axis are significant to transient instabilities that are typically resolved with pseudospectra, as discussed in Sect. 4.2. Nevertheless, the existence of branch-cut modes that infect the spectrum with noise, and subsequently the pseudospectrum at the imaginary axis, forbids us from drawing any solid conclusions regarding transient phenomena in extremal spacetimes, since the crossing of the contour lines to the side of the complex plane where unstable modes reside may be tainted by the branch-cut modes.

For gravitoelectric QNMs, the $\ell - 1$ electromagnetic-led and ℓ gravitational-led modes of extremal RN geometries are isospectral [131, 134]. Figure 12 shows that not only the spectra coincide, but also the pseudospectra for such perturbations at extremality. Thus, the isospectrality taking place here has a stronger nature and occurs because not only the poles, but the Green's function as a whole coincide in the two cases.

5.3.3 Asymptotic Universality of Pseudospectra

We finally discuss the asymptotic behavior of pseudospectra for different kinds of fluctuations. Figure 13 shows the asymptotic structure of three spectral problems, namely the $\ell = 2$ scalar, electromagnetic-led and gravitational-led QNMs of a subextremal RN BH. It is evident that the lower pseudospectra contours of the three distinct

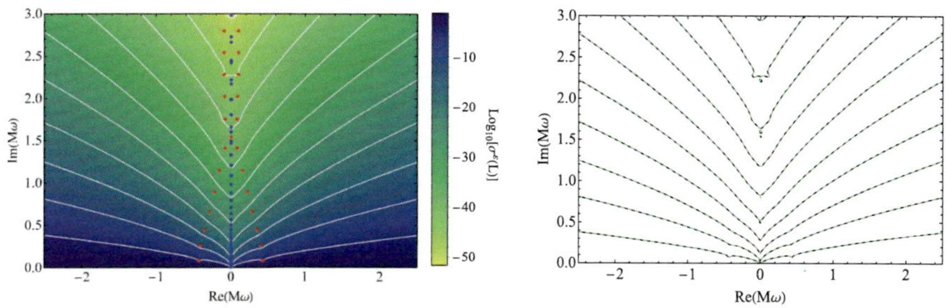

Fig. 12 Left: $\ell = 2$ gravitational-led QNMs (red dots) and ϵ-pseudospectra boundaries of a RN BH with $Q/M = 1$. Right: superimposed $\ell = 1$ electromagnetic-led (black lines) and $\ell = 2$ gravitational-led (green dashed lines) ϵ-pseudospectral contours of a RN BH with $Q/M = 1$. In both cases, the contour levels $\log_{10}(\epsilon)$ range from -50 (top level) to -5 (bottom level) in steps of 5. The blue dots designate branch-cut (non-convergent) modes. The figure has been taken from [54]

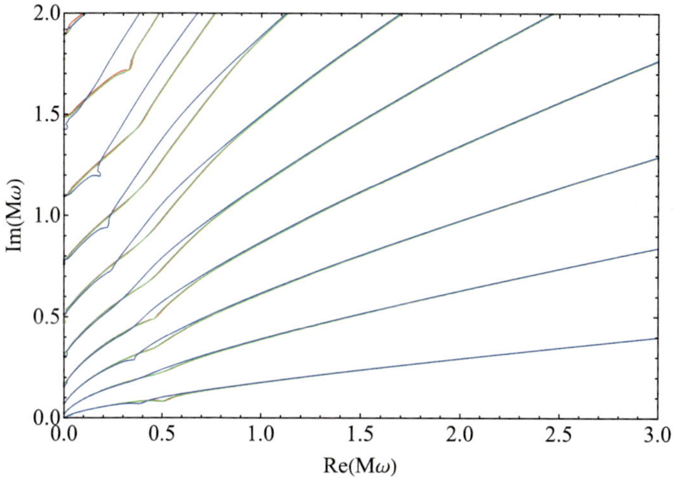

Fig. 13 Superimposed scalar (red), electromagnetic-led (green) and gravitational-led (blue) pseudospectral levels for a RN BH with $Q/M = 0.5$. All perturbations share the same angular index $\ell = 2$. The contour levels $\log_{10}(\epsilon)$ range from -50 (top level) to -5 (bottom level) with steps of 5. The figure has been taken from [54]

effective potentials coincide already at small frequencies and almost from the respective QNM for each level set, while higher contours begin to coincide at frequencies with larger real parts. This poses as a numerical evidence of the asymptotic universality of pseudospectra shared by a whole class of BH potentials, with similar logarithmic patterns.

6 Pseudospectrum of Horizonless Exotic Compact Objects

Our current understanding of stellar evolution suggests that astrophysical compact objects can experience the process of gravitational collapse when the interaction of stellar matter cannot counterbalance the star's self-gravity. The most natural outcome is a collapse to a BH. Nevertheless, the singularity that appears through this classical process may hint at an incomplete model of gravitational collapse. In order to avoid singularities, we can consider either localized corrections around the singularity, leaving the exterior of the BH effectively classical, or slightly larger scale corrections that are still localized but can extend slightly outside the event horizon, thus removing the event horizon. Currently, a plethora of horizonless ECOs has been constructed through the latter aforementioned technique, such as classical boson stars [138–140], gravastars [141], or semi-classical ECOs with exotic near-horizon structure [142, 143] such as firewalls [144] and fuzzballs [145–147].

ECOs are interesting not only due to their composition, but also their dynamical behavior under perturbations. Their QNMs can differ dramatically from those of BHs, due to the different boundary condition that has to be imposed at their surface. In particular, QNMs of BHs require purely ingoing waves at the event horizon while for horizonless ECOs that possess a centrifugal barrier, slightly outside the would-be event horizon, the boundary condition may become (at least partially) reflective. This in turn, can lead to fundamental modes that are extremely long-lived, when the ECO is ultracompact, which are clearly distinguishable from those of BHs [12, 38, 39, 148]. These long-lived QNMs, which are obtained with BH perturbation theory at the linear level, can source instabilities in ECOs when non-linear effects are taken into account [149–151]. Thus, constructing stable ultracompact horizonless objects that can mimick BHs seem to be a non-trivial task.

The motivation for studying the pseudospectrum of spherically-symmetric horizonless ECOs stems from the fact that they both exhibit arbitrarily long-lived QNMs (that reside close to the real axis of the complex plane) and, at the non-linear level, they become unstable due to various instability mechanisms [151, 152]. Such a problem is in total analogy to the results discussed in Sect. 4.2 regarding hydrodynamic flows, where a linear analysis fails to capture the phenomena found in experiments, and the pseudospectrum resolved the inconsistency between linear analyses and full non-linear experiments.

6.1 Spherically-Symmetric Horizonless Compact Objects: A Simplistic Approach

Even though there are plenty of ECO configurations that can be constructed for the purpose of studying their QNMs and pseudospectra, such a task is highly demanding. As in neutron stars, a complete ECO model requires knowledge of the object's internal structure and how its matter content interacts with external fluctuations.

In [56], we have instead taken a simplified approach to construct a spherically-symmetric ECO effectively by hand that reproduces the whole phenomenology of realistically constructed ECOs. Our analysis assumes a Schwarzschild BH spacetime on the domain $r \in [r_0, \infty)$, where $r = r_0 = (1 + \mathcal{E})r_+$ is the ECO surface and $\mathcal{E} \ll 1$ for ultracompact objects. At $r = r_0$ we impose a purely reflective boundary condition, i.e. for a perturbation field Ψ we assume the Dirichlet condition $\Psi(t, r_0) = 0$. Even though our assumption of pure reflection of waves at the ECO surface is very simplistic, there are other choices where one can introduce partial reflection and absorption [152–154]. In the interest of studying the spectrum and pseudospectrum on a static background that does not absorb energy, we will assume a purely reflective ECO surface, though the results that follow can be extrapolated to other choices of partially reflective boundary conditions. In the frequency domain, where our results will be interpreted, the boundary condition at infinity remains the same as that of an asymptotically-flat BH, i.e. purely outgoing waves, while at the surface of the ECO we impose a Dirichlet boundary condition $\Psi(r_0) = 0$.

The majority of the hyperboloidal approach and the energy norm is used exactly as described in Sect. 4.4. Specifically, at null infinity, the outgoing boundary condition is imposed automatically. At the ECO surface though, the purely reflective boundary condition is achieved by removing the row and column of the discretized operator L that corresponds to $r = r_0$, or equivalently $x = x_0$ at the compactified coordinate system utilized for numerical calculations. For a detailed analysis of the equations involved to properly set up the wave equation of perturbations of ECOs in hyperboloidal coordinates, we refer the reader to [56, 155].

6.2 Pseudospectrum of Horizonless Objects with a Pöschl-Teller Potential

Following the steps of [51], we first analyze a simple case where the ECO potential is replaced with the Pöschl-Teller function (17), where at the right of the potential, that corresponds to infinity, the boundary condition is kept outgoing and at the left of the potential we use a Dirichlet boundary at a finite distance $x = x_0$ where the hypothetical ECO surface resides. In Fig. 14, we show the spectra and pseudospectra resulting from the aforementioned system. It is apparent that the QNMs of the Pöschl-Teller ECO are much more long-lived than those of the Pöschl-Teller potential itself, which is an illustration of stable trapping that occurs in ECOs. Moreover, ECO modes suffer from spectral instability, like any other dissipative system discussed. Most importantly, the pseudospectrum suggests that the fundamental QNM and first overtone are able to cross the real axis, into the unstable part of the complex plane, for $\epsilon \simeq 0.02$. This is suggestive of a perturbation-triggered instability, where when one adds a random or deterministic perturbation into the effective potential, the long-lived modes may migrate to the lower half of the complex plane where unstable QNMs reside. Nevertheless, by employing the aforementioned strategy, we reached

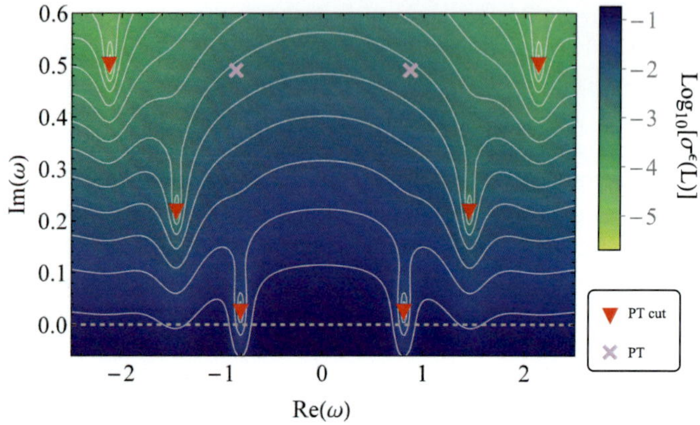

Fig. 14 QNMs (red triangles) and pseudospectrum (white curves) of the L operator for the Pöschl-Teller potential with a Dirichlet boundary condition at $\mathcal{E} = 10^{-3}$ (PT cut) and $N = 200$ spatial discretization points. Contour lines correspond to values between -4.7 (uppermost contour) and -1.1 (lowest contour) with steps of 0.3 in the log scale. The horizontal dashed line indicates the position where $\text{Im}(\omega) = 0$. For comparison, we also include the QNMs of the Pöschl-Teller with outgoing boundaries on the right and ingoing on the left of the domain (PT, pink crosses). The figure has been adapted from [56]

the same conclusions as those for BHs: the fundamental mode is spectrally stable as long as the asymptotic structure of the potential remains intact.

Despite that, the protrusion of the pseudospectral contour lines into the unstable part of the complex plane indicates a tendency with which pseudoresonances can be triggered easily, such as those found in hydrodynamics [95], if the evolution operator had a time-dependent external source, whose Fourier transform had a real frequency close to the pseudospectrum protrusion regime [53].

6.3 *Pseudospectrum of Horizonless Compact Objects*

The analysis in Sect. 6.2 has been repeated for the scalar and axial gravitational perturbation potentials of a Schwarzschild BH, where instead of ingoing waves at the event horizon, we impose a Dirichlet boundary at the surface of the ECO $r = r_0$. In this case, $\mathcal{E}$ controls the compactness of the ECO. The smaller it gets, the closer the ECO surface gets to the would-be event horizon, and the object becomes more compact. This allows for perturbations to be trapped between the photon sphere and the ECO surface, leading to long-lived modes in the frequency domain and echoes in the ringdown signal [37, 38, 156, 157]. The systematics of calculating the QNMs and pseudospectra are detailed in [56] and in the previous sections.

Figure 15 demonstrates, as expected, similar results as those found in the Pöschl-Teller case discussed above. Scalar and axial gravitational QNMs of ECOs with

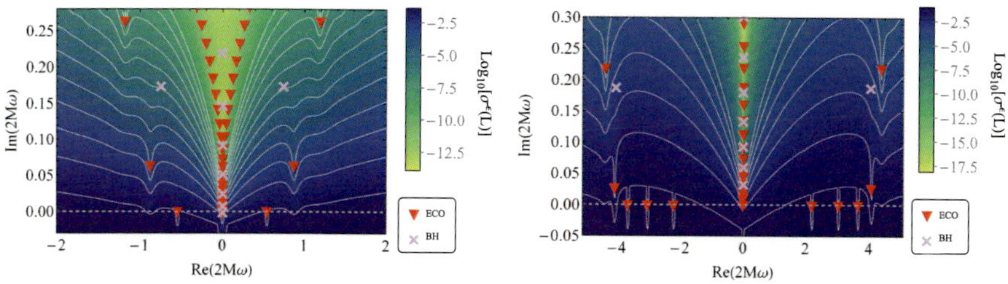

Fig. 15 Left: ECO axial gravitational $\ell = 2$ QNMs (ECO, red triangles) with compactness $\mathcal{E} = 10^{-3}$. The contour lines range from -6.5 (uppermost contour) to -1.7 (lowest contour) with steps of 0.6 in the log scale. Right: ECO scalar $\ell = 10$ QNMs (ECO, red triangles) with compactness $\mathcal{E} = 10^{-2}$. The contour lines range from -6.5 (uppermost contour) to -1.7 (lowest contour) with steps of 0.6 in the log scale. In both figures, we also include the respective field QNMs of a Schwarzschild BH (BH, pink crosses). The figure is adapted from [56]

varying compactness $\mathcal{E}$ suffer from spectral instabilities, as their pseudospectra suggest, and the long-lived QNMs close to the real axis protrude to the unstable side of the complex plane. Therefore, with a similar mechanism as in the Pöschl-Teller potential, a time-dependent source with a real frequency in the vicinity of the protrusion of the contour lines could trigger a pseudoresonance that can, in turn, lead to an instability, in a similar manner as it is found for ultracompact objects when higher than linear-order perturbations are taken into consideration [149–152].

Nonetheless, when adding random perturbations to the effective potentials of scalar and axial gravitational perturbations of the ECO we find interesting results. For less compact objects, such as $\mathcal{E} = 1$ (see Fig. 16), the ECO QNMs have an imaginary part that is larger than the fundamental BH mode with the same mass. As the compactness increases (decreasing $\mathcal{E}$), the imaginary parts of the ECO QNMs become significantly smaller than those of a BH. We find the same results for both types of field perturbations. Once more, stable trapping, and the formation of quasibound states [158–165], is the mechanism in hand that leads to the change of the fundamental mode when comparing a BH and an ECO of the same mass. The protrusion of the cases shown in Fig. 15 occurs for $\epsilon \lesssim 10^{-3}$, for the fundamental QNM, and $\epsilon \lesssim 10^{-2}$ for the first overtone.

In order to further understand the protrusion's capability for an instability of the ECO under study, we calculate the perturbed QNMs of the system by adding random and deterministic perturbations δL (with energy norm $\|\delta L\| \simeq \epsilon$) to the ECO potential, as was performed in [51]. The fundamental mode always appears to be spectrally stable up to perturbations of energy norm $\epsilon \lesssim 10^{-1}$ and higher, suggesting that a migration to the unstable side of the complex plane cannot be triggered. When the ECO is less compact and its surface reaches the photon sphere, then the perturbation response is significantly different. Pseudospectra are much harder to protrude into the unstable side of the complex plane for sufficiently small values of ϵ. Nevertheless, the ECO QNMs are still spectrally unstable. From Fig. 16, it is apparent that the spectral stability of the fundamental ECO QNMs against random

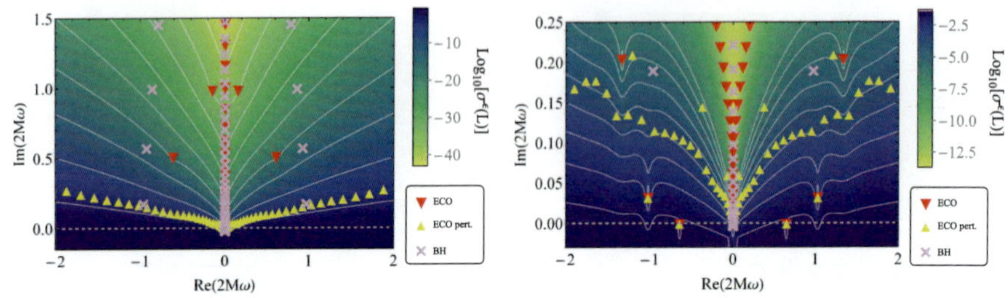

Fig. 16 Left: ECO scalar $\ell = 2$ QNMs (ECO, red triangles) with compactness $\mathcal{E} = 1$, and the perturbed QNMs (ECO pert., yellow triangles) resulting by adding a random perturbation to the effective potential of energy norm 10^{-2}. The contour lines range from -44 (uppermost contour) to -3.5 (lowest contour) with steps of 4.5 in the log scale. Right: Same as left with compactness $\mathcal{E} = 10^{-3}$, and a random perturbation to the effective potential of energy norm 10^{-1}. The contour lines range from -6.5 (uppermost contour) to -1.7 (lowest contour) with steps of 0.6 in the log scale. On both figures, we also include the scalar $\ell = 2$ QNMs of a Schwarzschild BH for comparison (BH, pink crosses). The figure is adapted from [56]

and deterministic perturbations is present only when their imaginary part is of the same, or smaller, order than the imaginary part of the fundamental Schwarzschild BH QNM.

Finally, it is important to notice the case where $\ell \gg 1$, specifically the case of scalar $\ell = 10$ QNMs in Fig. 15, and how it connects to the fact that ECO instabilities are triggered in the parametric region of high angular modes. In the mentioned figure, we observe at least four QNMs (the fundamental and the first three overtones) that protrude to the unstable part of the complex plane. This suggests a stronger tendency to trigger pseudoresonances, which may make the ultracompact horizonless object unstable when nonlinearities are considered, as already shown to occur [149, 150].

This further justifies the strength of the pseudospectrum to potentially predict nonlinear transient effects even though the operator in hand is still linear. In particular, the transitive protrusion of pseudospectral contours through the real axis, along with the notion of pseudoresonances, can be combined with BH perturbation theory in GR at second order [166–168] to give rise to a possible bootstrap instability mechanism, discussed in [56] and exhaustively detailed in [53]. Such a mechanism seems perfectly suited for ultracompact objects since pseudoresonances may lead to the growth of external time-dependent sources close to protrusions, which break down the perturbative expansion and leads to nonlinear dynamical instabilities. Therefore, in the gravitational cases studied in this section, the bootstrap instability mechanism is provided by the GR perturbative structure that sources second-order gravitational perturbations through first-order gravitational ones [53, 56], thus the characterization as a *bootstrap mechanism*.

7 Destabilizing the Fundamental Quasinormal Mode

In the previous sections, the pseudospectra of BHs and ECOs were studied, together with the migration of QNMs when the effective potential of the aforementioned compact objects was perturbed with random or deterministic (sinusoidal) perturbations. Random perturbations still do not possess any reasoning on why they may be important physically, but sinusoidal ones can be interpreted as quantum-gravity oriented, though this aspect is still up for discussion. Nevertheless, both perturbations introduced to the BH or ECO potentials led to the same conclusion: even though the pseudospectrum suggests that the fundamental mode can be destabilized by generic perturbations, actually perturbing only the effective potential would lead to the migration of overtones, while the fundamental mode was always remaining intact. Therefore, the fundamental QNM seems to admit a special status when it comes to perturbing the BH potential; a characteristic that is quite reassuring for BH spectroscopy.

The simplistic approaches that regarded BHs as isolated objects in vacuum or in binaries made them perfect laboratories for tests of GR, as long as their environment was ignored. Recent works [14, 15, 169, 170] have shown that not taking into account environmental effects, as well as other astrophysical uncertainties, may compromise precision tests of GR, especially when the shadow detections of dark compact objects in the galactic core of M87 [171] and the Milky way [172] solidified the existence of extremely rich environments around BHs, such as accretion disks [169, 173–176]. Moreover, the suggestion that large length-scale, galactic dark matter halos act as "galactic glue" and contribute further non-luminous mass to the galaxy [177–179] made even more clear the need for taking into account both short and large-scale environmental effects when performing precision GW astrophysics. In this section, we will explore the possibility of not ignoring environmental effects, and in particular if they can lead to spectral instabilities in Schwarzschild BHs surrounded by matter shells.

To model the ringdown signal of a BH merger, by using QNMs, we usually take into account a part of the spectrum obtained in vacuum. In order to consider also surrounding matter, we will add a bump in the effective potential beyond the photon sphere. For simplicity, we consider the effective potential of a Schwarzschild BH and add a small bump to it of the form [12, 55]

$$V_\epsilon = V + \epsilon V_{\text{bump}}, \tag{24}$$

with $\epsilon \ll 1$ and V_{bump} a generic matter shell that induces a bump located at $r_* = a$, such that the bump decays faster that V. By solving the frequency domain Eq. (3) with QNM boundary conditions, and replacing its potential with that of Eq. (24), we obtain the perturbed QNMs $\omega_n^{(\epsilon)}$. The bump drives the original, unperturbed, fundamental mode $\omega_0^{(0)}$ to migrate continuously in the complex plane along a curve that depends on the distance between the BH potential peak and the bump a. To avoid confusion, we will drop the subscript 0 from the perturbed QNMs since we do not

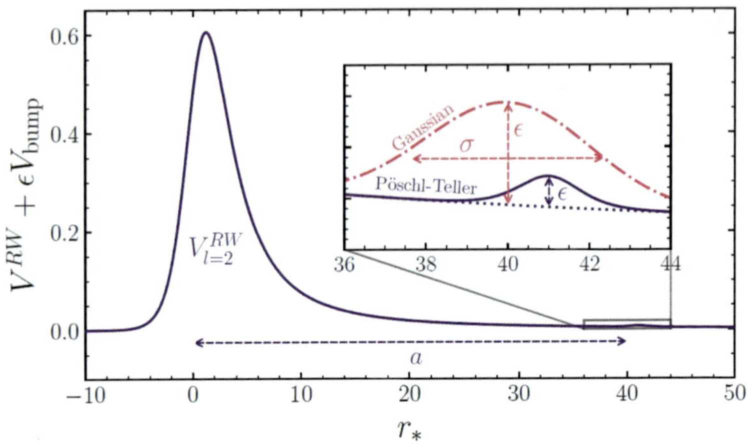

Fig. 17 Schematic illustration of the $\ell = 2$ axial gravitational (Regge-Wheeler) potential V^{RW} perturbed by a Pöschl-Teller bump (solid line in the inset) and by a Gaussian bump (dot-dashed line). The figure has been taken from [55]

necessarily know if they are fundamental anymore after their migration. We further define the variation $\Delta\omega^{(\epsilon)} = |\omega^{(\epsilon)} - \omega^{(0)}|$. When the quantity $|\Delta\omega^{(\epsilon)}/\omega^{(0)}| \gg \epsilon$ then we either observe a destabilization of the fundamental mode due to disproportional migration in the complex plane or due to the phenomenon of *mode overtaking*, where unperturbed overtones become fundamental modes as a increases. In what follows, to avoid confusion we will denote the perturbed fundamental mode as ϖ, no matter if it is the original, the overtaken or the destabilized QNM.

We consider the potential V to be that of $\ell = 2$ axial gravitational perturbations on a Schwarzschild BH with mass $M = 1/2$ and two types of bumps added in distance a, namely the Pöschl-Teller potential

$$V_{\text{PT}}(r_* - a) = \text{sech}^2(r_* - a), \tag{25}$$

and a Gaussian peak

$$V_{\text{G}}(r_* - a) = \exp\left(-\frac{(r_* - a)^2}{2\sigma}\right), \tag{26}$$

where σ is the Gaussian distribution's width. An illustration of the "toy model" potential is demonstrated in Fig. 17[2].

The migration of QNMs due to a Pöschl-Teller bump with $\epsilon = 10^{-6}$ is shown in Fig. 18. The left panel depicts the displacements of QNMs in the complex plane as

[2] We stress that even though here we add the bump by hand in the BH potential, a matter shell can induce a bump with the same properties as those we observe here. See the Appendix A of Ref. [55] for a more detailed discussion.

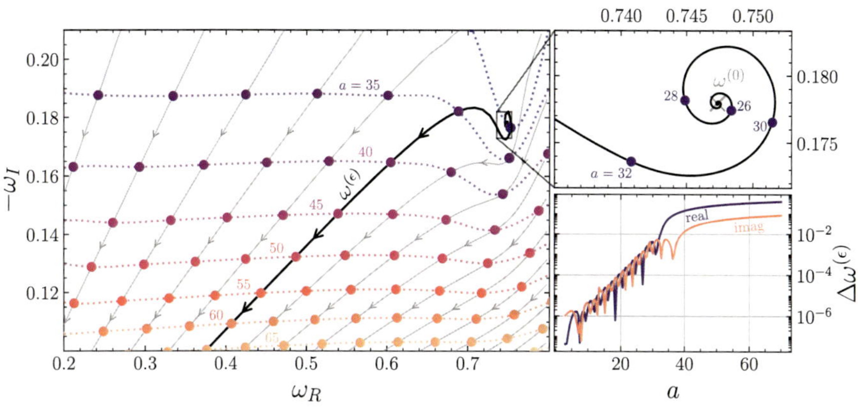

Fig. 18 Migration of $\ell = 2$ axial gravitational QNMs as a function of the bump position a for a Pöschl-Teller perturbation with $\epsilon = 10^{-6}$. Left: Individual modes migrate along the black lines. The mode $\omega^{(\epsilon)}$ (bold line) reduces to the Schwarzschild fundamental mode when $\epsilon \to 0$, and the arrows indicate the direction of migration as a increases. Modes with the same value of a have the same color, and are connected with dotted lines. Top right: Zoom-in around the unperturbed fundamental QNM $\omega^{(0)}$. Bottom right: Real and imaginary parts of the migration distance $\Delta\omega^{(\epsilon)}$ of the perturbed fundamental QNM. We use units such that $2M = 1$. The figure has been taken from [55]

we increase a. Specifically, the bold continuous curve traces the original fundamental mode's migration. In the top right panel, we zoom in on the trajectory in the close vicinity of $\omega^{(0)}$ for small to moderate a, while the bottom right panel shows the change in the variation of the real and imaginary parts of the perturbed QNM. The variation exceeds the perturbation scale by at least four order of magnitude, so we can conclude that even the fundamental mode of BHs can be destabilized, as conjectured by the pseudospectrum, by including appropriate and astrophysically-relevant perturbations to the effective potential.

Figure 19 presents the whole parametric analysis performed in [55], where not only a but also ϵ and σ is varied, together with a different perspective on mode migration and overtaking by overtones. The top left panel details the discontinuous overtaking of the original fundamental mode. For small a, the fundamental mode remains the same even though it continuously migrates in the complex plane. Around $a \sim 30M$, an overtone $\omega^{(\epsilon)}$ takes over discontinuously and becomes the new fundamental mode $\omega^{(0)}$. After some further overtakings, the original fundamental mode has been completely destabilized and migrated up to order $\mathcal{O}(10^{-1})$. The bottom left panel demonstrates this behavior of migrating QNMs and how the overtaking occurs at the real part of the QNM frequency. The first regime, shown in white, consists of a quasi-stable region where the original fundamental mode has not been overtaken yet. In the grey region, three consecutive overtakings take place, where overtones become the new perturbed fundamental modes until the migration is so large that leads to a completely destabilized fundamental mode in the orange region. In the latter regime, the separation between the real parts of two consecutive modes

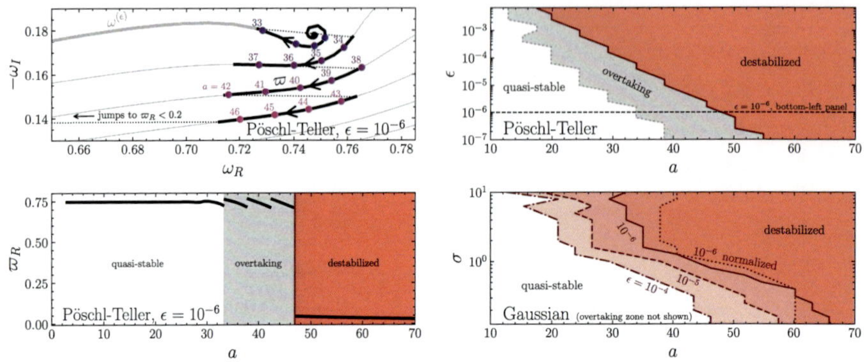

Fig. 19 Top left: Migration of the fundamental mode $\omega^{(\epsilon)}$ in the complex plane when a is increased, for a Pöschl-Teller bump with $\epsilon = 10^{-6}$. Discontinuous jumps in ω (in the plot denoted as ϖ which designates the new perturbed fundamental QNM) are shown with dotted lines. Bottom left: Variation of the real part of ϖ_R from the top left panel as a function of a. Top right: Phase diagram of ϵ vs a for a Pöschl-Teller bump. Bottom right: "Phase diagram" of σ vs a for a Gaussian bump with different values of ϵ. The figure has been taken from [55]

is given to a very good approximation by the equality $\omega_{n+1,R} - \omega_{n,R} = \pi/a$. This is an expected characteristic of trapped modes between two potential barriers located at distance a from each other and leads to repetitions of photon sphere excitations in the ringdown signal, i.e. echoes. Therefore, the overtaking and eventual destabilization of the fundamental mode is due to stable trapping and the appearance of longer-lived modes (with large decay timescales) that overtake the unperturbed QNMs, which decay faster than the trapped, long-lived ones.

We have repeated the analysis in a significant volume of the parameter space for Pöschl-Teller and Gaussian bumps by varying ϵ, σ and a in order to sketch phase diagrams that characterize the destabilization. In the right column of Fig. 19 we demonstrate that the larger ϵ is (for the Pöschl-Teller case), the easiest is to form quasibound states between the two potentials since the destabilization takes place for shorter distances a. It is clear that no matter how small we choose the bump height to be, there is always a critical a beyond which the destabilization of the fundamental QNM occurs. Similar results are shown for the Gaussian bump perturbation, where the wider the bump is, the easier is to destabilize the fundamental mode. From the above results, it is quite clear that the fundamental mode does not hold a distinguished status in the spectrum, as the analyses in [51, 56] suggest, but can also migrate disproportionately as the pseudospectrum recommends. The question that now follows is if a destabilized fundamental mode, due to environmental effects, is present and dominant in the ringdown waveform of such a perturbed potential. We discuss this aspect in the following section.

8 Destabilized Fundamental Mode Detection in Ringdown Signals

Until this point, we have been discussing the stability of the QNM spectrum from a *formal* perspective, using the powerful tool of the pseudospectrum. We have concluded that, in general, the QNM spectra of BHs and ECOs in GR are spectrally unstable. However, what is the actual consequence of this for GW astronomy, and in particular for the BH spectroscopy program? In the presence of a perturbation in the effective BH potential, is the ringdown signal also destabilized in some sense? A key point in this question is that the QNM spectral instability is obtained in the frequency domain, while GW interferometers measure a strain in time. The two pictures should be equivalent and related by a simple Fourier transformation. However, this is only true if we have access to the full signal in time *ad eternum*. In practice, this is impossible in an experiment due to systematics, such as instrumental noise.

This situation is reminiscent to what happens in ECOs. Their QNM spectra are extremely different from a BH in GR, in particular, they have long-lived trapped modes. However, by causality, the time-domain response of a BH and an ECO is the same for the time necessary for a perturbation to travel to the surface of the ECO, which is close to the BH horizon limit, and then be reflected back [38, 180].

Here, we solve the equation governing axial gravitational perturbations in the time domain ($s = 2$ in Eq. (2)), using a two-step Lax-Wendroff method with hyperboloidal layers [111, 181–184] (we refer the reader to the references mentioned for more details on this numerical scheme). We focus on the case where the perturbation to the effective potential is described by a Gaussian bump, as in Eq. (26). However, our qualitative conclusions are independent of the actual shape of the perturbation. We considered other ones, such as a concave bump with negative ϵ; and a bump decaying as r^{-3} at large distances, which is exactly zero for small enough r, mimicking a perturbation due to a thin shell of matter. The only relevant aspect of the perturbation is that it introduces a peak in the effective potential, otherwise, the instability is not present as seen in the previous section. As initial data for the perturbation, we prescribe a localized Gaussian pulse

$$\Psi\Big|_{t=0} = 0 \ , \quad \frac{\partial \Psi}{\partial t}\Big|_{t=0} = e^{-(r_*-5)^2/2} \,, \tag{27}$$

but, again, our qualitative conclusions are independent of this particular choice.

Finally, in order to extract the ringdown spectra, we fit the time-domain waveform to a superposition of exponentially damped sinusoids

$$\Psi(t) = \mathrm{Re} \sum_{n=0}^{N-1} A_n e^{-i(\omega_n t - \phi_n)} = \sum_{n=0}^{N-1} A_n e^{\omega_{nI} t} \cos(\omega_{nR} t - \phi_n) \,, \tag{28}$$

where the index n labels the different modes we find by fitting, and it does not necessarily coincide with the overtone number. Each mode is characterized by four parameters: an amplitude A_n, a phase ϕ_n, and the real and imaginary parts of the QNM frequency $\omega_n = \omega_{nR} + i\omega_{nI}$. We will find that several QNMs could have similar decay times and hence comparable amplitudes. When this happens, a good fit of the waveform requires a relatively large number of modes N. The largest number of modes we will look for is $N = 8$, corresponding to $8 \times 4 = 32$ fitting parameters.

8.1 *Ringdown Waveform Stability*

First, let us clarify what we mean by destabilization here. We will say that *destabilization* occurs when a quantity that characterizes the BH's behavior under an external perturbation, such as its waveform amplitude, changes by an amount much larger than the magnitude of the perturbation (in this case introduced in the effective potential).

Figure 20 shows the resulting ringdown waveform from the scattering of Gaussian pulses for different bumps. We define $t_{\max}$ as the instant at which $|\Psi|$ is maximum. In Schwarzschild ($\epsilon = 0$, black curve), one obtains the typical exponentially damped quasinormal ringing followed by the power-law tail due to backscaterring of radiation on the background curvature [75]. The larger the ϵ, the larger the difference in the waveform with respect to vacuum. For earlier times, when $t - t_{\max} \lesssim a$, the

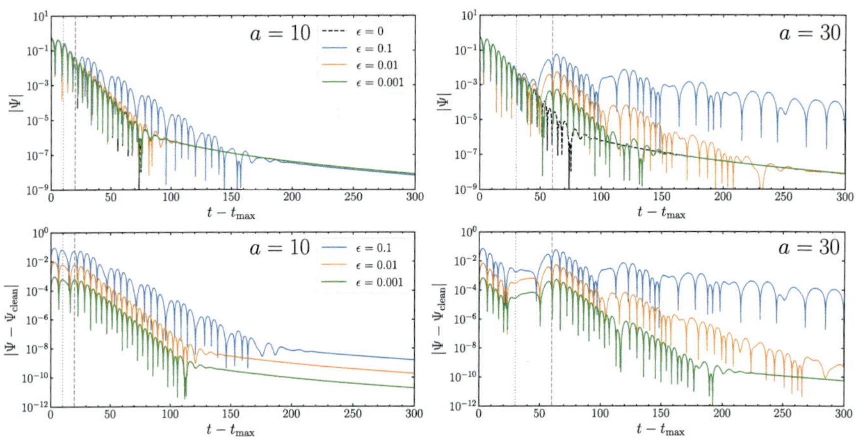

Fig. 20 Top panel: absolute value of the waveform arising from the scattering of the Gaussian pulse of Eq. (27) for bumps with different amplitudes ϵ, located at two selected distances a from the main peak. The bump's width σ in Eq. (26) is fixed at $\sigma = 0.5$. Echoes are apparent when the bumps has a large separation from the photon sphere ($a = 30$). The dotted and dashed vertical gray lines correspond to $t - t_{\max} = a$ and $2a$, and they illustrate how the delay between echoes is related to the size of the "cavity" between the two maxima in the perturbed potential. Bottom panels: Absolute value of the difference between the ringdown waveforms shown in the top panel and the unperturbed clean waveform without a bump ($\epsilon = 0$). The figure has been adapted from [57]

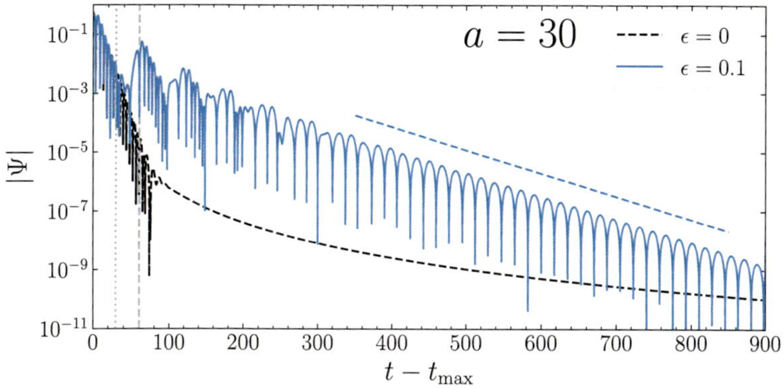

Fig. 21 Ringdown waveform for $a = 30$ and $\epsilon = 0.1$ over long time windows, where the new perturbed fundamental QNM dominates the late-time behavior of the signal. The dashed blue line represents its expected decay, which corresponds to the bottom blue cross with smallest $|\omega_I|$ in the top panel of Fig. 23. The figure was taken from [57]

absolute difference on the waveform with respect to the unperturbed potential case scales linearly with ϵ. The *prompt* ringdown signal close to the waveform peak is thus not destabilized in the sense defined above. This is the portion of the ringdown with the most interest to GW astronomy, meaning such differences are not expected to be observable with the singal-to-noise ratios (SNRs) achievable by current interferometers [12, 148].

After this initial regime, for $t - t_{\max} \gtrsim 2a$, we observe the repetition of the original prompt ringdown. These are echoes of waves trapped between the two peaks of the effective potential that "leak" out to infinity. At late times, the signal's frequency content can be drastically different. Here, the waveform is well described by a superposition of the long-lived QNMs of the perturbed spectra, as illustrated in Fig. 21. For smaller ϵ, the power-law tail can hide the difference in the late-time behavior with respect to the Schwarzschild waveform.

Typically, larger bumps lead to QNM spectra with longer damping times, which thus decay slower and survive longer before being dominated by the power-law tail. For large a there is a hierarchy in the timescales between the prompt ringdown, which is excited at the photon sphere, and the travel time of waves characterizing the *cavity* between the two peaks of the effective potential, located at the photon sphere and at the bump. Thus, a pulse bounces back and forth within the cavity and gradually loses its high-frequency component, which tunnels out more easily. This produces a sequence of echoes repeating at a characteristic frequency defined by the cavity size and damped on a timescale defined by the transmission coefficient of the small peak, as shown in Fig. 20 (see also Refs. [38, 39, 148, 157] for similar behavior when the bump is arbitrarily close to the horizon). These two scales determine the QNM spectrum of the bumpy potential, which is unstable and therefore can be non-perturbatively different from the $\epsilon = 0$ case.

The prompt ringdown is excited mainly at the peak of the effective potential, which is close to the photon sphere. If we place the bump closer to it, it changes the shape of the peak and, consequently the frequency content of the prompt ringdown also. If the bump is farther, the spectrum also changes because the QNMs are sensitive to the entire potential. However, the wave train excited at the peak should be similar to the one in the case of the unperturbed potential. Part of it will be reflected as it meets the bump, and the other will tunnel out to far-away distances. The reflection coefficient is of order $\mathcal{O}(\epsilon)$ for any frequency, and consequently, the change in the prompt ringdown also scales with this factor, designating ringdown stability at early times.

In the prompt ringdown, the relative difference in the waveform $|\Psi - \Psi_{\text{clean}}| / |\Psi|$ scales linearly with ϵ, while in the echo-dominated regime the difference is larger. The reason for this is the following: the original ringdown signal decays as $e^{-\omega_I t}$. Each reflection of the waves in the cavity reduces their order of magnitude by $\sim \epsilon$. On the other hand, each back-and-forth bounces inside the cavity occurs on timescales of $t_{\text{bounce}} \sim 2a$. This means that the amplitude of the n-th echo will be larger than the ringdown by a factor of $\left(\epsilon/e^{-\omega_I t_{\text{bounce}}}\right)^n$, or $\epsilon^n e^{n2a|\omega_I|}$. When the power-law tail starts dominating, the modification returns to order ϵ because the tail can be formally seen as a "direct" zero-frequency signal.

8.2 *Implications on Black-Hole Spectroscopy*

Having understood the stability problem regarding the ringdown, we focus on the core procedure of BH spectroscopy, which is to compare QNM spectra of compact objects with the frequencies extracted by fitting the ringdown with damped sinusoids. However, there are two different questions:

1. How the spectral instability affects the prompt ringdown, which is the louder portion of the signal and consequently of more interest for GW astronomy.
2. If the full waveform, in particular the late-time content, is well described by the destabilized QNM spectrum, including the long-lived trapped modes.

To answer these, we use two different portions of the ringdown signal when fitting the time-domain waveforms. These are highlighted in Fig. 22 . In both cases, we discard times $t - t_{\text{max}} \lesssim 5$, where there is contamination from the direct signal coming from the initial data [63]. When studying the full signal (shaded in blue), we discard the portion of the waveform dominated by the power-law tail. For the prompt ringdown analysis (shaded in green), we only consider the portion of the signal before the appearance of the first echo ($t - t_{\text{max}} \lesssim a$). In every fit, the starting time is varied to ensure convergence of the extracted frequencies.

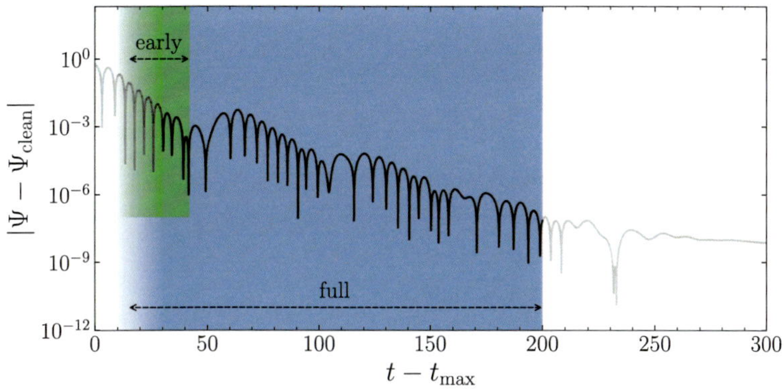

Fig. 22 The portion of the waveform used for the damped-sinusoid fitting in the two different regimes of interest. For completeness, this waveform corresponds to the case $a = 30$, $\epsilon = 0.01$, but the same procedure applies to other examples. The figure has been taken from [57]

8.2.1 The Full Time-Domain Signal

We begin by addressing the problem of the full signal. Figure 23 demonstrates results for bumps located at $a = 10$ and $a = 30$. The fitted (extracted) frequencies from the time-domain signals are shown with dots with different shades, where darker dots correspond to later starting times. The crosses, instead, are the QNM frequencies computed in the frequency domain, with the black cross denoting the fundamental axial QNM of Schwarzschild ($\epsilon = 0$) and the rest of the crosses denote the QNMs for specified bump sizes ϵ. We observe that the fitted frequencies for the bump closer to the potential peak ($a = 10$) are in very good agreement with the ones predicted by the frequency-domain computations for all values of ϵ presented, when the fitting time starting point is larger than the prompt response time, where we avoid initial data contamination. The minor discrepancies can be attributed to numerical error and contamination by the initial data and the power-law tail. For the further bump ($a = 30$), the destabilization of the spectrum is more noticeable. In this case, the new perturbed QNMs decay slower. After one echo, the waveform transitions to a combination of new QNMs until their amplitude becomes so small that they are masked by the late-time power-law tail. For $\epsilon = 0.1$ the ringdown can even transition to a clean exponential decay controlled by the new perturbed fundamental mode, as illustrated in Fig. 21. This allowed us to extract multiple slowly decaying modes from the fit when $\epsilon \gtrsim 0.01$.

The fitting procedure becomes more complicated when $\epsilon \lesssim 0.001$. For these parameters, we can only confidently claim the presence of a mode close to the clean fundamental mode of the unperturbed potential. This is a consequence of the very short decay time of the QNMs for small perturbations and therefore the waveform does not transition to the new set of QNM spectra before the power-law tail dominates the signal. In an attempt to try removing the contribution of the tail, in the bottom panel of Fig. 23, we subtract the clean ($\epsilon = 0$) waveform from the signal

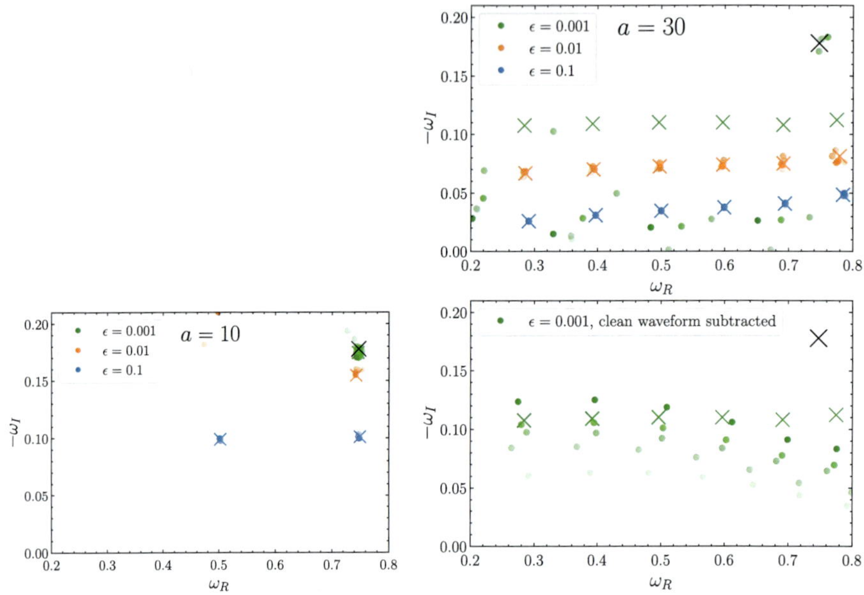

Fig. 23 Left panel: Comparison between the QNM frequencies computed with the shooting method in the frequency domain (crosses) and those extracted by fitting the full time domain waveform (dots). The black cross corresponds to the unperturbed clean fundamental QNM of Schwarzschild, where $M\omega = 0.374 - 0.089i$. The starting times used are $t - t_{\max} = 10,\ 15,\ 20,\ 25,\ 30M$. Darker shaded dots correspond to later initial fitting times. Right top panel: Same as in the left panel but with $a = 30$. For $\epsilon = 0.001$, the full time-domain fits can only confidently detect a mode close to the fundamental mode of Schwarzschild, because the power-law tail dominates the signal before it transitions to the new set of perturbed QNMs. Right bottom panel: Same analysis as in the right top panel for $\epsilon = 0.001$, but subtracting the unperturbed clean waveform from the full signal before performing the fit. The frequencies extracted do not converge exactly to the QNMs, but the structure is more similar, in particular for the real part. The figure is adapted from [57]

and repeat the fit using the green curve in the bottom-right panel of Fig. 20. There are additional QNM oscillations that were previously hidden by the power-law tail. The fitted modes do not converge as when $\epsilon \geq 0.01$, but their structure is in better agreement with the predicted QNM spectrum. The upshot is that if the perturbation is strong enough (typically for larger ϵ/smaller a), we can safely recover the unstable spectrum by fitting the full time-domain signal to a collection of modes, even though as we demonstrated in Sect. 8.1, the waveform itself is stable under perturbations.

8.2.2 The Prompt Ringdown

In a real GW detector, we do not have access to the full signal due to the presence of noise. In addition, for astrophysical systems, the perturbation bump should be small. Matter typically introduces corrections in the effective potential of amplitude $\epsilon V_{\text{bump}} \sim \rho$, where ρ is the matter density [14, 28, 55]. This corresponds to

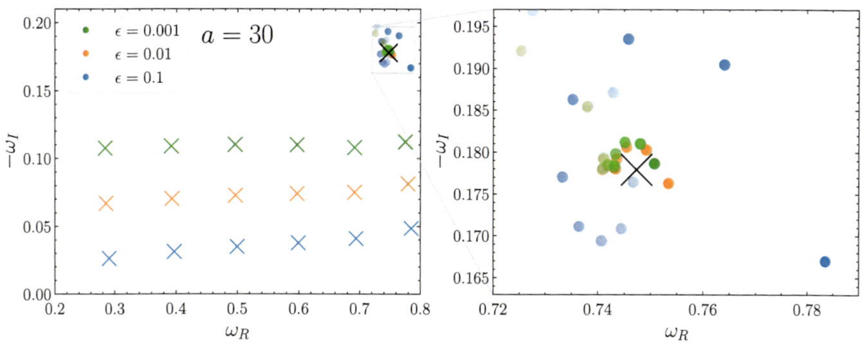

Fig. 24 Same as Fig. 23, but we only fit the first train of the initial ringdown without echoes. The starting times are $t - t_{\max} = 10,\ 11,\ 12,\ \ldots,\ 20M$. All the dots obtained using the fitting method now cluster around the clean fundamental mode of the unperturbed potential. The zoom-in in the right panel shows that a perturbation of order ϵ can induce systematic errors (approximately of order ϵ) in the measurement of the fundamental mode's frequency and damping time. The figure has been taken from [57]

$$\rho M^2 = 1.6 \times 10^{-18} \frac{\rho}{\rho_{\text{water}}} \frac{M^2}{M_\odot^2}\,, \tag{29}$$

where ρ_{water} is the density of water, M is matter's mass and $M_\odot$ the solar mass. Therefore, we expect ϵV_{bump} to be small in most astrophysical scenarios. Thus, with both current and future SNRs, we should only have access to the prompt ringdown signal.

Because of this, in Fig. 24 we repeat the analysis of the previous section but restrict it to the *prompt* portion of the signal before the appearance of the first echo, for times $t - t_{\max} \lesssim a$, shaded in green in Fig. 22. Remarkably, the prompt ringdown is well described by a *single* mode with a frequency close to the unperturbed fundamental QNM of Schwarzschild, in contrast with the vast spectrum recovered when fitting the full signal. Thus, we are led to conclude that the observed frequency associated with the prompt ringdown is spectrally stable, in a similar spirit to what happened with the waveform. The combination of our results for the full and prompt signal implies that the BH spectroscopy program is not jeopardized by the instability of the QNM spectra of very compact objects in GR. In order to recover the *correct* destabilized QNM spectrum obtained in the frequency domain, one needs to consider the late-time portion of the signal, which is not expected to be observable with both current and future GW detectors. Even this might not be enough, as the power-law tail can dominate the signal's amplitude, which would hide the presence of new perturbed modes.

Our conclusions highlight the necessity of always complementing computations of the QNM spectra in modified gravity with time-domain studies in order to verify that changes are present in the prompt ringdown. We have to note, however, that the analysis presented here was restricted to the spectral instability of the fundamental

QNM. As discussed in the previous sections, overtone destabilization is more severe for short-wavelength deterministic perturbations that we have focused on in this discussion, and these overtones generally can affect the signal at early times and the predictions of BH spectroscopy [52]. Yet, as we also mentioned, the extraction of overtones from waveforms is highly nontrivial, even at the linear level, and is a subject of current research and debate [35, 59, 63]. Any future investigation of the impact of overtone spectral instabilities in GW signals must be complemented by a better understanding on how to fit them accurately.

9 Conclusions and Outlook

BH spectroscopy is the pinnacle of GW ringdown physics. By taking into account recent improvements concerning nonlinearities at the early ringdown stage, and by eventually finding a common ground regarding the status of overtone detection, we will soon be able to perform unprecedented tests regarding the validity of GR, the existence and properties of environments and galactic halos, as well as the quantum nature of spacetime. Nevertheless, there are still many aspects of the BH spectroscopy program that need improvement in order to form a trustworthy tool, when future ground- and space-based interferometers begin their observation runs, in order to analyze GWs from a huge variety of radiation sources in the cosmos.

In this chapter, we have portrayed yet another interesting aspect of QNMs, and how it may affect BH spectroscopy, namely the fact that they suffer from spectral instabilities. This practically means that considering BH and ECO solutions only in vacuum, which is currently being done with numerical relativity simulations to evolve compact object binaries, may spoil precision GW astronomy. In order to have more accurate models of waveforms from coalescing binaries to perform data analysis and parameter estimation, environmental effects must be taken into account. Considering non-vacuum binaries comes with a drawback; the environment may taint QNMs and ringdown signals in accord with its nature and how strongly it affects the vacuum results.

We have described the current literature of spectral instabilities that occur in BH and ECO QNMs at the frequency domain with the mathematical notion of the pseudospectrum. We demonstrated that static, spherically-symmetric, and asymptotically-flat BHs exhibit an overtone instability, even including the fundamental mode when matter shell environments are taken into account in the effective potential of linear field perturbations. At the level of the spectrum itself, these results show a fragility in modal analysis, and if we want to further comprehend the underlying structure of QNMs we need non-modal tools, like the pseudospectrum, to gain more intuition into the limitations of linear predictions and transient phenomena that are obscured with typical eigenvalue investigations. Specifically for horizonless ECOs, we showed that the pseudospectrum may point towards a linear mechanism that could possibly lead to nonlinear instabilities when the ECO is ultracompact. This nonlinear instability mechanism is extremely interesting when one considers that the QNM operator is linear.

As mentioned above, if only the overtones suffer from spectral instabilities, then BH spectroscopy is safe at the moment since the fundamental mode is stable and plays a crucial role in analyzing ringdown signals. Unfortunately, it has been found that at the frequency domain, even the fundamental mode can be significantly destabilized with the addition of a tiny matter shell at some distance away from the photon sphere. This result may prove highly problematic for BH spectroscopy if such spectral instability is seeded into the ringdown waveform. Luckily, a time-domain analysis of linear gravitational perturbations on a Schwarzschild BH with a matter shell bump proves that the fundamental mode remains unsullied. The prompt ringdown waveform and the frequency fitted from this early part of the signal has an error proportional to the perturbation introduced in the potential. Only at very late times the spectrally-unstable fundamental mode appears, though our current detectors do not possess the SNR needed to probe such a long-lived mode.

There is still plenty of work to be done for a spectroscopy program to be unequivocally successfully, especially when LISA begins its observation run of novel targets that can last for years and give much cleaner ringdown waveforms. Specifically, when the overtone dispute is solved, one must also perform analyses that include not only the stability of the fundamental mode in ringdowns, but also the overtones which play a crucial role in tests of GR and the no-hair theorems. Ultimately, the pseudospectrum should be vigorously added to the spectroscopy program while the holy grail of spectral (in)stability, namely the pseudospectrum of Kerr BHs, has to be examined in order to further understand if QNM instabilities are attributes of spherical symmetry or are omnipresent to all compact objects in the Universe without any symmetry assumptions.

Acknowledgements The authors warmly thank the organization committee of the 11th Aegean Summer School that took place in Syros, Greece. The authors also want to warmly thank all the collaborators and colleagues that, even though not present at the Summer School, contributed significantly on the works discussed in this chapter, namely Emanuele Berti, Valentin Boyanov, Vitor Cardoso, Mark H. Y. Cheung, Edgar Gasperin, Jose-Luis Jaramillo, Rodrigo Panosso Macedo and Lamis Al Sheikh. K.D. acknowledges financial support provided by the DAAD program for the "Promotion of the exchange and scientific cooperation between Greece and Germany IKYDAAD 2022" (57628320). K.D. also acknowledges financial support under the European Union's H2020 ERC, Starting Grant agreement no. DarkGRA-757480 and the MIUR PRIN and FARE programmes (GW-NEXT, CUP: B84I20000100001). F.D. acknowledges financial support by FCT under project No. 2022.01324.PTDC, and financial support provided by FCT/Portugal through the grant No. SFRH/BD/143657/2019.

References

1. L. Barack et al., Black holes, gravitational waves and fundamental physics: a roadmap. Class. Quant. Grav. **36**(14), 143001 (2019)
2. P. Mészáros, D.B. Fox, C. Hanna, K. Murase, Multi-messenger astrophysics. Nat. Rev. Phys. **1**, 585–599 (2019)
3. R. Abbott et al., GWTC-2: compact binary coalescences observed by LIGO and Virgo during the first half of the third observing run. Phys. Rev. X **11**, 021053 (2021)
4. R. Abbott et al., GWTC-3: compact binary coalescences observed by LIGO and Virgo during the second part of the third observing run. 11 (2021)
5. P. Amaro-Seoane et al., Laser Interferometer Space Antenna. 2 (2017)
6. P. Amaro-Seoane et al., Astrophysics with the laser interferometer space antenna. 3 (2022)
7. H. Wen-Rui, W. Yue-Liang, The Taiji Program in Space for gravitational wave physics and the nature of gravity. Natl. Sci. Rev. **4**(5), 685–686 (2017)
8. J. Luo et al., TianQin: a space-borne gravitational wave detector. Class. Quant. Grav. **33**(3), 035010 (2016)
9. Y. Gong, J. Luo, B. Wang, Concepts and status of Chinese space gravitational wave detection projects. Nat. Astron. **5**(9), 881–889 (2021)
10. K.G. Arun et al., New horizons for fundamental physics with LISA. Living Rev. Relativ. **25**(1), 4 (2022)
11. P.T. Leung, Y.T. Liu, W.M. Suen, C.Y. Tam, K. Young, Quasinormal modes of dirty black holes. Phys. Rev. Lett. **78**, 2894–2897 (1997)
12. E. Barausse, V. Cardoso, P. Pani, Can environmental effects spoil precision gravitational-wave astrophysics? Phys. Rev. D **89**(10), 104059 (2014)
13. J. Bamber, O.J. Tattersall, K. Clough, P.G. Ferreira, Quasinormal modes of growing dirty black holes. Phys. Rev. D **103**(12), 124013 (2021)
14. V. Cardoso, K. Destounis, F. Duque, R.P. Macedo, A. Maselli, Black holes in galaxies: Environmental impact on gravitational-wave generation and propagation. Phys. Rev. D **105**(6), L061501 (2022)
15. V. Cardoso, K. Destounis, F. Duque, R.P. Macedo, A. Maselli, Gravitational waves from extreme-mass-ratio systems in astrophysical environments. Phys. Rev. Lett. **129**(24), 241103 (2022)
16. K. Destounis, A. Kulathingal, K.D. Kokkotas, G.O. Papadopoulos, Gravitational-wave imprints of compact and galactic-scale environments in extreme-mass-ratio binaries. 10 (2022)
17. T. Clifton, P.G. Ferreira, A. Padilla, C. Skordis, Modified gravity and cosmology. Phys. Rept. **513**, 1–189 (2012)
18. N. Yunes, X. Siemens, Gravitational-wave tests of general relativity with ground-based detectors and pulsar timing-arrays. Living Rev. Rel. **16**, 9 (2013)
19. B. Carter, Axisymmetric black hole has only two degrees of freedom. Phys. Rev. Lett. **26**, 331–333 (1971)
20. L. Lehner, Numerical relativity: a review. Class. Quant. Grav. **18**, R25–R86 (2001)
21. L. Barack, A. Pound, Self-force and radiation reaction in general relativity. Rep. Prog. Phys. **82**(1), 016904 (2019)
22. B. Wardell, A. Pound, N. Warburton, J. Miller, L. Durkan, A. Le Tiec, Gravitational waveforms for compact binaries from second-order self-force theory. 12 (2021)
23. A. Albertini, A. Nagar, A. Pound, N. Warburton, B. Wardell, L. Durkan, J. Miller, Comparing second-order gravitational self-force, numerical relativity, and effective one body waveforms from inspiralling, quasicircular, and nonspinning black hole binaries. Phys. Rev. D **106**(8), 084061 (2022)
24. K. Destounis, A.G. Suvorov, K.D. Kokkotas, Testing spacetime symmetry through gravitational waves from extreme-mass-ratio inspirals. Phys. Rev. D **102**(6), 064041 (2020)
25. K. Destounis, A.G. Suvorov, K.D. Kokkotas, Gravitational-wave glitches in chaotic extreme-mass-ratio inspirals. Phys. Rev. Lett. **126**(14), 141102 (2021)

26. K. Destounis, K.D. Kokkotas, Gravitational-wave glitches: resonant islands and frequency jumps in nonintegrable extreme-mass-ratio inspirals. Phys. Rev. D **104**(6), 064023 (2021)
27. K. Destounis, G. Huez, K.D. Kokkotas, Geodesics and gravitational waves in chaotic extreme-mass-ratio inspirals: the curious case of Zipoy-Voorhees black-hole mimickers. 1 (2023)
28. K.D. Kokkotas, B.G. Schmidt, Quasinormal modes of stars and black holes. Living Rev. Relativ. **2**, 2 (1999)
29. E. Berti, V. Cardoso, A.O. Starinets, Quasinormal modes of black holes and black branes. Class. Quant. Grav. **26**, 163001 (2009)
30. R.A. Konoplya, A. Zhidenko, Quasinormal modes of black holes: from astrophysics to string theory. Rev. Mod. Phys. **83**, 793–836 (2011)
31. E. Berti, V. Cardoso, C.M. Will, On gravitational-wave spectroscopy of massive black holes with the space interferometer LISA. Phys. Rev. D **73**, 064030 (2006)
32. M. Giesler, M. Isi, M.A. Scheel, S. Teukolsky, Black hole ringdown: the importance of overtones. Phys. Rev. X **9**(4), 041060 (2019)
33. M. Isi, M. Giesler, W.M. Farr, M.A. Scheel, S.A. Teukolsky, Testing the no-hair theorem with GW150914. Phys. Rev. Lett. **123**(11), 111102 (2019)
34. M. Isi, W.M. Farr, M. Giesler, M.A. Scheel, S.A. Teukolsky, Testing the black-hole area law with GW150914. Phys. Rev. Lett. **127**(1), 011103 (2021)
35. M. Isi, W.M. Farr, Analyzing black-hole ringdowns. 7 (2021)
36. N. Franchini, S.H. Völkel, Testing general relativity with black hole quasi-normal modes. 5 (2023)
37. J. Abedi, N. Afshordi, N. Oshita, Q. Wang, Quantum black holes in the sky. Universe **6**(3), 43 (2020)
38. V. Cardoso, E. Franzin, P. Pani, Is the gravitational-wave ringdown a probe of the event horizon? Phys. Rev. Lett. **116**(17), 171101 (2016). [Erratum: Phys. Rev. Lett. **117**, 089902 (2016)]
39. V. Cardoso, S. Hopper, C.F.B. Macedo, C. Palenzuela, P. Pani, Gravitational-wave signatures of exotic compact objects and of quantum corrections at the horizon scale. Phys. Rev. D **94**(8), 084031 (2016)
40. V. Cardoso, P. Pani, Testing the nature of dark compact objects: a status report. Living Rev. Relativ. **22**(1), 4 (2019)
41. J.L. Blázquez-Salcedo, F.S. Khoo, J. Kunz, Quasinormal modes of Einstein-Gauss-Bonnet-dilaton black holes. Phys. Rev. D **96**(6), 064008 (2017)
42. K. Destounis, G. Panotopoulos, Á. Rincón, Stability under scalar perturbations and quasinormal modes of 4D Einstein–Born–Infeld dilaton spacetime: exact spectrum. Eur. Phys. J. C **78**(2), 139 (2018)
43. F. Moulin, A. Barrau, K. Martineau, An overview of quasinormal modes in modified and extended gravity. Universe **5**(9), 202 (2019)
44. J.L. Blázquez-Salcedo, D.D. Doneva, S. Kahlen, J. Kunz, P. Nedkova, S.S. Yazadjiev, Polar quasinormal modes of the scalarized Einstein-Gauss-Bonnet black holes. Phys. Rev. D **102**(2), 024086 (2020)
45. L. Pierini, L. Gualtieri, Quasi-normal modes of rotating black holes in Einstein-dilaton Gauss-Bonnet gravity: the first order in rotation. Phys. Rev. D **103**, 124017 (2021)
46. C. Vlachos, E. Papantonopoulos, K. Destounis, Echoes of compact objects in scalar-tensor theories of gravity. Phys. Rev. D **103**(4), 044042 (2021)
47. N. Chatzifotis, C. Vlachos, K. Destounis, E. Papantonopoulos, Stability of black holes with non-minimally coupled scalar hair to the Einstein tensor. Gen. Relativ. Gravit. **54**(6), 49 (2022)
48. R. Ghosh, S. Sk, S. Sarkar, Hairy black holes: Nonexistence of short hairs and a bound on the light ring size. Phys. Rev. D **108**(4), L041501 (2023)
49. H.-P. Nollert, About the significance of quasinormal modes of black holes. Phys. Rev. D **53**, 4397–4402 (1996)
50. H.-P. Nollert, R.H. Price, Quantifying excitations of quasinormal mode systems. J. Math. Phys. **40**, 980–1010 (1999)

51. J.L. Jaramillo, R.P. Macedo, L. Al Sheikh, Pseudospectrum and black hole quasinormal mode instability. Phys. Rev. X **11**(3), 031003 (2021)
52. J.L. Jaramillo, R.P. Macedo, L. Al Sheikh, Gravitational wave signatures of black hole quasinormal mode instability. Phys. Rev. Lett. **128**(21), 211102 (2022)
53. J.L. Jaramillo, Pseudospectrum and binary black hole merger transients. Class. Quant. Grav. **39**(21), 217002 (2022)
54. K. Destounis, R.P. Macedo, E. Berti, V. Cardoso, J.L. Jaramillo, Pseudospectrum of Reissner-Nordström black holes: Quasinormal mode instability and universality. Phys. Rev. D **104**(8), 084091 (2021)
55. M.H.-Y. Cheung, K. Destounis, R.P. Macedo, E. Berti, V. Cardoso, Destabilizing the fundamental mode of black holes: the elephant and the flea. Phys. Rev. Lett. **128**(11), 111103 (2022)
56. V. Boyanov, K. Destounis, R.P. Macedo, V. Cardoso, J.L. Jaramillo, Pseudospectrum of horizonless compact objects: a bootstrap instability mechanism. Phys. Rev. D **107**(6), 064012 (2023)
57. E. Berti, V. Cardoso, M.H.-Y. Cheung, F. Di Filippo, F. Duque, P. Martens, S. Mukohyama, Stability of the fundamental quasinormal mode in time-domain observations against small perturbations. Phys. Rev. D **106**(8), 084011 (2022)
58. R. Ghosh, N. Franchini, S.H. Völkel, E. Barausse, Quasinormal modes of nonseparable perturbation equations: the scalar non-Kerr case. Phys. Rev. D **108**(2), 024038 (2023)
59. R. Cotesta, G. Carullo, E. Berti, V. Cardoso, Analysis of ringdown overtones in GW150914. Phys. Rev. Lett. **129**(11), 111102 (2022)
60. M. Isi, W.M. Farr, Revisiting the ringdown of GW150914. 2 (2022)
61. M.H.-Y. Cheung et al., Nonlinear effects in black hole ringdown. 8 (2022)
62. K. Mitman et al., Nonlinearities in black hole ringdowns. 8 (2022)
63. V. Baibhav, M.H.-Y. Cheung, E. Berti, V. Cardoso, G. Carullo, R. Cotesta, W. Del Pozzo, F. Duque, Agnostic black hole spectroscopy: quasinormal mode content of numerical relativity waveforms and limits of validity of linear perturbation theory. 2 (2023)
64. J. Redondo-Yuste, G. Carullo, J.L. Ripley, E. Berti, V. Cardoso, Spin dependence of black hole ringdown nonlinearities (2023)
65. P.J. Nee, S.H. Völkel, H.P. Pfeiffer, Role of black hole quasinormal mode overtones for ringdown analysis. Phys. Rev. D **108**(4), 044032 (2023)
66. R.A. Konoplya, A. Zhidenko, First few overtones probe the event horizon geometry. 9 (2022)
67. A. Courty, K. Destounis, P. Pani, Quasinormal mode (in)stability and strong cosmic censorship (2023). In preparation
68. L. Trefethen, M. Embree, *Spectra and Pseudospectra: The Behavior of Nonnormal Matrices and Operators*. 01 (2005)
69. T. Regge, J.A. Wheeler, Stability of a Schwarzschild singularity. Phys. Rev. **108**, 1063–1069 (1957)
70. F.J. Zerilli, Effective potential for even parity Regge-Wheeler gravitational perturbation equations. Phys. Rev. Lett. **24**, 737–738 (1970)
71. S.A. Teukolsky, W.H. Press, Perturbations of a rotating black hole. III. Interaction of the hole with gravitational and electromagnetic radiation. Astrophys. J. **193**, 443–461 (1974)
72. S. Chandrasekhar, *The Mathematical Theory of Black Holes* (1985)
73. C.V. Vishveshwara, Scattering of gravitational radiation by a Schwarzschild black-hole. Nature **227**, 936–938 (1970)
74. S. Chandrasekhar, S.L. Detweiler, The quasi-normal modes of the Schwarzschild black hole. Proc. R. Soc. Lond. A **344**, 441–452 (1975)
75. E.W. Leaver, Spectral decomposition of the perturbation response of the Schwarzschild geometry. Phys. Rev. D **34**, 384–408 (1986)
76. P. Pani, Advanced methods in black-hole perturbation theory. Int. J. Mod. Phys. A **28**, 1340018 (2013)
77. C. Gundlach, R.H. Price, J. Pullin, Late time behavior of stellar collapse and explosions: 1. Linearized perturbations. Phys. Rev. D **49**, 883–889 (1994)

78. B.P. Abbott et al., Observation of gravitational waves from a binary black hole merger. Phys. Rev. Lett. **116**(6), 061102 (2016)
79. G. Carullo, W. Del Pozzo, J. Veitch, Observational black hole spectroscopy: a time-domain multimode analysis of GW150914. Phys. Rev. D **99**(12), 123029 (2019). [Erratum: Phys. Rev. D **100**, 089903 (2019)]
80. G. Carullo, R. Cotesta, E. Berti, V. Cardoso, Reply to comment on "analysis of ringdown overtones in GW150914". In preparation (2023)
81. E. Finch, C.J. Moore, Searching for a ringdown overtone in GW150914. Phys. Rev. D **106**(4), 043005 (2022)
82. M. Crisostomi, K. Dey, E. Barausse, R. Trotta, Neural posterior estimation with guaranteed exact coverage: the ringdown of GW150914. 5 (2023)
83. A. Kehagias, D. Perrone, A. Riotto, F. Riva, Explaining nonlinearities in black hole ringdowns from symmetries. Phys. Rev. D **108**(2), L021501 (2023)
84. A. Kehagias, A. Riotto, On nonlinear black hole ringdowns from gauge-invariance and measurements. 2 (2023)
85. D. Perrone, T. Barreira, A. Kehagias, A. Riotto, Non-linear black hole ringdowns: an analytical approach. 8 (2023)
86. B. Bucciotti, A. Kuntz, F. Serra, E. Trincherini, Nonlinear quasi-normal modes: uniform approximation. 9 (2023)
87. H. Zhu, J.L. Ripley, A. Cárdenas-Avendaño, F. Pretorius, Challenges in quasinormal mode extraction: perspectives from numerical solutions to the Teukolsky equation. 9 (2023)
88. E.W. Leaver, Spectral decomposition of the perturbation response of the Schwarzschild geometry. Phys. Rev. D **34**(2), 384–408 (1986)
89. M. Casals, A.C. Ottewill, Analytic investigation of the branch cut of the Green function in Schwarzschild space-time. Phys. Rev. D **87**(6), 064010 (2013)
90. K.D. Kokkotas, B.F. Schutz, W-modes: a new family of normal modes of pulsating relativistic stars. Mon. Not. R. Astron. Soc. **255**, 119 (1992)
91. R.G. Daghigh, M.D. Green, J.C. Morey, Significance of black hole quasinormal modes: a closer look. Phys. Rev. D **101**(10), 104009 (2020)
92. L.N. Trefethen, *Spectral Methods in Matlab* (Society for Industrial and Applied Mathematics, Philadelphia, PA, 2000)
93. Edgar Gasperin and José Luis Jaramillo, Energy scales and black hole pseudospectra: the structural role of the scalar product. Class. Quant. Grav. **39**(11), 115010 (2022)
94. T.A. Driscoll, L.N. Trefethen, Pseudospectra for the wave equation with an absorbing boundary. J. Comput. Appl. Math. **69**(1), 125–142 (1996)
95. L.N. Trefethen, A.E. Trefethen, S.C. Reddy, T.A. Driscoll, Hydrodynamic stability without eigenvalues. Science **261**(5121), 578–584 (1993)
96. D. Krejcirik, P. Siegl, M. Tater, J. Viola, Pseudospectra in non-Hermitian quantum mechanics. J. Math. Phys. **56**(10), 103513 (2015)
97. S.A. Orszag, L.C. Kells, Transition to turbulence in plane Poiseuille and plane Couette flow. J. Fluid Mech. **96**(1), 159–205 (1980)
98. V.C. Patel, M.R. Head, Some observations on skin friction and velocity profiles in fully developed pipe and channel flows. J. Fluid Mech. **38**(1), 181–201 (1969)
99. D.R. Carlson, S.E. Widnall, M.F. Peeters, A flow-visualization study of transition in plane Poiseuille flow. J. Fluid Mech. **121**, 487–505 (1982)
100. S. Sarkar, M. Rahman, S. Chakraborty, Perturbing the perturbed: stability of quasi-normal modes in presence of a positive cosmological constant. 4 (2023)
101. D. Arean, D.G. Fariña, K. Landsteiner, Pseudospectra of holographic quasinormal modes. 7 (2023)
102. V. Cardoso, M. Kimura, A. Maselli, E. Berti, C.F.B. Macedo, R. McManus, Parametrized black hole quasinormal ringdown: decoupled equations for nonrotating black holes. Phys. Rev. D **99**(10), 104077 (2019)
103. R. McManus, E. Berti, C.F.B. Macedo, M. Kimura, A. Maselli, V. Cardoso, Parametrized black hole quasinormal ringdown. II. Coupled equations and quadratic corrections for nonrotating black holes. Phys. Rev. D **100**(4), 044061 (2019)

104. M. Kimura, Note on the parametrized black hole quasinormal ringdown formalism. Phys. Rev. D **101**(6), 064031 (2020)
105. S.H. Völkel, N. Franchini, E. Barausse, Theory-agnostic reconstruction of potential and couplings from quasinormal modes. Phys. Rev. D **105**(8), 084046 (2022)
106. S.H. Völkel, N. Franchini, E. Barausse, E. Berti, Constraining modifications of black hole perturbation potentials near the light ring with quasinormal modes. Phys. Rev. D **106**(12), 124036 (2022)
107. N. Franchini, S.H. Völkel, Parametrized quasinormal mode framework for non-Schwarzschild metrics. Phys. Rev. D **107**(12), 124063 (2023)
108. Marcus Ansorg and Rodrigo Panosso Macedo, Spectral decomposition of black-hole perturbations on hyperboloidal slices. Phys. Rev. D **93**(12), 124016 (2016)
109. R.P. Macedo, J.L. Jaramillo, M. Ansorg, Hyperboloidal slicing approach to quasi-normal mode expansions: the Reissner-Nordström case. Phys. Rev. D **98**(12), 124005 (2018)
110. Rodrigo Panosso Macedo, Hyperboloidal framework for the Kerr spacetime. Class. Quant. Grav. **37**(6), 065019 (2020)
111. A. Zenginoglu, Hyperboloidal foliations and scri-fixing. Class. Quant. Grav. **25**, 145002 (2008)
112. A. Zenginoglu, A Geometric framework for black hole perturbations. Phys. Rev. D **83**, 127502 (2011)
113. K. Destounis, R.D.B. Fontana, F.C. Mena, Accelerating black holes: quasinormal modes and late-time tails. Phys. Rev. D **102**(4), 044005 (2020)
114. V. Ferrari, B. Mashhoon, New approach to the quasinormal modes of a black hole. Phys. Rev. D **30**, 295–304 (1984)
115. H.R. Beyer, On the completeness of the quasinormal modes of the Poschl-Teller potential. Commun. Math. Phys. **204**, 397–423 (1999)
116. P. Bizoń, T. Chmaj, P. Mach, A toy model of hyperboloidal approach to quasinormal modes. Acta Phys. Pol. B **51**, 1007 (2020)
117. R.H. Price, Nonspherical perturbations of relativistic gravitational collapse. 1. Scalar and gravitational perturbations. Phys. Rev. D **5**, 2419–2438 (1972)
118. R.H. Price, Nonspherical perturbations of relativistic gravitational collapse. II. Integer-spin, zero-rest-mass fields. Phys. Rev. D **5**, 2439–2454 (1972)
119. C. Gundlach, R.H. Price, J. Pullin, Late time behavior of stellar collapse and explosions: 2. Nonlinear evolution. Phys. Rev. D **49**, 890–899 (1994)
120. S.W. Hawking, G.F.R. Ellis, in *The Large Scale Structure of Space-Time*, vol. 2. Cambridge Monographs on Mathematical Physics (Cambridge University Press, 2023)
121. V. Cardoso, J.L. Costa, K. Destounis, P. Hintz, A. Jansen, Quasinormal modes and strong cosmic censorship. Phys. Rev. Lett. **120**(3), 031103 (2018)
122. Y.-W. Kim, Y.S. Myung, Y.-J. Park, Quasinormal modes and hidden conformal symmetry in the Reissner-Nordström black hole. Eur. Phys. J. C **73**, 2440 (2013)
123. A. Zimmerman, H. Yang, Z. Mark, Y. Chen, L. Lehner, Quasinormal Modes Beyond Kerr. Astrophys. Space Sci. Proc. **40**, 217–223 (2015)
124. M. Richartz, Quasinormal modes of extremal black holes. Phys. Rev. D **93**(6), 064062 (2016)
125. V. Cardoso, J.L. Costa, K. Destounis, P. Hintz, A. Jansen, Strong cosmic censorship in charged black-hole spacetimes: still subtle. Phys. Rev. D **98**(10), 104007 (2018)
126. K. Destounis, Charged fermions and strong cosmic censorship. Phys. Lett. B **795**, 211–219 (2019)
127. H. Liu, W.-L. Qian, Y. Liu, W. Jian-Pin, B. Wang, R.-H. Yue, Alternative mechanism for black hole echoes. Phys. Rev. D **104**(4), 044012 (2021)
128. S. Aretakis, Stability and instability of extreme Reissner-Nordström black hole spacetimes for linear scalar perturbations I. Commun. Math. Phys. **307**, 17–63 (2011)
129. S. Aretakis, Stability and instability of extreme Reissner-Nordström black hole spacetimes for linear scalar perturbations II. Ann. Henri Poincare **12**, 1491–1538 (2011)
130. Y. Angelopoulos, S. Aretakis, D. Gajic, Horizon hair of extremal black holes and measurements at null infinity. Phys. Rev. Lett. **121**(13), 131102 (2018)

131. H. Onozawa, T. Okamura, T. Mishima, H. Ishihara, Perturbing supersymmetric black hole. Phys. Rev. D **55**, 4529–4531 (1997)
132. T. Okamura, On scattering off the extreme Reissner-Nordström black hole in N = 2 supergravity. Phys. Rev. D **56**, 4927–4933 (1997)
133. R. Kallosh, J. Rahmfeld, W.K. Wong, One loop supergravity corrections to the black hole entropy and residual supersymmetry. Phys. Rev. D **57**, 1063–1067 (1998)
134. E. Berti, Black hole quasinormal modes: hints of quantum gravity? Conf. Proc. C **0405132**, 145–186 (2004)
135. V. Moncrief, Odd-parity stability of a Reissner-Nordström black hole. Phys. Rev. D **9**, 2707–2709 (1974)
136. V. Moncrief, Stability of Reissner-Nordström black holes. Phys. Rev. D **10**, 1057–1059 (1974)
137. V. Moncrief, Gauge-invariant perturbations of Reissner-Nordström black holes. Phys. Rev. D **12**, 1526–1537 (1975)
138. S.L. Liebling, C. Palenzuela, Dynamical boson stars. Living Rev. Rel. **26**(1), 1 (2023)
139. M. Vaglio, C. Pacilio, A. Maselli, P. Pani, Multipolar structure of rotating boson stars. Phys. Rev. D **105**(12), 124020 (2022)
140. K. Destounis, F. Angeloni, M. Vaglio, P. Pani, Extreme-mass-ratio inspirals into rotating boson stars: nonintegrability, chaos, and transient resonances. 5 (2023)
141. P.O. Mazur, E. Mottola, Gravitational condensate stars: an alternative to black holes. Universe **9**(2), 88 (2023)
142. R. Carballo-Rubio, Stellar equilibrium in semiclassical gravity. Phys. Rev. Lett. **120**(6), 061102 (2018)
143. J. Arrechea, C. Barceló, R. Carballo-Rubio, L.J. Garay, Semiclassical constant-density spheres in a regularized Polyakov approximation. Phys. Rev. D **104**(8), 084071 (2021)
144. D.E. Kaplan, S. Rajendran, Firewalls in general relativity. Phys. Rev. D **99**(4), 044033 (2019)
145. S.D. Mathur, The Fuzzball proposal for black holes: an Elementary review. Fortschr. Phys. **53**, 793–827 (2005)
146. S.D. Mathur, Fuzzballs and the information paradox: a summary and conjectures. 10 (2008)
147. T. Ikeda, M. Bianchi, D. Consoli, A. Grillo, J.F. Morales, P. Pani, G. Raposo, Black-hole microstate spectroscopy: ringdown, quasinormal modes, and echoes. Phys. Rev. D **104**(6), 066021 (2021)
148. V. Cardoso, P. Pani, Testing the nature of dark compact objects: a status report. Living Rev. Rel. **22**(1), 4 (2019)
149. V. Cardoso, L.C.B. Crispino, C.F.B. Macedo, H. Okawa, P. Pani, Light rings as observational evidence for event horizons: long-lived modes, ergoregions and nonlinear instabilities of ultracompact objects. Phys. Rev. D **90**(4), 044069 (2014)
150. J. Keir, Slowly decaying waves on spherically symmetric spacetimes and ultracompact neutron stars. Class. Quant. Grav. **33**(13), 135009 (2016)
151. P.V.P. Cunha, C. Herdeiro, E. Radu, N. Sanchis-Gual, Exotic compact objects and the fate of the light-ring instability. Phys. Rev. Lett. **130**(6), 061401 (2023)
152. Z. Zhong, V. Cardoso, E. Maggio, On the instability of ultracompact horizonless spacetimes. 11 (2022)
153. E. Maggio, L. Buoninfante, A. Mazumdar, P. Pani, How does a dark compact object ringdown? Phys. Rev. D **102**(6), 064053 (2020)
154. V. Vellucci, E. Franzin, S. Liberati, Echoes from backreacting exotic compact objects. 5 (2022)
155. Z. Jackiewicz, R. Renaut, A note on stability of pseudospectral methods for wave propagation. J. Comput. Appl. Math. **143**(06), 127–139 (2002)
156. V. Cardoso, L. Gualtieri, Testing the black hole 'no-hair' hypothesis. Class. Quant. Grav. **33**(17), 174001 (2016)
157. V. Cardoso, P. Pani, Tests for the existence of black holes through gravitational wave echoes. Nat. Astron. **1**(9), 586–591 (2017)
158. S.R. Dolan, Instability of the massive Klein-Gordon field on the Kerr spacetime. Phys. Rev. D **76**, 084001 (2007)

159. J.G. Rosa, S.R. Dolan, Massive vector fields on the Schwarzschild spacetime: quasi-normal modes and bound states. Phys. Rev. D **85**, 044043 (2012)
160. V. Cardoso, S. Chakrabarti, P. Pani, E. Berti, L. Gualtieri, Floating and sinking: the imprint of massive scalars around rotating black holes. Phys. Rev. Lett. **107**, 241101 (2011)
161. P. Pani, V. Cardoso, L. Gualtieri, E. Berti, A. Ishibashi, Perturbations of slowly rotating black holes: massive vector fields in the Kerr metric. Phys. Rev. D **86**, 104017 (2012)
162. K. Destounis, Superradiant instability of charged scalar fields in higher-dimensional Reissner-Nordström-de Sitter black holes. Phys. Rev. D **100**(4), 044054 (2019)
163. H.S. Vieira, K.D. Kokkotas, Quasibound states of Schwarzschild acoustic black holes. Phys. Rev. D **104**(2), 024035 (2021)
164. H.S. Vieira, K. Destounis, K.D. Kokkotas, Slowly-rotating curved acoustic black holes: Quasinormal modes, Hawking-Unruh radiation, and quasibound states. Phys. Rev. D **105**(4), 045015 (2022)
165. H.S. Vieira, K. Destounis, K.D. Kokkotas, Analog Schwarzschild black holes of Bose-Einstein condensates in a cavity: Quasinormal modes and quasibound states. Phys. Rev. D **107**(10), 104038 (2023)
166. C.T. Cunningham, R.H. Price, V. Moncrief, Radiation from collapsing relativistic stars. III - Second order perturbations of collapse with rotation. ApJ **236**, 674–692 (1980)
167. R.J. Gleiser, C.O. Nicasio, R.H. Price, J. Pullin, Colliding black holes: how far can the close approximation go? Phys. Rev. Lett. **77**, 4483–4486 (1996)
168. J. Miller, B. Wardell, A. Pound, Second-order perturbation theory: the problem of infinite mode coupling. Phys. Rev. D **94**(10), 104018 (2016)
169. L. Speri, A. Antonelli, L. Sberna, S. Babak, E. Barausse, J.R. Gair, M.L. Katz, Probing accretion physics with gravitational waves. Phys. Rev. X **13**(2), 021035 (2023)
170. N. Speeney, A. Antonelli, V. Baibhav, E. Berti, Impact of relativistic corrections on the detectability of dark-matter spikes with gravitational waves. Phys. Rev. D **106**(4), 044027 (2022)
171. K. Akiyama et al., First M87 event horizon telescope results. I. The shadow of the supermassive black hole. Astrophys. J. Lett. **875**, L1 (2019)
172. K. Akiyama et al., First Sagittarius A* event horizon telescope results. I. The shadow of the supermassive black hole in the center of the milky way. Astrophys. J. Lett. **930**(2), L12 (2022)
173. D. Lynden-Bell, Galactic nuclei as collapsed old quasars. Nature **223**, 690 (1969)
174. V.G. Gurzadian, L.M. Ozernoi, Accretion on massive black holes in galactic nuclei. Nature **280**(5719), 214–215 (1979)
175. P.J. Armitage, Theory of disk accretion onto supermassive black holes. Astrophys. Space Sci. Libr. **308**, 89–126 (2004)
176. A.J. Dittmann, M.C. Miller, Star formation in accretion discs and SMBH growth. Mon. Not. R. Astron. Soc. **493**(3), 3732–3743 (2020)
177. Y. Sofue, M. Honma, T. Omodaka, Unified Rotation Curve of the Galaxy - Decomposition into de Vaucouleurs Bulge, Disk, Dark Halo, and the 9-kpc Rotation Dip -. Publ. Astron. Soc. Jap. **61**, 227 (2009)
178. R. Catena, P. Ullio, A novel determination of the local dark matter density. JCAP **08**, 004 (2010)
179. M. Weber, W. de Boer, Determination of the local dark matter density in our galaxy. Astron. Astrophys. **509**, A25 (2010)
180. Z. Mark, A. Zimmerman, S.M. Du, Y. Chen, A recipe for echoes from exotic compact objects. Phys. Rev. D **96**(8), 084002 (2017)
181. W. Krivan, P. Laguna, P. Papadopoulos, N. Andersson, Dynamics of perturbations of rotating black holes. Phys. Rev. D **56**, 3395–3404 (1997)
182. E. Pazos-Ávalos, C.O. Lousto, Numerical integration of the Teukolsky equation in the time domain. Phys. Rev. D **72**(8) (2005)
183. A. Zenginoglu, G. Khanna, Null infinity waveforms from extreme-mass-ratio inspirals in Kerr spacetime. Phys. Rev. X **1**, 021017 (2011)
184. Rodrigo Panosso Macedo, Hyperboloidal framework for the Kerr spacetime. Class. Quant. Grav. **37**(6), 065019 (2020)

WKB Method and Quasinormal Modes of String-Theoretical D-Dimensional Black Holes

Filipe Moura and João Rodrigues

Abstract After a brief introduction to quasinormal modes in dissipative systems, we review the WKB formalism in the context of the analytical calculation of quasinormal frequencies. We apply these results to the calculation of quasinormal frequencies associated with gravitational perturbations of d-dimensional spherically symmetric black holes with string corrections. We do this for two distinct limits: the eikonal limit and the asymptotic limit.

1 Introduction and Summary

Gravitational wave detectors can directly measure quasinormal ringing frequencies, which carry unique information about parameters of the black hole in the ringdown phase resulting from a binary black hole coalescence. This feature turns quasinormal modes into preferential probes for testing theories of gravity beyond Einstein, since the ringing frequencies represent a universal part of the gravitational wave signals. With the advent of gravitational wave astronomy, therefore, interest in the study of black hole quasinormal modes has raised, as their measurement can provide a test to modified gravity theories [1].

We provide an introductory review to this topic of great current interest. In Sect. 2 we start by introducing quasinormal modes as solutions to field equations of dissipative systems, and we verify that quasinormal frequencies arise as the poles of a frequency domain Green's function. We then define black hole quasinormal modes and the boundary conditions they must verify. In Sect. 3 we briefly review WKB

F. Moura (✉)
Departamento de Matemática, Escola de Tecnologias e Arquitetura and Instituto de Telecomunicações, ISCTE—Instituto Universitário de Lisboa, Av. das Forças Armadas 1649-026 Lisboa, Portugal
e-mail: fmoura@lx.it.pt

J. Rodrigues
Centro de Análise Matemática, Geometria e Sistemas Dinâmicos, Departamento de Matemática, Instituto Superior Técnico, Av. Rovisco Pais, 1049-001 Lisboa, Portugal
e-mail: joao.carlos.rodrigues@tecnico.ulisboa.pt

© The Author(s), under exclusive license to Springer Nature Switzerland AG 2024

E. Papantonopoulos and N. Mavromatos (eds.), *Compact Objects in the Universe*,
https://doi.org/10.1007/978-3-031-55098-0_7

theory in order to obtain approximations to general solutions of Schrödinger–like differential equations, like the ones describing black hole perturbations. In Sect. 4 we introduce a d-dimensional spherically symmetric black hole solution with leading string corrections obtained by Callan, Myers and Perry, and briefly review the formalism for gravitational perturbations applied to it. We then compute analytical expressions for quasinormal frequencies, associated with tensorial gravitational perturbations of the black hole solution we introduced. We do this for two distinct limits. In Sect. 5, we compute these frequencies in the eikonal limit, resorting to the WKB approximations we addressed. Finally in Sect. 6 we compute these frequencies in the asymptotic highly damped limit, studying two different monodromies of the perturbation, when analytically continued to the complex r-plane.

2 Quasinormal Modes

2.1 Dissipative Systems and Quasinormal Modes

We start with a digression on physical systems described by differential equations whose spatial variables are defined on non compact subsets of $\mathbb{R}^n$. Here we closely follow [2], conveniently adapted to our context. As an example, let us picture a general 1-dimensional physical system, defined by the wave equation

$$\frac{\partial^2 \psi}{\partial t^2} + \left(V(x) - \frac{\partial^2}{\partial x^2} \right) \psi = 0 \tag{1}$$

where V is a continuous positive function of compact support in $\mathbb{R}$, meaning that

$$V(x) = 0 \tag{2}$$

for all $x \in \mathbb{R}$ such that $|x| > L$ with $L \in \mathbb{R}^+$. Let us choose general initial data of compact support

$$\psi(x, 0) = \Gamma(x), \tag{3}$$

$$\frac{\partial \psi}{\partial t}(x, 0) = \Omega(x) \tag{4}$$

for some $\Gamma \in C^2(\mathbb{R})$ and $\Omega \in C^1(\mathbb{R})$.

Now, we consider the unique solution ψ of the differential Eq. (1), with initial data as above. The Laplace transform of this solution takes the form

$$\mathcal{L}(\psi)(x, s) = \int_0^{+\infty} e^{-st} \psi(x, t) dt. \tag{5}$$

The differential equation satisfied by $\mathcal{L}(\psi)$ is

$$s^2\mathcal{L}(\psi) - \frac{\partial^2 \mathcal{L}}{\partial x^2}(\psi) + V\mathcal{L}(\psi) = s\Gamma + \Omega. \tag{6}$$

Because we are working with initial data of compact support, we know ψ is bounded [2]. Hence, the Laplace transform $\mathcal{L}(\psi)$ is analytic for $s \in]0, +\infty[$ and admits an analytic continuation [2] onto the complex half plane $\Re(s) > 0$[1].

Now, let us write the Green function of the differential Eq. (6). Such function, takes the well known form

$$G(s,x,x') = \begin{cases} \frac{f_-(s,x')f_+(s,x)}{W(s)} \; ; \; x' \leq x \\ \frac{f_-(s,x)f_+(s,x')}{W(s)} \; ; \; x' > x \end{cases} \tag{7}$$

where f_- and f_+ are two linearly independent solutions of the homogeneous differential equation, associated with (6), and W denotes the respective Wronskian. In order to simplify the notation, we define the function

$$\Lambda := s\Gamma + \Omega. \tag{8}$$

Using the previous definition, we can write the particular solution of the differential Eq. (6) as

$$\mathcal{L}(\psi)(s,x) = \int_{-\infty}^{+\infty} G(s,x,x')\Lambda(s,x')dx'. \tag{9}$$

Considering the Laplace transform definition, we know $\mathcal{L}(\psi)$ is bounded as a function of x. Indeed, since ψ is bounded, we know that $|\psi|$ has a majorant $M \in \mathbb{R}^+$. Thus, for some s in the complex half plane $\Re(s) > 0$, we have

$$|\mathcal{L}(\psi)| = \left| \int_0^{+\infty} e^{-st}\psi(x,t)dt \right| \leq \int_0^{+\infty} \left|e^{-st}\right| |\psi(x,t)|\, dt \leq M \int_0^{+\infty} e^{-\Re(s)t}dt = \frac{M}{\Re(s)}. \tag{10}$$

Moreover, for a fixed s in the complex half plane $\Re(s) > 0$ and for $|x| > L$, the general solution of the homogeneous differential equation, associated with (6), is a linear combination of the functions $e^{\pm xs}$. Since $\mathcal{L}(\psi)$ is bounded as a function of x, we are forced to choose a Green function G such that

$$f_+(s,x) \propto e^{-sx} \; ; \; x \geq L, \tag{11}$$

$$f_-(s,x) \propto e^{sx} \; ; \; x \leq -L. \tag{12}$$

The above conditions uniquely define the Green function of (6) and the respective particular solution through (9).

[1] We denote by $\Re(s)/\Im(s)$ the real/imaginary parts of the complex number s..

It is within this context that quasinormal frequencies emerge. Quasinormal frequencies are defined as the complex numbers $s_n \in \mathbb{C}$ such that

$$f_+(s_n, x) = c(s_n) f_-(s_n, x) \tag{13}$$

for all $x \in \mathbb{R}$, where $c(s_n) \in \mathbb{C}$. For these values of s, the solutions f_+ and f_- become linearly dependent, with respect to x. When this happens, the Wronskian W vanishes and the chosen Green function becomes singular. The corresponding functions $f_-(s_n, x)$ and $f_+(s_n, x)$ are called quasinormal modes. We did not prove that such frequencies exist, but in fact they do. In [3], for the conditions we are considering, a countable family of such frequencies, in the complex half plane $\Re(s) < 0$, was proven to exist. We know $\mathcal{L}(\psi)$ admits an analytic continuation onto the complex half plane $\Re(s) > 0$. Thus, there should not exist any quasinormal frequency with positive real component.

The quasinormal frequencies and quasinormal modes come into play when we reconsider the solution to the original differential Eq. (1). Indeed, we can use Mellin's inverse formula to write ψ as

$$\psi(x, t) = \frac{1}{2\pi i} \int_{a-i\infty}^{a+i\infty} e^{st} \mathcal{L}(\psi)(s, x) ds = \frac{1}{2\pi i} \int_{a-i\infty}^{a+i\infty} e^{st} \int_{-\infty}^{+\infty} G(s, x, x') \Lambda(s, x') dx' ds \tag{14}$$

for some $a > 0$. Now, we can compute the integral over s, using the contour depicted in Fig. 1. To do this, we define

$$\mathcal{F}(s, x) := \frac{e^{st}}{2\pi i} \int_{-\infty}^{+\infty} G(s, x, x') \Lambda(s, x') dx' \tag{15}$$

and rewrite (14) as

$$\psi(x, t) = \int_{a-i\infty}^{a+i\infty} \mathcal{F}(s, x) ds. \tag{16}$$

Because we chose initial data of compact support, we know Λ is compactly supported on x' as well. Therefore, there can be no singularities of $\mathcal{F}$, arising from the integration. On the other hand, we know G is singular for quasinormal frequencies s_n. As our contour encloses these frequencies, the residue theorem yields

$$\oint_{\mathcal{D}} \mathcal{F}(s, x) ds = \int_{a-i\infty}^{a+i\infty} \mathcal{F}(s, x) ds + \int_C \mathcal{F}(s, x) ds = 2\pi i \sum_{n \in I} \text{Res}\left(\mathcal{F}, s_n\right)(x) \tag{17}$$

where we denoted the blue contour and the associated arc shaped portion by $\mathcal{D}$ and C respectively. Furthermore, we denoted the countable index set of quasinormal frequencies by I. Rewriting the equation above, using (16), yields

$$\psi(x, t) = 2\pi i \sum_{n \in I} \text{Res}\left(\mathcal{F}, s_n\right)(x) - \int_C \mathcal{F}(s, x) ds. \tag{18}$$

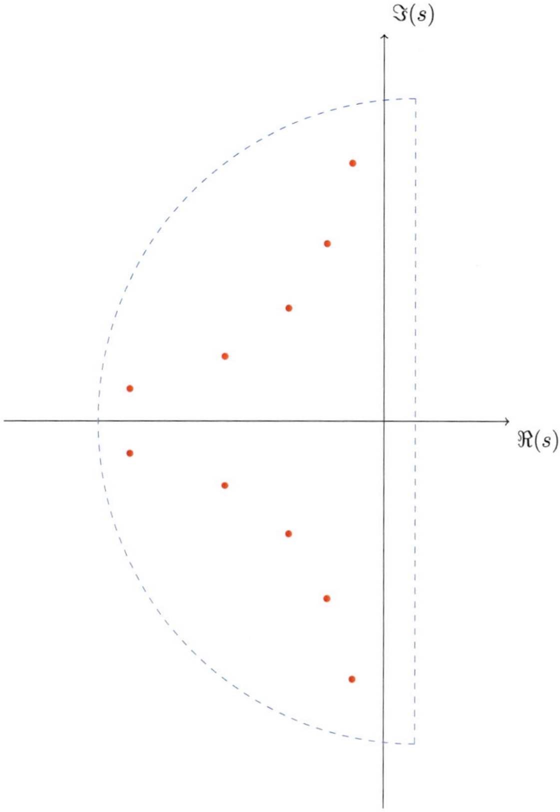

Fig. 1 Pictorial representation of the closed contour needed to compute the integral over s in (14) as the blue dashed line. Furthermore, possible quasinormal frequencies are depicted by conjugated red dots. These conjugation pairs arise from conjugation of (13)

We arrived at an expression relating ψ to the quasinormal frequencies.

The second contribution to ψ, in the equation above, is due to the contour integration along $\mathcal{C}$. We notice that $\Re(s) \ll -1$ in most of $\mathcal{C}$. Thus, the exponential factor in the definition of $\mathcal{F}$ is bound to decay rapidly as t increases and so is the contribution as a whole [4]. This renders the contribution negligible for the late time behaviour of the system. Hence, we end up with the asymptotic behaviour

$$\psi(x,t) \sim 2\pi i \sum_{n\in I} \text{Res}\,(\mathcal{F}, s_n) \tag{19}$$

in this limit. Finally, the expression above can be recast in the form

$$\psi\,(x,t) \sim \sum_{n\in I} C_n e^{s_n t} f_+\,(s_n, x) \tag{20}$$

for finite values of x and for some $C_n \in \mathbb{C}$ [2].

As we can see from (20), the unique solution of our physical problem will decay exponentially fast in time, for spatially bounded regions. In this sense, equations such as (1), defined on non compact spatial domains, model physical dissipative systems. Solutions to these systems, as we just saw, cannot be fully described by normal modes. Instead, in the late time behaviour of the system, they can be well approximated by quasinormal modes. In this way, they play the analogous role of normal modes in systems of compact spatial domain. A major distinction between the two sides is that normal frequencies are real while quasinormal frequencies may be complex.

2.2 *Quasinormal Modes of Black Holes*

A general static spherically symmetric metric in d dimensions has the form

$$ds^2 = -f dt \otimes dt + \frac{1}{f} dr \otimes dr + r^2 d^2\Omega_{d-2}, \tag{21}$$

where $d\,\Omega^2_{d-2}$ is the canonical metric tensor field of the unit (d-2)-sphere.

We consider a complex massless scalar test field ψ, minimally coupled to a background of the form (21). In such background, this scalar field can be decomposed as [1]

$$\psi(t, r, \theta, \phi) = \sum_{l=0}^{+\infty} \sum_{m=-l}^{l} \frac{e^{i\omega t}}{r} \psi_r(r) Y_{lm}(\theta, \phi). \tag{22}$$

Each component ψ_r obeys a second order field equation of the form

$$\frac{d^2\psi_r}{dx^2} + \left(\omega^2 - V\right)\psi_r = 0 \tag{23}$$

where x is the tortoise coordinate defined as

$$dx := \frac{dr}{f}. \tag{24}$$

The potential V in (23) can be expressed, with respect to r, as

$$V[f(r)] = f(r)\left(\frac{l(l+d-3)}{r^2} + \frac{(d-2)(d-4)f(r)}{4r^2} + \frac{(d-2)f'(r)}{2r}\right). \tag{25}$$

This potential (25) is valid for minimally coupled massless scalar fields in the background of any metric of the type (21), in Einstein gravity, but also in the presence of higher derivative corrections (which appear implicitly through the λ' corrections of $f(r)$).

Now we admit the background metric (21) to be a black hole with horizon radius R_h, and we concern ourselves with boundary conditions associated to (23). We are interested in a solution of the perturbation ψ, propagating in the spatial region outside of the black hole. That is to say, we only seek a solution for values of r in the region $r > R_h$. The tortoise coordinate x, defined as we did, takes values in $\mathbb{R}$ and has the asymptotic behaviours $x \to +\infty$ when $r \to +\infty$ and $x \to -\infty$ when $r \to R_h^+$. It is also easy to see that

$$\lim_{r\to+\infty} V[f(r)] = 0, \quad \lim_{r\to R_h^+} V[f(r)] = 0. \tag{26}$$

From the limits above, we gather that, in the boundaries of the region we are interested in, the solutions to the master Eq. (23) behave as

$$\psi_r(x) = A_+ e^{i\omega x} + A_- e^{-i\omega x} \tag{27}$$

for some $A_\pm \in \mathbb{C}$. Because nothing should leave the event horizon, we impose the boundary condition

$$\psi_r(x) \propto e^{i\omega x}, \quad x \to -\infty. \tag{28}$$

On the other hand, as we are describing an isolated system, we want to disregard unphysical waves coming from spacial infinity. Thus, we impose the boundary condition

$$\psi_r(x) \propto e^{-i\omega x}, \quad x \to +\infty. \tag{29}$$

Conditions (28) and (29), independent of initial data, define the asymptotic behaviour of the dominant radial contribution to ψ in the late time behaviour of the system: the *quasinormal modes*. These modes are defined as the solutions of Eq. (23), obeying the boundary conditions above for some associated quasinormal frequency ω.

In Sect. 2.1, we defined quasinormal frequencies as values of s. However, it is common to write the variable associated with the Laplace transform as $s = i\omega$. Thus, quasinormal frequencies are usually defined as values of ω instead. Using this notation, it is easy to see that our latest definitions of quasinormal modes and frequencies match those previously introduced.

As an observer, far away from the black hole, we are interested in the limit of spatial infinity, for the late time behaviour of ψ. Looking at (20), we see that in order to know ψ, we first need to know the quasinormal frequencies and the associated complex constants C_n, also called quasinormal excitation coefficients. How do we compute these? Regarding the excitation coefficients, the only way to compute them is by providing the problem with initial data [1]. On the other hand, quasinormal frequencies only depend on the metric tensor field of the background black hole space time.

We are left with the task of computing quasinormal frequencies. In general, it is very complicated (if not impossible) to obtain an analytical expressions of them.

However, there are two limiting cases where we might have an easier time looking for analytical expressions. These limiting cases target existing quasinormal frequencies with, operationally wise, desirable properties. The first limiting case targets quasinormal frequencies obeying the master equation for arbitrarily large values of the azimuthal number l[2]. This limit is usually called the eikonal limit. The second limiting case targets quasinormal frequencies ω whose imaginary part is much larger (in magnitude) than the real part: $|\Im(\omega)| \gg |\Re(\omega)|$. This limit is usually called the asymptotic limit. We will be working with both these limits.

Finally, while we postpone to later sections the introduction and discussion of actual analytical methods to compute quasinormal frequencies in both limiting cases introduced, we take one last note. In Sect. 2.1, we saw that quasinormal frequencies were restricted to the complex half plane $\Re(s) < 0$. It is clear that this restriction translates to $\Im(\omega) > 0$ in the new notation.

3 WKB Theory: A Review

We briefly review the WKB theory and explain some basic results which will be used extensively. The name WKB comes from the founders Gregor Wentzel, Hendrik Anthony Kramers, and Léon Brillouin.

3.1 *Defining Purpose and Schrödinger Like Equations*

The broad purpose of the WKB theory is to provide approximations to general solutions of differential equations whose highest order derivative is multiplied by some small real parameter. In this work, we will only deal with ordinary homogeneous linear differential equations. Hence, we restrict ourselves to this class of differential equations. Let us write a general differential equation of this class as

$$\epsilon \frac{d^n y}{dx^n} + \sum_{m=0}^{n-1} a_i \frac{d^m y}{dx^m} = 0 \tag{30}$$

where $a_i \in C^k(\mathbb{R})$, for some $k \in \mathbb{N}$. The WKB method proposes an asymptotic series expansion of the form

$$y \sim \prod_{n=0}^{+\infty} e^{S_n \delta^{n-1}}. \tag{31}$$

[2] In our example, this number arises from the spherical harmonic decomposition (22). Naturally, this limiting case is restricted to background black hole space times whose metric tensor field allows for similar decompositions.

where $S_n \in C^{k_n}(\mathbb{R})$ for some $k_n \in \mathbb{N}$ and where δ is a small real parameter. The scaling of δ with ϵ is determined by substitution along with the functions S_n.

As an example, we consider the Schrödinger like equation

$$\epsilon^2 \frac{d^2 y}{dx^2} = Qy \tag{32}$$

where ϵ^2 is a small real parameter and $Q \in C^\infty(\mathbb{R})$. Replacing the asymptotic series (31) in the differential equation above yields

$$\epsilon^2 \left[\frac{1}{\delta^2} \left(\sum_{n=0}^{+\infty} \delta^n \frac{dS_n}{dx} \right)^2 + \frac{1}{\delta} \sum_{n=0}^{+\infty} \delta^n \frac{d^2 S_n}{dx^2} \right] = Q. \tag{33}$$

We fix $\delta = \epsilon$ and proceed to solve the differential equation perturbatively in powers of ϵ. The equation of leading order is

$$\left(\frac{dS_0}{dx} \right)^2 = Q \tag{34}$$

whose solutions are (up to an additive constant, which we will ignore since (32) is linear)

$$S_0(x) = \pm \int \sqrt{Q(x)} dx. \tag{35}$$

The differential equation of first order in ϵ is

$$2 \frac{dS_0}{dx} \frac{dS_1}{dx} + \frac{d^2 S_0}{dx^2} = 0 \tag{36}$$

whose solution is (also up to an additive constant)

$$S_1(x) = -\frac{1}{4} \log (Q(x)). \tag{37}$$

We can follow this procedure iteratively, solving differential equations of higher order in ϵ. These equations take the generic form (for $n \geq 2$)

$$2 \frac{dS_0}{dx} \frac{dS_n}{dx} + \frac{d^2 S_{n-1}}{dx^2} + \sum_{j=1}^{n-1} \frac{dS_j}{dx} \frac{dS_{n-j}}{dx} = 0. \tag{38}$$

Because (31) is an asymptotic series, there is usually an accuracy limit associated as a representation of the solution to the differential equation considered. Hence, only a finite amount of functions S_n are of interest to us.

Up to zeroth order in ϵ, the general solution of the differential Eq. (32) can be expressed as

$$y(x) \sim \frac{\mathcal{C}_+}{\sqrt[4]{Q(x)}} \exp\left(\frac{1}{\epsilon}\int \sqrt{Q(x)}dx\right) + \frac{\mathcal{C}_-}{\sqrt[4]{Q(x)}} \exp\left(-\frac{1}{\epsilon}\int \sqrt{Q(x)}dx\right) \quad (39)$$

for some $\mathcal{C}_\pm \in \mathbb{C}$.

3.2 Analytic Continuation and Stokes Lines

Now, we study the validity domain of the solution (39), when analytically continued to complex x-plane. In order to do that, we rederive the solution (39), using a different approach. First, let us assume Q can be analytically continued onto the complex x-plane. Furthermore, let us rewrite Eq. (32) as

$$\frac{d^2y}{dx^2} - \frac{Q}{\epsilon^2}y = 0. \quad (40)$$

We introduce the new dependent variable

$$\rho(x) := y(x)\sqrt{q(x)} \quad (41)$$

and the new independent variable

$$w(x) := \int q(x)dx + \mathcal{C}_w \quad (42)$$

where q is an analytical function and $\mathcal{C}_w \in \mathbb{C}$. Rewriting our differential equation with respect to these variables yields [5]

$$\frac{d^2\rho}{dw^2} + (1+\delta)\rho = 0 \quad (43)$$

where we defined the function

$$\delta(x) := -\frac{Q(x)}{q^2(x)\epsilon^2} - 1 + q^{-\frac{3}{2}}(x)\frac{d^2}{dx^2}\left(\frac{1}{\sqrt{q(x)}}\right). \quad (44)$$

The idea is to choose a function q such that $|\delta| \sim 0$ in the region of the complex x-plane where we want to approximate the solution. Assuming we made such a choice, we can approximate our differential equation by

$$\frac{d^2\rho}{dw^2} + \rho = 0 \quad (45)$$

whose general solution is

$$\rho(w) = C_+ e^{iw} + C_- e^{-iw}. \tag{46}$$

for some $C_\pm \in \mathbb{C}$. Now, we seek a function q such that $|\delta| \sim 0$. If we choose

$$q(x) = \frac{i\sqrt{Q(x)}}{\epsilon} \tag{47}$$

we can rewrite δ as

$$\delta = \frac{\epsilon^2}{4}\left(\frac{1}{Q^2}\frac{d^2Q}{dx^2} - \frac{5}{4}\frac{1}{Q^3}\frac{dQ}{dx}\right). \tag{48}$$

Because ϵ^2 is a small real parameter, the choice (47) yields a small δ everywhere in the complex x-plane, except near roots of Q.

Plugging (47) in (46) yields the approximated solution

$$y(x) = \frac{C'_-}{\sqrt[4]{Q(x)}} \exp\left(-\frac{1}{\epsilon}\int \sqrt{Q(x)}dx\right) + \frac{C'_+}{\sqrt[4]{Q(x)}} \exp\left(\frac{1}{\epsilon}\int \sqrt{Q(x)}dx\right) \tag{49}$$

for some $C'_\pm \in \mathbb{C}$. We notice the solution above is identical to (39)!

Now, we want to know how good the expression above is, as an approximation to the true solution of (32).

We already know the expression above does not accurately represent the solution near roots of Q. In the remaining zones of the complex x-plane, δ is small. We will now argue that this condition alone is not enough to ensure the expression above to be a good representation of the true solution. We start by expressing a true solution of the differential Eq. (43) as

$$\rho(w) = a_+(w)e^{iw} + a_-(w)e^{-iw} \tag{50}$$

for some analytical functions a_+, a_-. The expression above comes with an extra degree of freedom, for we are expressing the function ρ using two arbitrary analytical functions a_+, a_-. To eliminate this freedom, we impose the condition

$$\frac{da_+}{dw}e^{iw} + \frac{da_-}{dw}e^{-iw} = 0. \tag{51}$$

Using the condition above, we can write

$$\frac{d\rho}{dw} = ia_+e^{iw} - ia_-e^{-iw}. \tag{52}$$

The equations above allow us to rewrite the second order ordinary differential Eq. (43) as the system of two first order ordinary differential equations [5]

$$\frac{da_+}{dw} = \frac{i\delta}{2}\left(a_+ + a_- e^{-2iw}\right) \tag{53}$$

$$\frac{da_-}{dw} = -\frac{i\delta}{2}\left(a_- + a_+ e^{2iw}\right). \tag{54}$$

Furthermore, we can rewrite the system above as the matrix differential equation

$$\frac{da}{dw} = Ma \tag{55}$$

where we defined the matrices

$$M(w) := \frac{i\delta}{2}\begin{bmatrix} 1 & e^{-2iw} \\ -e^{2iw} & -1 \end{bmatrix} \qquad a(w) := \begin{bmatrix} a_+(w) \\ a_-(w) \end{bmatrix}. \tag{56}$$

Looking at Eqs. (48) and (55), we notice the relative change of a with respect to w is proportional to ϵ^2. Thus, as long as the off diagonal elements of M are of the order of unity, this change is negligible. In particular, we can ask for the condition

$$\left|e^{-2iw}\right| \sim \left|e^{2iw}\right| \sim 1. \tag{57}$$

We found the validity domain we were looking for. Indeed, the solutions (46) and consequently (39) are good approximations of the respective true solutions, for values of x where the condition above is fulfilled.

The condition above naturally leads to the definition of Stokes lines. Stokes lines are the curves in the complex x-plane where w is real:

$$\Im(w) = 0. \tag{58}$$

Naturally, from the definition (42) the topology of these lines depends on the constant of integration $\mathcal{C}_w$. Moreover, along these lines (57) holds true!

With this loose argument, we reason the solution (39) is a good approximation of the true solution of (32), along Stokes lines, sufficiently far away from roots of Q. A much more formal proof of this statement can be found in [5].

A final note should be taken regarding cases where an analytic continuation of Q onto the complex x-plane cannot be made. As a general rule, we do not expect the solution (39) to be a good approximation of the true solution, near singularities of Q. This is so, because δ might share the same singularities [5].

4 A String-Corrected Black Hole and Its Perturbations

In this section, the d-dimensional spherically symmetric black hole solution with leading string-theoretical higher derivative corrections obtained by Callan, Myers and Perry is introduced along with some results concerning gravitational and scalar field perturbations.

4.1 Physical Origin

String theories require corrections in powers of the inverse string tension α', which manifest themselves in the form of higher-derivative terms in the effective action, and obviously in the corresponding field equations and their solutions.

We are focusing, in particular, in corrections to first order in α', which are present in bosonic and heterotic string effective actions (but not on type II superstring). The effective action we are thus considering is [6]

$$\frac{1}{16\pi G}\int\sqrt{-g}\left(\mathcal{R}-\frac{4}{d-2}\left(\partial^{\mu}\phi\right)\partial_{\mu}\phi+\mathrm{e}^{\frac{4}{d-2}\phi}\frac{\lambda}{2}\,\mathcal{R}^{\mu\nu\rho\sigma}\mathcal{R}_{\mu\nu\rho\sigma}\right)\mathrm{d}^{d}x \tag{59}$$

where $\lambda=\frac{\alpha'}{2},\frac{\alpha'}{4},0$ for bosonic, heterotic and type II strings respectively. We are only considering gravitational terms: we can consistently settle all fermions and gauge fields to zero. That is not the case of the dilaton, as it can be seen from the corresponding field equations.

The stringy correction present in the action above is bound to manifest itself as a correction term proportional to λ, and to affect measurable quantities associated with the black hole solution (e.g. the quasinormal modes we consider in this work). Thus, studying how these quantities change according to the stringy correction might be a viable way to provide experimental verifications of string theory.

4.2 The Metric

The Callan-Myers-Perry black hole we will be considering is a d-dimensional static spherically symmetric black hole of the form (21). It is an asymptotically flat solution of the field equations from (59)—a λ-corrected Tangherlini black hole with $f(r)$ given by [6]

$$f(r) := f_0(r)\left(1 + \lambda'\delta f\right), \tag{60}$$

$$f_0(r) := 1 - \left(\frac{R_h}{r}\right)^{d-3}, \tag{61}$$

$$\delta f(r) := -\frac{(d-3)(d-4)}{2}\left(\frac{R_h}{r}\right)^{d-3}\frac{1-\left(\frac{R_h}{r}\right)^{d-1}}{1-\left(\frac{R_h}{r}\right)^{d-3}}. \tag{62}$$

In the equations above, λ' stands for an adimensional version of λ, defined as $\lambda' := \frac{\lambda}{R_h^2}$. Moreover, R_h stands for the radius of the black hole event horizon, expressed as $R_h = (2\mu)^{\frac{1}{d-3}}$ where the parameter μ can be related to the black hole mass m by

$$m = \left(1 + \frac{(d-3)(d-4)}{2}\lambda'\right)\frac{(d-2)\Omega_{d-2}}{8\pi G}\mu, \ \Omega_{d-2} = \frac{2\pi^{\frac{d-1}{2}}}{\Gamma\left(\frac{d-1}{2}\right)}. \tag{63}$$

We notice R_h remained invariant under the stringy correction. This is a consequence of the coordinates used to express the metric (21).

The Hawking temperature of this black hole is given, from (60), by

$$T_{\mathcal{H}} = \frac{1}{4\pi}\frac{df}{dr}(R_h) = \frac{d-3}{4\pi R_h}\left(1 - \lambda'\frac{(d-1)(d-4)}{2}\right). \tag{64}$$

4.3 Gravitational Perturbations and Quasinormal Modes

Here, we introduce, very briefly, gravitational perturbations and present some associated results in the context of the Callan-Myers-Perry black hole (60).

Linear perturbations of the metric are called gravitational perturbations. If we consider a spherically symmetric and static d-dimensional black hole space time of the form (21), in the process of simplifying the equations of motion associated with the perturbation components one ends up with a simplified system of (decoupled) master equations [7]. The independent variables associated with these equations are functions that describe distinct types of gravitational perturbations. These types, scalar, vector and (for $d > 4$) tensor types, are all needed in order to fully describe general gravitational perturbations. The master equations associated with these perturbations are second order Schrödinger-like differential equations of the form (23), with ψ_r replaced by ψ, a function describing the gravitational perturbations. In this context, x is the tortoise coordinate corresponding to (60).

This framework is valid in Einstein gravity and, as shown in different articles [8–11], also in the presence of quadratic (Gauss-Bonnet) gravitational corrections.

In this work, we address gravitational perturbations associated with d-dimensional black holes with leading string gravitational corrections, like (59). We will be specifically concerned with tensor type gravitational perturbations of the black hole (60), but our methods apply to other types of gravitational perturbations and other analogous black hole solutions [12, 13]. In [14] and more generally in [15], it was shown that the corresponding potential V can be expressed, with respect to r, as

$$V[f(r)] = f(r)\left(\frac{l(l+d-3)}{r^2} + \frac{(d-2)(d-4)f(r)}{4r^2} + \frac{(d-2)f'(r)}{2r}\right) + \lambda'$$
$$\left(\frac{R_h}{r}\right)^2 f(r)\Bigg[\left(\frac{2l(l+d-3)}{r} + \frac{(d-4)(d-5)f(r)}{r} + (d-4)f'(r)\right)\left(2\left(\frac{1-f(r)}{r}\right) + f'(r)\right)$$
$$+ (4(d-3) - (5d-16)f(r))\frac{f'(r)}{r} - 4\left(f'(r)\right)^2 + (d-4)f(r)f''(r)\Bigg] \tag{65}$$

where the apostrophe stands for derivation with respect to r.

We notice the potential above is identical to (25) to zeroth order in λ'. Indeed, in the classical ($\lambda = 0$) limit, linear perturbations of complex massless scalar fields and tensor type gravitational perturbations share the same quasinormal frequency spectrum [16].

5 Quasinormal Modes in the Eikonal Limit

In this section, we analytically compute the quasinormal frequencies, in the eikonal limit, of tensor type gravitational perturbations of the black hole (60). For a general discussion see [17, 18].

We recall the eikonal limit targets quasinormal frequencies ω obeying the master Eq. (23) for large values of the azimuthal number l.

5.1 Definition and Analysis of Q

From the potential (65) we define the function

$$Q(x) := \omega^2 - V(x). \tag{66}$$

For large values of l, the asymptotic expansion of Q can be written, with respect to r, as

$$Q(r) \sim \omega^2 - h(r)\left(\frac{l}{r}\right)^2 \tag{67}$$

where we defined

$$h(r) := f(r)\left(1 + 2\lambda'\left(\frac{R_h}{r}\right)^2\left[2(1-f(r)) + rf'(r)\right]\right). \tag{68}$$

Because we are working in the eikonal limit, we redefine Q as the respective large l asymptotic expansion (67), for convenience purposes.

Up to zeroth order in λ', we have

$$Q(r) = \omega^2 - \left[1 - \left(\frac{R_h}{r}\right)^{d-3}\right]\left(\frac{l}{r}\right)^2. \tag{69}$$

We notice the expression above only has one critical point, besides $r = 0$, corresponding to one minimizer. Indeed, taking the derivative yields

$$\frac{dQ}{dr}(r) = l^2\left(\frac{2}{r^3} - (d-1)R_h^{d-3}r^{-d}\right) \tag{70}$$

Computing the root of the expression above, we get

$$r_g := \left(\frac{d-1}{2}\right)^{\frac{1}{d-3}} R_h. \tag{71}$$

Furthermore, taking the second order derivative yields

$$\frac{d^2Q}{dr^2}(r_g) = \frac{2^{\frac{d+1}{d-3}}(d-3)(d-1)^{-\frac{4}{d-3}}l^2}{R_h^4}. \tag{72}$$

Because the expression above is strictly positive for $d \geq 4$, we gather r_g is a minimizer. When we consider (67), up to first order in λ', a correction of this minimizer emerges. To compute this correction, we redefine r_g as the corrected minimizer

$$r_g \mapsto r_g + \lambda'\delta r_g \tag{73}$$

for some unknown $\delta r_g \in \mathbb{R}$ and replace it in the derivative of Q. The resulting expression, up to first order in λ', is

$$\lambda'\frac{32^{\frac{1}{d-3}}(d-1)^{-\frac{2(d+2)}{d-3}}l^2}{R_h^4}\left(\delta r_g 2^{\frac{4-d}{3-d}}(d-3)(d-1)^{\frac{2d}{d-3}} + 4((d-11)d+16)(d-1)^{\frac{d+2}{d-3}}R_h + \left(4(d+1) - 4^{\frac{1}{3-d}}(d-4)(d-3)(d-1)^{\frac{2}{d-3}}\right)(d-1)^{\frac{2d-1}{d-3}}R_h\right). \tag{74}$$

In order to find δr_g, we equate the expression above to zero and solve for δr_g. After some algebraic manipulation, we get

$$\delta r_g = \left(\frac{d-1}{2}\right)^{\frac{1}{d-3}}\left(\frac{d-4}{2} - (2d-5)\left(\frac{2}{d-1}\right)^{\frac{d-1}{d-3}}\right)R_h. \tag{75}$$

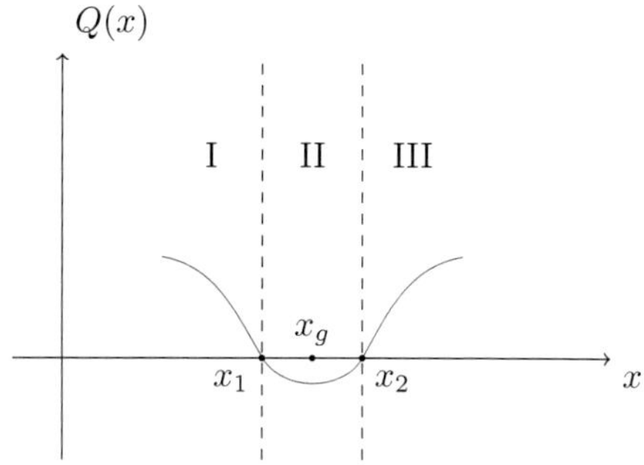

Fig. 2 Graphical depiction of Q as the blue line. Possible roots of Q are denoted by x_1 and x_2. Moreover, the x-axis is subdivided in three regions, denoted by I, II and III

Finally, we can write the corrected minimizer as

$$r_g = \left(\frac{d-1}{2}\right)^{\frac{1}{d-3}} R_h \left[1 + \lambda' \left(\frac{d-4}{2} - (2d-5)\left(\frac{2}{d-1}\right)^{\frac{d-1}{d-3}}\right)\right]. \tag{76}$$

5.2 *WKB Approximation*

We now use the WKB theory to find an analytical relation between quasinormal frequencies, in the eikonal limit, and the potential (65). To this end, we resort to the previously found minimum of Q. We start by making the assumption that quasinormal modes ω, in the eikonal limit, are such that

$$\omega^2 \sim V(r_g). \tag{77}$$

If this is the case, Q can be graphically represented near $x_g = x(r_g)$, as depicted in Fig. 2.

In this figure, we notice the x-axis is subdivided in three regions, denoted by I, II and III.

Looking at (67), we notice $|Q|$ is very large far enough from x_g and decreases abruptly, approaching the latter. Thus, we can use WKB theory to approximate the solution of the master Eq. (23) in regions I and III. Using (39) yields

$$\psi_{\mathrm{I}}(x) \sim \frac{C_1^+}{\sqrt[4]{Q(x)}} \exp\left(i \int_x^{x_1} \sqrt{Q(y)} dy\right) + \frac{C_1^-}{\sqrt[4]{Q(x)}} \exp\left(-i \int_x^{x_1} \sqrt{Q(y)} dy\right) \tag{78}$$

$$\psi_{\mathrm{III}}(x) \sim \frac{C_2^+}{\sqrt[4]{Q(x)}} \exp\left(i \int_{x_2}^{x} \sqrt{Q(y)} dy\right) + \frac{C_2^-}{\sqrt[4]{Q(x)}} \exp\left(-i \int_{x_2}^{x} \sqrt{Q(y)} dy\right) \tag{79}$$

for some $C_1^\pm, C_2^\pm \in \mathbb{C}$. To compute an approximation for ψ_{II}, we start by considering the Taylor expansion

$$Q(x) = Q(x_g) + \frac{dQ}{dx}(x_g)(x - x_g) + \frac{d^2Q}{dx^2}(x_g)\frac{(x - x_g)^2}{2} + \mathcal{O}\left(\left(x - x_g\right)^3\right). \tag{80}$$

Noting that

$$\frac{dQ}{dx}(x_g) = 0 \tag{81}$$

we can rewrite (80) as

$$Q(x) = Q(x_g) + \frac{d^2Q}{dx^2}(x_g)\frac{(x - x_g)^2}{2} + \mathcal{O}\left(\left(x - x_g\right)^3\right). \tag{82}$$

Furthermore, we can neglect terms of higher orders in $(x - x_g)$ and write

$$Q(x) \sim Q(x_g) + \frac{d^2Q}{dx^2}(x_g)\frac{(x - x_g)^2}{2}. \tag{83}$$

in region II. This approximation is valid because region II is very slim, under the assumption (77). Now, defining the constants

$$k := \frac{1}{2}\frac{d^2Q}{dx^2}(x_g) \qquad \nu := -\frac{1}{2} - iQ(x_g)\left(2\frac{d^2Q}{dx^2}(x_g)\right)^{-\frac{1}{2}} \tag{84}$$

and the variable

$$t := \sqrt[4]{4k}e^{\frac{i\pi}{4}}(x - x_g) \tag{85}$$

allow us to rewrite the master Eq. (23) as

$$\frac{d^2\psi_{\rm II}}{dt^2} + \left(\nu + \frac{1}{2} - \frac{t^2}{4}\right)\psi_{\rm II} = 0 \tag{86}$$

in region II. The general solution of the differential equation above is

$$\psi_{\rm II}(t) = C_1 D_\nu(t) + C_2 D_{-\nu-1}(it) \tag{87}$$

for some $C_1, C_2 \in \mathbb{C}$, where

$$D_\nu(w) = \frac{2^{\frac{\nu}{2}}e^{-\frac{w^2}{4}}\sqrt[4]{-iw}\sqrt[4]{iw}}{\sqrt{w}}\,{}_1F_1\left(-\frac{\nu}{2}, \frac{1}{2}, \frac{w^2}{2}\right) \tag{88}$$

denotes the parabolic cylinder function and

$${}_1F_1\,(a, b, w) = \sum_{k=0}^{+\infty}\frac{\Gamma\,(a + k)}{\Gamma(a)}\frac{\Gamma(b)}{\Gamma(b + k)}\frac{w^k}{k!} \tag{89}$$

stands for the confluent hypergeometric function of the first kind.

The asymptotic expansions of the parabolic cylinder function yield [17]

$$\psi_{\mathrm{II}}(x) \sim C_2 e^{-3i\pi\frac{\nu+1}{4}}(4k)^{-\frac{\nu+1}{4}}(x-x_g)^{-(\nu+1)}e^{i\sqrt{k}\frac{(x-x_g)^2}{2}} + \left(C_1 + C_2\frac{\sqrt{2\pi}e^{-i\nu\frac{\pi}{2}}}{\Gamma(\nu+1)}\right)e^{i\pi\frac{\nu}{4}}(4k)^{\frac{\nu}{4}}\left(x-x_g\right)^{\nu}e^{-i\sqrt{k}\frac{(x-x_g)^2}{2}} \tag{90}$$

$$\psi_{\mathrm{II}}(x) \sim C_1 e^{-3i\pi\frac{\nu}{4}}(4k)^{\frac{\nu}{4}}(x_g-x)^{\nu}e^{-i\sqrt{k}\frac{(x-x_g)^2}{2}} + \left(C_2 - iC_1\frac{\sqrt{2\pi}e^{-i\nu\frac{\pi}{2}}}{\Gamma(-\nu)}\right)e^{i\pi\frac{\nu+1}{4}}(4k)^{-\frac{\nu+1}{4}}\left(x_g-x\right)^{-(\nu+1)}e^{i\sqrt{k}\frac{(x-x_g)^2}{2}} \tag{91}$$

for $te^{-i\frac{\pi}{4}} \gg 1$ and $te^{-i\frac{\pi}{4}} \ll -1$ respectively. Looking at (72), we see that $k \gg 1$. Hence, the asymptotic expansions above are valid for values of x near the respective points x_1 and x_2, where it still makes sense to match them with ψ_{I} and ψ_{III}.

Imposing the boundary conditions (28) and (29) on solutions ψ_{I} and ψ_{III} respectively yields

$$C_1^+ = 0 \qquad C_2^+ = 0. \tag{92}$$

Thus, we can rewrite ψ_{I} and ψ_{III} as

$$\psi_{\mathrm{I}}(x) \sim \frac{C_1^-}{\sqrt[4]{Q(x)}}\exp\left(-i\int_x^{x_1}\sqrt{Q(y)}dy\right) \tag{93}$$

$$\psi_{\mathrm{III}}(x) \sim \frac{C_2^-}{\sqrt[4]{Q(x)}}\exp\left(-i\int_{x_2}^{x}\sqrt{Q(y)}dy\right). \tag{94}$$

In the matching region, we still can use the approximation (83) and write

$$\psi_{\mathrm{I}}(x) \sim \frac{C_1^-}{\sqrt[4]{Q(x)}}\exp\left(-i\int_x^{x_1}\sqrt{Q(x_g)+k(y-x_g)^2}dy\right) \tag{95}$$

$$\psi_{\mathrm{III}}(x) \sim \frac{C_2^-}{\sqrt[4]{Q(x)}}\exp\left(-i\int_{x_2}^{x}\sqrt{Q(x_g)+k(y-x_g)^2}dy\right). \tag{96}$$

Under the assumption (77), we know that

$$Q(x_g) \sim 0. \tag{97}$$

Thus, we can write

$$\int_{x}^{x_1} \sqrt{Q(x_g) + k(y - x_g)^2} dy \sim \int_{x}^{x_1} \sqrt{k(y - x_g)^2} dy = \sqrt{k} \int_{x}^{x_1} |y - x_g| dy$$
$$= -\sqrt{k} \int_{x}^{x_1} (y - x_g) dy = \sqrt{k} \frac{(x - x_g)^2}{2} - \sqrt{k} \frac{(x_1 - x_g)}{2}, \tag{98}$$

$$\int_{x_2}^{x} \sqrt{Q(x_g) + k(y - x_g)^2} dy \sim \int_{x_2}^{x} \sqrt{k(y - x_g)^2} dy = \sqrt{k} \int_{x_2}^{x} |y - x_g| dy =$$
$$\sqrt{k} \int_{x_2}^{x} (y - x_g) dy = \sqrt{k} \frac{(x - x_g)^2}{2} - \sqrt{k} \frac{(x_2 - x_g)}{2}. \tag{99}$$

Using the approximations above, we can rewrite (95) and (96) as

$$\psi_{\mathrm{I}}(x) \sim \frac{\mathcal{D}_1}{\sqrt[4]{Q(x)}} e^{-i\sqrt{k}\frac{(x-x_g)^2}{2}}, \tag{100}$$

$$\psi_{\mathrm{III}}(x) \sim \frac{\mathcal{D}_2}{\sqrt[4]{Q(x)}} e^{-i\sqrt{k}\frac{(x-x_g)^2}{2}} \tag{101}$$

for some $\mathcal{D}_1, \mathcal{D}_2 \in \mathbb{C}$. We notice the exponential terms of the asymptotic expansions (90) and (91) match those of (101) and (100) respectively if and only if $\mathcal{C}_2 = 0$ and

$$\frac{1}{\Gamma(-\nu)} \sim 0. \tag{102}$$

The last condition demands ν to be very close to singular points of Γ. Thus, we end up with the restriction

$$\nu \in \mathbb{N}_0. \tag{103}$$

Using definition (84), we can rewrite the condition above as

$$Q(x_g) \left(2 \frac{d^2 Q}{dx^2}(x_g) \right)^{-\frac{1}{2}} = i \left(n + \frac{1}{2} \right) \tag{104}$$

for $n \in \mathbb{N}_0$. This is the condition we use to analytically compute the quasinormal frequencies.

5.3 Quasinormal Frequencies Computation

We now explicitly compute an analytical expression for the quasinormal frequencies in the eikonal limit. Looking at the asymptotic expression (67), we easily see that

$$2h(r_g) = r_g \frac{dh}{dr}(r_g). \tag{105}$$

Using (67), we can rewrite the Eq. (104) as

$$\left(\omega^2 - h(r_g)\left(\frac{l}{r_g}\right)^2\right) = i\left(n + \frac{1}{2}\right)\sqrt{2\frac{d^2Q}{dx^2}(x_g)} \tag{106}$$

for $n \in \mathbb{N}_0$. Rearranging some terms yields

$$\omega = \pm l\sqrt{\frac{h(r_g)}{r_g^2}}\sqrt{1 + i\left(n + \frac{1}{2}\right)\frac{r_g^2}{l^2 h(r_g)}\sqrt{2\frac{d^2Q}{dx^2}(x_g)}}. \tag{107}$$

Once more, using (67) yields

$$\begin{aligned}\frac{d^2Q}{dx^2}(x_g) &= -l^2\frac{d^2}{dx^2}\left(\frac{h}{r^2}\right)(x_g) = -l^2\frac{d}{dx}\left[\left(\frac{dh}{dr}\frac{1}{r^2} - \frac{2}{r^3}h\right)\frac{dr}{dx}\right](x_g)\\ &= -l^2\frac{dr}{dx}\left[\frac{d^2r}{drdx}\left(\frac{dh}{dr}\frac{1}{r^2} - \frac{2}{r^3}h\right) + \frac{dr}{dx}\left(\frac{d^2h}{dr^2}\frac{1}{r^2} - \frac{2}{r^3}\frac{dh}{dr} + \frac{6}{r^4}h - \frac{2}{r^3}\frac{dh}{dr}\right)\right](x_g).\end{aligned} \tag{108}$$

Using definition (24) and Eq. (105), we can rewrite the previous equation as

$$\frac{d^2Q}{dx^2}(x_g) = -l^2\frac{f^2(r_g)}{r_g^4}\left(\frac{d^2h}{dr^2}(r_g)r_g^2 - 2h(r_g)\right). \tag{109}$$

Hence, we can rewrite Eq. (107) as

$$\omega = \pm l\sqrt{\frac{h(r_g)}{r_g^2}}\sqrt{1 + \frac{i}{l}\left(n + \frac{1}{2}\right)\sqrt{2\left(\frac{f(r_g)}{h(r_g)}\right)^2\left(2h(r_g) - \frac{d^2h}{dr^2}(r_g)r_g^2\right)}} \tag{110}$$

for $n \in \mathbb{N}_0$. Since we are in the eikonal limit, we can Taylor expand the expression above yielding

$$\omega = l\sqrt{\frac{h(r_g)}{r_g^2}} + \frac{i}{\sqrt{2}}\left(n + \frac{1}{2}\right)\sqrt{\frac{f^2(r_g)}{r_g^2 h(r_g)}\left(2h(r_g) - \frac{d^2h}{dr^2}(r_g)r_g^2\right)} \tag{111}$$

where we chose the positive root, assuming stability of the black hole solution. In order to simplify the notation, we define the constants

$$\Gamma_r := \sqrt{\frac{h(r_g)}{r_g^2}}, \qquad \Gamma_i := \sqrt{\frac{f^2(r_g)}{2r_g^2 h(r_g)}\left(2h(r_g) - \frac{d^2h}{dr^2}(r_g)r_g^2\right)}. \tag{112}$$

After some algebraic manipulation, we can write Γ_r and Γ_i, up to first order in λ', as

$$\Gamma_r = \sqrt{\frac{d-3}{d-1}} \left(\frac{2}{d-1}\right)^{\frac{1}{d-3}} \frac{1}{R_h} \left[1 + \frac{\lambda'}{2}\left(3(d-2)\left(\frac{2}{d-1}\right)^{\frac{d-1}{d-3}} - (d-4)\right)\right], \tag{113}$$

$$\Gamma_i = \frac{d-3}{\sqrt{d-1}} \left(\frac{2}{d-1}\right)^{\frac{1}{d-3}} \frac{1}{R_h} \left[1 - \lambda' \frac{d-4}{2}\left(1 + (d-2)\left(\frac{2}{d-1}\right)^{\frac{d-1}{d-3}}\right)\right]. \tag{114}$$

Rewriting the expressions above with respect to the black hole Hawking temperature $T_{\mathcal{H}}$, given by (64), yields

$$\Gamma_r = \frac{4\pi T_{\mathcal{H}}}{\sqrt{(d-1)(d-3)}} \left(\frac{2}{d-1}\right)^{\frac{1}{d-3}} \left[1 + \lambda' \left(\frac{d-2}{2}\right)\left(d-4+3\left(\frac{2}{d-1}\right)^{\frac{d-1}{d-3}}\right)\right], \tag{115}$$

$$\Gamma_i = \frac{4\pi T_{\mathcal{H}}}{\sqrt{d-1}} \left(\frac{2}{d-1}\right)^{\frac{1}{d-3}} \left[1 + \lambda' \frac{(d-2)(d-4)}{2}\left(1 - \left(\frac{2}{d-1}\right)^{\frac{d-1}{d-3}}\right)\right]. \tag{116}$$

Finally, we can rewrite the quasinormal frequencies (111) as

$$\omega = l\Gamma_r + i\left(n + \frac{1}{2}\right)\Gamma_i \tag{117}$$

for $n \in \mathbb{N}_0$. We notice they obey the starting assumption (77).

This concludes our derivation of the quasinormal frequencies for tensor type gravitational perturbations of the black hole (60) in the eikonal limit. This derivation can also be found in [19], together with an analogous calculation for test scalar fields in the same background. Quasinornal frequencies in the eikonal limit for generic gravitational perturbations of black holes with Gauss Bonnet corrections can be found in [20].

6 Quasinormal Modes in the Asymptotic Limit

In this section, we analytically compute the quasinormal frequencies, in the asymptotic limit, of tensor type gravitational perturbations of the black hole (60). We recall the asymptotic limit targets quasinormal frequencies ω such that $|\Im(\omega)| \gg |\Re(\omega)|$. The corresponding quasinormal modes are therefore highly damped.

First, we address the master Eq. (23) by making a crucial variable change followed by a standard perturbation theory approach.

Secondly, we set up some of the WKB theory formalism and study the associated Stokes lines topology.

Finally, we introduce the monodromy method to analytically compute the quasinormal frequencies. The monodromy method, introduced in [21], is a very powerful method allowing one to analytically compute quasinormal frequencies, in the asymptotic limit, associated with linear perturbations of fields coupled to gravity in a black hole space time [16].

In this section we allow r to take complex values. Thus, we assume an analytical continuation of functions of r to the complex plane.

6.1 Variable Change and Perturbation Theory

For future convenience, we want to rewrite the master Eq. (23) with respect to a simpler independent variable. The variable we seek is the tortoise coordinate of the d-dimensional Tangherlini black hole space time. Recalling (61), this variable is defined as

$$dz = \frac{dr}{f_0} \tag{118}$$

up to a complex integration constant. We allow the integration constant to be complex, because we are assuming an analytical continuation of the functions of r to the complex plane.

Looking at definitions (24) and (118), we can write

$$\frac{dz}{dx} = 1 + \lambda' \delta f. \tag{119}$$

We can also rewrite the second order derivative of the master Eq. (23) as

$$\frac{d^2\psi}{dx^2} = \frac{d}{dx}\left(\frac{d\psi}{dz}\frac{dz}{dx}\right) = \frac{d^2\psi}{dz^2}\left(\frac{dz}{dx}\right)^2 + \frac{d\psi}{dz}\frac{d^2z}{dx^2} = \frac{d^2\psi}{dz^2}\left(\frac{dz}{dx}\right)^2 + \frac{d}{dr}\left(\frac{dz}{dx}\right)\frac{dr}{dx}\frac{d\psi}{dz}. \tag{120}$$

Thus, we can rewrite the master Eq. (23), with respect to z, as

$$\left(\frac{dz}{dx}\right)^2 \frac{d^2\psi}{dz^2} + \frac{d}{dr}\left(\frac{dz}{dx}\right)\frac{dr}{dx}\frac{d\psi}{dz} + \left(\omega^2 - V\right)\psi = 0. \tag{121}$$

Using (24) and (119), we can write, with respect to r, the expressions

$$\frac{d}{dr}\left(\frac{dz}{dx}\right)\frac{dr}{dx}(r) = \lambda' \frac{(d-4)(d-3)r^{2-2d}\left((d-1)r^4 R_h^{2d} - 2(d-2)R_h^{d+3}r^{d+1} + (d-3)R_h^4 r^{2d}\right)}{2R_h^{7-d}r^d - 2R_h^4 r^3} \tag{122}$$

$$\left(\frac{dz}{dx}\right)^2 (r) = 1 + 2\lambda'\delta f = 1 - \lambda'(d-3)(d-4)\left(\frac{R_h}{r}\right)^{d-3} \frac{1 - \left(\frac{R_h}{r}\right)^{d-1}}{1 - \left(\frac{R_h}{r}\right)^{d-3}} \tag{123}$$

up to first order in λ'.

Now, we use perturbation theory to address the differential Eq. (121). We start by considering the expansions

$$\psi = \psi_0 + \lambda' \psi_1 \tag{124}$$

$$V = V_0 + \lambda' V_1. \tag{125}$$

Here, V_0 stands for the potential (65) in the limit of λ' set to zero, i.e. the potential (25) taken with $f_0(r)$. Moreover, V_1 contains all the corrections of first order in λ'. This includes those coming from the explicit correction present in (65) and those coming from the correction present in f in (60).

Replacing these expansions in (121) and solving perturbatively in powers of λ' yields two distinct linear ordinary differential equations. The first one, of zeroth order in λ', is

$$\frac{d^2\psi_0}{dz^2} + \left(\omega^2 - V_0\right)\psi_0 = 0. \tag{126}$$

The second one, of first order in λ', is the non homogeneous differential equation

$$\frac{d^2\psi_1}{dz^2} + \left(\omega^2 - V_0\right)\psi_1 = \xi. \tag{127}$$

The non homogeneous term is

$$\xi = \xi_1 \frac{d^2\psi_0}{dz^2} + \xi_2 \frac{d\psi_0}{dz} + \xi_3 \psi_0 \tag{128}$$

where we defined the functions

$$\xi_1(r) := -2\delta f(r) \tag{129}$$

$$\xi_2(r) := -f(r)\,[\delta f]'(r) \tag{130}$$

$$\xi_3(r) := V_1(r) \tag{131}$$

From now on, we restrict ourselves to the asymptotic limit. Hence, we assume that $|\Im(\omega)| \gg |\Re(\omega)|$. In the asymptotic limit, some of the WKB theory formalism developed in Sect. 3 can be used to give a viable approximation of the general solutions of the differential Eqs. (126) and (127). In the next section, we set up this formalism and proceed to compute this approximation for the general solution of (126).

6.2 WKB Theory Set up

We now set up the WKB theory formalism associated with the differential Eq. (126).

We note it is preferable to rewrite (126) with respect to the coordinate r, instead of z. This is so, mainly because, as we will see in (146), z is a multivalued function of r. Following [22], we define a new dependent variable

$$\Psi := \sqrt{f_0}\psi_0. \tag{132}$$

Now, we can write

$$\frac{d^2\psi_0}{dz^2} = \frac{d}{dz}\left(\frac{d\psi_0}{dr}\frac{dr}{dz}\right) = \frac{d^2\psi_0}{dr^2}\left(\frac{dr}{dz}\right)^2 + \frac{d\psi_0}{dr}\frac{d^2r}{drdz}\frac{dr}{dz}. \tag{133}$$

Using definition (118) allows us to rewrite the previous equation as

$$\frac{d^2\psi_0}{dx^2} = \frac{d^2\psi_0}{dr^2}f_0^2 + f_0' f_0 \frac{d\psi_0}{dr} \tag{134}$$

where the apostrophe stands for derivation with respect to r. Using definition (132), we can write

$$\frac{d\psi_0}{dr} = -\frac{1}{2}f_0^{-\frac{3}{2}} f_0' \Psi + \frac{1}{\sqrt{f_0}}\frac{d\Psi}{dr} \tag{135}$$

$$\frac{d^2\psi_0}{dr^2} = \frac{3}{4}f_0^{-\frac{5}{2}}\left(f_0'\right)^2\Psi - \frac{1}{2}f_0^{-\frac{3}{2}}f_0''\Psi - f_0^{-\frac{3}{2}}f_0'\frac{d\Psi}{dr} + \frac{1}{\sqrt{f_0}}\frac{d^2\Psi}{dr^2}. \tag{136}$$

Replacing these expressions in Eq. (134) allow us to rewrite (126) as

$$\frac{d^2\Psi}{dr^2} + R\Psi = 0. \tag{137}$$

where we defined the function

$$R := \left(\frac{\omega}{f_0}\right)^2 - \frac{V_0}{f_0^2} - \frac{1}{2}\frac{f_0''}{f_0} + \frac{1}{4}\left(\frac{f_0'}{f_0}\right)^2. \tag{138}$$

In the asymptotic limit, quasinormal frequencies are such that $|\omega| \gg 1$. Hence, we can use (39) to write the WKB approximation of the general solution of (137) for some $C_\pm \in \mathbb{C}$ as

$$\Psi(r) \sim C_+\sqrt[4]{\frac{\omega^2}{R(r)}}\exp\left(i\omega\int\sqrt{\frac{R(r)}{\omega^2}}dr\right) + C_-\sqrt[4]{\frac{\omega^2}{R(r)}}\exp\left(-i\omega\int\sqrt{\frac{R(r)}{\omega^2}}dr\right). \tag{139}$$

Looking at (138), we notice R has singularities in roots of f_0 and in $r = 0$. Considering the definition (61), we figure these roots are solutions of the polynomial equation

$$1 - \left(\frac{R_h}{r}\right)^{d-3} = 0 \tag{140}$$

such as the event horizon $r = R_h$. The remaining d-4 solutions, usually called "fictitious horizons", can be found equally distributed among the circle $|r| = R_h$. Furthermore, because we are considering the asymptotic limit and consequently assuming that $|\omega| \gg 1$, we know the roots of R are very close to $r = 0$. Thus, we expect (139) to provide a good approximation of the general solution of the differential Eq. (137), in Stokes lines, far enough from $r = 0$ and from the real and fictitious horizons. In these regions, the approximation

$$R \sim \left(\frac{\omega}{f_0}\right)^2 \tag{141}$$

holds true. Hence, we can rewrite the WKB approximation (139) as

$$\Psi(r) \sim C_+\sqrt{f_0(r)}\exp\left(i\omega\int\frac{dr}{f_0(r)}\right) + C_-\sqrt{f_0(r)}\exp\left(-i\omega\int\frac{dr}{f_0(r)}\right) \tag{142}$$

in these regions. Using definition (118), we can rewrite the approximation above as

$$\Psi(r) \sim C_+\sqrt{f_0(r)}e^{i\omega z(r)} + C_-\sqrt{f_0(r)}e^{-i\omega z(r)}. \tag{143}$$

Finally, using definition (132) and omitting the r dependence for simplicity purposes, we can rewrite the approximation above as

$$\psi_0(z) \sim C_+e^{i\omega z} + C_-e^{-i\omega z}. \tag{144}$$

We will use this approximation several times during this work.

6.3 Stokes Lines Topology

Now, we want to know the regions of the complex r-plane where the WKB approximation (144) is effective. As we saw in Sect. 3, these regions will generally coincide with Stokes lines associated with the approximation. Therefore, in order to know these regions well, we need to start by studying the Stokes lines topology associated with the WKB approximation (144).

Considering the approximation (141) and definition (58), Stokes lines, far enough from $r = 0$ and from the real and fictitious horizons, are such that

$$\Im(\omega z) = 0 \Rightarrow \Re(z) = 0. \tag{145}$$

In the first equality above, we used the fact that we are in the asymptotic limit and consequently assuming ω to be approximately imaginary.

It is rather easy to see the topology of these lines in the surroundings of $r = 0$. Indeed, integrating (118) yields [16]

$$z(r) = r + \frac{1}{2}\sum_{n=0}^{d-4}\frac{1}{k_n}\log\left(1 - \frac{r}{R_n}\right) \tag{146}$$

where we defined

$$k_n := \frac{1}{2}f_0'(R_n) \tag{147}$$

and where we chose the complex integration constant, defining z, to be such that $z(r = 0) = 0$. In the expressions above, R_n denotes fictitious horizons for $1 \leq n \leq d - 4$ and the real horizon for $n = 0$.

For values of r sufficiently close to the origin of the complex r-plane, the logarithm in (146) is given by the Taylor series

$$\log\left(1 - \frac{r}{R_n}\right) = -\sum_{k=1}^{+\infty}\frac{1}{k}\left(\frac{r}{R_n}\right)^k. \tag{148}$$

The lowest order contribution from the Taylor series above to the sum in (146) is

$$-\frac{1}{2}\sum_{n=0}^{d-4}\frac{r}{k_n R_n}. \tag{149}$$

Using definitions (61) and (147), we can rewrite the sum above as

$$-\frac{1}{2}\sum_{n=0}^{d-4}\frac{r}{k_n R_n} = -\sum_{n=0}^{d-4}\frac{r}{(d-3)} = -r. \tag{150}$$

Thus, we notice the leading order contribution from the truncated Taylor expansion of the logarithm is cancelled by the first term in (146). As a consequence, the dominant term of z, near the origin of the complex r-plane, is provided by the next nonzero contribution from the Taylor expansion of (146). It is easy to see that

$$\sum_{n=0}^{d-4}\frac{1}{k_n}\left(\frac{r}{R_n}\right)^k = \sum_{n=0}^{d-4}\frac{R_n^{d-2}}{(d-3)R_h^{d-3}}\left(\frac{r}{R_n}\right)^k = \frac{r^k}{d-3}\sum_{n=0}^{d-4}R_n^{1-k} = 0 \tag{151}$$

for $1 < k < d-2$. Indeed, noting the horizons are equally distributed along the circle $|r| = R_h$, the last equality above follows easily. For $k = d-2$, we have the contribution

$$-\frac{1}{d-2}\sum_{n=0}^{d-4}\frac{r^{d-2}}{d-3}R_h^{3-d} = -\frac{R_h^{3-d}}{d-2}\sum_{n=0}^{d-4}\frac{r^{d-2}}{d-3} = -\frac{R_h^{3-d}r^{d-2}}{d-2}. \tag{152}$$

Thus, we can write

$$z(r) \sim -\frac{R_h^{3-d}r^{d-2}}{d-2}. \tag{153}$$

near the surroundings of $r = 0$.

Now, we parametrize r as

$$r = \rho e^{i\theta} \tag{154}$$

for $\rho \in \mathbb{R}_0^+$ and $\theta \in [0, 2\pi[$. We know that

$$\Re\left(r^{d-2}\right) = \rho^{d-2}\cos\left((d-2)\theta\right). \tag{155}$$

Equating the expression above to zero yields

$$\theta = \frac{\pi}{d-2}\left(n + \frac{1}{2}\right) \tag{156}$$

for $1 \leq n \leq 2d-4$. Thus, we know there are $2d$-4 Stokes lines emerging from the origin of the complex r-plane, all equally distributed and separated by an angle of $\frac{\pi}{d-2}$ radians. In order to show the topology of Stokes lines, far from the origin of the complex r-plane, we display, in Fig. 3, several numerical plots of the these lines, for different dimensions.

By looking at this figure, we first notice the topology of these lines, near the origin of the complex r-plane, is as we predicted. Moreover, for every dimension displayed in the figure, it appears to always exist two Stokes lines emerging from the origin of the complex r-plane, encircling the event horizon R_h. Furthermore, these lines are always followed, clockwise and counter clockwise, by two unbounded Stokes lines. This feature does not restrict itself to the dimensions displayed in Fig. 3, being a general feature for every dimension $d \geq 4$. In [16], a good justification of this fact is presented.

Finally, for purposes that will become clear later, we also want to know the topology of the Stokes lines associated with the WKB approximation of the master Eq. (23). Although we do not show this here, it should be clear that if we were to set up the WKB formalism around (23), just as we did in the previous section with Eq. (126), we would end up with a WKB approximation analogous to (144), with z replaced by x. Hence, the Stokes lines associated with this approximation are defined by the condition

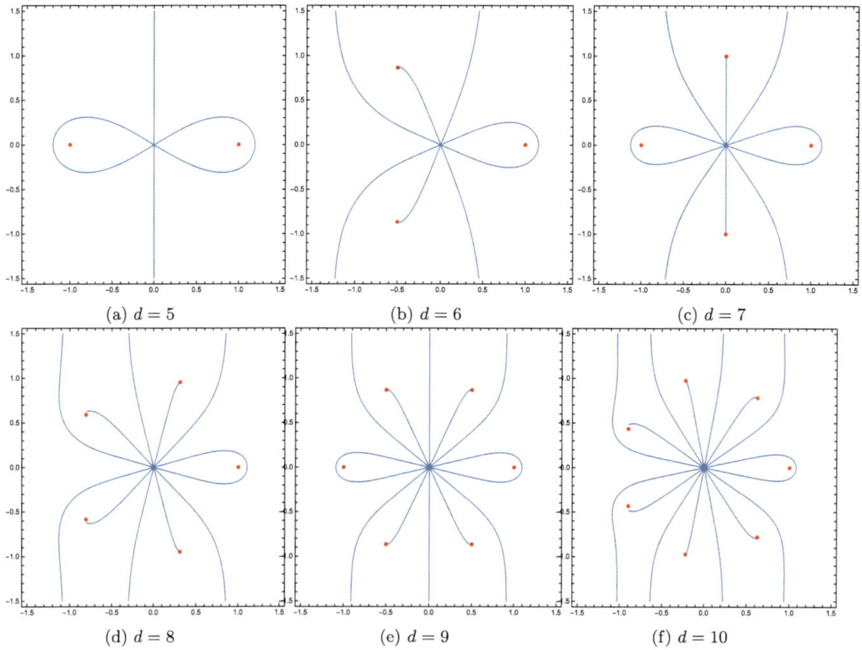

Fig. 3 Numerical plot of the Stokes lines topology for different dimensions. The horizontal axis stands for $\Re(r/R_h)$ and the vertical axis stands for $\Im(r/R_h)$. We denoted the positions of the physical horizon R_h and of the fictitious horizons by red dots

$$\Re\left(x(r)\right) = 0. \tag{157}$$

Integrating Eq. (119), yields the asymptotic expansion

$$z(r) \sim x(r) + \mathcal{C} \tag{158}$$

as $|r| \to +\infty$, for some $\mathcal{C} \in \mathbb{C}$. We set the complex integration constant, defining x, to be such that $\mathcal{C} = 0$. Hence, the Stokes lines topologies, associated with the WKB approximations of (23) and (126), match, far enough from the origin of the complex r-plane. In particular, the two previously mentioned unbounded Stokes lines are also present in the Stokes lines topology of the WKB approximation of (23), only differing near the origin of the complex r-plane. For a more detailed discussion of this point see [23].

6.4 Boundary Conditions

Here, we discuss the boundary conditions, in spatial infinity and in the event horizon R_h.

First, we introduce a common issue in the quasinormal mode problem. It should be clear that any method devised to compute quasinormal frequencies will have to make use of the defining boundary conditions (28) and (29). Because quasinormal frequencies are complex, gathering information of these boundary conditions amounts to distinguish between an exponentially small and an exponentially large term. Clearly, a numerical approach will face problems with such task. Moreover, any kind of analytical approximate approach will also fail to some extent. Indeed, many lower order terms of the approximation are needed in order to make sense of the exponentially decreasing term, otherwise this term might be much smaller than the approximation error and consequently needs to be disregarded.

In the previous section, this problem was not too worrisome because quasinormal frequencies, in the eikonal limit, are approximately real. However, as we know, this is not true in general.

An elegant solution to half of this issue emerges from the moment we allow r to take complex values and consequently assume an analytic continuation of functions of r to the complex plane. Indeed, we can distinguish the two exponential terms near the event horizon, by computing the respective monodromies around it. These monodromies are non trivial because x has a branch point in the event horizon.

So, we are left with the task of imposing the boundary condition (29). As it turns out, the solution to this problem also comes with the analytic continuation of functions of r to the complex plane. Indeed, imposing the boundary condition (29) is equivalent to imposing the boundary condition

$$\psi(x) \propto e^{-i\omega x} \tag{159}$$

for $|r| \to +\infty$, in any of the two unbounded Stokes lines. Looking at (157), we know that

$$|e^{i\omega x}| \sim |e^{-i\omega x}| \sim 1 \tag{160}$$

in these lines. Thus, imposing the boundary condition (159), no longer poses a challenge to an approximate analytical method.

Looking at Eq. (158), we notice we can rewrite the boundary condition (159) as

$$\psi(z) \propto e^{-i\omega z} \tag{161}$$

for $|r| \to +\infty$, in the same Stokes lines. This boundary condition applies to the differential Eq. (121) as the latter is simply (23), rewritten with respect to the independent variable z. Finally, the boundary condition (161 also applies to (126) and (127), for both arise from a perturbative approach of (121). In fact, the differential

Eq. (126) is the master equation associated with tensor type gravitational perturbations, in the d-dimensional Tangherlini black hole space time [16].

6.5 The Monodromy Method

The monodromy method uses both solutions, introduced in the previous subsection, to impose the appropriate boundary conditions. The general idea goes as follows:

- We pick two closed homotopic contours on the complex r-plane. Both these contours enclose only one horizon: the real one R_h. Moreover, neither one of them encloses the origin of the complex r-plane.
- One of these contours, which we name the big contour, seeks to encode information of the boundary condition (161) on the monodromy of ψ, associated with a full loop around it.
- The other contour, which we name the small contour, seeks to encode information of the boundary condition (28) on the monodromy of ψ, associated with a full loop around it.
- As both contours are homotopic, the monodromy theorem asserts that the respective monodromies must be the same. Thus, equating them hopefully yields a restriction on the values of the quasinormal frequencies ω, from the complex plane to an infinite but countable subset.

6.6 The Big Contour

In order to build the big contour, we make use of the feature we found, when studying the topology of the Stokes lines, associated with the WKB approximation (144). More precisely, for every dimension d, there will be two Stokes lines, emerging from the origin of the complex r-plane, encircling the event horizon R_h. Furthermore, these lines are followed, counter clockwise and clockwise, by two unbounded Stokes lines. The big contour will follow these unbounded Stokes lines, reaching the condition $|r| \to +\infty$ twice. There, the contour abandons the Stokes lines and follows an arc shaped path enclosing it.

Overall, the big contour is well represented as depicted in Fig. 4 (we stress that in this figure the proportions may not be right). However, the topology of the big contour is well represented by the blue dashed line for every dimension $d \geq 5$. The boundary condition (161) is to be imposed in the regions marked by D or U.

After defining the big contour, we want to compute the monodromy of ψ, associated with a full clockwise loop around it. In order to attain reliable information on how ψ changes around the big contour, we resort to WKB theory. Indeed, we can use (144) in portions of the big contour, matched with Stokes lines, far enough from the

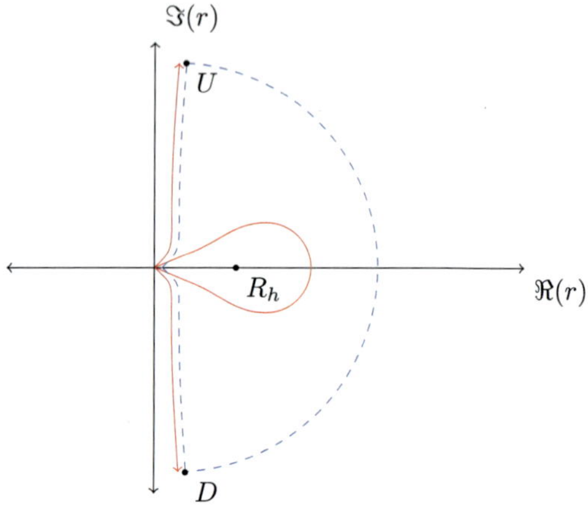

Fig. 4 Schematic depiction of the big contour, as the blue dashed line. The Stokes lines are depicted as red curves. Naturally, not all Stokes lines are depicted. Furthermore, we marked by D and U the regions where the boundary condition (161) may be imposed

the origin of the complex r-plane. Near the origin, the WKB approximation fails as $r = 0$ is a singular point of (138). Thus, we need to analytically solve the differential Eqs. (126) and (127) in this region.

The full calculation of the monodromy of ψ around the big contour is a long and complex task, full of subtleties and technical details, whose reproduction is outside the scope of this review. We refer the reader to [23, 24], where it is explained in detail.

6.7 The Small Contour

The small contour is remarkably simple, when compared to the big contour. Indeed, we build an arbitrarily small closed contour around the event horizon R_h. Such contour can be represented as depicted in Fig. 5. In this contour, we won't need to perturbatively approach the master Eq. (23) nor change the independent variable. Indeed, we can simply consider the solution (27) to (23), together with the boundary condition (28), and (as the contour is arbitrarily small) take a Taylor expansion around $r = R_h$. The tortoise (146) has a branch point at $r = R_h$, from which we can compute the monodromy of ψ around the small contour. The calculation of the monodromy is therefore considerably simpler when compared to the case of the big contour. Again we refer the reader to [23, 24], where it can be found in detail.

6.8 Equating Monodromies: The Final Result

We notice the big contour is homotopic to the small one. Thus, by the monodromy theorem we know the monodromies of ψ, associated with the full clockwise loops

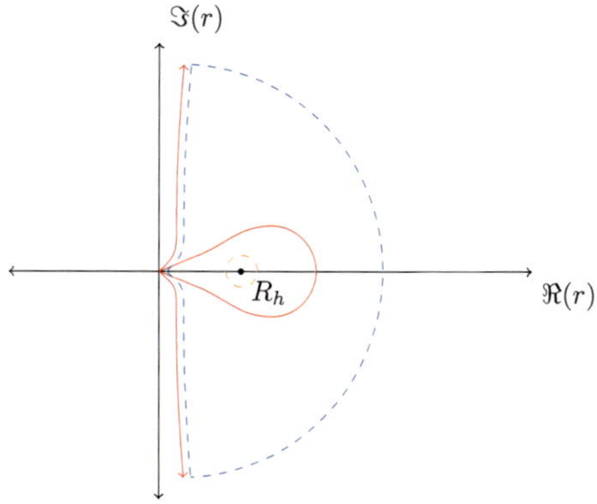

Fig. 5 Schematic depiction of the small and big contours as the orange and blue dashed lines respectively. The orange contour is to be interpreted as arbitrarily close to R_H. The Stokes lines are depicted by red curves. Naturally, not all Stokes lines are depicted

around the big and the small contours, are equal. Hence, equating the two monodromies we obtain an algebraic condition restraining the possible values of the asymptotic quasinormal frequencies ω, written in terms of the black hole temperature (64):

$$\frac{\omega}{T_{\mathcal{H}}} = \ln(3) + (2k+1)\pi i + \lambda \left(\frac{4\pi}{d-3}\right)^2 T_{\mathcal{H}}^2 \left[\frac{d-3}{d-2}\frac{(2k+1)}{4}\right]^{\frac{d-1}{d-2}} \Pi_{\mathsf{T}}(d)\, e^{\frac{d-5}{2(d-2)}\pi i}, \tag{162}$$

with $k \in \mathbb{N}$ and $\Pi_{\mathsf{T}}(d)$ given by

$$\Pi_{\mathsf{T}}(d) := \frac{2}{3}\frac{(d-4)\,(d(d-5)+2)}{d-1}\sqrt{\pi}\sin\left(\frac{\pi}{d-2}\right)\frac{\Gamma\left(\frac{1}{2(d-2)}\right)\Gamma\left(\frac{d-3}{2(d-2)}\right)}{\left[\Gamma\left(\frac{d-1}{2(d-2)}\right)\right]^2}. \tag{163}$$

The $\lambda = 0$ limit corresponds to the result obtained in [16, 21]: the real part is a universal value $\ln(3)$; the imaginary part depends only on the mode number k. In contrast, the λ corrections depend strongly on the dimension d, both for the real and the imaginary parts. The result we derived here for tensorial perturbations of the Callan-Myers-Perry black hole is also valid for gravitational perturbations of any type (tensorial, vectorial and scalar) of d-dimensional Gauss-Bonnet corrected black holes, as shown in [23]. In this same reference, and in [24], one can also find the calculations of asymptotic quasinormal modes of test scalar fields.

Acknowledgements This work has been supported by Fundação para a Ciência e a Tecnologia under grants IT (UIDB/50008/2020 and UIDP/50008/2020), CAMGSD/IST-ID (UIDB/04459/2020 and UIDP/04459/2020) and projects CERN/FIS-PAR/0023/2019, 2022.08368.PTDC. João Rodrigues is supported by Fundação para a Ciência e a Tecnologia through the doctoral fellowship UI/BD/151499/2021.

References

1. E. Berti, V. Cardoso, A.O. Starinets, Quasinormal modes of black holes and black branes. Class. Quant. Grav. **26**, 163001 (2009). ([arXiv:0905.2975 [gr-qc]])
2. K.D. Kokkotas, B.G. Schmidt, Quasinormal modes of stars and black holes. Living Rev. Relativ. **2**, 2 (1999). ([arXiv:gr-qc/9909058 [gr-qc]])
3. A. Bachelot, A. Motet-Bachelot, Les résonances d'un trou noir de Schwarzschild. Ann. Inst. H. Poincare Phys. Theor. **59**, 3–68 (1993)
4. E.S.C. Ching, P.T. Leung, W.M. Suen, K. Young, Wave propagation in gravitational systems: late time behavior. Phys. Rev. **D52**, 2118–2132 (1995). [arXiv:gr-qc/9507035 [gr-qc]]
5. N. Froman, P. Froman, *JWKB Aproximation: Contributions to the theory* (North-Holland publishing company, Amsterdam, 1965)
6. C.G. Callan, R.C. Myers, M.J. Perry, Black holes in string theory. Nucl. Phys. B **311**, 673 (1989)
7. A. Ishibashi, H. Kodama, A master equation for gravitational perturbations of maximally symmetric black holes in higher dimensions. Prog. Theor. Phys. **110**, 701 (2003). ([arXiv:hep-th/0305147])
8. T. Takahashi, J. Soda, Instability of small Lovelock black holes in even-dimensions. Phys. Rev. D **80**, 104021 (2009). ([arXiv:0907.0556 [gr-qc]])
9. T. Takahashi, J. Soda, Master equations for gravitational perturbations of static Lovelock black holes in higher dimensions. Prog. Theor. Phys. **124**, 911 (2010). ([arXiv:1008.1385 [gr-qc]])
10. G. Dotti, R.J. Gleiser, Linear stability of Einstein-Gauss-Bonnet static spacetimes. Part I. Tensor perturbations. Phys. Rev. **D72**, 044018 (2005). [arXiv:gr-qc/0503117]
11. R.J. Gleiser, G. Dotti, Linear stability of Einstein-Gauss-Bonnet static spacetimes. Part II: Vector and scalar perturbations. Phys. Rev. **D72**, 124002 (2005). [arXiv:gr-qc/0510069]
12. J.T. Wheeler, Symmetric solutions to the Gauss-Bonnet extended Einstein equations. Nucl. Phys. B **268**, 737 (1986)
13. F. Moura, String-corrected dilatonic black holes in d dimensions. Phys. Rev. D **83**, 044002 (2011). ([arXiv:0912.3051 [hep-th]])
14. F. Moura, R. Schiappa, Higher-derivative corrected black holes: perturbative stability and absorption cross-section in heterotic string theory. Class. Quant. Grav. **24**, 361 (2007). (arXiv:hep-th/0605001 [hep-th/0605001])
15. F. Moura, Tensorial perturbations and stability of spherically symmetric d-dimensional black holes in string theory. Phys. Rev. D **87**, 044036 (2013). ([arXiv:1212.2904 [hep-th]])
16. J. Natário, R. Schiappa, On the classification of asymptotic quasinormal frequencies for d-dimensional black holes and quantum gravity. Adv. Theor. Math. Phys. **8**(6), 1001 (2004). (arXiv:hep-th/0411267 [hep-th/0411267])
17. B.F. Schutz, C.M. Will, Black hole normal modes - a semianalytic approach. Astrophys. J. Lett. **291**, L33–L36 (1985)
18. V. Cardoso, A.S. Miranda, E. Berti, H. Witek, V.T. Zanchin, Geodesic stability, Lyapunov exponents and quasinormal modes. Phys. Rev. D **79**, 064016 (2009). ([arXiv:0812.1806 [hep-th]])
19. F. Moura, J. Rodrigues, Eikonal quasinormal modes and shadow of string-corrected d-dimensional black holes. Phys. Lett. B **819**, 136407 (2021). ([arXiv:2103.09302 [hep-th]])
20. R.A. Konoplya, Z. Stuchlík, Are eikonal quasinormal modes linked to the unstable circular null geodesics? Phys. Lett. B **771**, 597 (2017). ([arXiv:1705.05928 [gr-qc]])
21. L. Motl, A. Neitzke, Asymptotic black hole quasinormal frequencies. Adv. Theor. Math. Phys. **7**(2), 307–330 (2003). ([arXiv:hep-th/0301173])
22. N. Andersson, C.J. Howls, The Asymptotic quasinormal mode spectrum of nonrotating black holes. Class. Quant. Grav. **21**, 1623 (2004). ([arXiv:gr-qc/0307020])
23. F. Moura, J. Rodrigues, The isospectrality of asymptotic quasinormal modes of large Gauss-Bonnet d-dimensional black holes. Nucl. Phys. B **993**, 116255 (2023). ([arXiv:2206.11377 [hep-th]])
24. F. Moura, J. Rodrigues, Asymptotic quasinormal modes of string-theoretical d-dimensional black holes. JHEP **08**, 078 (2021). ([arXiv:2105.02616 [hep-th]])

Observational Signals of Compact Objects

Spacetime Geometry of Rotating Boson Stars and KBHsSH

Lucas G. Collodel

Abstract In this article we give an overview of the proprieties of rotating boson stars and Kerr black holes with scalar hair. A fluid description of the scalar matter is provided, and by adopting a frame co-moving with the fluid we find a structural connection between both objects. We show that the unusual distribution of the scalar matter warps spacetime in a very unique way, such that some metric functions on the equatorial plane behave nonmonotonically. The implications it bears to circular orbits is explored.

1 Introduction

Boson stars (BSs) are among the most studied hypothetical exotic objects. They were first theorized half a century ago [1, 2] by minimally coupling a complex scalar field theory to gravity (general relativity). Shortly after, it was shown that the quantized real scalar field gives rise to the same equations of motion in a semi-classical approach [3]. These are special objects in many different ways. Their gravitational collapse is prevented due to the highly dispersive character of the complex field. Although the field might interact with itself, its only other channel of energy exchange is via gravity and therefore common matter might orbit freely in its interior. These stars lack a surface where the pressure goes smoothly to zero. Instead, the field extends all the way to infinity, albeit with exponential decay. They are nevertheless localized and might be thought of as compact since they lack any hard layers and might host lightrings. Its stability is linked to the existence of a conserved Noether charge representing the particle number, Q. As long as the system finds itself in the lowest energy level for a particular charge, it could be stable. Upon rotation, the topology of the field changes and it distributes itself in a toroidal shape. The angular momentum is quantized as $J = mQ$, as found in the first investigations to solve the fully nonlinear rotating system [4, 5]. Consequently, BSs cannot rotate slowly in a perturbative manner and earlier attempts in this direction rendered fruitless, raising

L. G. Collodel (✉)
Theoretical Astrophysics, Eberhard Karls University of Tübingen, 72076 Tübingen, Germany
e-mail: lucas.gardai-collodel@uni-tuebingen.de

© The Author(s), under exclusive license to Springer Nature Switzerland AG 2024

E. Papantonopoulos and N. Mavromatos (eds.), *Compact Objects in the Universe*,
https://doi.org/10.1007/978-3-031-55098-0_8

the question whether BSs could rotate at all [6]. A recent review on the topic is found at [7]. The inner structure of BSs and its connection to perfect fluids received some attention in the early works [1, 3] where it was shown that spherically symmetric configurations are anisotropic. In recent years, a similar approach was done to more involved case of rotating (charged and uncharged) BSs, which by loss of symmetry become completely anisotropic [8, 9].

The realization of Kerr black holes with scalar hair (KBHsSH) came only recently with the seminar paper [10], showing that the system indeed evades the no-hair theorems as the matter fields do not share the same isometries as the spacetime. Stationary modes of a complex scalar field on a Kerr background had already been found for rapid rotation [11], connecting the horizon's angular velocity to the scalar field's natural frequency. As it turns out in the backreacting case, this synchronization is a regularity condition at the level of the equations, which sets the appropriate boundary conditions. KBHsSH share traits of both black holes and rotating BSs and can be thought of as the combined system of both.

Unlike Schwarzschild, Kerr is not a unique vacuum solution of a rotating source. Nevertheless, the exterior geometry of spacetimes warped by rotating fermionic matter such as neutron stars is thought to deviate just marginally from Kerr. BSs make one of the best counter examples. Due to the off-centered distribution of energy density for rotating scalar fields, the deviations can be enormous for small radii. Notoriously, the metric functions in usual adapted spherical coordinates might behave nonmonotonically. One of the most prominent consequences is that these spacetimes might host *static rings* [12], i.e. a ring of point on the equatorial plane where a test particle at rest in the fiducial frame remains inertially at rest. This feature is present for configurations of rotating BSs and KBHsSH. Similarly, orbits in which the particle finds itself multiple times momentarily at rest along a full revolution were reported in [13].

In this article we aim at exploring the fluid description of rotating BSs and trace a direct link to KBHsSH by assuming the role of an observer co-moving with the scalar fluid. By considering equatorial circular orbits, we present the peculiar profiles the orbital velocity might have in different parts of the parameter space. Such might have great astrophysical implications that should help probe the existence of these objects. In this sense, investigations have been made such as the extraction of $K\alpha$ iron lines of some KBHsSH [14], the construction of thick disks around KBHsSH [15, 16] and rotating BSs [17–19] and thin disks around KBHsSH [20]. More recently, preliminary research on the impact of the orbital parameters on the fluxes of an inspiralling object towards KBHsSH due to gravitational radiation were reported in [21].

This article is organized as follows. In Sect. 2 we provide the general framework to study both BSs and KBHsSH and scrutinize the underlying theory. In Sect. 3 we focus on rotating BSs by specifying the appropriate boundary conditions and provide a fluid description to it. In Sect. 4 we present the domain of existence for KBHsSH, as well as a short description of their theoretical construction. The spacetime geometry of these objects and how they affect circular orbital parameters is the topic of Sect. 5. We conclude in Sect. 6. Throughout this paper, we adopt geometrical units where $c = 8\pi G = 1$.

2 Theory

2.1 *Action and Equations of Motion*

Boson stars and KBHsSH are both realizations of bound objects of scalar matter withing the same theory. Hence, their prescription is given by the the same set of differential equations, but with different boundary conditions. Fundamentally, they are given by a complex field minimally coupled to gravity, which we assume to be described by general relativity. Therefore, at the theory level the action reads

$$S = \int \left[\frac{R}{2} - g^{\mu\nu}\partial_\mu \Phi^* \partial_\nu \Phi - 2U(\Phi) \right] \sqrt{-g} d^4x, \tag{1}$$

where R is the Ricci scalar, Φ is the complex scalar field with self-interaction U, $g^{\mu\nu}$ is the contravariant metric tensor and g its determinant. Hence, the Lagrangian density for the gravitational sector is simply $\mathcal{L}_G \equiv R/2$, while that for the complex scalar field is given by

$$\mathcal{L}_\Phi \equiv -g^{\mu\nu}\partial_\mu \Phi^* \partial_\nu \Phi - 2U(\Phi). \tag{2}$$

Einstein's field equations are obtained by varying the action (1) with respect to the inverse of the metric $g^{\mu\nu}$, and read

$$R_{\mu\nu} - \frac{1}{2} R g_{\mu\nu} = T_{\mu\nu}, \tag{3}$$

where the energy momentum tensor is spelled out as

$$\begin{aligned} T_{\mu\nu} &\equiv g_{\mu\nu}\mathcal{L}_\Phi - 2\frac{\partial \mathcal{L}_\Phi}{\partial g^{\mu\nu}} \\ &= -g_{\mu\nu}\left(\partial^\sigma \Phi^* \partial_\sigma \Phi + 2U\right) + 2\partial_\mu \Phi^* \partial_\nu \Phi. \end{aligned} \tag{4}$$

These field equations for the metric components are to be solved simultaneously with the Einstein-Klein-Gordon (EKG) equations, which arises from the variation of the action (1) with respect to the scalar field and its complex conjugate, giving

$$\left(\Box - \frac{\partial U}{\partial |\Phi|^2}\right)\Phi^* = 0, \qquad \left(\Box - \frac{\partial U}{\partial |\Phi|^2}\right)\Phi = 0. \tag{5}$$

2.2 Charges

In this work, we wish to investigate spacetimes which are exclusively stationary and axisymmetric, possessing therefore two Killing vector fields associated with them, $\eta = \partial_t$ and $\chi = \partial_\varphi$. Another crucial assumption we make is that these spacetimes are circular, i.e. $\eta^\mu R_{\mu[\alpha}\chi_\beta \eta_{\gamma]} = 0 = \chi^\mu R_{\mu[\alpha}\chi_\beta \eta_{\gamma]}$. Then, it is possible to find a basis for our coordinate system that can everywhere accommodate our Killing fields. It also implies the discrete isometry $t \rightarrow -t$ and $\varphi \rightarrow -\varphi$, which simultaneously reverses the direction of time and of rotation. In order to describe rotating BSs and KBHsSH we will adopt different metric parametrizations, but in both cases we will employ the Boyer-Lindquist coordinates (t, r, θ, φ).

The Komar integrals are conserved quantities associated with each vector field and are overall depleted of a clear physical meaning. The situation changes however, when these are Killing fields. Stationarity is directly related to conservation of energy, and a global asymptotic mass

$$M = 2 \int_\Sigma n^\mu \nabla_\nu \nabla_\mu \eta^\nu dV, \tag{6}$$

while axisymmetry implies in the conservation of angular momentum which is also a global charge

$$J = -\int_\Sigma n^\mu \nabla_\nu \nabla_\mu \chi^\nu dV. \tag{7}$$

In the Komar expressions above, dV is the volume element enclosed by the asymptotically flat, spacelike hypersurface Σ whose unitary normal vector is n^μ.

The scalar field theory (2) is endowed with a global $U(1)$ symmetry and, for any constant α, the transformation $\Phi \rightarrow e^\alpha \Phi$ leaves the system unchanged. Thus, there is a conserved Noether current

$$j^\mu = -i \left(\Phi^* \partial^\mu \Phi - \Phi \partial^\mu \Phi^* \right), \tag{8}$$

whose projection onto the future directed vector field n^ν can be integrated over Σ to give a conserved Noether charge, Q, corresponding to the particle number of the ensemble

$$Q = \int_\Sigma j_\mu n^\mu dV. \tag{9}$$

2.3 Scalar Field Ansatz

The main motivation to work with complex scalar fields lies on the fact that static real scalar field bound configurations should be unstable [22]. In fact, if we think of the system as an ensemble of particles as we will in the next section in order to assign a fluid description to it, it is imperative that we have a well defined four-momentum for each particle or element of volume. The canonical way to do that is via the Lagrangian density (2),

$$\Pi^{\mu} \equiv \frac{\partial \mathcal{L}_{\Phi}}{\partial\left(\partial_{\mu} \Phi\right)}=-\partial^{\mu} \Phi^{*}, \qquad \Pi^{* \mu} \equiv \frac{\partial \mathcal{L}_{\Phi}}{\partial\left(\partial_{\mu} \Phi^{*}\right)}=-\partial^{\mu} \Phi, \tag{10}$$

whence we conclude that the scalar particles cannot be timelike once the field is static. Similarly, in order to describe a rotating object and have a source for the dragging of the spacetime, these particles must have a nonzero axial component in their four-momentum. As discussed above, we are nevertheless solely focused on stationary and axisymmetric spacetimes, i.e. even though the scalar field must be given as a function of t and φ, the Lagrangian density (and therefore the equations of motion) must remain independent. This is only possible with an explicit harmonic dependence, and thus the complex scalar field must be written as

$$\Phi=\phi(r, \theta) e^{i\left(\omega_{s} t+m \varphi\right)}, \tag{11}$$

where ω_s is a real number that gives the field's natural frequency, while m is the winding number and must be an integer due to the identification $\Phi(\varphi=0)=\Phi(\varphi=2\pi)$. It corresponds to the number of nodes the field has in a closed curve along the axial direction. Furthermore, ϕ is a real-valued function of the remaining coordinates (r and θ), much in the same fashion as the metric function.

2.4 Note on Self-interaction

The explicit form of the potential U is overall not important for our purposes. We wish to investigate the inner structure of rotating self-gravitating bosonic matter and how it affects its surrounding spacetime. The way the bosons interact with one another has no qualitative impact on this topic. Rather, the potential has greater implications on the maximum of the total mass and compactness of the resulting star. However, it shall not be completely arbitrary and therefore some considerations are still required.

The Lagrangian density is a real quantity and we wish to avoid absolute valued terms at the level of the theory. Thus, the potential should only contain terms involving the product of the scalar field with its complex conjugate $\Phi\Phi^*$, i.e. only even powers of ϕ.

The complex field makes the Einstein-Klein-Gordon equation (5) ever more dispersive. Therefore, while a free field theory is perfectly fine, we require that the scalar field have a mass so that it drops to zero at infinity, otherwise we could not fulfill the requirement of an asymptotically flat spacetime. Hence, the leading order term of the self-interaction must be a mass term, and for the rest of this paper we establish

$$U(\phi) = \frac{1}{2}\mu^2\phi^2 + O(4), \tag{12}$$

where μ is the mass of the bosonic particle.

3 Rotating Boson Stars

3.1 Metric Ansatz and Boundary Conditions

To study rotating BSs we adopt the quasi-isotropic Lewis-Papapetrou metric, whose line element is

$$ds^2 = -f dt^2 + \frac{l}{f}\left[g\left(dr^2 + r^2 d\theta^2\right) + r^2 \sin^2\theta \left(d\varphi - \frac{\omega}{r}dt\right)^2\right], \tag{13}$$

where f, l, g and ω are all functions of r and θ. Plugging this metric into Eq. (3) together with Eq. (5) leads to a system of five coupled nonlinear partial differential equations (PDEs) of elliptical character. In order to solve them, one needs to prescribe boundary conditions that accurately describe the system at hand. Our domain lies in the intervals $r \in [0, \infty)$ and $\theta \in [0, \pi/2]$, due to even parity with respect to reflections on the equatorial plane.

At the center of the star, $r = 0$, regularity requires that

$$\partial_r f|_{r=0} = 0, \qquad \partial_r l|_{r=0} = 0, \qquad g|_{r=0} = 1, \qquad \omega|_{r=0} = 0, \qquad \phi|_{r=0} = 0. \tag{14}$$

The last condition comes in contrast with spherically symmetric BSs in which the scalar field has a nonzero value at its center and decreases monotonically to zero when not excited. In fact, expanding Eq. (5) around the origin yields

$$\phi(r \sim 0) \sim \phi_m r^m \sin^m\theta + O\left(m + 1\right), \tag{15}$$

where m in the winding number defined in the ansatz for the scalar field and ϕ_m is a real number. Hence, spherically symmetric BSs do indeed profile a non-trivial scalar field at the origin while any rotating BSs have a nonmonotonically varying field starting from zero at the origin. The first fundamental mode of rotation, $m = 1$ profiles a scalar field linearly increasing with r from the star's core. Hence, upon rotation the field's contour surfaces undergo a topology change from the two sphere to a torus, $S^2 \to S^1 \times S^1$. Consequently, there is an off-center distribution of energy density which in turn warps the spacetime in an unusual way. As we will see below, some metric functions behave nonmonotonically resulting in very large deviations of orbital dynamics when compared to Kerr black holes.

At infinity, the spacetime is Minkowski and the field is zero so that the boundary conditions read

$$f|_{r\to\infty} = 1, \qquad l|_{r\to\infty} = 1, \qquad g|_{r\to\infty} = 1, \qquad \omega|_{r\to\infty} = 0, \qquad \phi|_{r\to\infty} = 0. \tag{16}$$

An asymptotic expansion of the EKG Eq. (5) at radial infinity gives

$$\phi(r \to \infty) \sim \frac{\phi_\infty}{r} \exp\left(-r\sqrt{\mu^2 - \omega_s^2}\right), \tag{17}$$

confirming the necessity of a nonzero mass term in the potential so that the field indeed drops to zero at infinity. This expansion remains true for KBHsSH. In fact, it is completely oblivious to the central object and yields the same result in an asymptotically flat spacetime for any complex scalar field whose leading term in the potential is the mass term. This expansion also unveils a particularity of boson stars, namely the lack of a well defined surface. In fact, the field extends all the way to infinity where it finally becomes trivial. However, the exponential decay allows for localized compact lumps of energy density.

Along the polar axis the conditions are set by axisymmetry in the north pole,

$$\partial_\theta f|_{\theta=0} = 0, \qquad \partial_\theta l|_{\theta=0} = 0, \qquad g|_{\theta=0} = 1, \qquad \partial_\theta \omega|_{\theta=0} = 0, \qquad \phi|_{\theta=0} = 0, \tag{18}$$

and on the equatorial plane by requiring even parity among the hemispheres,

$$\partial_\theta f|_{\theta=\pi/2} = 0, \quad \partial_\theta l|_{\theta=\pi/2} = 0, \quad \partial_\theta g|_{\theta=\pi/2} = 0, \quad \partial_\theta \omega|_{\theta=\pi/2} = 0, \quad \partial_\theta \phi|_{\theta=\pi/2} = 0. \tag{19}$$

The set of PDE (3) and (5) together with the ansätze (11) and (13) and the boundary conditions above allow one to search for families of solution. Of course one needs to define the form of the self-interaction potential and fix the free parameters. The complete process of solving the PDE is out of the scope of the present paper, but we give a very short description of how the solutions used below were found. We first compactify the radial coordinate as to accommodate the infinity at a real number, defining $x := r/(r+1)$. A Cartesian grid of finite points is then built in the domain $x \in [0, 1]$ and $\theta \in [0, \pi/2]$. The PDE are discretized onto this grid by a finite difference method. A good initial guess is given before seeking each solution and a relaxed Newton method is employed to solve the algebraic nonlinear system, in which a linear solver is used to invert the Jacobian. After a solution is found, it is usually used as the initial guess for the subsequent solution where one of the parameters (say ω_s) has a slightly different input.

3.2 Charges

Now that we have our system in explicit form, meaning the metric and boundary condition, let us turn back to the Komar expression of Sect. 2.2. Because the spacetime is regular everywhere and contains no null hypersurface, we may keep these as volume integrals for the whole domain. The future-directed timelike unitary vector is given by $n^\mu = (\eta^\mu + \omega/r\chi^\mu)/\sqrt{f}$, as its covariant form is only nonzero for the time component which is always minus the lapse, $n_\mu = \left(-\sqrt{f}, 0, 0, 0\right)$. The element of volume read $dV = \sqrt{-g/f}drd\theta d\varphi$. Then, Eqs. (6) and (7) read, in terms of the energy-momentum tensor,

$$M = \int_\Sigma \left(2T^t{}_t - T\right)\sqrt{-g}drd\theta d\varphi; \qquad J = -\int_\Sigma T^t{}_\varphi \sqrt{-g}drd\theta d\varphi. \tag{20}$$

Similarly, the Noether charge (9) becomes

$$Q = -\int_\Sigma j^t \sqrt{-g}drd\theta d\varphi, \tag{21}$$

and from the ansatz (11) one can easily observe that $T^t{}_\varphi = mj^t$, implying that $J = mQ$. This remarkable result, first reported in [4], unravels a macroscopic quantum property of rotating BSs: their angular momentum is quantized. Such a trait makes it impossible for BSs to have slow rotation (perturbatively) as long as stationarity and axisymmetry is assumed. For this reason, perturbed solutions could not be obtained in the first studies, leading to questioning whether BSs could rotate at all [6].

3.3 Fluid Description and Co-moving Frame

In the previous subsection we gave an overview on how we find solutions for BSs. At the level of the equations, the system differs from that of usual fermionic matter in that, instead of having to solve a Tolman-Oppenheimer-Volkoff equation to which one prescribes an equation of state, one has to simultaneously solve the EKG equation with a prescribed self-interaction potential. The disperssive character of this equations is in fact what prevents further gravitational collapse. Nevertheless, in order to assess the energy density and pressures at any given volume element of the star we must match the energy-momentum tensor with that of a perfect fluid. This was first done for spherically symmetrical BSs by Kaup in the first paper to consider the action (1) [1]. A straightforward outcome is that static BSs are anisotropic, i.e. their tangential pressures are different from their radial one. The energy-momentum tensor for an anisotropic fluid is given by

$$T^\mu{}_\nu = (\rho + p_\perp)u^\mu u_\nu + (p_r - p_\perp)r^\mu r_\nu + p_\perp \delta^\mu{}_\nu, \tag{22}$$

where ρ is the fluids energy density, p_r and $p_\perp$ the radial and tangential pressures, respectively. The fluid's four-velocity is u^μ, while r^μ is the unitary spacelike vector in the radial direction. Upon matching the equation above with Eq. (4) we obtain

$$\rho = g^{rr}\,(\partial_r \phi)^2 - g^{tt}\phi^2\omega_s^2 + U, \qquad p_r = \rho - 2U, \qquad p_\perp = p_r - 2g^{rr}\,(\partial_r\phi)^2\,. \tag{23}$$

In the above equation, g^{tt} and g^{rr} are contravariant components of the metric tensor and were left purposely in this form for generality as most of the time isotropic coordinates are not the ones employed for spherically symmetric spacetimes.

Rotating BSs, however, are more involved. First, we must find the fluid's frame which might be a bit ambiguous as it differs from the frame of the scalar field Φ. We circumvent this by defining the co-moving frame of a fluid element as the one for which the energy-momentum tensor is diagonal. In other words, an observer in the co-moving frame measures zero momentum fluxes and shear effects.

The diagonalization is done with aid of the vierbein basis vectors $\hat{e}^{\mu}_{(a)}$,

$$T^{\mu}{}_{\nu}\hat{e}^{\nu}{}_{(a)} = \lambda_{(a)}\hat{e}^{\mu}{}_{(a)}, \quad a \in [1,4], \tag{24}$$

so that in this basis $T^{a}{}_{b}$ is diagonal and the eigenvalues $\lambda_{(a)}$ give the energy density and pressures as measured by the co-moving observer. A general expression for them is possible and it shows convenient to split the energy-momentum tensor in two parts as

$$T^{\mu}{}_{\nu} = {}^{(1)}\mathcal{T}^{\mu}{}_{\nu} + {}^{(2)}\mathcal{T}^{\mu}{}_{\nu}, \tag{25}$$

with

$${}^{(1)}\mathcal{T}^{\mu}{}_{\nu} = \begin{pmatrix} T^{t}{}_{t} & 0 & 0 & T^{t}{}_{\varphi} \\ 0 & 0 & 0 & 0 \\ 0 & 0 & 0 & 0 \\ T^{\varphi}{}_{t} & 0 & 0 & T^{\varphi}{}_{\varphi} \end{pmatrix}, \quad {}^{(2)}\mathcal{T}^{\mu}{}_{\nu} = \begin{pmatrix} 0 & 0 & 0 & 0 \\ 0 & T^{r}{}_{r} & T^{r}{}_{\theta} & 0 \\ 0 & T^{\theta}{}_{r} & T^{\theta}{}_{\theta} & 0 \\ 0 & 0 & 0 & 0 \end{pmatrix}. \tag{26}$$

Then the eigenvalues read

$$\lambda_{i\pm} = \frac{1}{2}\left[\pm\sqrt{2\,{}^{(i)}\mathcal{T}^{\mu}{}_{\nu}\,{}^{(i)}\mathcal{T}^{\nu}{}_{\mu} - {}^{(i)}\mathcal{T}^{2}} + {}^{(i)}\mathcal{T}\right], \tag{27}$$

where ${}^{(i)}\mathcal{T} = {}^{(i)}\mathcal{T}^{\mu}{}_{\mu}$.

And

$$\rho = -\lambda_{1+}; \qquad p_r = \lambda_{2+}; \qquad p_\theta = \lambda_{2-}; \qquad p_\varphi = \lambda_{1-}. \tag{28}$$

The results above are general and valid for any stationary, axisymmetric spacetime in adapted spherical coordinates. In the case of an isotropic fluid, $p_r = p_\theta = p_\varphi$, while an anisotropic fluid like the spherically symmetric BS yields $p_\theta = p_\varphi$. Upon rotation, BSs are described by a *completely anisotropic fluid* with three different pressures whose energy-momentum tensor can be written as

$$T^{\mu}{}_{\nu} = (\rho + p_\varphi)u^{\mu}u_{\nu} + (p_r - p_\varphi)r^{\mu}r_{\nu} + (p_\theta - p_\varphi)\theta^{\mu}\theta_{\nu} + p_\varphi\delta^{\mu}{}_{\nu}, \tag{29}$$

where θ^{μ} is the spacelike unitary vector in the polar direction.

Finally, the explicit form of the hydrodynamic quantities in the rotating scalar fluid are readily obtained by substituting Eq. (4) into (27), giving

$$\rho = \frac{f}{lgr^2}\left[(r\partial_r\phi)^2 + (\partial_\theta\phi)^2\right] + \frac{|m^2 f^2 - l\sin^2\theta(\omega_s r + m\omega)^2|}{flr^2\sin^2\theta}\phi^2 + U, \quad (30)$$

$$p_r = \frac{f}{lgr^2}\left[(r\partial_r\phi)^2 + (\partial_\theta\phi)^2\right] + \frac{m^2 f^2 - l\sin^2\theta(\omega_s r + m\omega)^2}{flr^2\sin^2\theta}\phi^2 - U, \quad (31)$$

$$p_\theta = p_r - 2\frac{f}{lgr^2}\left[(r\partial_r\phi)^2 + (\partial_\theta\phi)^2\right], \quad (32)$$

$$p_\varphi = \rho - 2\frac{f}{lgr^2}\left[(r\partial_r\phi)^2 + (\partial_\theta\phi)^2\right] - 2U. \quad (33)$$

The expressions above, in particular for ρ and p_φ, feature an absolute value term which at first seems quite unphysical. For the co-moving observer, the energy density and axial pressure of a rotating BS are continuous but not smooth, as the argument

$$\Delta := m^2 f^2 - l\sin^2\theta(\omega_s r + m\omega)^2, \quad (34)$$

switches sign at the surface $\Delta|_{S_c} = 0$. Hence, the observer measures a cusp in the profile of those hydrodynamic quantities. In Fig. 1 we depicted the profile of the energy density with respect to the compactified coordinate x as seen by an observer co-moving with the fluid on the equatorial plane for several different rotating BSs. In all of those configurations, $m = 1$ and ϕ_1 is the leading order term of the scalar field at the origin, as given by the expansion (15). For the sake of completeness, the self-interaction of the underlying theory is

$$U(|\Phi|) = \phi^2\left(\frac{\mu^2}{2} - \sigma\phi^2 + \gamma\phi^4\right), \quad (35)$$

with $\mu^2 = 0.055$, $\sigma = 2$ and $\gamma = 1$. This is the sixtic potential which allows for bound configurations even in the absence of gravity [23, 24]. Nevertheless, as previously stated and easily seen by the definition of Δ, the self-interaction potential has no implication in the existence of the cusp. Instead, one can note from the figure that this trait becomes more prominent as ϕ_1 increases.

In order to understand better what happens at the cusp we must turn our attention to the observer itself. According to our definition, this co-moving observer follows a worldline for which the energy-momentum tensor of the star's fluid is diagonal. Due to the symmetry of our system, we can think of each element of fluid (and consequently each observer of interest) performing a circular motion at fixed r and θ in our coordinate system. Hence, the four-velocity reads

$$u^\mu = u^t\left(\eta^\mu + \Omega\chi^\mu\right), \quad (36)$$

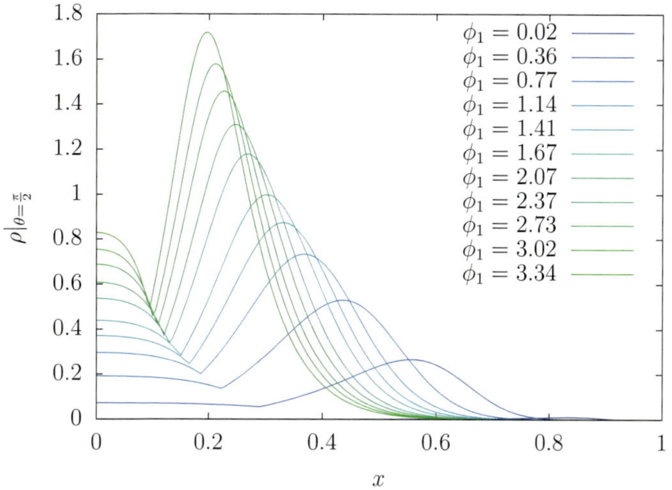

Fig. 1 Energy density of the scalar matter as seen by a co-moving observer on the equatorial plane for several different solutions in compactified radial coordinate

as usual. The angular velocity is simply $\Omega \equiv u^\varphi/u^t$. In Eq. (24), the energy density is obtained by contracting the mixed form of the energy-momentum tensor with the eingenvector representing the four-velocity, $e^\mu{}_{(1)} = u^\mu$. After working out the algebraic relations we obtain

$$u^t = \frac{\sqrt{2fl}m(\omega_s r + m\omega)\sin\theta}{\sqrt{\left[2l\sin^2\theta(\omega_s r + m\omega)^2 + m^2\Delta\right]|\Delta| - m^2\Delta^2}}, \tag{37}$$

$$\Omega = -\frac{\omega_s}{m} + \frac{m|\Delta|\,(1 - \text{sign}\,\Delta)}{2rl\sin^2\theta(\omega_s r + m\omega)}. \tag{38}$$

These expressions elucidate the rather unphysical emergence of the cusp. The timelike component of the four-velocity of our observer blows up at the cusp surface, and so does its energy and angular momentum per unit mass,

$$u^t \xrightarrow{\Delta\to 0} \infty; \qquad -u_t \xrightarrow{\Delta\to 0} \infty; \qquad u_\varphi \xrightarrow{\Delta\to 0} \infty. \tag{39}$$

The angular velocity, on the other hand, remains finite at the cusp

$$\Omega \xrightarrow{\Delta\to 0} -\frac{\omega_s}{m}. \tag{40}$$

In other words, there is no timelike observer who experiences zero momentum fluxes and shears at the surface that defines the cusp. Hence the apparent pathology.

As a matter of fact this surface is a very peculiar one and its description will help us transition to KBHsSH. Let us momentarily look at other quantities whose properties change at the cusp. By defining the parameter Killing vector

$$w^\mu := \eta^\mu - \frac{\omega_s}{m}\chi^\mu, \tag{41}$$

its contraction yields $w^\mu w_\mu = \Delta/m^2 f^2$. Hence, at the cusp it becomes a null vector. Note, however, that there is no null surface as the normal vector $\partial^\mu \Delta$ is not orthogonal to itself at S_c and therefore not parallel to w^μ in any frame. It would nevertheless be the case if both terms defining Δ became each trivial, i.e. $f = 0$ and $\omega/r = -\omega_s/m$, which is exactly what happens with the hairy BHs of the next section. The first term establishes the existence of a horizon at this surface, while the second one defines the dragging of the spacetime there, namely the horizon's angular velocity.

According to the ansatz of the scalar field (11), the parameter Killing vector is everywhere orthogonal to the complex scalar field's gradient, $w^\mu \partial_\mu \Phi = 0$.

The norm of the Noether current j^μ can also be written in terms of Δ, as

$$j^\mu j_\mu = \frac{4\phi^4}{f l r^2 \sin^2\theta}\Delta, \tag{42}$$

where the fraction is everywhere positive, so Δ dictates the nature of the current, which is then timelike outside S_c, null on the surface and spacelike in the volume enveloped by it. This illustrates another counter intuitive property of rotating BSs, i.e. if we wish to describe the system by an ensemble of timelike particles, their four velocity u^μ cannot be written everywhere in terms of the current j^μ. It also means that within the surface S_c, all timelike observers (including our co-moving one) measure zero charge density $\varrho \equiv u^\mu j_\mu$. In Fig. 2 we present again the profile of the energy density on the equatorial plane as seen by a co-moving observer, adding how the quantities w^2, j^2 and ϱ vary across S_c.

4 KBHsSH

KBHsSH are, in a sense, the combination of a Kerr BH with a rotating BS. Because general relativity is a nonlinear theory, those objects do not simply superpose but warp spacetime in their own way. Nevertheless, this simplistic picture is very useful to visualize the system: an event horizon surrounded by a toroidal lump of scalar matter, all rotating. As such, KBHsSH share traits of both Kerr BHs and rotating BSs, weighting more on one or another depending on how hairy a solution is. These solutions were first found and reported less than a decade ago by Herdeiro and Radu [10].

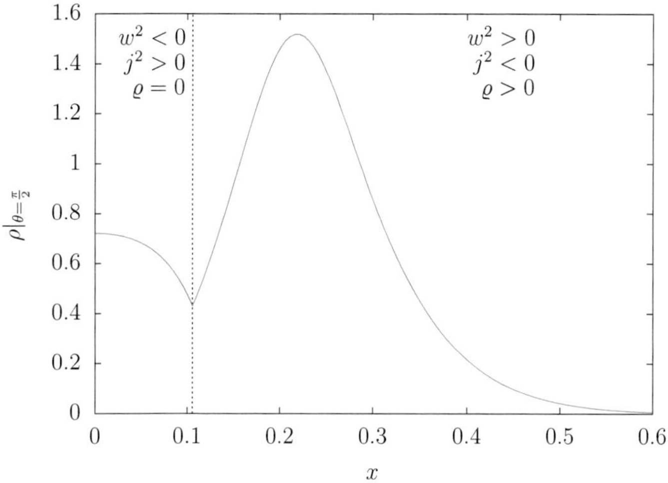

Fig. 2 Energy density of the scalar matter as seen by a co-moving observer on the equatorial plane in compactified radial coordinate. The norm of the vectors w^μ and j^μ changes signs across the surface S_c. Since the current is spacelike inside the envelope, all timelike observers measure zero charge density ϱ therein

4.1 Metric Ansatz and Boundary Conditions

The metric employed before is not the most suitable one to describe this system. In particular, for technical reasons working with numerics, it pays out to factor out the poles related to the event horizon in the metric components. We adopt,

$$ds^2 = -\mathcal{N} e^{2F_0} dt^2 + e^{2F_1}\left(\frac{dr^2}{\mathcal{N}} + r^2 d\theta^2\right) + e^{2F_2} r^2 \sin^2\theta \left(d\varphi - \frac{\omega}{r} dt\right)^2, \quad (43)$$

where, similarly as before, F_0, F_1, F_2 and ω are the functions of r and θ we need to solve for, while $\mathcal{N} \equiv 1 - r_H/r$ and r_H is the horizon radius and an input parameter. In what follows, we are only interested in the exterior region $r \geq r_H$. Again we choose to work with a compactified coordinate but need to redefine it. We choose $x = \tilde{r}/(\tilde{r}+1)$ where $\tilde{r} = \sqrt{r^2 - r_H^2}$. This transformation brings things much closer to the strong regime of the event horizon's vicinity when compared to Kerr BHs in usual Boyer-Lindquist coordinates. It also forces most of the boundary conditions at the horizon to be trivial Neumann,

$$\partial_x F_0|_{x=0} = 0, \qquad \partial_x F_1|_{x=0} = 0, \qquad \partial_x F_2|_{x=0} = 0, \qquad \partial_x \phi|_{x=0} = 0. \quad (44)$$

The function ω, which relates to the dragging of the spacetimes as $-g_{t\varphi}/g_{\varphi\varphi} = \omega/r$ has a Dirichlet type boundary condition required by regularity,

$$\omega|_{x=0} = -r_H \frac{\omega_s}{m}. \tag{45}$$

Meaning, the angular velocity of the event horizon is set to $\Omega_H = -\omega_s/m$, the same a co-moving observer would have had at the energy-density cusp of a rotating BS, Eq. (38). Because this condition cannot be set freely in order to obtain KBHsSH, they are also referred to as *black holes with synchronized hair*. The Killing vector field, $w^\mu = \eta^\mu + \Omega_H \chi^\mu$ is null at the horizon and generates the flow along which the scalar field is dragged, and its action on it $w^\mu \partial_\mu \Phi = 0$ implies in there being no flux across the horizon.

At infinity, asymptotic flatness demands

$$F_0|_{x=1} = 0, \quad F_0|_{x=1} = 0, \quad F_1|_{x=1} = 0, \quad F_2|_{x=1} = 0, \quad \omega|_{x=1} = 0, \quad \phi|_{x=1} = 0, \tag{46}$$

and the scalar field falls off at large distances exactly like for rotating BSs (16).

At the rotation axis, axisymmetry and regularity dictates that

$$\partial_\theta F_0|_{\theta=0} = 0, \quad \partial_\theta F_1|_{\theta=0} = 0, \quad \partial_\theta F_2|_{\theta=0} = 0, \quad \partial_\theta \omega|_{\theta=0} = 0, \quad \phi|_{\theta=0} = 0, \tag{47}$$

while on the equatorial plane we require even parity

$$\begin{aligned} &\partial_\theta F_0|_{\theta=\pi/2} = 0, \quad \partial_\theta F_1|_{\theta=\pi/2} = 0, \quad \partial_\theta F_2|_{\theta=\pi/2} = 0, \\ &\partial_\theta \omega|_{\theta=\pi/2} = 0, \quad \partial_\theta \phi|_{\theta=\pi/2} = 0. \end{aligned} \tag{48}$$

4.2 Charges

Similarly to what was done in the previous section, we shall now calculate the global charges of KBHsSH. As the system is again stationary and axisymmetric, the Komar expressions associated to the Killing vectors yield the total mass and angular momentum, while a Noether charge arises from the $U(1)$ global invariance of the scalar-field theory.

Mass and angular momentum are stored partially in the hole itself, which requires us to evaluate its contribution in the Komar expressions (6) and (7) as a surface integral. Splitting the equations into two parts, we obtain

$$M = \underbrace{2\int_{\mathcal{H}} n^\mu \sigma^\nu \nabla_\mu \eta_\nu dA_H}_{M_H} + \underbrace{2\int_{\Sigma\backslash\mathcal{H}} n^\mu \nabla_\nu \nabla_\mu \eta^\nu dV}_{M_\phi},$$

$$J = \underbrace{-\int_{\mathcal{H}} n^\mu \sigma^\nu \nabla_\mu \chi_\nu dA_H}_{J_H} \underbrace{- \int_{\Sigma\backslash\mathcal{H}} n^\mu \nabla_\nu \nabla_\mu \chi^\nu dV}_{J_\phi}, \tag{49}$$

where $\mathcal{H}$ is the horizon surface, dA_H its area element and σ^μ a spacelike vector perpendicular to it such that $\sigma^\mu \sigma_\mu = 1$. Similarly, $\Sigma\backslash\mathcal{H}$ is a spacelike asymptotically flat hypersurface bounded by the horizon, dV is its volume element and n^μ is the future-oriented timelike vector normal to it. The line element (43) implies in $\sigma^\mu = (0, \sqrt{\mathcal{N}}e^{-F_1}, 0, 0), n^\mu = (\xi^\mu + \omega/r\eta^\mu)/\left(\sqrt{\mathcal{N}}e^{F_0}\right), dA_H = e^{F_1+F_2}r^2 \sin\theta d\theta d\phi$ and $dV = \sqrt{-g/\mathcal{N}}e^{-F_0}drd\theta d\phi$.

$$M_\phi = \frac{1}{2\pi}\int_{r_H}^{\infty}\int_0^{\pi}\int_0^{2\pi} (\omega_s \mathrm{j} - U)\sqrt{-g}drd\theta d\phi, \qquad J_\phi = \frac{1}{4\pi}\int_{r_H}^{\infty}\int_0^{\pi}\int_0^{2\pi} m\mathrm{j}\sqrt{-g}drd\theta d\phi, \tag{50}$$

where we define the density

$$\mathrm{j} = \frac{(m\omega + \omega_s r)\,\phi^2}{\mathcal{N}e^{2F_0}r}. \tag{51}$$

The conserved Noether charge, differently from the quantities above, has no counterpart stored in the hole thanks to the no-flux condition. Hence, it can be obtained by a volume integral in the exterior region, similar to Eq. (21)

$$Q = \int_{\Sigma\backslash\mathcal{H}} j_\mu n^\mu dV, \tag{52}$$

and by applying the same substitutions we did for the Komar integrals, we arrive at

$$Q = \frac{1}{4\pi}\int_{r_H}^{\infty}\int_0^{\pi}\int_0^{2\pi} \mathrm{j}\sqrt{-g}drd\theta d\phi, \tag{53}$$

noticing that even in the presence of a horizon, the angular momentum due to the scalar field remains quantized as it happens for rotating BSs, $J_\phi = mQ$. Following the approach done in [10], we define a normalized charge $q = J_\phi/J \in [0, 1]$ which serves as a measure of how hairy the black hole is.

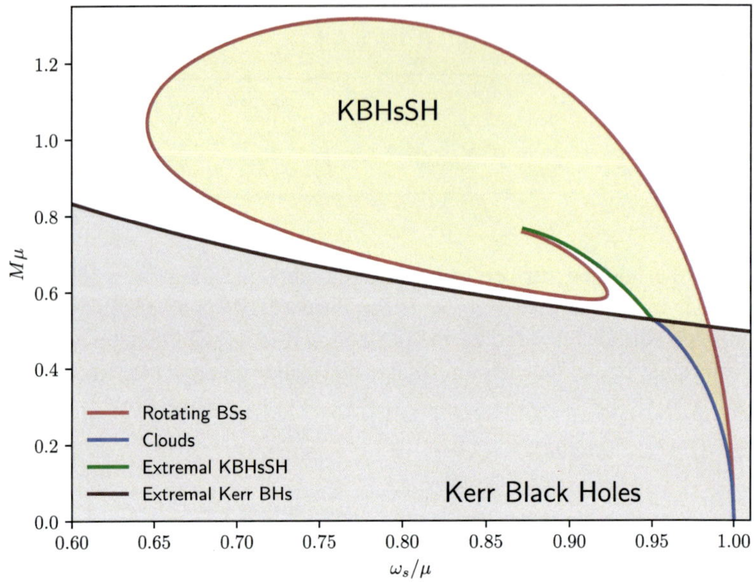

Fig. 3 Domain of existence of KBHsSH in the mass versus ω_s plane. There are three curves that delimit the region where hairy black holes dwell: rotating BSs (red), clouds (blue) and extremal hairy black holes (green). KBHsSH co-exist with bald Kerr BHs below the extremal Kerr BHs line near $\omega_s/\mu \sim 1$. More than one solution might exist for a given pair of parameters $\{M\mu, J\mu^2\}$

4.3 Domain of Existence

In this subsection, we aim to provide a comprehensive overview of KBHsSH solutions. To that end, we must prescribe which scalar-field theory is used as our basis. Specifically, we present results for the free theory, which only consists of a mass term in its potential, $U(\Phi) = \frac{1}{2}\mu^2\Phi\Phi^*$. Like the BSs mentioned earlier, all solutions observed belong to the first rotation mode with $m = 1$.

The domain of existence of KBHsSH is enclosed by three curves consisting of limiting solutions. In Fig. 3 we depict the mass versus frequency diagram for hairy and Kerr BHs together with its boundaries. We mentioned earlier that one can think of KBHsSH as the combined system of a hole and a rotating BS. This is better visualized in this figure where one can appreciate how these objects approach BSs on one end and Kerr BHs on another. Thus, rotating BSs form the first boundary and are characterized by not having a hole at its center, so all angular momentum and mass are stored in the scalar matter ($r_H = 0$, $q = 1$). Second are the clouds. Those are stationary solutions of the EKG Eq. (5) on a Kerr background and are therefore not backreacting ($q = 0$). Finally, extremal KBHsSH connect BSs to the clouds. Those are saturated BHs with zero Hawking temperature (and in our parametrization $r_H = 0$). A triple point appears in the figure where extremal Kerr, clouds and extremal

KBHsH meet. In this case they are all the same solution, but the overlapping region of Kerr and KBHsSH is not unique.

Recall that the scalar field's frequency is related to the horizon angular velocity through the synchronization condition (45), which allows us to draw Kerr in this diagram even in the absence of bosonic matter. These solutions obey

$$M - \frac{s}{2\omega_s\left(\sqrt{1-s^2}+1\right)} = 0, \qquad s \equiv \frac{J}{M^2}. \tag{54}$$

5 Equatorial Circular Orbits

Circular orbits, albeit seemingly simple, provide great insight to more complicated processes and lie at the core of astrophysical phenomenology such as accretion discs and extreme-mass-ratio inspirals. In this section we shall look at particular traits of the spacetime geometry of some KBHsSH (and their limiting rotatings BSs) and how they impair the dynamics of orbiting particles. On the equatorial plane, $\theta = \pi/2, d\theta = 0$ and the line element reads

$$ds^2_{\pi/2} = g_{tt}(r,\pi/2)dt^2 + g_{rr}(r,\pi/2)dr^2 + 2g_{t\varphi}(r,\pi/2)dtd\varphi + g_{t\varphi}(r,\pi/2)d\varphi^2. \tag{55}$$

The first integrals of the equations of motion come from the spacetime isometries. Due to stationary, the total energy per unit mass of the test particle $E = -u_t$ is conserved, while axisymmetry grants conservation of its angular momentum per unit mass $L = u_t$. Then, in terms of E and L, the components of the four-velocity, $u^\mu = (\eta^\mu + \Omega\chi^\mu)$, of the orbiting particle read

$$u^t = -\frac{g_{t\varphi}L + g_{\varphi\varphi}E}{g_{t\varphi}^2 - g_{tt}g_{\varphi\varphi}}; \qquad u^\varphi = \frac{g_{tt}L + g_{t\varphi}E}{g_{t\varphi}^2 - g_{tt}g_{\varphi\varphi}}. \tag{56}$$

The norm $u^\mu u_\mu = -1$ can be used to derive the equation of motion for the radial coordinate, and it gives

$$V_{eff} \equiv g_{rr}\dot{r}^2 = \frac{g_{\varphi\varphi}}{g_{t\varphi}^2 - g_{tt}g_{\varphi\varphi}}\left(E - V_+\right)\left(E - V_-\right), \tag{57}$$

where we define the effective potential V_{eff} in terms of the potentials $V_\pm$

$$V_\pm \equiv L\frac{g_{t\varphi}}{g_{\phi\phi}} \pm \frac{\sqrt{L^2\left(g_{t\varphi}^2 - g_{tt}g_{\varphi\varphi}\right)}}{g_{\phi\phi}}. \tag{58}$$

Circular orbits are characterized by $\dot{r} = \ddot{r} = 0$. This implies in $E = V_\pm$ and $\partial_r V_\pm = 0$, which fixes the values of E and L for each radius where the orbit is to take place. Since $\Omega \equiv u^\varphi/u^t$, the angular velocity is also set. In general, for any stationary, axisymmetric and circular spacetime, the circular orbital parameters on the equatorial plane can be written

$$E = -\frac{g_{tt} + g_{t\varphi}\Omega}{\sqrt{-g_{tt} - 2g_{t\varphi}\Omega - g_{\varphi\varphi}\Omega^2}}, \tag{59}$$

$$L = \frac{g_{t\varphi} + g_{\varphi\varphi}\Omega}{\sqrt{-g_{tt} - 2g_{t\varphi}\Omega - g_{\varphi\varphi}\Omega^2}}, \tag{60}$$

$$\Omega_\pm = \frac{-\partial_r g_{t\varphi} \pm \sqrt{(\partial_r g_{t\varphi})^2 - \partial_r g_{tt}\partial_r g_{\varphi\varphi}}}{\partial_r g_{\varphi\varphi}}. \tag{61}$$

The innermost-stable-circular-orbit (ISCO) lies at the smallest radius (r_{ISCO}) for which the orbit is marginally stable, i.e. where $\partial_r^2 V_{eff} = 0$.

On the equatorial plane of most of the studied stationary spacetimes, specially in general relativity, and constraining the analysis to the regions exterior to a horizon (if there is any), the three metric functions that appear in these equations are monotonic, with the clear exception for manifolds with naked singularities. As a consequence, one might intuitively and naively take for granted that

- For all $r > r_{ISCO}$ there exists a circular orbit and it is stable.
- For all circular orbits, $\Omega_+ > 0$ while $\Omega_- < 0$ and, therefore, they correspond to the angular velocities of co-rotating and counter-rotating orbits, respectively.

Rotating BSs, as we have seen in Sect. 3, feature a distribution of energy density which is off-centered due to the nontrivial topology that the scalar field assumes. This trait is carried along to KBHsSH which have a significant part of its mass and angular momentum stored in the hair. As a consequence, the metric functions g_{tt} and $g_{t\varphi}$ are no longer monotonic, i.e. their slopes are positive in a region and negative in another. In Fig. 4 we illustrate the profile of these metric functions on the equatorial plane of a particular KBHSH. The line element is given as Eq. (43). Note that such metric parametrization requires a larger resolution near the horizon where the ergoregion lies.

The angular velocity of a circularly orbiting particle, Eq. (61) depends directly on the slope of the metric functions, and therefore these have great impact in the overall dynamics. Next, we shall explore some specific proprieties of Ω that arise as a consequence of the off-center distribution of matter that makes it very discernible from what is obtained from most objects in general relativity. In Fig. 5, we display once again the domain of existence of KBHsSH but highlight the areas that contain the solutions we are interested in. These shaded regions overlap as many solutions feature more than one of the peculiar properties we shall now analyze.

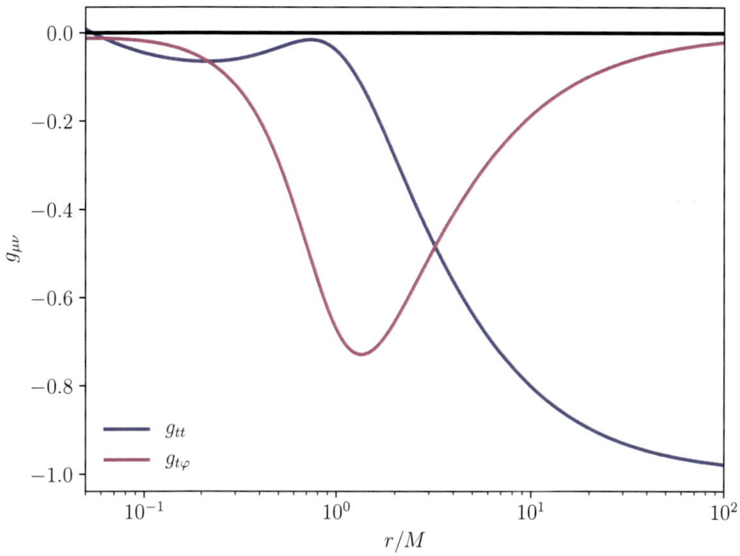

Fig. 4 Metric function profiles on the equatorial plane of a particular KBHsSH. Unlike for bald Kerr BHs, these can vary nonmonotonically

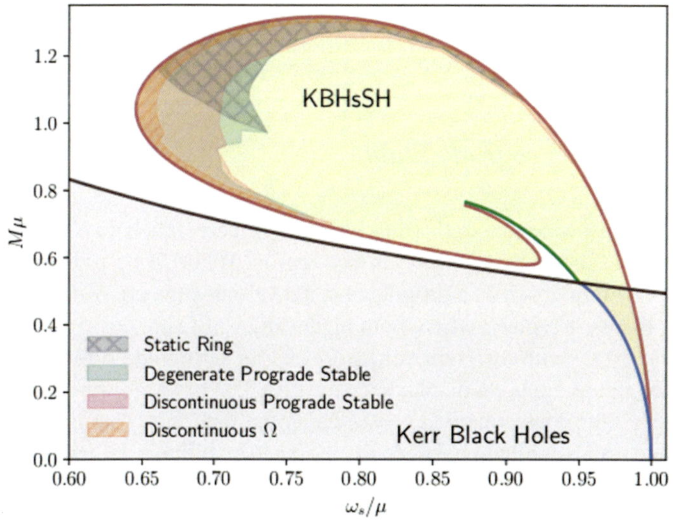

Fig. 5 Domain of existence of KBHsSH in the mass versus ω_s plane. The shaded areas encompass solutions for which the circular orbital velocity features peculiarities. Static rings: $\Omega_- = 0$. Degenerate prograde stable: $\Omega_- > 0$ and stable. Discontinuous prograde stable: disjoint regions of unstable orbits. Discontinuous Ω: regions where inertial circular motion is not possible

5.1 Static Ring

The static ring is simply a ring a points on the equatorial plane where a test particle momentarily at rest with respect to a zero angular momentum observer (ZAMO) at infinity remains at rest. Its dynamics is characterized by a circular orbit geodesic with zero angular velocity. According to Eq. (61), if the slope of g_{tt} disappears at a certain point where its value is still negative, then $\Omega_- = 0$. Such property, together with several spacetime examples were reported in [12]. Note that the requirement $g_{tt} < 0$ at this point simply means that it must not be contained within an ergoregion where no timelike particle can stand still. In Fig. 5, the lower boundary for the shaded region depicting solution featuring static rings marks the onset of ergoregions where $\partial_r g_{tt} = 0$. Note that all KBHsSH have one topological S^2 ergoregion in the vicinity of the horizon, while a second, disjoint, topological $S^1 \times S^1$ ergoregion appears for solutions lying below the aforementioned boundary. Configurations on top of the boundary for which $g_{tt} = \partial_r g_{tt} = 0$ at the same radial coordinate host the so-called light points [25]. Note also that, unless $\partial_r g_{tt} = 0$ is a saddle point, the solution will portray two static rings. One at the local minimum (closer to the center) and another one at the local maximum (further from the center). While at the local maximum the static orbit is always stable, at the local minimum it can be either stable or unstable. Finally, we remark that this phenomenon is purely inertial (the particle is indeed following a geodesic). In essence, there are two gravitational wells that counterbalance each other where the particle finds itself at rest.

5.2 Degenerate Prograde Stable

Configurations featuring two static rings have the property that between them $\partial_r g_{tt} > 0$. Then, Eq. (61) shows that $\Omega_- > 0$ in this region. While $\Omega_+ > 0$ everywhere and can therefore be thought of as a prograde orbital velocity, the opposite is not true for Ω_-. Hence, there are regions where both eigenvalues of Ω are positive and there are two prograde orbits with different velocities and no retrograde orbit. Furthermore, both might be stable. Figure 6 displays the positive axis of $\Omega_\pm$ for one such example. There are two static rings where $\Omega_- = 0$, being the innermost one unstable and the outermost one stable, while in between $\Omega_- > 0$ and stability is gained. Furthermore, although $\Omega_+ > 0$ everywhere, unlike for the exterior region of Kerr BHs and stars in general, it varies nonmonotonically.

5.3 Discontinuous Prograde Stable

At the beginning of the section we stated that one may, naively, assume that every circular orbit at a radius $r > r_{ISCO}$ is stable. Figure 6 also gives a counter example to

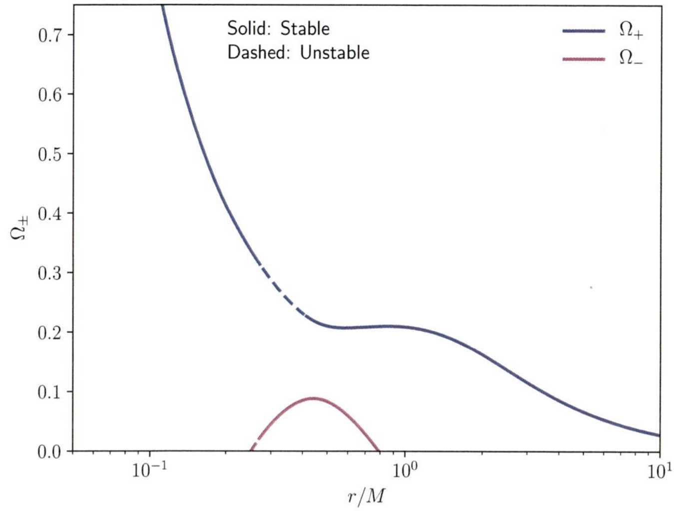

Fig. 6 Profiles of the equatorial circular orbital velocities (positive axis only) for a particular solution of a KBHSH. Solid line for stable orbits and dashed for unstable. This configurations hosts two static rings (where $\Omega_- = 0$), and a region where both orbital velocities are positive but different, and both orbits are stable

this assumption. KBHsSH might feature disjoint regions of (un)stable circular orbits. As seen in Fig. 5 this encompasses a significant amount of solutions in the parameter space. Such feature is not exclusive of these objects but has also been reported for axionic solitons and Kerr BHs with axionic hair [26, 27]. However, in all known cases the scalar field is topologically nontrivial and its energy density distribution is off-centered.

5.4 *Discontinuous Circular Orbits*

The final shaded region contains solutions for which there is a discontinuity in the existence of circular orbits. In the region between two static rings, the slope of g_{tt} can be steep enough that $\left(\partial_r g_{t\varphi}\right) - \partial_r g_{tt} \partial_r g_{\varphi\varphi} < 0$. When that happens, the eigenvalues of Ω are both complex and inertial circular orbits cease to exist. Such phenomenon has nothing to do with the timelike character of the orbiting particle, which is more closely related to the square root appearing in Eqs. (59) and (60). Circular motion is still possible as long as the particle is accelerated towards the black hole. A rather simple mechanical analogy that helps illustrate the problem is given by a particle sliding a potential of particular shape. Let there be an upside-down cone. Given a large enough angular velocity, a particle can be made orbit the center of the cone at a certain height, sliding on the inner surface such that the centrifugal force causes the vertical component of the normal force to perfectly balance the downward gravitational pull.

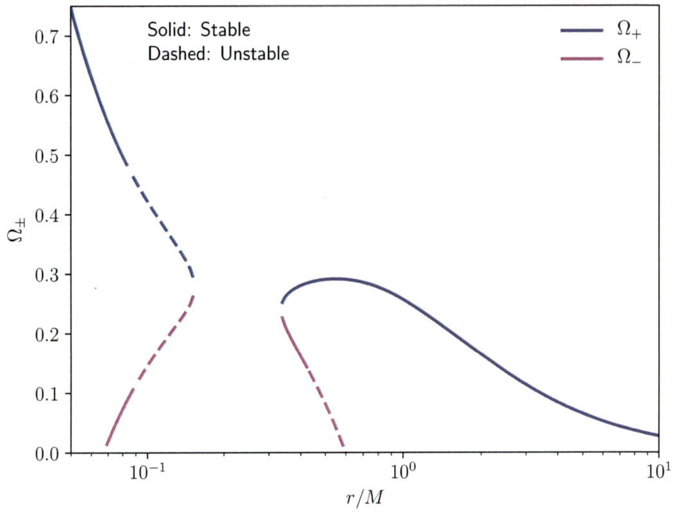

Fig. 7 Profiles of the equatorial circular orbital velocities (positive axis only) for a particular solution of a KBHSH. Solid line for stable orbits and dashed for unstable. This solutions hosts two static rings, degenerate prograde stable orbits and the interruption of the existence of circular orbits. In the region where no Ω is found, only non-inertial circular motion is possible

This trick is obviously impossible to be achieved if the cone is placed upwards and the particle made orbit its outer surface. Figure 7 shows the profile of $\Omega_\pm$ for a solution with discontinuous circular orbits. Another interesting feature is that at the points limiting the nonexistence region, $\Omega_- = \Omega_+ = -\partial_r g_{t\varphi}/\partial_r g_{\varphi\varphi}$ which is simply the dragging of the spacetime. At these two points, particles in circular orbits have zero angular momentum, namely they are ZAMOs. This sheds a different light on the phenomenon. The dragging of the spacetime is the maximum possible value for Ω_- and the minimum possible value for Ω_+.

6 Conclusions

In this article we have laid out particular features of rotating BSs and KBHsSH that have its roots in the topologically nontrivial distribution of energy density due to the scalar matter. The general theory underling these rotating compact objects has been presented, as well as an outline on how the solutions are constructed. We provided the boundary conditions which arise from regularity requirements and naturally give the overall behavior of the scalar field, as well as the synchronization condition for hairy black holes. We have shown that, by employing a frame co-moving with the stellar fluid for rotating BSs an interesting surface can be defined, where no timelike observer experiences zero shear or momentum fluxes as its four-velocity diverges while its orbital velocity remains finite and equals the angular velocity a horizon would have if the background metric were given by a hairy black hole instead. Hence, rotating BSs naturally carry the structure to host a stationary black hole.

The existence domain of KBHsSH was explored in terms of peculiar traits of circular orbits. We have shown that, unlike for exterior Kerr and its small deviations, the metric functions vary nonmonotonically. This creates disjoint regions of stable (and unstable) orbits. Some solutions feature static rings, where circular orbits have zero orbital velocity, as well as regions where for a given radial distance both eigenvalues of the angular velocity are positive and the designation *retrograde* for Ω_- loses meaning. Furthermore, both prograde orbits at a same radial coordinate might be stable. Some solutions contain a region where no inertial circular motion is possible, limited by points where both angular velocities equal the dragging of the spacetime itself. Moreover, while Ω_- might become positive, Ω_+ is never negative but still behaves nonmonotonically.

While these interesting discrepancies with exterior Kerr happen at the vicinity of the horizon making it hard for the direct observation of orbiting stars, the impact on astrophysical phenomenology should not be negligible. As of yet, it is unclear how this unusual profile of the orbital parameter might affect accretion discs, but the existence of disjoint regions of unstable orbits might cause violent events as the inflow is no longer adiabatic once the orbits become unstable. If the objects hosts static rings, it is reasonable to ask whether it will even accrete at its center (BSs) or through the horizon (KBHsSH) as infalling matter might just pile up at the ring. Similarly, the quadruple formula for gravitational waves show that the energy and angular momentum fluxes stop once the orbital velocity becomes zero. Therefore, at leading order, an inspiralling object should indeed stop once it reaches the static ring. Preliminary research on these topics has been done in the references, but more in-depth investigations are still lacking in the literature, leaving many open questions to the consequences of having nonmonotonically varying metric functions.

References

1. D.J. Kaup, Klein-gordon geon. Phys. Rev. **172**, 1331–1342 (1968)
2. D.A. Feinblum, W.A. McKinley, Stable states of a scalar particle in its own gravational field. Phys. Rev. **168**, 1445–1450 (1968)
3. R. Ruffini, S. Bonazzola, Systems of self-gravitating particles in general relativity and the concept of an equation of state. Phys. Rev. **187**, 1767–1783 (1969)
4. E.W. Mielke, F.E. Schunck, Rotating boson stars, pp. 391–420 (1997)
5. S. Yoshida, Y. Eriguchi, Rotating boson stars in general relativity. Phys. Rev. D **56**, 762–771 (1997)
6. Y. Kobayashi, M. Kasai, T. Futamase, Does a boson star rotate? Phys. Rev. D **50**, 7721–7724 (1994)
7. S.L. Liebling, C. Palenzuela, Dynamical Boson stars. Living Rev. Rel. **15**, 6 (2012)
8. L.G. Collodel, B. Kleihaus, J. Kunz, Excited Boson stars. Phys. Rev. D **96**(8), 084066 (2017)
9. L.G. Collodel, B. Kleihaus, J. Kunz, Structure of rotating charged boson stars. Phys. Rev. D **99**(10), 104076 (2019)
10. C.A.R. Herdeiro, E. Radu, Kerr black holes with scalar hair. Phys. Rev. Lett. **112**, 221101 (2014)

11. S. Hod, Stationary scalar clouds around rotating black holes. Phys. Rev. D **86**, 104026 (2012). [Erratum: Phys. Rev. D **86**, 129902 (2012)]
12. L.G. Collodel, B. Kleihaus, J. Kunz, Static orbits in rotating spacetimes. Phys. Rev. Lett. **120**(20), 201103 (2018)
13. P. Grandclement, C. Somé, E. Gourgoulhon, Models of rotating boson stars and geodesics around them: new type of orbits. Phys. Rev. D **90**(2), 024068 (2014)
14. Y. Ni, M. Zhou, A. Cárdenas-Avendaño, C. Bambi, C.A.R. Herdeiro, E. Radu, Iron kα line of kerr black holes with scalar hair. J. Cosmol. Astroparticle Phys. **2016**, 049 (2016)
15. S. Gimeno-Soler, J.A. Font, C. Herdeiro, E. Radu, Magnetized accretion disks around kerr black holes with scalar hair: constant angular momentum disks. Phys. Rev. D **99**, 043002 (2019)
16. M.C. Teodoro, L.G. Collodel, D. Doneva, J. Kunz, P. Nedkova, S. Yazadjiev, Thick toroidal configurations around scalarized kerr black holes. Phys. Rev. D **104**, 124047 (2021)
17. Z. Meliani, F.H. Vincent, P. Grandclément, E. Gourgoulhon, R. Monceau-Baroux, O. Straub, Circular geodesics and thick tori around rotating boson stars. Classical Quantum Gravity **32**, 235022 (2015)
18. M.C. Teodoro, L.G. Collodel, J. Kunz, Retrograde polish doughnuts around boson stars. J. Cosmol. Astroparticle Phys. **2021**, 063 (2021)
19. K. Gjorgjieski, J. Kunz, M.C. Teodoro, L.G. Collodel, P. Nedkova, Magnetized thick disks around boson stars (2023)
20. L.G. Collodel, D.D. Doneva, S.S. Yazadjiev, Circular orbit structure and thin accretion disks around kerr black holes with scalar hair. Astrophys. J. **910**(1), 52 (2021)
21. L.G. Collodel, D.D. Doneva, S.S. Yazadjiev, Equatorial extreme-mass-ratio inspirals in kerr black holes with scalar hair spacetimes. Phys. Rev. D **105**, 044036 (2022)
22. G.H. Derrick, Comments on nonlinear wave equations as models for elementary particles. J. Math. Phys. **5**, 1252–1254 (1964)
23. M.S. Volkov , E. Wöhnert, Spinning Q-balls. Phys. Rev. D **66**(8), 085003 (2002) https://doi.org/10.1103/PhysRevD.66.085003
24. B. Kleihaus, J. Kunz, M. List, Rotating boson stars and Q-balls. Phys. Rev. D **72**(6), 064002 (2005) https://doi.org/10.1103/PhysRevD.72.064002
25. P. Grandclément, Light rings and light points of boson stars. Phys. Rev. D **95**, 084011 (2017)
26. J.F. Delgado, C.A. Herdeiro, E. Radu, Rotating axion boson stars. J. Cosmol. Astroparticle Phys. **2020**, 037 (2020)
27. J.F.M. Delgado, C.A.R. Herdeiro, E. Radu, Kerr black holes with synchronized axionic hair. Phys. Rev. D **103**, 104029 (2021)

Seven Years of LIGO-Virgo Science

Giancarlo Cella

Abstract I review the first seven years of gravitational wave detections. Gravitational waves are ripples in spacetime generated by the acceleration of astrophysical objects. A direct consequence of general relativity, they were first directly observed in 2015 by the twin Laser Interferometer Gravitational Wave Observatory (LIGO) observatories. The LIGO and Virgo gravitational–wave detectors carried out the third observing run in 2020. Several new signals where detected, including peculiar events not seen before such as coalescences with a large mass or large mass asymmetry, and coalescences of a neutron star-black hole pair. The analysis of these events is providing information about stellar compact object populations, the local expansion rate of the universe, the nature of black holes and the validity of general relativity. They can constrain the behavior of matter at densities higher than an atomic nucleus, and mergers that emit both gravitational and electromagnetic waves probe the formation of short gamma ray bursts and the nucleosynthesis of heavy elements. Other searches are constraining more and more the emission of isolated sources and the possible background of gravitational waves from unresolved sources of astrophysical or cosmological nature.

1 Signal and Noises

The first direct observation of gravitational wave signals opened a new frontier for astrophysics. Since then, we can look at the universe using an additional channel which carries on information directly related to the mass–energy distribution and dynamics. Gravitational waves are perturbation of the spacetime metric which are described by the Einstein equation

$$R_{\mu\nu} - \frac{1}{2} g_{\mu\nu} R = \frac{8\pi G}{c^4} T_{\mu\nu} \tag{1}$$

G. Cella (✉)
I.N.F.N. Pisa, Largo B. Pontecorvo 3, Pisa, Italy
e-mail: giancarlo.cella@pi.infn.it

© The Author(s), under exclusive license to Springer Nature Switzerland AG 2024

E. Papantonopoulos and N. Mavromatos (eds.), *Compact Objects in the Universe*,
https://doi.org/10.1007/978-3-031-55098-0_9

linearized around a given background solution ${}^0g_{\mu\nu}$. In the simplest case where the background is the flat spacetime described by the Minkowski metric, ${}^0g_{\mu\nu} = \eta_{\mu\nu}$, it is possible to choose coordinates where

$$g_{\mu\nu} = \eta_{\mu\nu} + h_{\mu\nu} \tag{2}$$

and $h_{\mu\nu}$ is supposed to be a small perturbation, in the sense that $|h_{\mu\nu}| \ll 1$. The linearized version of Riemann tensor reads

$$R_{\mu\alpha\nu\beta} = \frac{1}{2}\left(h_{\mu\beta,\nu\alpha} + h_{\alpha\nu,\beta\mu} - h_{\mu\nu,\alpha\beta} - h_{\alpha\beta,\mu\nu}\right) \tag{3}$$

and it is invariant under "small" coordinate transformations

$$x^\mu \to x^\mu + \xi^\mu(x) \tag{4}$$

with $\xi = O(h)$. Now it is possible to rewrite the Einstein equation (1) as

$$\Box \overline{h}_{\mu\nu} = -\frac{16\pi G}{c^4} T_{\mu\nu} \,, \tag{5}$$

$$\overline{h}_{\mu\nu}{}^{,\mu} = 0 \tag{6}$$

where the possibility of performing the small transformation (4) has been used to impose the additional constraint (6) and the trace reversed perturbation

$$\overline{h}_\mu{}^\nu \equiv h_{\mu\nu} - \frac{1}{2}\eta_{\mu\nu} h \tag{7}$$

has been introduced for convenience. Equations (5) and (6) have the same structure of the wave equation in electromagnetism and can be solved in the same way, given the knowledge of the source term, obtaining

$$\overline{h}_{\mu\nu}(\vec{x}, t) = \frac{4G}{c^4} \int d^3x' \frac{T_{\mu\nu}\left(\vec{x}', t - \frac{|\vec{x}-\vec{x}'|}{c}\right)}{|\vec{x} - \vec{x}'|} \,. \tag{8}$$

It is possible to decompose in an unique way the perturbation to separate their radiative degrees of freedom [1] by writing

$$h_{tt} = 2\phi \,, \tag{9}$$

$$h_{ti} = \beta_i + \gamma_{,i} \,, \tag{10}$$

$$h_{ij} = h_{ij}^{TT} + \frac{1}{3} H \delta_{ij} + \epsilon_{i,j} + \epsilon_{j,i} + \left(\partial_i \partial_j - \frac{1}{3}\delta_{ij}\nabla^2\right)\lambda \,, \tag{11}$$

where all the vector fields are transverse and h_{ij}^{TT} is transverse and traceless. This is a classification based on the properties of transformation under *spatial* rotations. Some combination of the fields introduced are invariant with respect to the small coordinate transformations (4), namely

$$\Phi = -\phi + \dot{\gamma} - \frac{1}{2}\ddot{\lambda}\,, \tag{12}$$

$$\Theta = \frac{1}{3}\left(H - \nabla^2\lambda\right)\,, \tag{13}$$

$$\Xi_i = \beta_i - \frac{1}{2}\dot{\epsilon}_i\,, \tag{14}$$

and h_{ij}^{TT}. This transverse and traceless part is the only one that satisfy a wave equation, namely

$$\Box\bar{h}_{ij}^{TT} = -\tfrac{16\pi G}{c^4}T_{ij}^{TT} \tag{15}$$

where T_{ij}^{TT} is the transverse and traceless part of the spatial part of energy momentum tensor. We can immediately conclude that the radiative degrees of freedom are just two, the number of independent component of a transverse and traceless tridimensional symmetric tensor. In particular Eq. (15) has plane wave solution with two independent polarizations. The other invariant fields Φ, Θ and Ξ_i are governed by Poisson–like equations, in other words they represent non propagating Coulomb–like contributions. For example, in the Newtonian limit Φ reduces to the Newtonian potential. Note that this decomposition is unique only in the full spacetime, not in a finite region of it.

We see that a gravitational wave carry information about energy and momentum distribution inside the source, complementary to the information carried by electromagnetic waves.

A convenient coordinate system to discuss in simple terms the basic peculiarities of a gravitational wave detector is the *local proper reference frame*, which is defined as follows. First, a reference mass in geodesic motion is chosen, and its world line is defined to be $x^\mu = (c\tau, \vec{0})$. Next, the coordinates are chosen in such a way that

$$g_{\mu\nu}(\tau,\vec{0}) = \eta_{\mu\nu}\,, \tag{16}$$

$$\Gamma^\mu{}_{\nu\rho}(\tau,\vec{0}) = 0\,. \tag{17}$$

In the neighborhood of $\vec{x} = \vec{0}$ the metric is different from $\eta_{\mu\nu}$ by terms $O(|\vec{x}|^2/\rho^2)$ where ρ is the typical curvature length scale. Looking at Eq. (3) we see that $\rho = O(\lambda_{GW}/\sqrt{h})$.

In the large wavelength limit $\lambda_{GW} \gg L$, which is a decent approximation for earthbound gravitational wave interferometric detectors (GWID), we can describe the evolution of the separation between nearby free falling test masses using the geodesic deviation equation. The relevant part can be written as

$$\frac{d^2L^i(t)}{dt^2} = -R_{i0j0}(t,\vec{0})L^j(t) \tag{18}$$

where

$$R_{i0j0} = -\frac{1}{2}\ddot{h}_{ij}^{TT} + \Phi_{,ij} + \left(\dot{\Xi}_{i,j} + \dot{\Xi}_{j,i}\right) - \frac{1}{2}\ddot{\Theta}\delta_{ij} \tag{19}$$

and L^i is the proper distance between two test masses. By writing the variation of L^i as a small perturbation, $L^i = L_0^i + \delta L^i$ and by integrating Eq. (18) we get

$$\delta L^i(t) = \frac{1}{2}h_{ij}^{TT}(t)L_0^j \tag{20}$$

where we retained only the radiative part of (19). Note that non–radiative contributions could also play a role: for example the term $\Phi_{,ij}$ is related to the *gravity gradient noise* that we will shortly discuss later.

Equation (20) give us an intuition of the coupling of gravitational waves to the detector in term of a tidal force. In GWID as Virgo and LIGO the variations δL^i are measured with an interferometric technique. A simplified optical scheme of the Virgo detector is depicted in Fig. 1. Basically, this is a Michelson interferometer with Fabry–Perot resonant cavities inserted in the arms. A laser beam is injected in the apparatus and separated in two by a semitransparent mirrors (the *beam splitter*). Each splitted part propagates back and forth along an interferometer's arm for a time $\delta\tau$, and before being recombined get a phase shift that can be written, again in the large wavelength approximation, as $\delta\phi_i = 2\pi\mathcal{F}\delta L_i\lambda_\ell^{-1}$ where $\mathcal{F}$ is the finesse of the Fabry–Perot cavity and λ_ℓ the wavelength of the laser.

The measured signal is given by the interference of the two beams, so it is proportional to the difference between the δL_i of the two arms. It is always possible to calibrate the interferometer's output $s(t)$ in such a way that

$$s(t) = D^{ij}h_{ij}^{TT} + n(t) \tag{21}$$

where $n(t)$ is the additive strain noise and

$$D^{ij} = u^iu^j - v^iv^j \,, \tag{22}$$

where $\vec{u}$ and $\vec{v}$ are unitary vectors parallel to interferometer's arms, is called *detector tensor*. We can introduce an *angular sensitivity pattern*

$$F^A(\vec{n}) = D^{ij}\epsilon_{ij}^A(\vec{n}) \tag{23}$$

which gives information about the directional sensitivity of the GWID. In Eq. (23) the index A labels the polarization of a gravitational wave coming from the direction $\vec{n}$, and ϵ_{ij}^A is the corresponding polarization tensor. In Fig. 2 the amplitude $|F^A(\vec{n})|$ is represented as a function of the direction $\vec{n}$. We see that a GWID has some directionality, however it is not a detector that can by itself be *pointed* to a given source.

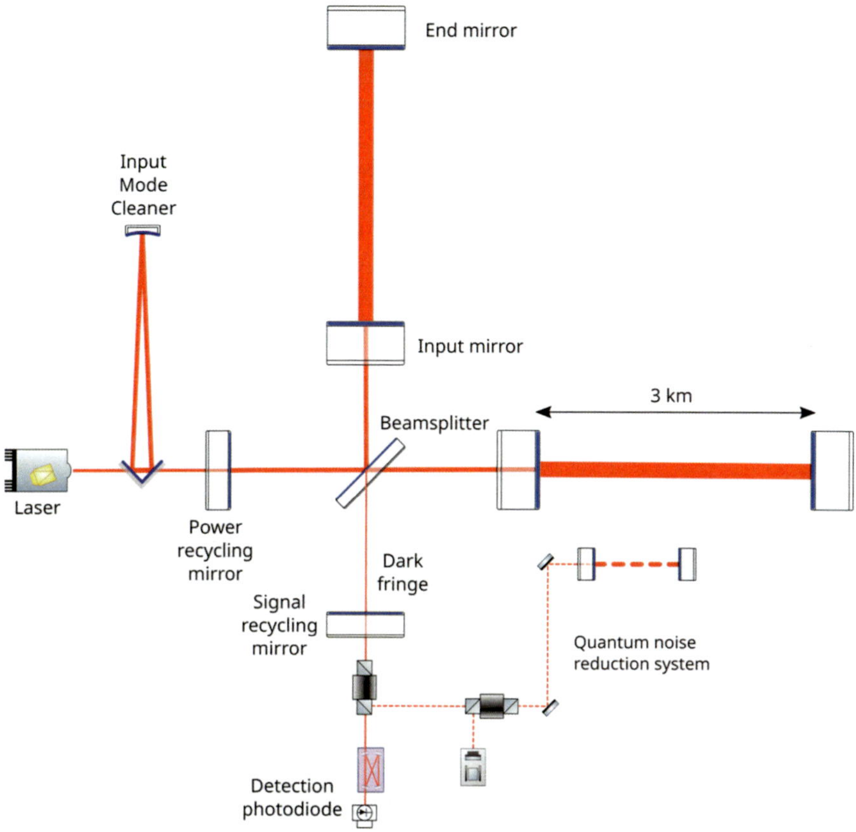

Fig. 1 A simplified optical scheme of the Advanced Virgo GWID, a laser interferometer with 3 kms long perpendicular arms. The laser path is folded hundreds of times before the light recombines at the beam splitter by Fabry-Perot cavities along each arm. Advanced Virgo is designed to be a dual recycled Michelson interferometer, for reasons that are explained in the text. Two additional optical systems are the *input mode cleaner*, which is used to improve the quality of the laser beam circulating the interferometer, and the *quantum noise reduction* which is used to reduce the noises due to the quantum nature of light

Another important point is that the angular pattern depends on the polarization: this opens up the possibility of disentangle the polarization components of a signal.

The main effect of Fabry–Perot cavities is to introduce the finesse proportionality factor between δL^i and $\delta\phi^i$. We could think that this should produce an increase of the sensitivity proportional to $\mathcal{F}$, however we should remind that the same amplification factor is also applied to all the disturbances which induces a motion of the test masses (the mirrors of the cavity). This means that only noises which are not connected to mirror motion are reduced.

Other relevant elements depicted in Fig. 1 are the *power recycling mirror* and the *signal recycling mirror*. To understand properly their effect we look at a semiclas-

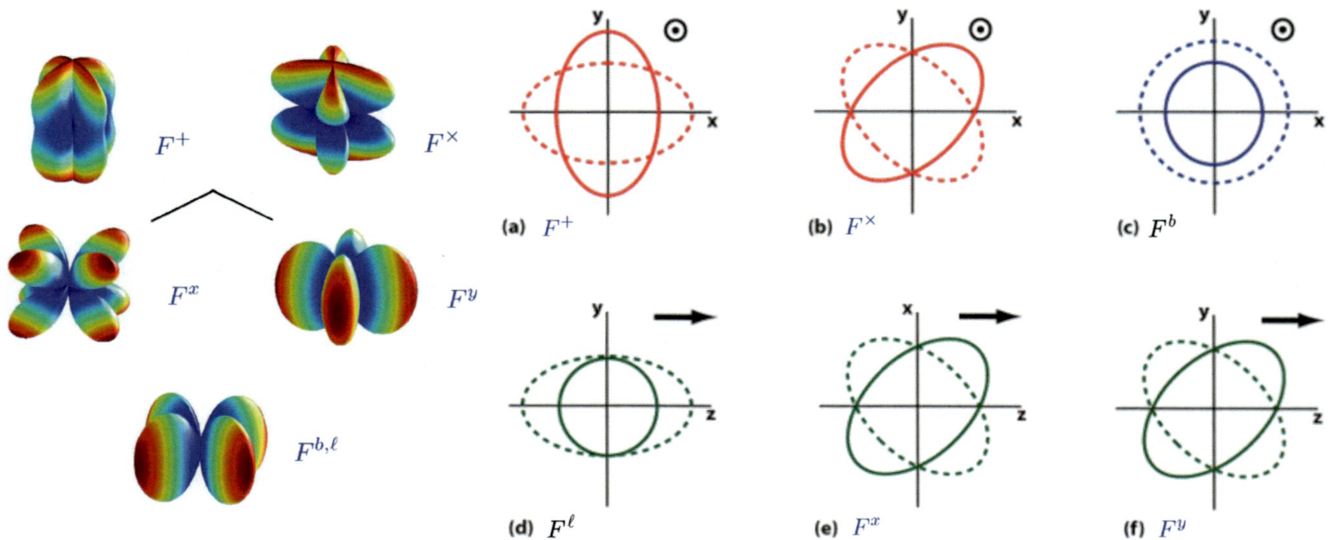

Fig. 2 Left: angular patterns F^A for a given polarization of a gravitational wave. In general relativity only polarizations $A = +$ and $A = \times$ correspond to physical degrees of freedom. The orientation of the interferometer's arms are represented in black. Right: possible polarization for a metric theory. The tidal action on a circular ring of test masses is represented, the arrow corresponds to the propagation direction

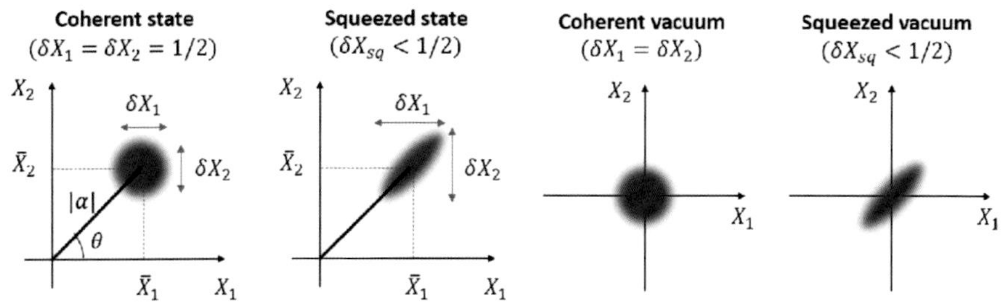

Fig. 3 Semiclassical representation of some beam quantum states relevant for the physics of GWID. From the left: a *coherent state*, which is a good approximation of a laser beam. A *squeezed state*, with a reduced phase fluctuation. A *coherent vacuum*, which is a good approximation of what enters the interferometer through the dark port. A *squeezed vacuum*, which can be produced and injected in the dark port to mitigate the quantum noise

sical description of the laser light depicted in Fig. 3. The electromagnetic mode of the laser injected in the interferometer can be represented in the quadrature space as a coherent state, which is sketched as a vector with modulus proportional to the classical amplitude α with a cloud representing the quantum uncertainty. The motion of the mirrors (the test masses) generate a variation of the phase of the mode, so the gravitational wave signal is encoded in the quadrature orthogonal to the classical field direction (the *phase quadrature*). The quantum fluctuations in the same direction generate a *quantum phase noise*, which can be visualized as the fluctuation $\delta\theta$ associated to the quantum cloud. Quantitatively, the spectral noise amplitude of the phase fluctuations of a coherent state, commonly called *shot noise*, is given by

$$S_h = \frac{\lambda_\ell}{4\pi L\mathcal{F}}\sqrt{\frac{h\nu_\ell}{W}\left[1+\left(\frac{\omega_{GW}L\mathcal{F}}{c}\right)^2\right]}\,. \tag{24}$$

From the representation in Fig. 3 it is easy to understand that a practical way to reduce the quantum phase fluctuation is to increase the laser power, which is proportional to $|\alpha|^2$.

The *power recycling mirror* is used to recycle the laser light injected in, and almost completely reflected by the interferometer. In this way a resonant cavity between the recycling mirror and the interferometer is formed, which amplify the effective power injected on it.

It turns out that this is a too naive picture: the quantum fluctuations parallel to the classical field are amplitude fluctuations, and generate a fluctuating radiation pressure on the mirrors. The mirrors move reacting to this pressure, and this generate an additional phase fluctuation with spectral amplitude

$$S_h = \frac{\mathcal{F}}{LcM\omega_{GW}^2}\sqrt{Wh\nu_\ell\left[1+\left(\frac{\omega_{GW}L\mathcal{F}}{c}\right)^2\right]^{-1}} \tag{25}$$

which is called *radiation pressure noise*. Note that while *shot noise* decreases with the laser power W, the opposite is true for *radiation pressure noise*. It turns out that, whatever W we choose, we always obtain a quantum noise which is larger than the *standard quantum limit* (SQL)

$$S_h = \frac{2}{\omega L}\sqrt{\frac{\hbar}{M}} \,. \tag{26}$$

It is important to realize that the quantum fluctuations in the output beam are not generated by the quantum fluctuations of the laser: as we said, the laser is almost completely reflected by the interferometer. They enter the interferometer through the dark port, as quantum fluctuation of an electromagnetic mode in the vacuum state [2]. This opens up the possibility of evading the SQL by manipulating the vacuum state entering in the dark port. In particular it is possible to inject a *squeezed vacuum state* [3] (as the one depicted in Fig. 3). If the squeezing is done in such a way that phase quadrature fluctuations are reduced, a reduction of the quantum noise at high frequency (where it is dominated by the shot noise) is obtained, at a fixed laser power. This is currently done with an appropriate optical system (schematically represented in Fig. 1). A more complex possibility is to tune the squeezing angle as a function of the frequency: in this way the effect of the radiation pressure could be completely removed in principle [2] and the SQL can be evaded.

The *signal recycling mirror* recycles the output beam exiting from the interferometer, which contains the signal in its phase quadrature. By an appropriate tuning of the signal recycling cavity which is formed between the interferometer and the mirror, and taking advantage by the peculiarity of the beam exiting the interferometer, which as we said has been modified by the radiation pressure through ponderomotive effects, we obtain a further mitigation of quantum noise, see [4] and reference therein for details. In Fig. 4 the typical noise budget of an advanced detector is represented. Beside the already discussed quantum noise, we see that several other fundamental noises limit the sensitivity.

1.1 Thermal Noise

This is a consequence of the coupling of the detector to the environment. It is the main sensitivity limitation in the middle frequency range of advanced detectors. As a consequence of the fluctuation–dissipation theorem, thermal fluctuations are proportional to dissipative effects, and the main mitigation strategy is to reduce these. The most important case is about thermal noise generated by optical coating of mirrors, which is needed to guarantee the extreme optical properties of them but introduce dissipation. The search for better coatings is a very active field of research [6]. Another handle that can be used to reduce thermal noise is to increase

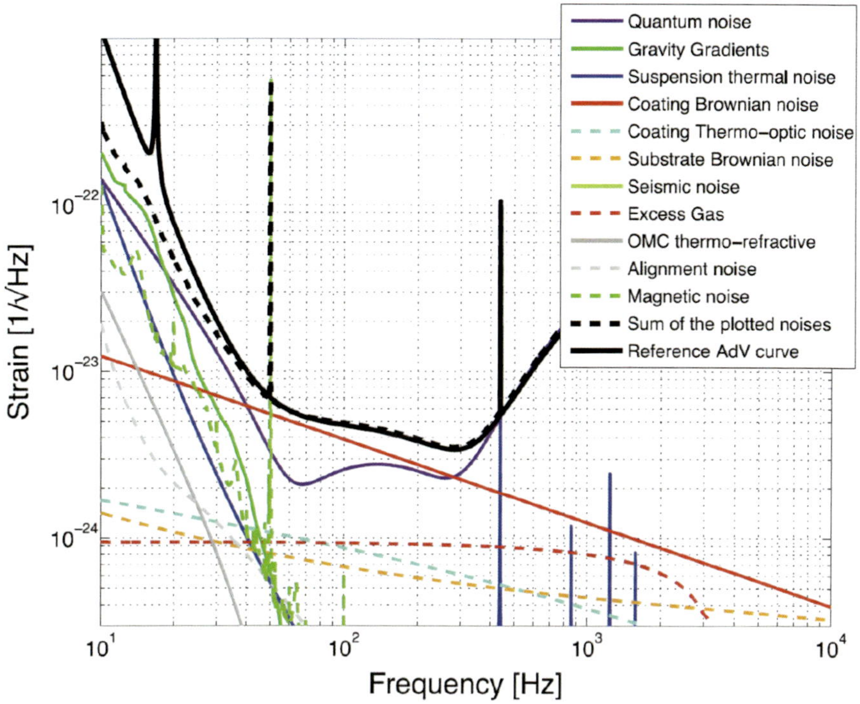

Fig. 4 Virgo Advanced noise budget. Figure taken from [5]: © The Authors, published by EDP Sciences. This is an open access article distributed under the terms of the Creative Commons Attribution License 4.0 (http://creativecommons.org/licenses/by/4.0/ (accessed on 10 August 2021))

the transverse size of the beams on the mirror surface, which results in an improved average of surface fluctuations. Finally, a direct approach could be the reduction of the temperature. Neither Virgo or LIGO are cryogenic detectors: KAGRA is attempting to explore this solution.

1.2 Seismic Noise

This is the effect of the mirrors' motion induced by movement of the ground. These fluctuations are so large that to not spoil completely the sensitivity sophisticated isolators are needed to isolate the mirrors. This noise set the lower frequency of the sensitivity band [7].

1.3 Gravity Gradient Noise

This is the effect of the direct coupling of test masses to local gravitational fluctuations induced by mass density fluctuations of seismic or atmospheric origin. These couples to the detector as described in Eq. (19) (the term $\Phi_{,ij}$ there is just the gradient of gravitational force in the Newtonian approximation). This kind of noise can't be directly screened, and mitigation strategies are actively developed and studied especially in anticipation of third generation detectors [8].

In addition there are several "technical" sources of noise that needs to be controlled, for example scattered light noise, residual gas noise, control noise, electromagnetic noise and so on.

1.4 Signals

During these first years of observations many kinds of signals have been searched by LIGO and Virgo. They can be roughly classified in four different categories.

1.4.1 Bursts

These are short duration transient, that can be generated during the core collapse of massive stars, or by cusps and kinks in cosmic strings. These are typically searched by doing only minimal assumptions about the waveform's shape.

1.4.2 Compact Binary Coalescences

These are gravitational waves generated by a coalescence between two compact objects, black holes and neutron stars. They are searched using specific, theoretically motivated templates for the waveform.

1.4.3 Continuous Waves

These are continuous signals generated by rotating neutron stars.

1.4.4 Stochastic Backgrounds

These are stochastic signals, with an astrophysical (superposition of many unresolved events) or cosmological (inflation, phase transitions, cosmic strings) origin. The noise spectral amplitude $\sqrt{S(f)}$ of the noise is the most relevant information needed to quantify the performances of a GWID. Currently, coalescences for all the the three possible pairs (see Fig. 6) have been detected. The search has been done by applying to the data s a Wiener filter

$$W(p_i, s) = \int_0^\infty \frac{\tilde{T}(p_i, f)^* \tilde{s}(f)}{S(f)} df \tag{27}$$

which is basically a scalar product, weighted with the noise power spectrum $S(f)$, between a set of theoretical templates $\tilde{T}(p_i, f)$ and s. The maximum of W over the set of physical parameters p_i is compared to a threshold which sets the significance of the detection. The integral (27) is dominated by the frequency range where $|\tilde{T}(p_i, f)|^2/S(f)$ is large. Roughly, $|\tilde{T}(p_i, f)|^2 \sim f^{-7/3}$ for $f < f_{cut}$ and is negligible for $f > f_{cut}$, where

$$f_{cut} \simeq 4.4\text{kHz} \left(\frac{M_\odot}{m_1 + m_2} \right) . \tag{28}$$

The coalescence will be difficult to detect for large total mass systems. On the other hand the signal–to–noise ratio for the statistics (27) scales as $\mathcal{M}^{5/6}$, where the chirp mass is defined by

$$\mathcal{M} = \frac{(m_1 m_2)^{3/5}}{(m_1 + m_2)^{1/5}} \tag{29}$$

so low–mass systems will be also difficult to detect. A typical figure of merit which is used to quantify GWID performance is the *sight distance*, defined as the maximum distance where a coalescence can be detected with a given signal to noise ratio. It can be written as (see [9, Sect. 7.7.2])

$$d_{sight} = \frac{2}{5}\sqrt{\frac{5}{6}} \frac{c}{\pi^{2/3}} \left(\frac{G\mathcal{M}}{c^3} \right)^{5/6} \sqrt{\int_0^{f_{cut}} \frac{df}{f^{7/3}} \frac{1}{S(f)}} SNR^{-1} \tag{30}$$

where an average over the source position in the sky and on the orbit orientation is assumed.

In the next two sections we will discuss the scientific results of the LIGO–Virgo–KAGRA (LVK) collaboration. LIGO collaboration manages two identical 4-km arm size GWID [10]: the LIGO Hanford Advanced detector is situated in southeastern Washington State and the LIGO Livingston Advanced detector is in Livingston, Louisiana. The Advanced Virgo detector [11] is a 3-km interferometer in Cascina, Italy. It is operated by the European Gravitational Observatory and funded by INFN (Italy), CNRS (France) and Nikhef (Netherlands). The KAGRA observatory is an underground 3-km interferometer in Kamioka, Japan, who joined the collaboration during the third observing run of LIGO and Virgo. Past and future observing plans of LVK are represented in Fig. 5.

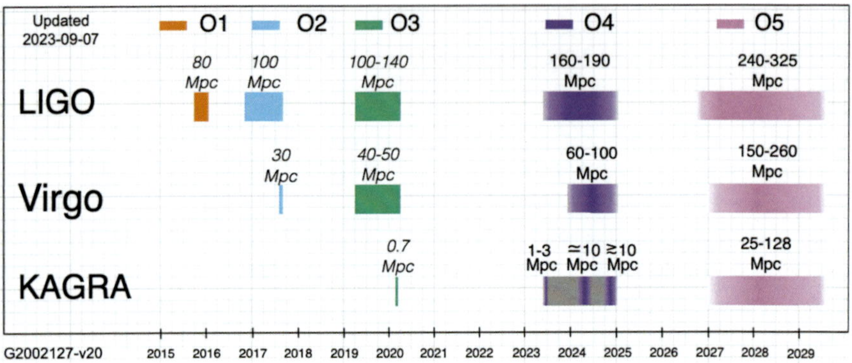

Fig. 5 Timeline of the observing plans for the LVK collaboration. The obtained and foreseen sensitivity of the involved GWID is quantified by the *BNS horizon*, which is the sight distance (30) for a binary neutron star coalescence and $SNR = 8$. Updated information about observing plans are available at https://observing.docs.ligo.org/plan/#live-status

2 The First Discoveries

The first two scientific runs O1 (from September 2015 to January 2016) and O2 (from November 2016 to August 2017) have been highly successful.

In O1 the first direct detection of gravitational waves was obtained, from the observation of a coalescence of two black holes [12] with mass of about $30 M_\odot$ each. The signal lasted less than 0.5 s in the sensitivity band and was detected by the two LIGO interferometers [13], as Virgo was not yet commissioned at the time. The event, named GW150914, had a significativity greater than 5.3σ, with a false alarm probability of about 5×10^{-6} yr^{-1}. Another event, GW151226 [14], was detected in the following. It was interpreted once again as a BBH coalescence [15].

Another BBH coalescence was detected by the two LIGO interferometers during the O2 run [16]. In all cases a detailed study of the signal shape allowed for an estimate of the parameters of the systems [15]: the masses and the spins of the two progenitor black holes and of the final one, the distance of the source and its position in the sky. The accuracy of these estimates depends on the parameter considered. For example, the determination of the chirp mass $\mathcal{M}$ (see Eq. (29)) which is the combination of the progenitor masses m_i which dominates the signal's shape was quite accurate, $28.1^{+1.8}_{-1.5} M_\odot$ for GW150914 and $8.3^{+0.3}_{-0.3} M_\odot$ for GW151226. On the other hand the accuracy on the mass ratio m_1/m_2, which contributes only at higher order in the post Newtonian expansion, was lower (around 20% for GW150914) [15].

2.1 Localization

Toward the end of O2 run Virgo was commissioned and it joined the network. Immediately after, the first three interferometers detection of a BBH coalescence, GW170814, was obtained [17].

The inclusion of Virgo produced a qualitative improvement for the accuracy of the localization of the source in the sky. This can be understood easily: the largest amount of information about the direction of propagation of the gravitational wave signal is encoded in the relative delays between the signal detected by the different interferometers in the network. With two detectors, we have just a single delay. GW150914 was seen by the two LIGO interferometers with a delay of about 7 ms, while the light time of flight between them was 10 ms. This gave a constraint for the angle between the propagation direction and the line which connects the two detectors, which allowed for a determination of a ring shaped region in the sky where the source must be. The width of the ring was determined by the accuracy on the delay estimate and additional, weaker constraints about the compatibility of the two signal shapes allowed for some reduction of the region.

With a three detector network two independent delays can be used: in principle this allows for a complete determination of the source position, modulo a reflection degeneracy. Once again the final accuracy depends on the error on the delays and on additional constraint, but a large improvement was expected and found [17]: the typical determination for the initial LIGO only network was about 1160 deg^2, to be compared with about 60 deg^2 for the LIGO-Virgo case.

This improvement was not really important for GW170814, as the only signal expected from a BBH coalescence is a gravitational one [18]. But it was the demonstration that the era of multi messenger observations could start.

2.2 GW170817

In the very final part of O2 run a different kind of coalescence was jointly detected by the LIGO-Virgo collaboration, with a false alarm probability of about 10^{-5} yr^{-1}. While previous observations remained inside the sensitivity window for less than two seconds, in this case the signal last around a minute, and reached much higher frequencies.

The largest instantaneous frequency of the signal, and the time interval of the observation, depends mainly on the dimension of progenitors. The frequency of emitted gravitational waves are two times the orbital one, so the progenitors need to reach a small orbital period. This is possible only (think about the third Kepler law for a rough estimate) if they can orbit one very near to the other before merger. Black holes with reasonable mass can't be compact enough to allow for this, so the simplest interpretation is that the observed signal is the result of a coalescence between two neutron stars [19]. This interpretation was confirmed by the detailed determination of the parameters, which gives for the progenitor masses $1.36M_\odot < m_1 < 1.60M_\odot$ and $1.17M_\odot < m_2 < 1.36M_\odot$.

Another support to the hypothesis came from the observation of several electromagnetic counterparts. Fermi/GBM detected a gamma radiation peak 1.7 s after the merger. The localization done with the gravitational detectors was consistent with the one provided by Fermi and INTEGRAL, and the signal was interpreted as a short GRB (GRB 170817A) generated by the merger [20]. The observed delay between the coalescence and the GRB counterparts allowed for the best determination of the difference between the speed of light and the gravitational wave speed,

$$-3 \times 10^{-15} < \frac{v_{GW}}{c} - 1 < 7 \times 10^{-16}\,. \tag{31}$$

2.2.1 Nuclear Matter Equation of State

Looking at the gravitational signal only, it was possible for the first time to measure the tidal deformability of the progenitors. Information about that are contained in the waveform: a high deformability can enhance the production of gravitational radiation in the final inspiral phase, and introduce a early cut off during the merger. The tidal deformability gives information about the nuclear matter equation of state, and this opens the possibility of ruling out some of the existing models [21]. With the current sensitivity in the high frequency region which is sensitive to tidal effects few model (with very low stiffness) have been excluded [22].

2.2.2 Hubble Parameter Determination

Another interesting result was the independent estimation of the Hubble parameter, using only the gravitational signal. Gravitational wave provides information about distance, while the red shift can be inferred from the electromagnetic counterpart [23]. The constraint is currently not competitive with the ones from Planck and SHoES [24, 25], but is conceptually important because it does not need the use of a distance ladder. When n similar events will be available, it will be possible to combine the results of all the analysis reducing the error. As we will see, O3 run provided a larger number of detections, but for none of them an electromagnetic counterparts has been found. It has been shown however that it is possible to infer information about Hubble parameter also for this kind of sources [26]: the idea is to localize statistically the host galaxy to infer the red shift and average over many events.

2.2.3 Electromagnetic Counterparts

Beyond the mentioned gamma signal from Fermi, GW170817 was followed by an extensive set of electromagnetic emissions in the X, UV, optical, infrared and radio band. These were extensively observed by a large number of collaborations [27], owing to the good localization provided by LIGO-Virgo and Fermi+INTEGRAL. An important results obtained in this way was the detailed observation of the evolution of the spectra, which are consistent with the kilonova scenario [28]: matter ejected in the post-merger phase undergoes r-process in a neutron rich environment, providing a mechanism for the production of elements heavier than iron.

3 O3 Run and Beyond

The third LVK observing run started in April 1, 2019 and ended in March 27, 2020. The total number of significant GW detections, all compact binary coalescences, found by LVK analyses is 90 [29–31], compared with 3 candidates found after the end of O1 and 11 after the end of O2. Additional candidates have also been reported by non–LVK searches on public data [32]. The number of detections scales as expected with the sensitivity of the LVK network, in proportion to the spacetime volume explored (see Fig. 1 in [31]). The detections are summarized in Fig. 6. With the currently available number of detected events we are entering in a regime driven

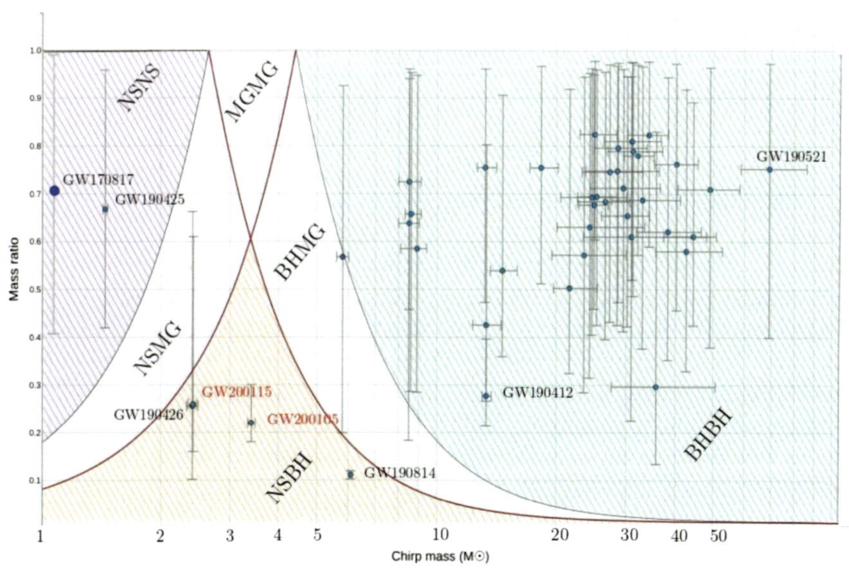

Fig. 6 Detections in the O1, O2 and O3 runs. On the horizontal axis, the chirp mass of the coalescing pair. On the vertical axis, its mass ratio. On the basis of the pair masses the coalescing system is classified as a NSNS hone (two neutron stars), BHNS (a neutron star and a black hole), BHBH (two black holes) and MG (mass gap). Credits: figure prepared with http://catalog.cardiffgravity.org/, web app maintained by Chris North (Cardiff University)

by statistical information. In other words, it start to be possible to extract parameters connected with astrophysical populations. Examples are mass and other parameters distributions [33] or cosmological parameters, as the already mentioned Hubble parameter.

3.1 Tests of General Relativity

The detection of compact binary coalescence offers an unique possibility of testing general relativity in the strong field, relativistic regime. A large number of tests has been done by LVK for finding possible violations of general relativity prediction. All these tests are *generic*, in the sense that they do not look at a specific alternative model.

No evidence of general relativity violations have been found currently, but several upper limits have been established and improved. This is an example of a precision study, which will improve with the LVK detector network sensitivity, both for an increased SNR of the best detections which will allow for a better estimation of parameters, and for a larger statistic that will be obtained with a big number of events. A very careful application of statistical inference method is mandatory to reduce as much as possible small statistical biases which, not relevant for the single events, can build up when lare averages are done.

In the following subsections we shortly describe and comments the most important tests performed at the current time. Detailed results and discussions can be found in [34–38].

3.1.1 Residual and Parametrized Tests

In principle the simpler general relativity test that can be done consist in subtracting the theoretical waveform which best fit the data, and check the hypothesis that what remains is compatible with the expected noise. This require a good enough theoretical model, and an accurate understanding of the detector noise. Another option is to parameterize general relativity violations in a post–Newtonian expansion, and infer the parameters value from the data. This is a standard and well established method, that it is now possible to apply to very relativistic and strong gravity systems. No violation of general relativity has been found, but the constraint of several post–Newtonian parameters have been improved by a factor 2.

3.1.2 Compatibility Tests

The signal coming from a coalescence can be partitioned in three different phases. In the first one, called *inspiral*, the two progenitors can be described as orbiting point masses, and the evolution of the system can be evaluated with an accurate perturbative expansion including general relativity effects of high enough order. The perturbative expansion is expected to be accurate because the system is not yet really in the strong gravity regime. In the second phase, the *merger*, perturbative expansions break down, together with the point mass approximation. The nonlinear general relativity effects become dominant and the evolution of the system can be obtained by solving numerically the Einstein equations. During this phase the two progenitor black holes merge together, radiating a significant fraction of their mass (about $3.0M_\odot$ for GW150914) as gravitational waves. In the third phase, called *ringdown*, the final compact object, which can be described as a perturbed black hole, radiates its excitation energy evolving toward its final stationary Kerr configuration. An estimate of the mass and final spin of the source can be done using only the inspiral signal (with a signal to noise ratio of about 19.5 for GW150914), or only the post-inspiral signal (with $SNR \simeq 16$ for GW150914). The two probability distributions can be tested for compatibility. In this way strong gravity effects not explained by the Einstein's theory can be put in evidence. Results show agreement with general relativity predictions.

3.1.3 Quasi Normal Modes and Echoes

In GR, no-hair theorem predicts that BHs are described only by their mass and spin (and charge). When a BH is perturbed, it reacts (in GR) in a very specific manner, relaxing to its stationary configuration by oscillating in a superpositions of quasi-normal modes, which are damped by the emission of GWs. In the *ringdown* phase the excitation of the final black hole can be described as a superposition of *quasi normal modes*, namely harmonic damped oscillations, each parameterized by a frequency ω_m and a decay time τ_m. A relevant fact is that the values of ω_m and Q_m are expected to depend only on the mass and spin of the black hole. This prediction can be tested by comparing the measured values of ω_m, τ_m with the ones expected from the estimated final mass and spin. This is a non trivial procedure. If we analyze the signal starting from a time t_0 too near to the merger we can't be sure that we entered the linear regime. Also if this is true, we expect a complicate expression

$$h(t) \simeq \sum_m A_m \theta(t - t_0) e^{-(t-t_0)/\tau_m} \cos[\omega_m (t - t_0) + \phi_m] \tag{32}$$

with relevant contributions coming from several modes. Moving the start time t_0 far from the merger we expect only the contribution of the mode with the largest τ_m to survive. But the amplitude of the signal will decay giving a low signal to noise ratio. Anyway tests using only the dominant mode has been attempted, obtaining results consistent once again with general relativity. Quantum gravity corrections could in principle change the structure of the horizon. We can model this as a phenomenological horizon reflection amplitude different from zero. In this case the ringdown waveform is no more parametrizable as in Eq. (32), and *echoes* can be generated. It become possible to test the horizon structure. No evidence of these echoes has been found.

3.1.4 Modified Dispersion Relations

General relativity predict that gravitational waves are non dispersive. When dispersion is present the gravitational wave speed becomes frequency dependent, and the waveform shape changes during the propagation. This can happen if we suppose a non zero mass for the graviton, or a modified dispersion relation like

$$E^2 = p^2c^2 + A_\alpha p^\alpha c^\alpha \tag{33}$$

where A_α parametrize possible Lorenz–violating terms. No evidence has been found for this kind of effects, and a stringent upper limit for the graviton mass ($m_g < 1.76 \times 10^{-23}$eV) has been obtained.

3.1.5 Polarization Contents

As we said in Sect. 1, in general relativity only two polarization degrees of freedom exists for gravitational waves. For a general metric theory up to six polarizations can be physical: their tidal effect is represented on the right side of Fig. 2. As they generally have a different angular pattern, it is possible in principle to infer their presence with a large enough network of detectors with different orientations. No evidence of non standard polarizations has been found.

3.2 "Exceptional" Events

Among all the events detected by the LVK collaboration there are some that were worth of a dedicated study for their peculiarities. We discuss shortly two of them.

3.2.1 GW190814

It is be best localised source in the sky (about 23 deg^2) and the most mass asymmetric GW event detected with high significance [39] currently ($m_1 = 23M_\odot, m_2 = 2.6M_\odot$, $q = m_1/m_2 \simeq 8.8$). The interest of mass asymmetry is that an asymmetric mass distribution amplify higher order modes in a multipole expansion of the waveform

$$h_+ - ih_\times = \sum_{\ell\geq 2}\sum_{-\ell\leq m\leq\ell} \frac{h_{\ell m}(t,q)}{D_L} {}_{-2}Y_{\ell m} \tag{34}$$

Besides the $\ell = 2$ dominant mode the $\ell = 3$ contribution has been clearly detected, allowing for a validation of the theory in this regime. For its very good localization it has been used to demonstrate the possibility of infer the Hubble parameter without an electromagnetic counterpart [26].

3.2.2 GW190521

This is a very massive BBH merger, with $m_1 = 142^{+28}_{-16} M_\odot$ and $m_2 = 66^{+17}_{-16} M_\odot$ [40]. There are several reasons of interest for this event: the primary mass is in mass gap for pair–instability SN theory, and the final remnant is the first IMBH detected ($m_f = 142^{+28}_{-16} M_\odot$). The evidence of a black hole in the pair–instability mass gap require an explanation for its formation. Several possibilities have been suggested [41]: the result of a stellar merger in young star clusters, the result of hierarchical merger, the result of dynamical phenomena inside AGN disks.

3.2.3 NSBH Coalescences

Two detections (GW200105 and GW200115) have been interpreted as BHNS coalescences, on the basis of the estimated masses [42]. In principle an electromagnetic counterpart can be expected from events of this kind, but none was found in spite of extensive searches. There are two possible reasons for this. First, both events happened at a large luminosity distance $D_L \sim 280^{+110}_{-110}$Mpc for GW200105 and $D_L \sim 310^{+150}_{-110}$Mpc for GW200115, to be compared with $D_L \sim 40^{+8}_{-14}$Mpc of GW170817 NSNS event. At these distances the possibility of an electromagnetic detection is not high. Additionaly, localization had large uncertainities: this reduced the chances of a positive follow up. Another reason is connected to the fact that electromagnetic counterpart are expected when the NS member of the pair is disrupted, but this is not expected to happen with the observed (too large) mass asymmetry [43].

3.3 *Other Searches*

Several other kind of searches have been done during these years of observations. They did not lead already to detection claims, but in several cases interesting upper limits and parameter constraint have been obtained. An extensive discussion is not possible here, and we will give only a quite fast description of the main activities.

3.3.1 Gravitational Lensing

Gravitational lensing of gravitational wave is analogous to gravitational lensing of light. Interacting with the spacetime metric modified by large concentration of mass a gravitational wave can be focused, with an observed increase of intensity. Lensing can also produce multiple "images" of the same source: in the gravitational wave context, this means that several signal can be detected, at different times and coming from different direction, but originated by the same source. Frequency dependent deformations of the waveform can be observed too. This opens the possibility of testing fundamental physics, but also to do precision cosmology and population studies with microlensing. The searches done up to now did not allow for specific claims [44, 45].

3.3.2 Unmodeled Searches

It is important to search for unmodeled or poorly–modeled sources. These can give rise to short or long transient that need specific methods for their detection. Short burst of gravitational wave radiation can be generated by massive BBH, CCSNe, cosmic strings, pulsar glitches [46]. Longer transient could be generated by fallback accretion, accretion disk instabilities, newborn neutron stars from BNS merger or core-collapse supernovae, eccentric compact binary coalescences [47].

3.3.3 CW Searches

Neutron stars can form from the remnant of stellar collapse. Their typical size is 10 km, and their mass is about $1.4M_{\odot}$. Some of these stars are observed as pulsars. Our galaxy might contain $\sim 10^9$ NS, of which $\sim 10^3$ have been identified. The gravitational wave signal is continuous, nearly monochromatic but modulated in amplitude and phase by the motion of the detectors. Spin down effect, environmental effects and glitches can encode interesting physical information. A large number of searches, both targeted (known source) [48] and all sky [49–52] have been performed in LVK, obtaining more and more constraining upper limits.

3.3.4 Stochastic Background

The observation of many CBC events allows to put constraint on their rates, and indirectly on black hole and neutron star populations. The gravitational wave luminosity of a BNS event is much lower (about $0.025M_{\odot}$ for GW170817) compared with the BBH one, so the range of detection is also smaller. Both for BBH and BNS coalescences distant events can be too weak to be detected with a good enough statistical confidence. But if we look far, we expect to see an increasing number of events. An interesting possibility is that the statistical superposition of a large number of events, too overlapped and weak to be resolvable, can generate a *stochastic background* of gravitational waves. This could be detected in principle by looking at the correlations between outputs of interferometers in the network. Estimate based on the detected events and population models shows that in principle this kind of background could be seen with the current generation of detectors, at their design sensitivity [53]. Stochastic backgrounds of BBH and BNS coalescences are just a promising example of a large set of *astrophysical stochastic background*. A really exciting perspective is the one of detecting a stochastic background of cosmological origin. A background of this kind can be produced in principle by a large number of mechanisms in the primordial universe, such as parametric amplification during inflation, phase transitions, cosmic string and others (see [54] for a short review). Their direct observation will allow to test for fundamental physics models, and is starting to give constraints for them in synergy with CMB and pulsar timing array observations [55]. In some cases there is no hope to see the expected background: for example the spectrum produced by a standard slow roll inflationary scenario is several order of magnitude below the current sensitivity and the one foreseen for third generation detectors. Some models are more promising, for example inflationary ones with coupling to an axion field, pre–big bang string motivated models, cosmic strings networks. The LIGO-Virgo collaboration produced upper limits on gravitational wave stochastic background. The current one [56] is the best bound on the interferometric observation frequency window.

3.3.5 Gravitational Waves and Dark Matter

As dark matter candidates interact gravitationally, we can wonder if gravitational waves can give interesting information about them. There are several possible connections. Dark matter can influence a compact object with environmental effects: the compact object structure can be changed (accretion disks, spin down effects, formation of a dark matter core, superradiance effects [57]). The gravitational wave mechanism itself can be changed, and finally dark matter can impact on the propagation of the generated waves. Primordial black holes can be a candidate for a fraction of the expected dark matter. The direct signature for this would be a detection of a subsolar mass black hole: currently no candidates have been found [58]. Some exotic objects can be the smoking gun of the existence of specific dark matter candidates: for example, it has been claimed that GW190521 signal is compatible with an head–on collision of two vector complex boson stars [59]. Finally, dark matter fields can couple directly to GWID [60].

4 Conclusions

In the last years the LIGO-Virgo collaboration obtained several important scientific discoveries, such as the first direct detection of gravitational waves, the first evidences of BBH and BNS coalescences, the first evidence of an association between a BNS coalescence, short GRB and kilonova. The detection of a BNS coalescence opened the era of multi-messenger astronomy.

Several other sources are expected in the future, for example rotating neutron stars and supernovae. Event with a very high signal to noise ratio will become available, and will allow for better parameter estimation. For example with an improved high frequency sensitivity it will be possible to obtain severe constraints on the nuclear matter equation of state.

The detection of a cosmological stochastic background is probably something that we can expect in third generation detectors, but in the meantime for sure several upper limits and constraint on models will be obtained. A new way of looking at the universe is born.

References

1. É.É. Flanagan, S.A. Hughes, The basics of gravitational wave theory. New J. Phys. **7**(1), 204 (2005)
2. H.J. Kimble, Y. Levin, A.B. Matsko, K.S. Thorne, S.P. Vyatchanin, Conversion of conventional gravitational-wave interferometers into quantum nondemolition interferometers by modifying their input and/or output optics. Phys. Rev. D **65**, 022002 (2001)
3. J. Aasi et al., Enhancing the sensitivity of the LIGO gravitational wave detector by using squeezed states of light. Nature Photon **7**, 613–619 (2013)
4. A. Buonanno, Y. Chen, Scaling law in signal recycled laser-interferometer gravitational-wave detectors. Phys. Rev. D **67**, 062002 (2003)
5. F. Acernese et al., Status of advanced virgo. EPJ Web Conf. **182**, 02003 (2018)
6. J. Steinlechner, Development of mirror coatings for gravitational-wave detectors. Phil. Trans. Roy. Soc. Lond. A **376**(2120), 20170282 (2018)
7. T. Accadia et al., The seismic Superattenuators of the Virgo gravitational waves interferometer. Low Freq. Noise Vibrat. Act. Control **30**(1), 63–79 (2011)
8. J. Harms, Terrestrial gravity fluctuations. Living Rev. Rel. **22**(1), 6 (2019)
9. M. Maggiore, *Gravitational Waves: Volume 1: Theory and Experiments*. Gravitational Waves (OUP Oxford, 2008)
10. J. Aasi et al., Advanced ligo. Class. Quantum Gravity **32**(7), 074001 (2015)
11. F. Acernese et al., Advanced virgo: a second-generation interferometric gravitational wave detector. Class. Quantum Gravity **32**(2), 024001 (2014)
12. B.P. Abbott et al., GW150914: first results from the search for binary black hole coalescence with Advanced LIGO. Phys. Rev. D **93**(12), 122003 (2016)
13. B.P. Abbott et al., GW150914: the advanced LIGO detectors in the era of first discoveries. Phys. Rev. Lett. **116**(13), 131103 (2016)
14. B.P. Abbott et al., GW151226: observation of Gravitational Waves from a 22-Solar-Mass Binary Black Hole Coalescence. Phys. Rev. Lett. **116**(24), 241103 (2016)
15. B.P. Abbott et al., Binary black hole mergers in the first advanced LIGO observing run. Phys. Rev. X **6**(4), 041015 (2016). [Erratum: Phys. Rev. X **8**, 039903 (2018)]
16. B.P. Abbott et al., GW170104: observation of a 50-solar-mass binary black hole coalescence at redshift 0.2. Phys. Rev. Lett. **118**(22), 221101 (2017). [Erratum: Phys. Rev. Lett. **121**, 129901 (2018)]
17. B.P. Abbott et al., GW170814: a three-detector observation of gravitational waves from a binary black hole coalescence. Phys. Rev. Lett. **119**(14), 141101 (2017)
18. B.P. Abbott et al., Localization and broadband follow-up of the gravitational-wave transient GW150914. Astrophys. J. Lett. **826**(1), L13 (2016)
19. B.P. Abbott et al., GW170817: observation of gravitational waves from a binary neutron star inspiral. Phys. Rev. Lett. **119**(16), 161101 (2017)
20. B.P. Abbott et al., Multi-messenger observations of a binary neutron star merger. Astrophys. J. Lett. **848**(2), L12 (2017)

21. W.D. Pozzo, T.G.F. Li, M. Agathos, C. Van Den Broeck, S. Vitale, Demonstrating the feasibility of probing the neutron-star equation of state with second-generation gravitational-wave detectors. Phys. Rev. Lett. **111**, 071101 (2013)
22. B.P. Abbott et al., GW170817: measurements of neutron star radii and equation of state. Phys. Rev. Lett. **121**(16), 161101 (2018)
23. D.E. Holz, S.A. Hughes, B.F. Schutz, Measuring cosmic distances with standard sirens. Phys. Today **71**(12), 34–40 12 (2018)
24. B.P. Abbott et al., A gravitational-wave standard siren measurement of the Hubble constant. Nature **551**(7678), 85–88 (2017)
25. B.P. Abbott et al., A gravitational-wave measurement of the hubble constant following the second observing run of advanced LIGO and virgo. Astrophys. J. **909**(2), 218 (2021)
26. M. Soares-Santos et al., First measurement of the hubble constant from a dark standard siren using the dark energy survey galaxies and the LIGO/Virgo binary-black-hole merger GW170814. Astrophys. J. Lett. **876**(1), L7 (2019)
27. B.P. Abbott et al., Gravitational waves and gamma-rays from a binary neutron star merger: GW170817 and GRB 170817A. Astrophys. J. Lett. **848**(2), L13 (2017)
28. E. Pian et al., Spectroscopic identification of r-process nucleosynthesis in a double neutron star merger. Nature **551**, 67–70 (2017)
29. R. Abbott et al., GWTC-2: compact binary coalescences observed by LIGO and virgo during the first half of the third observing run. Phys. Rev. X **11**, 021053 (2021)
30. R. Abbott et al., GWTC-2.1: deep extended catalog of compact binary coalescences observed by LIGO and virgo during the first half of the third observing run, 8 (2021). arXiv:2108.01045
31. R. Abbott et al. GWTC-3: compact binary coalescences observed by LIGO and virgo during the second part of the third observing run, 11 (2021). arXiv:2111.03606
32. A.H. Nitz, C.D. Capano, S. Kumar, Y.-F. Wang, S. Kastha, M. Schäfer, R. Dhurkunde, M. Cabero, 3-ogc: Catalog of gravitational waves from compact-binary mergers. Astrophys. J. **922**(1), 76 (2021)
33. R. Abbott et al., Population of merging compact binaries inferred using gravitational waves through GWTC-3. Phys. Rev. X **13**(1), 011048 (2023)
34. B.P. Abbott et al., Tests of general relativity with GW150914. Phys. Rev. Lett. **116**(22), 221101 (2016). [Erratum: Phys. Rev. Lett. **121**, 129902 (2018)]
35. B.P. Abbott et al., Tests of general relativity with GW170817. Phys. Rev. Lett. **123**(1), 011102 (2019)
36. B.P. Abbott et al., Tests of general relativity with the binary black hole signals from the LIGO-Virgo catalog GWTC-1. Phys. Rev. D **100**(10), 104036 (2019)
37. R. Abbott et al., Tests of general relativity with binary black holes from the second LIGO-Virgo gravitational-wave transient catalog. Phys. Rev. D **103**(12), 122002 (2021)
38. R. Abbott et al. Tests of general relativity with GWTC-3, 12 (2021). arXiv:2112.06861
39. R. Abbott et al. GW190814: gravitational waves from the coalescence of a 23 solar mass black hole with a 2.6 solar mass compact object. Astrophys. J. **896**(2), L44 (2020)
40. R. Abbott et al., GW190521: a binary black hole merger with a total mass of 150 $M_{\odot}$. Phys. Rev. Lett. **125**, 101102 (2020)
41. R. Abbott et al., Properties and astrophysical implications of the 150 Msun binary black hole merger GW190521. Astrophys. J. Lett. **900**, L13 (2020)
42. R. Abbott et al., Observation of gravitational waves from two neutron star-black hole coalescences. Astrophys. J. Lett. **915**, L5 (2021)
43. F. Foucart, A brief overview of black hole-neutron star mergers (2020)
44. R. Abbott et al., Search for lensing signatures in the gravitational-wave observations from the first half of LIGO-virgo's third observing run. Astrophys. J. **923**(1), 14 (2021)
45. R. Abbott et al. Search for gravitational-lensing signatures in the full third observing run of the LIGO-Virgo network, 4 (2023). arXiv:2304.08393
46. R. Abbott et al., All-sky search for short gravitational-wave bursts in the third Advanced LIGO and Advanced Virgo run. Phys. Rev. D **104**(12), 122004 (2021)

47. R. Abbott et al., All-sky search for long-duration gravitational-wave bursts in the third Advanced LIGO and Advanced Virgo run. Phys. Rev. D **104**(10), 102001 (2021)
48. R. Abbott et al., Diving below the spin-down limit: constraints on gravitational waves from the energetic young pulsar PSR J0537–6910. Astrophys. J. **913**, L27 (2021)
49. B.P. Abbott et al., All-sky search for short gravitational-wave bursts in the second advanced LIGO and advanced virgo run. Phys. Rev. D **100**(2), 024017 (2019)
50. B.P. Abbott et al., All-sky search for continuous gravitational waves from isolated neutron stars using advanced LIGO O2 data. Phys. Rev. D **100**(2), 024004 (2019)
51. R. Abbott et al., All-sky search in early O3 LIGO data for continuous gravitational-wave signals from unknown neutron stars in binary systems. Phys. Rev. D **103**, 064017 (2021)
52. R. Abbott et al., All-sky search for continuous gravitational waves from isolated neutron stars using advanced LIGO and advanced virgo O3 data. Phys. Rev. D **106**(10), 102008 (2022)
53. B.P. Abbott et al., GW170817: implications for the stochastic gravitational-wave background from compact binary coalescences. Phys. Rev. Lett. **120**(9), 091101 (2018)
54. C. Caprini, Stochastic background of gravitational waves from cosmological sources. J. Phys: Conf. Ser. **610**(1), 012004 (2015). (Apr)
55. P.D. Lasky, C.M.F. Mingarelli, T.L. Smith, J.T. Giblin, E. Thrane, D.J. Reardon, R. Caldwell, M. Bailes, N.D. Ramesh Bhat, S. Burke-Spolaor, S. Dai, J. Dempsey, G. Hobbs, M. Kerr, Y. Levin, R.N. Manchester, S. Osłowski, V. Ravi, P.A. Rosado, R.M. Shannon, R. Spiewak, W. van Straten, L. Toomey, J. Wang, L. Wen, X. You, X. Zhu, Gravitational-wave cosmology across 29 decades in frequency. Phys. Rev. X **6**, 011035 (2016)
56. B.P. Abbott et al., Search for the isotropic stochastic background using data from advanced LIGO's second observing run. Phys. Rev. D **100**(6), 061101 (2019)
57. R. Abbott et al., All-sky search for gravitational wave emission from scalar boson clouds around spinning black holes in LIGO O3 data. Phys. Rev. D **105**(10), 102001 (2022)
58. R. Abbott et al., Search for subsolar-mass black hole binaries in the second part of advanced LIGO's and advanced Virgo's third observing run 12 (2022)
59. J.C. Bustillo, N. Sanchis-Gual, A. Torres-Forné, J.A. Font, A. Vajpeyi, R. Smith, C. Herdeiro, E. Radu, S.H.W. Leong, Gw190521 as a merger of proca stars: a potential new vector boson of 8.7×10^{-13} eV. Phys. Rev. Lett. **126**, 081101 (2021)
60. R. Abbott et al., Constraints on dark photon dark matter using data from LIGO's and Virgo's third observing run. Phys. Rev. D **105**(6), 063030 (2022)

Astrophysics of Radio Sources: Analysis of 21 cm Post-reionization Signals and Modeling of FRBs as PBHs with Magnetic Fields

Alessandro Marins, Elcio Abdalla, Jordany Vieira, Lucas Formigari, Filipe Abdalla, Amilcar R. Queiroz, and Bin Wang

Abstract Between the scientist's theorizing and the Universe, there is the human codified data. Actually, data is the representation of the communication between nature and human beings through a certain type of view and information. The hard task of interpreting and understanding what the data means requires a huge and collective effort from the scientific community in different fields. Sometimes, a new scientific discovery comes from a theoretical prediction and needs a real observation to validate it. However, it frequently happens that new discoveries come from incidental observations with instruments developed for other goals. In the latter case, more information with new and independent observations should be needed to deeply investigate this possible new phenomenon and its theoretical description. Radioastronomy, since its origin, has been a fruitful field with a mix-up of theoretical and observational new discoveries. It happened, for instance, for the Fast Radio Bursts discovery through pulsar investigations. Hence, versatile and accurate radiotelescopes have been operating or are under development to be able to map huge sky swathes

A. Marins (✉)
Department of Astronomy, School of Physical Science, University of Science and Technology of China, Hefei, China
e-mail: amarins@ustc.edu.cn

A. Marins · E. Abdalla · J. Vieira · L. Formigari · F. Abdalla
Departamento de Física Geral, Instituto de Física, Universidade de São Paulo, São Paulo, Brazil

F. Abdalla
Department of Physics and Astronomy, University College London, Gower Street, London WC1E 6BT, UK

A. R. Queiroz
Unidade Acadêmica de Física, Universidade Federal de Campina Grande, R. Aprígio Veloso, Campina Grande 58429-900, Brazil

B. Wang
School of Aeronautics and Astronautics, Shanghai Jiao Tong University, Shanghai 200240, China

Center for Gravitation and Cosmology, Yangzhou University, Yangzhou 224009, China

© The Author(s), under exclusive license to Springer Nature Switzerland AG 2024

E. Papantonopoulos and N. Mavromatos (eds.), *Compact Objects in the Universe*,
https://doi.org/10.1007/978-3-031-55098-0_10

for large-scale investigations at the same as very precise pinpointing in the sky for small-scale searches; or else, different projects to be able to operate in interferometric observations composing as large as possible baseline with earth-dimensions. In this work, we will focus on describing briefly different types of astrophysical information that are observed when a radiotelescope operates in a 21-cm intensity mapping on large scales and how we can handle the images to extract the hidden 21-cm signal. In the last part of this work, we will interested in providing a new candidate to describe Fast Radio Bursts and their magnetic characteristic in addition to one most currently accepted and likely magnetar nature.

1 Introduction

Almost a century after Karl Jansky discovered for the first time a sky radio signal outside of Earth, today's radio astronomy can be seen as a mosaic of discoveries and opportunities. Being one of the most ancient frequency bands explored to observe the sky, recently, radio observations have suffered an increasing development, allowing us to achieve state-of-art to explore the Universe from the ground. Projects around the world have different and agreeing sky-covered, observational methods, angular precision, and/or scientific goals, pointing towards synergistic and complementary projects. A radio telescope can be used to explore different sciences, from nature, morphology, and the environment of transient sources to the birth of the first lights in the Universe.

Regarding the Universe's evolution, one of the most challenging and promising observational tools is to use the well-known line of emissions. Mainly due to the abundance of hydrogen after the radiation-matter decoupling, hydrogen line emissions turn into an important way to explore the Universe in low and high redshifts. The 21 cm emission from the neutral hydrogen (HI) can be used to understand astrophysical phenomena and the Universe on large scales, even being a weak signal. However, the effective use of 21 cm emission to explore most parts of the Universe's history and composition depends on the efficiency in the (pre- and) post-processing of the observational data. An inaccurate treatment of the raw observational data leads to an impossibility in the detection of the paramount signatures on the underlying matter as, e.g., the baryonic acoustic oscillations (BAO) [19, 23] through the large scale observations, it would render the use of 21 cm emissions inefficient in helping to explain many important puzzles of the Universe. Therefore, it is mandatory to ensure the viability of the project of detecting the emission and improving the current scientific scenery. Simulations and tests before a project starts, should drive an important part of the project's workforce. For 21 cm intensity mapping projects, it is necessary to simulate the astrophysical, cosmological, and instrumental signals and effects until we achieve good-quality data.

Radio telescopes operating in an Intensity Mapping (IM) approach do not have an interest in taking individual source pinpointing but explore a high survey speed in an integrated collecting signal across an as large as possible swath in the sky. This results in observed data composed of different types of sources being observed in the same frequency range of operating. Then, pre-processing data needed to be handled with respect to the first data treatments in the time domain, calibrating the gain, removing the radio frequency interference (RFI), masking known strong point sources, converting the data from an analogic to digital description until the map-making step. The final map obtained is the representation of the radio telescope's observation in a given (range of) frequency. After that, the next and essential part is the 21 cm post-processing data, aiming at the accuracy of the 21 cm signal reconstructions. This part of the processing will be in charge of extracting the 21 cm emissions from their mix-up with other contaminants and observational effects.

There are basically three ways to lead with this part: assuming parametric modeling for each one of the signals and instrumental effects and then fitting them by some method such as Bayesian; *blind* methods, assuming no knowledge of the astrophysical and cosmological sources and a hybrid of both, modeling known sources and allowing a fit of unknown sources. Here, we will be concentrating on the blind method. Although the efficiency of the blind methods for 21 cm reconstruction has been proved in several works [24, 37, 45], there is no perfect method, and leftovers from inaccuracy in modeling always leak into the final reconstructed data.

Signals coming from Gaussian fields, as neutral hydrogen is expected to follow, are completely characterized by their power spectrum. Higher-order statistical analysis of the data can provide signatures of deviations from the Gaussian fields, opening room for new physics of the primordial Universe. Furthermore, higher-order statistical analysis can be used to identify the inaccuracy in modeling the target signals. This is because of the different levels of non-Gaussianities from the signals composing the observed one. The first order that is higher than the power spectrum, bispectrum, has been a growing interest in radio projects, and its use to improve the understanding of the underlying matter of the Universe and to enlarge the signal reconstruction could contain a richness in information.

On the other hand, projects such as BINGO [1, 21], CHIME [13, 14], FAST [40, 53, 56], and Tianlai [17, 52] are some projects involved to search more than 21 cm IM but fast and transient radio sources, mainly pulsars and *Fast Radio Bursts* (FRBs), mainly due to their large observation time and higher time resolution. Highlighted, FRBs are extremely bright unknown-origin bursts like pulsars with higher frequency-dependent dispersion delay and flux density. Characterized by its very short pulse duration, huge brightness, wide broadband coverage, and large dispersion measure, it was first found in the Parkes 64-m telescope data and archives [29, 34, 47].

Currently, a new generation of radiotelescopes is in operation or under development to investigate the signatures, characteristics, morphologies, emission mechanisms, and environments in which this astrophysical phenomenon is involved. Even though the number of identified FRBs and their host galaxies have increased quickly—skipping more than ten times in detection after CHIME started its observations—there is even a tiny sampling of observations[1] to towards us a bet-

[1] To date, using the Transient Name Sever, 805 FRBs in total with 206 been repeaters and 45 host galaxy redshifts identified.

ter understanding about its origin(s). As an example, it has been found that FRB repeaters are usually in star-forming regions in their host galaxies, whereas non-repeaters seem to be more common in regions of very low rate of star-forming [41]. Nevertheless, there are exceptions to these two behaviors. As much FRB 20200129E (an extragalactic and repeater FRB) as SGR 1935+2154 (a Galactic magnetar) points to a possible magnetic neutron star nature of the FRBs, it is not clear yet if the bursts have a unique mechanism of emission, and some works evoke that only magnetars are not enough to explain all observations thus far (see [6]). The lack of a significative sampling must be gradually overcome in the upcoming years with the current and new projects developed or adapted to observe all-sky in different frequency regimes. Conversely, a new and complementary candidate to describe FRB's magnetic feature rises up when we assume a primordial black hole with a magnetic field. This interesting candidate comes up from the Gutsannaev-Manko's solution [26] describing a mass with a magnetic dipole.

In this work, our desire is to give a brief discussion about some fascinating ways to handle radio signals for recovering 21-cm signals from far distances in the Universe and how different and undesirable astrophysical mechanisms of radio emission can hide the target information. In addition, we will be discussing a new candidate to describe the FRBs and their magnetic characteristics.

2 Extracting HI Information from Sky Maps

The HI emission from the post-reionization epoch has an intensity that is approximately four orders of magnitude lower than the strongest Galactic emission in the same frequency range [37]. Then, multiresolution maps of the observational intensity, that is, images of the intensity in different frequency channels, do not show the 21 cm intensity emission patterns over the covered sky, though they are there. The major part of the emissions shown is from the Galaxy and strong unresolved sources such as other galaxies (or even clusters and superclusters of galaxies) with observed sizes smaller than the telescope beam resolution.

In a *blind* perspective in relation to the sources of emissions, we do not specify or model any source but use some statistical information to split between what we want and what we do not. In general, because of the mechanisms of emission, it is possible to model the foreground intensity as a law of the frequency, which means that the intensity has a smooth spectral behavior for different directions in the sky [18, 39, 44]. For the radio frequency range, it is useful to use (brightness) temperature rather than intensity through the Rayleight-Jeans relation. Therefore, the foreground temperatures will be described as some potency of the frequency.

These smooth behaviors of the foregrounds are distinct from the expected (and often simulated) 21 cm intensity—or brightness temperature. In the standard and current most accepted cosmological model description, there was a beginning inflationary period of the Universe when quantum fluctuations resulted in the Gaussian distribution of the total matter as presented by today's observations [42]. In this way, if we point out to and observe a random position in the sky for different frequencies, the 21 cm signal will be expected to follow a Gaussian distribution, or, at least, almost one. This later statement is because either smooth deviations from the standard inflationary models or local mechanisms of emission can appear in statistical analysis.

To be able to use some algorithm of the blind type in order to estimate the desired and undesired signals, it is necessary to explore patterns of the signals. One possible procedure is by splitting spectral and spatial (or any other information in another mathematical space) behaviors. In general, the spectral information for each sky's direction is commonly used to explore and distinguish sources of different mechanisms of emission.

First of all, it is necessary to model the observational data for each direction $\hat{n}$ in the sky for each frequency (or channel) ν. The brightness temperature is expressed by

$$T_{\text{obs}}(\nu, \hat{n}) = \int \mathrm{d}\Omega_{\hat{n},\hat{n}'} b\left(\nu, \hat{n}; \hat{n}'\right) T_{\text{sky}}\left(\nu, \hat{n}'\right) + T_{\nu,\text{instr}}(\hat{n}), \qquad (1)$$

with b the beam profile of the radiotelescope and $\Omega_{\hat{n},\hat{n}'}$ a solid angle formed between $\hat{n}$ and another direction $\hat{n}'$ to consider the beam effect over the observation. The data collected from the sky is expressed by T_{sky} and the instrumental signal contribution by T_{instr}. For the T_{sky} modeling, we will assume 21 cm emissions and a collection of other important sky emissions for the BINGO frequency range; that is, $T_{\text{sky}} = T_{\text{FG}} + T_{21\,\text{cm}}$, where FG its a short for *foregrounds* and condense all considered important emissions.

2.1 Caracterizing the 21 cm Post-reionization Signals

After reionization is complete ($z \sim 6$), it is only possible to have HI—consequently 21 cm emissions—because there are certain extremely dense systems called *Damped Lyman-α Absorption systems* (DLAs) [50]—very well characterized in the high-z quasar spectra—that protects HI leftovers from background ionizing radiation. The DLAs—and consequently the HI—are mostly distributed over dark halos with masses within the range $10^9 - 10^{12} M_{\odot} h^{-1}$. Hydrodynamic simulations performed in [49, 55] showed that for $z < 3$, basically 90% of the HI is distributed in dark halos, and reportedly this percentage achieves 99% for $z \approx 0$. For this reason, we can use HI as a tracer of the total matter. We express the quantity of HI through the parameter density $\Omega_{\text{HI}}(z) = 8\pi G m_{\text{HI}} n_{\text{HI}}(z)/3H_0^2$, with m_{HI}, n_{HI}, H_0, and G be the HI atom mass, the density number of HI atoms, the Hubble constant, and the gravitational constant, respectively.

Following [37], the observed (average) 21 cm brightness temperature at low redshifts is described by

$$T_{21\,\mathrm{cm}}(z) = \left(\frac{9hc^3 A_{10}}{256\pi^2 G k_{\mathrm{B}} \nu_{10}^2 m_{\mathrm{HI}}}\right) \frac{\Omega_{\mathrm{HI}}(z)}{(1+z)^2} \frac{H_0^2}{\|dv_{\|}/d\chi\|} \tag{2}$$

where A_{10}, c, and h are the spontaneous emission coefficient of the 21 cm transition, speed of light, and Planck (fundamental) constants, respectively. $\|dv_{\|}/d\chi\|$ is the gradient of the specific velocity field along the line of sight. In comoving coordinates $\chi = a^{-1}s$, with the scale factor $a = a(z)$.

It can be observationally proved in two main types or from the global signal measured by taking the average of the signal within the observed sky region, which can be thought of as the simplest and straightforward observation way for handling the weakness of the signal,

$$\bar{T}_{21\,\mathrm{cm}}(\nu) = \int \frac{\mathrm{d}\Omega(\hat{\mathbf{n}})}{4\pi} T_{21\,\mathrm{cm}}(\hat{\mathbf{n}}, \nu),$$

either from the spatial signal fluctuations. This later is because the 21 cm brightness temperature observed can be around its average value because different mechanisms along the line of sight caused by changes in the mean HI distribution, which, to first order, can be described as $n_{\mathrm{HI}}(z, \hat{n}) = \overline{n}_{\mathrm{HI}}(z)\left(1 + \delta_{\mathrm{HI}}(z, \hat{n})\right)$ in the $\hat{n}$ direction. Consequently, the first-order correction of the brightness temperature will be given by $T_{21\,\mathrm{cm}}(z, \hat{n}) = \overline{T}_{21\,\mathrm{cm}}(z)\left(1 + \delta T_{21\,\mathrm{cm}}(z, \hat{n})\right)$. The (first-order) perturbation term in the conformal Newtonian gauge is a composition of the different physical effects, as described in [16, 27],

$$\delta T_{21\,\mathrm{cm}}(z, \hat{n}) = \delta_{\mathrm{n}} - \frac{1}{\mathcal{H}}\hat{\mathbf{n}} \cdot (\hat{\mathbf{n}} \cdot \nabla \hat{\mathbf{v}}) + \left(\frac{\mathrm{d}}{\mathrm{d}\eta} \ln\left(a^3 \bar{\mathrm{n}}_{\mathrm{HI}}\right) - \frac{\dot{\mathcal{H}}}{\mathcal{H}} - 2\mathcal{H}\right)\delta\eta + \frac{1}{\mathcal{H}}\dot{\Phi} + \Psi. \tag{3}$$

The function $\mathcal{H}$ is the Hubble function in the conformal time. The first term to the right of the above equality expresses the contribution from perturbation of the HI density, and the second term is related to the *redshift space distortion* from peculiar velocities of the sources. The third term between parenthesis is the zero-order brightness temperature calculated at the perturbed conformal time η of the observed redshift. The fourth and fifth terms come from the Newtonian gauge, where Ψ is the gravitational potential, and Φ is a local perturbation of the average scale factor. The first of the final term on the Eq. 3 is associated with the integrated Sachs-Wolf effect, and the later term is related to the conversion between increments in redshift to radial distances in the gas frame.

2.2 Foreground Signals

The frequency range to be operated by the BINGO project is low compared to the other projects, such as, for instance, the Planck [43],[2] with the main astrophysical and cosmological sources—other than 21 cm and often called generically as *foregrounds*—can be described in two groups: diffuse Galactic emissions and extragalactic emissions. In a general description, Galactic emissions come from the interstellar medium (ISM) of the Milky Way, which in turn, are constituted by cold molecular and atomic clouds, a partially ionized medium between the clouds, and presumably ionized regions formed by supernovae. These regions are strongly concentrated on the Galactic plane and content three important mechanisms of emission contributing to the BINGO observations: *synchrotron*, *free-free*, and *anomalous microwave emission*.

Galactic synchrotron emissions are non-thermic emissions by charged particles, typically electrons and positrons, spiraling in the galactic magnetic field [39]. For the Milky Way, this type of emission is highly concentrated in the Galactic plane as well as has really known loops of Galactic magnetic field where there are high rates of emissions [20, 48]. The Galactic synchrotron emission can be seen in Fig. 1 looking at the top-middle map, where it is clear the *North Polar Loop I* on the right of the figure is a behavior like an arc. In comparison with the summed-up foreground emission represented by the map on the top-left, we can note that the major contribution to the total foreground emission came exactly from the synchrotron mechanism at the Milky Way.

Free-free emissions, also known as *thermic bremsstrahlung*, are due to the electron-ion scattering in an ionized part of the Galactic ISM, and it has a strong presence in star-forming regions. This emission is higher in environments with temperatures higher than the bound hydrogen energy. In Fig. 2, we can see the free-free map for the range 1053–1063 MHz in galactic coordinates. This figure has highlighted two easily identifiable structures that are known as the Large and Small Magellanic Clouds (LMC and SMC, respectively). Even though these structures have strong free-free emission, the synchrotron contribution by both is still very strong, as can be seen in Fig. 3. In it, there is a comparison of the same region, with right ascension between –120° and –50°, and declination between –60° and 0°. The first image of the figure, on the left, is all foreground components' contribution, the second (middle image) represents the total emission of the foregrounds extracted from the free-free emission, and the last one represents only the free-free emission in the region. All images have the same intensity gradation. Therefore, it is possible to see that the biggest contribution of free-free in the total composition of the foregrounds is in LMC.

Anomalous Microwave Emissions (AME) are emissions first observed in CMB data in the 1990s and correlated with free-free emissions [30, 31]. The most accepted description today follows the Draine-Lazarian model [22] such that AME is a non-

[2] Planck satellite operated in nine channels between 30–857 GHz.

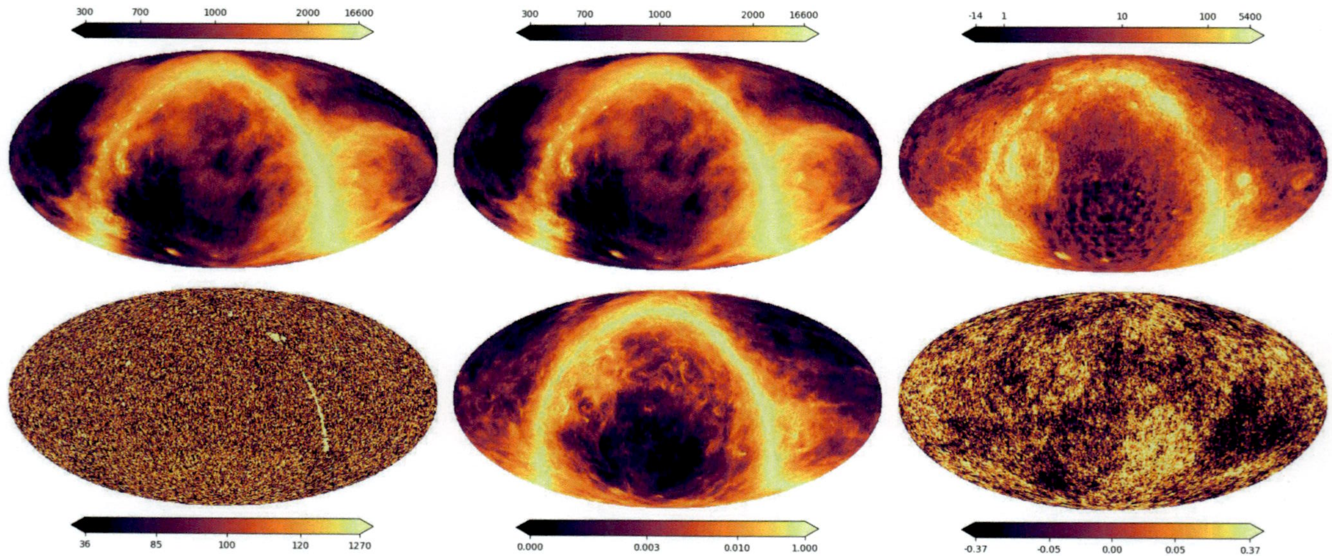

Fig. 1 Mollview projection of the foreground emissions through the HEALPix pixelization [25]. All brightness temperature maps are in units of mK and in celestial coordinates within the range 1053–1063 MHz. From top-left to bottom-right, the maps are: all foregrounds emission together, synchrotron, free-free, point-source, AME, and CMB. From [36]

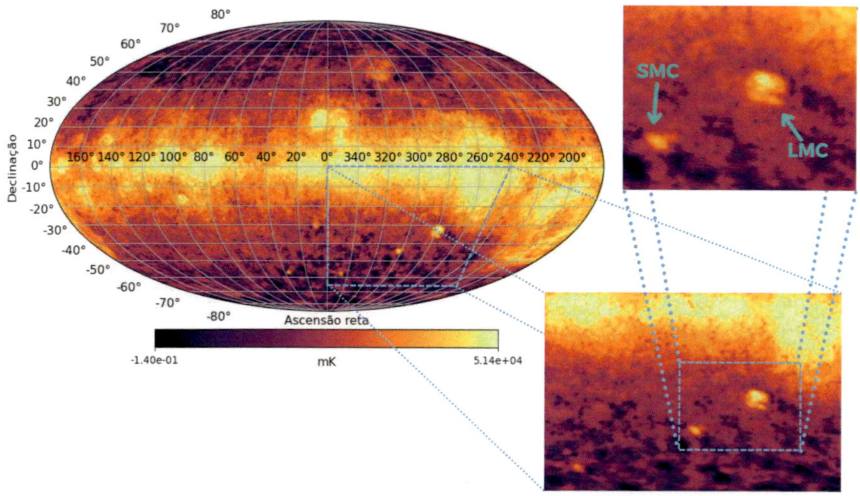

Fig. 2 Free-free emissions in galactic coordinates through the graduation color scale and expressing in declination and right ascension coordinates to focus on *Small Magellanic Cloud* (SMC) and *Large Magellanic Cloud* (LMC)

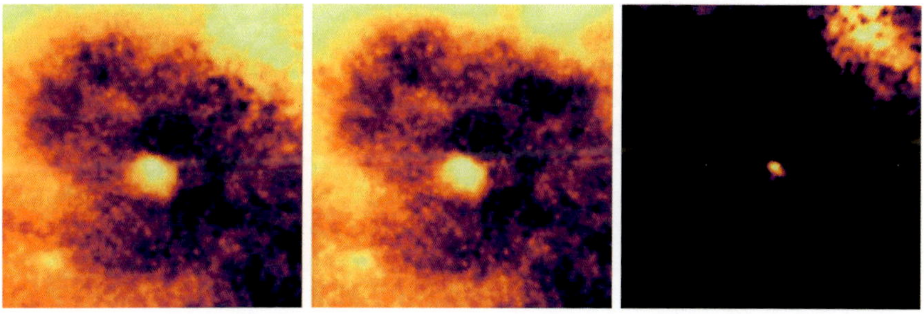

Fig. 3 Free-free emissions for the region with right ascension between –120° and –50°, and declination between –60° and 0°. The image on the left is the representation of the summed emission of all foreground components, the middle image is the total emission subtracted from the free-free emission, and the image on the right is the representation of only free-free emission. All images are with the scale limits of intensity, between $\sim 3.0 \times 10^2 - 1.6 \times 10^5$ mK. From [36]

thermic dust emission composed of spinning nanograins in the ISM. This rotation is to align the magnetic (or electric) dipole with the galactic magnetic field and has a total emission power described by the Larmor formula. We will not advance this topic and models trying to explain the origin of these grains, which is a vast and fruitful area of astrophysics and astrochemistry. We constrain us to comment that there is some explanation showing the viability and compatibility of their data with models based on silicate grains for different chemical compositions ($Mg_{1.48}Fe_{0.32}SiO_{3.79}$ as studied in [28]; and, SiO, Mg_2SiO_4, and $MgSiO_3$ in [35]).

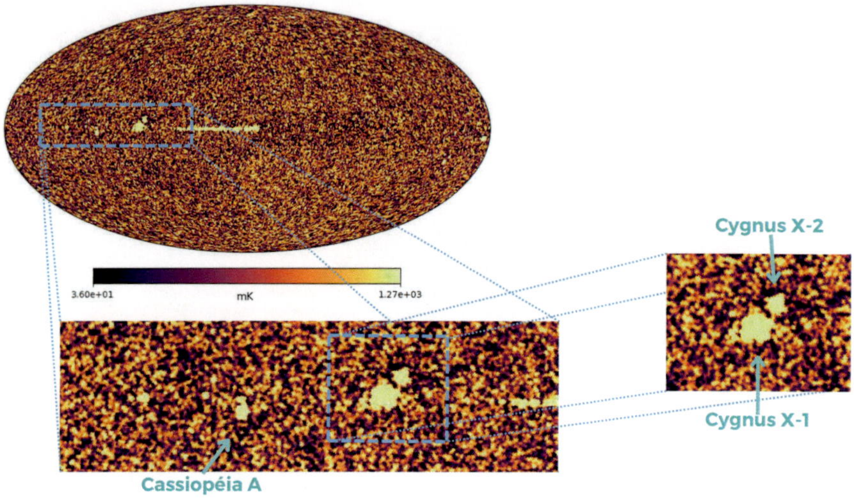

Fig. 4 Template representing both extragalactic and Galactic unresolved radio point sources within 1053–1063 MHz. In blue color, there is the identification of the highlighted regions. For instance, Cassiopea A, Cygnus X-1 e Cygnus X-2. On the right of the image, there is a distinction between Cygnus X-1 and X-2

Extragalactic emissions in radiofrequency are mainly emitted from active galactic nuclei (AGN), from star-forming galaxies (SFG) and are also composed of the cosmic microwave background (CMB). AGN emits synchrotron radiation from matter accretion by its central supermassive black holes, ejecting it in a beam perpendicular to the accretion plan. SFG produces synchrotron emissions like our galaxy and free-free from regions of ionized hydrogen [11].

Lastly, it is important to consider the CMB's influence on the radiofrequency regime. CMB is a radiation from electron-photon decoupling in the primordial Universe with an almost isotropic blackbody spectrum current with a mean thermodynamic temperature of 2.725 K [43]. The CMB thermodynamic and brightness temperature anisotropies are related by the expression

$$\delta T\left(\nu, p\right) \doteq \delta T^{\mathrm{B}}_{\mathrm{CMB}}\left(\nu, p\right) = \frac{x^2 e^x}{\left(e^x - 1\right)^2}\delta T^{\mathrm{th}}_{\mathrm{CMB}}\left(p\right), \quad \text{with } x = \frac{h\nu}{k_B T_{\mathrm{CMB}}}. \tag{4}$$

CMB emissions are easily obtained from open codes as CAMB [33] and CLASS [32].

For the BINGO beam resolution, several unresolved radio sources can contribute to the same localization observed in the sky. This leads to the existence of different source mechanisms contributing in the same direction toward the sky, which drives us to complex spectral descriptions. We can model these contributions by splitting them into two groups based on the specific intensity through a power law, $I_{\mathrm{ps}}(\nu) \sim \nu^{-\alpha}$ ($T_{\mathrm{ps}}(\nu) \sim \nu^{-\alpha-2}$): the first group is composed by sources following a steep spectral behavior ($\alpha > 0.5$) and the other group a smooth one ($\alpha < 0.5$) [5]. The spectrum must have a strong variation in the sky with a non-trivial description. The total brightness temperature of the radio point sources is obtained through a numerical distribution of the point sources (per solid angle) per specific intensity unit, $\mathrm{d}N/\mathrm{d}I_{\mathrm{ps}}$, such that $\int \mathrm{d}I_{\mathrm{ps}}(\mathrm{d}N/\mathrm{d}I_{\mathrm{ps}})I_{\mathrm{ps}}$ condense the contribution from several sources at the same sky positions. Thus, limiting the contributions up to a given maximum intensity (or flux),[3] $I_{\mathrm{ps}}^{\mathrm{max}}$, the brightness temperature for point sources can be approximated by

$$\bar{T}_{\mathrm{ps}}(\nu) \sim \nu^{-2} \int_0^{I_{\mathrm{ps}}^{\mathrm{max}}} \mathrm{d}I_{\mathrm{ps}} \frac{\mathrm{d}N}{\mathrm{d}I_{\mathrm{ps}}} I_{\mathrm{ps}}.$$

The expression above condenses a very important idea of trying to reconcile the different contributions in different frequencies. Very bright objects, such that their flux (or intensity) is greater than a certain upper limit, are seen as strong, compact sources of radio emission, whether they are of Galactic or extragalactic origin. The vast majority of the brightest fonts can be identified individually and looked up in catalogs. As there is no resolution, the other sources will make up a diffuse background composed of countless unresolved sources, such that the integrated contribution has a large effect on the data. In Fig. 4, we have a representation of point source emissions. In that figure, it is easy to identify in the central part (around the declination of 0°) stronger emission sources, with visual highlights for the cases: Cassiopeia A, which is a supernova remnant; Cygnus X-1, one of the most famous known black holes which led to the bet between Stephen Hawking and Kip Thorpe between the 70 s and 90 s about its real nature; and also Cygnus X-2, which is a low-mass X-ray binary.

2.3 Modeling the Foreground Removal Process

In the previous section, we characterized the most important (knowledge) sky emissions from the Universe to be considered by the BINGO project. Returning to Eq. 1, we can now split the sky brightness temperature into two distinct groups: the sum of the foregrounds and 21 cm,

$$T_{\mathrm{obs}}(\nu, \hat{n}) = \int \mathrm{d}\Omega_{\hat{n},\hat{n}'} b\left(\nu, \hat{n}; \hat{n}'\right) \left(\sum_j T_j^{\mathrm{FG}}\left(\nu, \hat{n}'\right) + T_{\nu,.21\,\mathrm{cm}}\left(\nu, \hat{n}'\right) \right) + T_{\mathrm{instr}}(\hat{n}). \tag{5}$$

[3] This uppercut in contributions can be thought of in a practical way as being such that, above it, the sources are so strong that they can be identified and extracted already in the pre-processing of the data, or even extracted from catalogs already available.

In a first approximation, to extract overall information and understand the process of recovering 21 cm information, we will assume that the BINGO beams follow a Gaussian shape, an axisymmetric function $b(\nu, \hat{n} - \hat{n}') = b(\nu, \|\hat{n} - \hat{n}'\|) = b(\nu, \sin\theta) \approx b(\nu, \theta)$,

$$b\left(\nu, \theta'\right) = e^{-4\ln 2\left(\theta'/\theta_{\text{FWHM}}(\nu)\right)^2}. \tag{6}$$

The θ_{FWHM} is the *Full Width at Half Maximum* (FWHM) value, and here we will assume it as 40 arcmin.[4] In practice, we can represent the observed map as the sum of the foreground map convolved with the beam, plus the 21 cm map also convolved with the beam, and then add the instrumental effects as an independent map. In addition, we do not have a continuous representation but a pixelized one, where the pointing direction $\hat{n}$ will be associated with the center of a pixel p. For the benefit of a general and easier description, we can rewrite each jth foreground pixelized and convolved map as y_j, the 21 cm map one as $x_{21\text{cm}}$, maps representing instrumental noise as n_{instr}, and the observed map as x. Considering N_S foreground components and because the blind methods only model foreground emissions, we will join 21 cm and instrumental noises in the same representation n $(= x_{21\,\text{cm}} + n_{\text{instr}})$ calling this simply as a (method) *residue*, as

$$x\left(\nu_i, p\right) = \sum_{j=1}^{N_S} y_j\left(\nu_i, p\right) + n\left(\nu_i, p\right) \quad . \tag{7}$$

The above expression can be rewritten in an even more convenient form exploring the fact that the foregrounds have smooth spectral information for each sky direction p. Thus, each $y_j\left(\nu_i, p\right)$ can be represented by a product of spectral law behavior at p-direction $a_j(\nu_i, p) = a_{ij}(p)$ with a spatial response at the same direction $s_j(p)$, that is, $y_j\left(\nu_i, p\right) = a_{ij}\left(p\right) s_j\left(p\right)$. If we express the different observations with the same frequency ν_i as $\mathbf{x}\left(\nu_i\right) = \left[x\left(\nu_i, p_1\right), \ldots, x\left(\nu_i, p_k\right), \ldots\right]^{\text{T}}$, this vector covers all positions in the sky. Each such vector corresponds to a HEALPix map [25], with m pixels, such that we can build an $\mathbb{X}$ matrix with each row corresponding to a map produced from a different frequency (or channel), that is, $\mathbb{X} = \left[\mathbf{x}\left(\nu_0\right), \mathbf{x}\left(\nu_1\right), \ldots, \mathbf{x}\left(\nu_{m-1}\right)\right]^{\text{T}}$. Taking the same idea for the foreground and residual contributions, we can write[5]

$$\mathbb{X} = \mathbb{A}\mathbb{S} + \mathbb{N}, \tag{8}$$

---

[4] This value of FWHM is a rough approximation for the BINGO beams. This is because the telescope is composed of 28 feed horns, each one in a specific location at the focal plane. Feed horns closer to the central region of the focal plane are really well described by a Gaussian shape; however, the further away the feed horn is, the greater the effects of aberrations and anisotropy. The first analysis of the BINGO beams can be seen in [3], and a complete analysis has been done and will be published soon [2].

[5] Note that $\mathbb{N} = \mathbb{X}_{21\text{cm}} + \mathbb{N}_{\text{Inst}}$, with multichannel data matrices for 21 cm and instrumental noise.

where $\mathbb{S}$ is called the *spatial response* and $\mathbb{A}$ the *mixing matrix*. As we previously mentioned, the former has in each row the spatial response of each foreground component, and the latter has each column corresponding to a spectral law.

The following step is handled with an inverse problem as above, trying to reconstruct the foreground maps from the mixing matrix estimated $\hat{\mathbb{A}}$. There are different methods to do that, and each one explores a way and a mathematical space (for instance, real, harmonic, wavelets, etc.). Here, we show results using GMCA algorithm [7, 8] in the starlet space [46], but there are many other interesting and efficient algorithms [37, 38, 45]. Generalized Morphological Component Analysis (GMCA) is an algorithm based on *morphological diversity* and *sparsity*. It uses the sparsity idea for signals as a linear instantaneous mixture from different sources through overcomplete dictionaries[6] for representation. The method takes advantage of the sparse representation to separate features based on their morphologies. For this work, we will use the dictionary called starlet[7] [46]. Starlet is well located in both harmonic and pixel domains, which is a known spherical wavelet transform in astronomical image analysis since it explores the features of astrophysical objects that are more or less isotropic (in most cases).

As the algorithm outcome, it is possible to build a *Moore-Penrose pseudoinverse*, $\mathbb{W} = (\hat{\mathbb{A}}^{\mathrm{T}}\hat{\mathbb{A}})^{-1}\hat{\mathbb{A}}^{\mathrm{T}}$, and then to obtain the estimation of the foreground maps, $\hat{\mathbb{X}}_{\mathrm{FG}} \doteq \hat{\mathbb{Y}} = (\hat{\mathbb{A}}\mathbb{W})\mathbb{X}$, and for the residual $\mathbb{R} = \mathbb{X} - \hat{\mathbb{X}}_{\mathrm{FG}}$.

Figure 5 shows the GMCA outcomes for one realization. The plots are for HEALPix resolution of NSIDE 256 ($\sim$ 14 arcmin) center at 0° and –13.4°. In the first row, the first plot on the left is the residue $\mathbb{R}$, which in the best case would be only 21 cm and noise. However, there are more than these two contributions, as we can see with the fourth plot, which is plotted $\mathbb{R} - \mathbb{N}$, and we can compare with the second plot, which is only the 21 cm emission used as input. Then, we can identify the 21 cm emission pattern, but there is more than 21 cm and noise (here, only white noise is assumed). The additional leftover blurring of the image is due to imperfect modeling of the algorithm, as presented in the fourth plot in the second row, and then for reconstructing the 21 cm information, we need more than extracting foreground and noise contribution. In the second row, the first plot represents all the foreground emissions used as input, and in the second plot, the foreground contribution is reconstructed by the GMCA algorithm. The difference between the input and the reconstructed is presented in the third plot, where there is a highlighted blue region at the top that represents the effects of the masking that does not represent a problem.

[6] A dictionary is a collection of parameterized waveforms [12].

[7] Also known as *Isotropic Undecimated Wavelet Transform* on the Sphere.

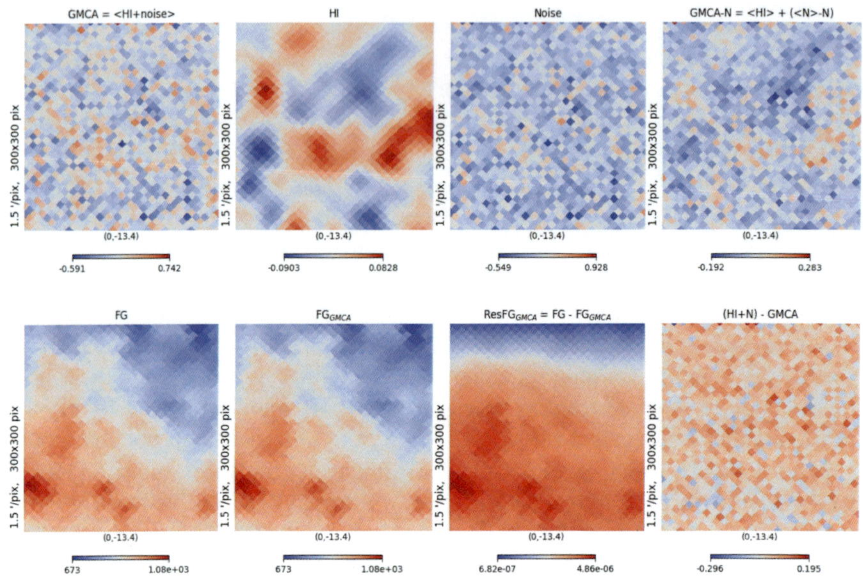

Fig. 5 Set of images representing the step of foreground removal from a 21-cm intensity mapping observation. The removal was done with the GMCA algorithm and the projection is centered at 0° right ascensions and –13.4° declination.The top left image represents the residues from the foreground removal process composed of HI, instrumental noise, and also foreground leftovers. The second left-top image represents the HI input data, the third-top image the noise input data, and the far right-top image the residues from GMCA minus the white noise input. On the far left-bottom, the image represents the foreground input map, the second left-bottom image is the foreground signal modeled by GMCA, the third-bottom image represents the difference between the input foreground map and the GMCA reconstructed one, and the far right-bottom image is the difference between HI plus noise input maps and GMCA residues

2.4 Statistical Estimation of the 21 cm Signals

Finally, the last step is to estimate the 21 cm signal itself. This estimation is realized through a set of realizations in the harmonic space. Each realization describing the observation is submitted to the GMCA process to remove foreground contribution and then used to estimate the aimed signal. Estimating the 21 cm power spectrum is done here using the angular power spectrum estimated from the ANAFAST function by HEALPix. We rely on the notation ${}^{(a)}\hat{C}_\ell^{(b)}$ where a and b denote which realization is and the type of map (for instance, 21 cm and/or noise N), respectively. The "hat"^ defines the estimated value of the angular power spectrum derived from the debiasing process. Therefore, for a specific L_i realization, the estimated 21 cm angular power spectrum is

$$
{}^{(L_i)}\hat{C}_\ell^{(21\,cm)} = \frac{{}^{(L_i)}C_\ell^{(\mathrm{R})}}{S_\ell} - \left\langle {}^{(L_j)}C_\ell^{(\mathrm{N})} \right\rangle_{L_j}^{j\neq i}
$$

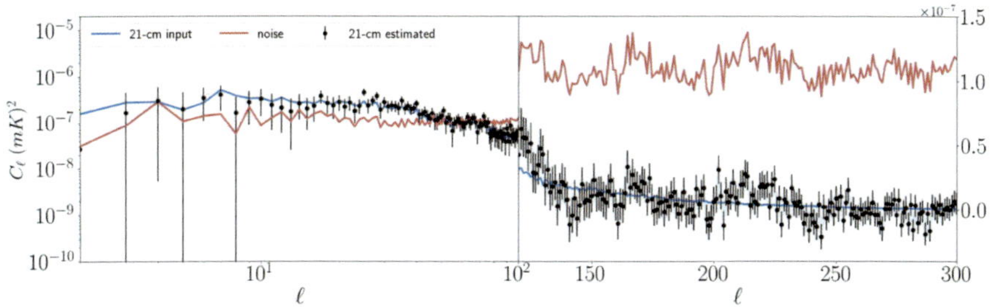

Fig. 6 The 21 cm input signals (blue color), described in the harmonic space, under instrumental noise contamination (red color), after removing foreground contamination and statistical analysis, is reconstructed (black color) in a high-accuracy order level until $\ell \sim 300$

where R is the residual, that is, the components of $\mathbb{R}$; and the suppression factor S_ℓ fixes the algorithm's error,

$$S_\ell = \left\langle \frac{^{(L_j)}C_\ell^{(\mathrm{R})}}{^{(L_j)}C_\ell^{(21\,\mathrm{cm}\,+\,\mathrm{N})}} \right\rangle_{L_j}^{j \neq i}, \tag{9}$$

with $\langle\rangle_{L_j}^{j \neq i}$ describing the average over all realizations except the L_i realization; that is, for $L_j \in \left(L_0, \ldots, L_{i-1}, L_{i+1}, \ldots, L_{n_r-1}\right)$, for n_r number of realizations. The result for this procedure for 400 realizations can be seen in Fig. 6 for the channel 1140–1150 MHz, with blue color describing the 21 cm input signal, red color the white noise angular power spectrum, and in black color the reconstructed 21-cm signal. The error bars are derived from the resampling method jackknife [37]. The plot on the left has a vertical axis in log scale and is linear on the right. The figure shows how highly accurate it is possible to recover the really faint 21 cm signal. The reconstruction accuracy tends to decrease the higher the signal-to-noise ratio.

3 Bispectrum in 21-cm Intensity Mapping

In the post-reionization epoch (z $\lesssim$ 6), the 21 cm brightness temperature depends, basically, on the HI fraction available and the gradient of the proper velocity along the line of sight because spin temperature suppresses the background radiation. As already in previous sections, as time goes by, more and more HI find themselves concentrated in Dark Matter halos until the case of $z < 3$, the amount of HI inside halos reaches 90%. That is, for far post-reionization epochs, most of the HI is associated with halos, and the approach of using HI to trace the total matter distribution is totally justified. In the standard scenario, in better agreement with the available data, there was a primordial inflationary period where quantum fluctuations resulted in the Gaussian distribution of the total matter. Therefore, it is straightforward that the HI distribution follows a Gaussian distribution, and any deviation from that, i.e., showing a primordial non-Gaussianity (PNG) distribution of the matter, may lead to a confrontation with the currently accepted scenario. Identifying any PNG signature in observational data is a challenge because of its possible tiny value of amplitude. In addition, after the pre- and post-processing of the observed (or simulated) data, non-Gaussianities can be introduced into the 21 cm reconstructed signal.

Studies of statistical analysis with order higher than two point functions (e.g., power spectrum) are important both for residual treatment in the data analysis and for investigating signatures of PNG. Bispectrum analysis of the BINGO data aims to look for non-Gaussiatinies of the HI distribution possibly presented in the observed maps, to investigate the Dark Sector, and to help in the recovery of the 21 cm maps. The bispectrum module has been implemented based on the algorithm presented in [10].

An (angular) bispectrum analysis through maps can be represented by

$$B_{\ell_1\ell_2\ell_3} = \int d\Omega M_{\ell_1}(\Omega) M_{\ell_2}(\Omega) M_{\ell_3}(\Omega) \tag{10}$$

where $M_\ell(\Omega) = \sum_m a_{\ell m} Y_{\ell m}(\Omega)$ is a map in the harmonic space and here it is representing the HI intensity maps. $B_{\ell_1\ell_2\ell_3}$ is the bispectrum arising from the maps. Writing the relation above in terms of spherical harmonics, we obtain

$$B_{\ell_1\ell_2\ell_3} = \sqrt{N_\Delta^{\ell_1\ell_2\ell_3}} \sum_{m_1,m_2,m_3} \begin{pmatrix} \ell_1 & \ell_2 & \ell_3 \\ m_1 & m_2 & m_3 \end{pmatrix} a_{\ell_1 m_1} a_{\ell_2 m_2} a_{\ell_3 m_3} \quad , \tag{11}$$

where the Wigner-3j symbol relation expressed as

$$\int d\Omega Y_{\ell_1 m_1}(\Omega) Y_{\ell_2 m_2}(\Omega) Y_{\ell_3 m_3}(\Omega) = \sqrt{N_\Delta^{\ell_1\ell_2\ell_3}} \begin{pmatrix} \ell_1 & \ell_2 & \ell_3 \\ m_1 & m_2 & m_3 \end{pmatrix} \quad , \tag{12}$$

has been used and the factor outside of the integral depends only on the ℓ's and is given by the expression

$$N_\Delta^{\ell_1\ell_2\ell_3} \equiv \frac{(2\ell_1+1)+(2\ell_2+1)+(2\ell_3+1)}{4\pi} \begin{pmatrix} \ell_1 & \ell_2 & \ell_3 \\ 0 & 0 & 0 \end{pmatrix} \quad . \tag{13}$$

Equation 11 provides a starting point for obtaining the bispectrum analysis. Due to the high complexity of the equation, it is necessary to do a huge number of calculations with enormous numbers that grow quickly with the ℓ values. Thus, an efficient algorithm is required in order to handle and calculate the bispectrum in a reasonable time.

In order to be able to use the bispectrum module in the simulations, we need to understand that a bispectrum is a statistical tool that computes the 3-point correlations from a dataset that, for our purpose, is an image (or up to three different ones). Hence, the ℓ multipoles of the bispectrum must create a triangle shape representing the triangular inequality $|\ell_1 - \ell_2| \leq \ell_3 \leq \ell_1 + \ell_2$. Other symmetries existing in Eq. 11 are the parity invariance due to the angular correlation function and the rotation invariance encoded in the Wigner-3j symbol. In the former, this symmetry happens because the sum of the three multipoles is an even number (i.e., $\ell_1 + \ell_2 + \ell_3 =$ even), and the latter symmetry, because the sum of m's is null (i.e., $m_1 + m_2 + m_3 = 0$).

For the purpose of testing the bispectrum module, simulated emissions and instrumental signals were used, and are expected to be considered by the telescope. Regarding observed sky signals, a composition among neutral hydrogen emissions and other radio emissions from astrophysical and cosmological phenomena, as described in the previous section, was considered the FLASK [51] and PSM codes [18], respectively. For the instrumental effects, only a beam convolution of the maps through a Gaussian beam with 40' angular resolution and white noise was assumed (see more about the construction of the dataset in the papers [24, 37]).

3.1 Bispectrum Patterns and Foreground Residues

We use here a triangle configuration called equisize, where the sum of the multipoles is constant, in the case example, 300 (i.e., $\ell_1 + \ell_2 + \ell_3 = \ell_0 = 300$). The analysis was done for all 30 redshift bins (consequently, 30 channels) here assumed (see [24]). In Fig. 7 is possible to see some bispectrum results for the total foreground contribution (mentioned in the figure as "total") in three distinct channels, which show that it is possible to identify the non-Gaussianity amplitude coming from foreground emissions and that a similar pattern is presented in the different channels. Figure 8 presents the bispectrum results considering only 21 cm and white noise signals. The tiny amplitude first represents the high level of Gaussianity of the 21 cm + white noise signals when compared with the foregrounds. Even though the 21 cm maps used here arise from simulations, the simulations have been carried out with a log-normal distribution. Hence, the Gaussianity of our HI distribution is almost zero. If the previous figure is compared with Fig. 10, where there is only the bispectrum of the white noise, it is possible to see that the contrast. The different bispectrum patterns between 21 cm and white noise signals can help to distinguish them and, therefore,

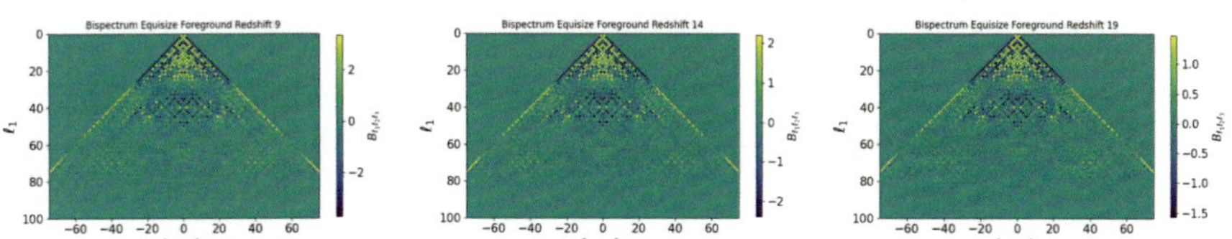

Fig. 7 Angular bispectrum plots in an equisize configuration for total foreground contribution in three different channels (with 10 MHz bandwidth). Beyond the high amplitude values, the plots present similar patterns

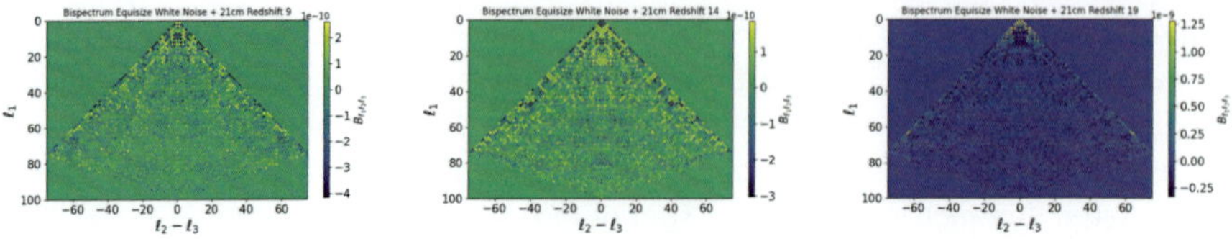

Fig. 8 Angular bispectrum plots in an equisize configuration for 21 cm + white noise signals in three different channels (with 10 MHz bandwidth). The tiny amplitude values of the outcome come from the high degree of Gaussianity from both signals

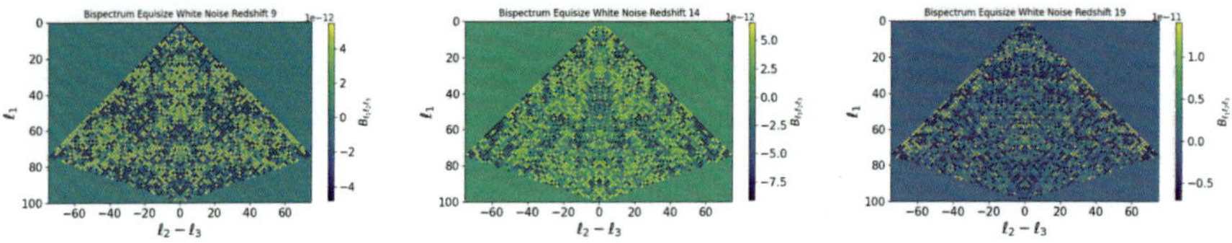

Fig. 9 Angular bispectrum plots in an equisize configuration for only white noise signal in three different channels (with 10 MHz bandwidth). Although the tiny amplitude values are also present, the white noise angular bispectrum pattern can be explored to distinguish in relation to the 21 cm

improve the quality and precision in the removal of the residual contaminations of the 21 cm signals recovered (Fig. 9).

To achieve the goal of helping in the process of 21 cm signal reconstruction and to quantify the strength of using bispectrum for identification of the foreground contamination, we carry out a foreground removal. The foreground removal procedure wcan be realized similarly to that described in the previous section but using GNILC algorithm [24], that as demonstrated by [37], has results equivalent to that of the GMCA.

Figure 10 presents the results of the equisize configuration ($\ell_1 + \ell_2 + \ell_3 = N$) for (N = 30) (as orange triangles on the left plot and green stars on the right plot) and 60 (as blue circles on the right plot and red dots on the left one) of the nonzero bispectrum values added, as given by the relation

$$B_{\ell_1+\ell_2+\ell_3=\ell_0} = \sum_{\ell_1=1}^{N}\sum_{\ell_2=1}^{N}\sum_{\ell_3=1}^{N} B_{\ell_1\ell_2\ell_3}\delta_{\ell_1+\ell_2+\ell_3,N} \quad . \tag{14}$$

The main difference between both plots is the number of degrees used to describe the foreground contribution. The m_{AIC} parameter represents the effective degree of freedom using the AIC criteria [4, 9]. Once the 21 cm+white noise bispectrum is suppressed by the foreground, significant amplitudes will be represented mainly from foreground signatures. The results on the left assume a sub-estimation of the foreground contribution, and on the right, use the real value provided by AIC. An underestimation of the foreground subspace leads to the foreground leakage into the 21 cm maps, and this fact is clearly represented on the left plot of Fig. 10 with

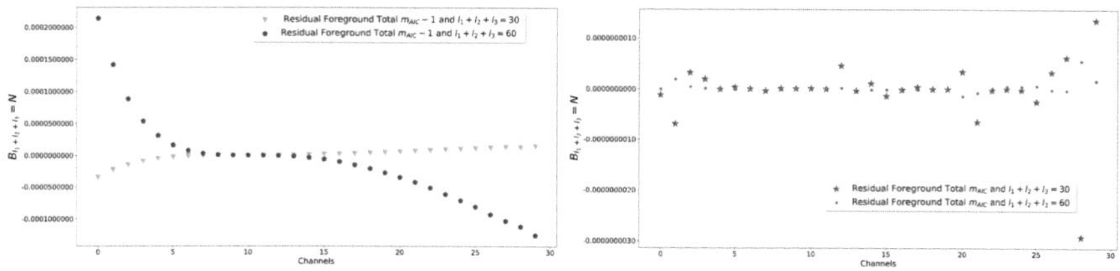

Fig. 10 The values of Eq. 14 per frequency, considering $N = 30$ (orange triangles on the left and green stars on the right) and 60 (blue circles on the left and red dots on the right). Images published in [24]

a higher (absolute) amplitude of the $B_{\ell_1+\ell_2+\ell_3=\ell_0}$ values, mainly for the extremes channels. The absolute value, $|B_{\ell_1+\ell_2+\ell_3=N}|$, is at most $\sim 3 \times 10^{-4}$ with $N = 60$, presenting the significant presence of the foregrounds as contaminants in the 21 cm maps. On the other hand, on the plot right, an optimal degree of freedom provided by AIC criteria makes it impossible to distinguish between foreground residuals and the tiny contribution from the 21 cm+white noise. The absolute bispectrum amplitude by $\mathcal{O}(10^{-9})$ is not significant enough to make an assertive distinction. Nonetheless, the results show GNILC is highly efficient in removing the foreground emissions if working in its optimal mode.

The higher challenge in the angular bispectrum calculation is the computational cost of the integral in the Eq. 12. Its time-consuming process has limited us from operating up to $N \sim 60$. Today, the development of higher-performance algorithms for calculations has concentrated the most effort from the workgroup to make available an efficient, precise, and suitable module operating with $N \sim 200 - 300$ in a feasible time. As it was possible to see, the capacity for detecting bispectrum signatures significantly increases as N. Therefore, values of $N \sim 200 - 300$ will potentially increase the potential in the detectability of foreground residues in the final maps and even, possibly, capturing PNG traces.

4 Primordial Black Holes with Magnetic Fields as a Candidate for Fast Radio Bursts

Fast Radio Bursts (FRBs) are astrophysical phenomena observed as short-duration transients in radio frequencies, whose origin, however, is not well understood. They were first discovered in 2007 by [34] from an event recorded by the 64-m Parkes Telescope in 2001. In the following years, the astrophysical nature of these events was not only confirmed but a series of other events were also detected, leading to the compilation of a catalog recording their occurrences.[8]

---

[8] Catalog available in https://www.herta-experiment.org/frbstats/catalogue.

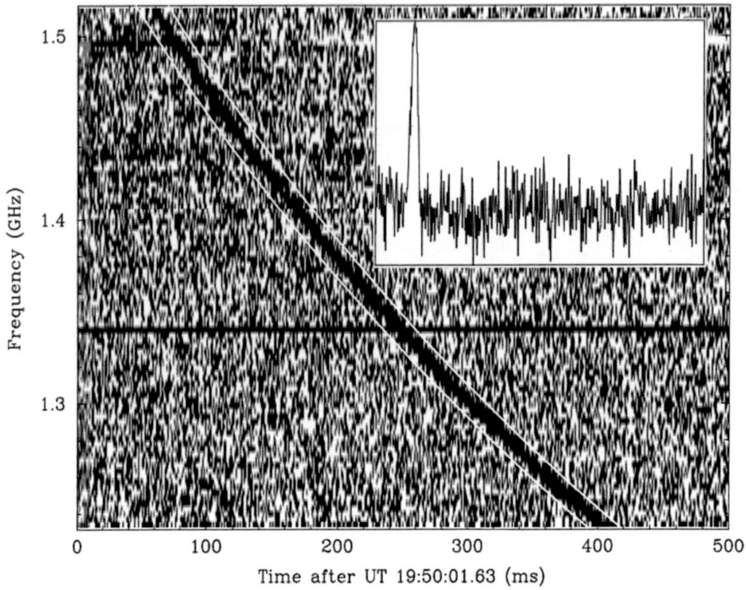

Fig. 11 Format of the observed Lorimer burst. Image extracted from [34]

Today, radiotelescope optical designs as that from BINGO, FAST, and Parkes are able to collect short-period signals compatible with FRBs, therefore perfect for the study of such transients, as well as pulsars. Today, one main aim of those projects is to observe FRBs. Moreover, with the planned outriggers, we are going to be able to determine the origin in the sky of the FRBs, permitting not only their study as astrophysical objects but also their role in cosmology once redshift as well as distance are determined (Figs. 11 and 12).

Two types of FRBs have been observed, repeating and non-repeating ones, the former showing a lasting semi-periodic behavior, while the latter has a short duration and high intensity. For the non-repeating kind, the typical duration of the pulses lies between 0.1 to 1 ms, during which time they emit an equivalent total isotropic energy from $10^{14}g$ to $10^{22}g$. Moreover, they have been found in redshifts ranging from 0 to 2.8 and present different pulse morphologies [54].[9]

In 2020, a detection by CHIME of an FRB from the direction of a galactic magnetar has further implied a possible connection between these objects and the radio bursts [14]. This also corroborates with the high linear polarisation, up to 50%, and points to some requirements for explanatory models.

These models should have, as aforementioned, intense magnetic fields and have the emissions formed around compact objects, such as black holes or neutron stars,

[9] It is not clear if all FRBs are in reality repeating, since we might, in fact, lose some, or even many, events. This is not an issue now.

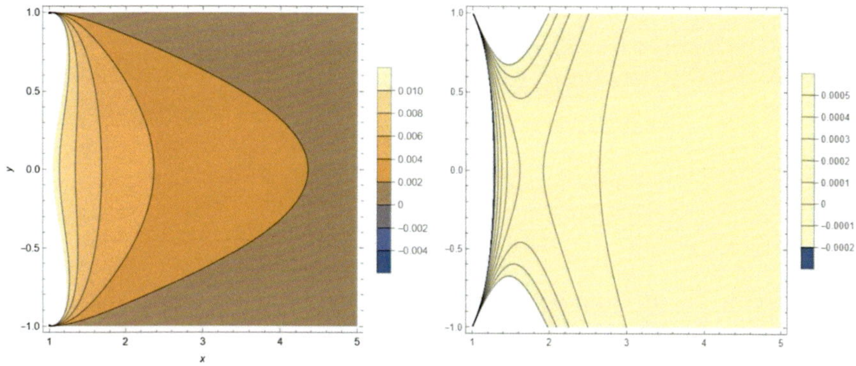

Fig. 12 (Left) Magnetic Potential, (right) Ricci Scalaer Curvature for $\alpha = 0.1$.Structure of the Metric for $\alpha = 0.1$. White regions demonstrate where these fields are too high for the numerical calculation at the defined precision, as they near the singularities

to allow the radiated energy. Furthermore, the possibility of directional emissions might lighten the energy budget required for such models.

5 Magnetized Space Times

A natural candidate as an explaining model concerns the perturbations of the Einstein-Maxwell equations around compact objects. Magnetars, and NS in general, have usually been treated perturbatively in the literature [15] by imposing multipole magnetic fields over static or stationary solutions of General Relativity. The proposed approach, conversely, uses Gutsannaev-Manko's solution [26] describing a mass with a magnetic dipole, on which the perturbations are calculated, and their quasi-normal modes are extracted. More general metrics can include angular moment and mass quadrupole moments at the cost of a more complex structure of the metric.

The solution is given by the metric

$$ds^2 = -\frac{(x-1)}{(x+1)} f^2 dt^2 + \left(\frac{g}{f}\right)^2 \left(\frac{x+1}{x-1} dx^2 + \frac{(x+1)^2}{1-y^2} dy^2\right) + \frac{(x+1)^2(1-y^2)}{f^2} d\varphi^2 \tag{15}$$

in which f and g are rational functions of x and y, which in turn are the prolate coordinates, related to the cylindrical coordinates by

$$\rho = k(x^2-1)^{1/2}(1-y^2)^{1/2}; \qquad z = kxy \tag{16}$$

and the functions in the metric are

$$f = \frac{(x^2 - y^2 + \alpha^2(x^2 - 1))^2 + 4\alpha^2 x^2(1 - y^2)}{(x^2 - y^2 + \alpha^2(x - 1)^2)^2 - 4\alpha^2 y^2(x^2 - 1)} \tag{17}$$

and

$$g = \frac{\left((x^2 - y^2 + \alpha^2(x^2 - 1))^2 + 4\alpha^2 x^2(1 - y^2)\right)^2}{(\alpha^2 + 1)^4 (x^2 - y^2)^4}. \tag{18}$$

This metric has an event horizon at $x = 1$; however, f has zeroes outside that range, what poses problems to the interpretation of this metric as a black hole solution, as it contains naked singularities, which can be checked in its Kretschman invariant. Thus one should construct a star solution from it, imposing an appropriate boundary condition that encompasses the singularity, leaving only the regular part of the space-time as the exterior metric.

Perturbations can be further imposed. Due to the lack of spherical symmetry, the perturbation will couple different spherical harmonics, which have to be solved as a matrix equation terminated at a final index chosen such that the coupling to the initial mode excited is low. These are being studied at the moment, both for charged and uncharged perturbations.

The new generation of the radiotelescope is suitable to work mapping large scale, in a poorly resolution but fast drifting, and—either alone (if it has a good resolution) or in an interferometric synergy—be a transient source's hunter and improve the FRBs catalog.

Regarding FRBS, these additional observations are a valuable tool not only to improve the understanding of the origin of these emissions but also to provide tools to analyze the dispersion measure to the host galaxy of the event. Thus, these measurements may allow for the estimation of free electron density in intergalactic media, providing another source of data for cosmological models to be tested.

References

1. E. Abdalla, E.G.M. Ferreira, R.G. Landim et al., A&A **664**, A14 (2022). https://doi.org/10.1051/0004-6361/202140883
2. Abdalla, F. B., Marins, A., Alberto, J., Santos, N., & Abdalla, E. 2023, (in preparation)
3. F.B. Abdalla, A. Marins, P. Motta et al., A&A **664**, A16 (2022). https://doi.org/10.1051/0004-6361/202141382
4. H. Akaike, IEEE Trans. Autom. Control. **19**, 716 (1974)
5. R.A. Battye, I.W.A. Browne, C. Dickinson et al., Mon. Not. R. Astron. Soc. **434**, 1239 (2013). https://doi.org/10.1093/mnras/stt1082
6. M. Bhardwaj, A.Y. Kirichenko, D. Michilli et al., ApJ **919**, L24 (2021). https://doi.org/10.3847/2041-8213/ac223b
7. J. Bobin, Y. Moudden, J.-L. Starck, J.M. Fadili, N. Aghanim, Astronomical data analysis ADA'06 (Marseille, France, 2006). https://hal.archives-ouvertes.fr/hal-01160933
8. J. Bobin, J.-L. Starck, J. Fadili, Y. Moudden, IEEE Trans. Image Process. **16**, 2662 (2007). https://doi.org/10.1109/TIP.2007.906256

9. H. Bozdogan, Psychometrika **52**, 345 (1987)
10. M. Bucher, B. Racine, B. van Tent, J. Cosmol. Astropart. Phys. **2016**, 055 (2016)
11. E. Chapman, V. Jelić (2019) arXiv preprint arXiv:1909.12369
12. Chen, S., & Donoho, D. L. 1995, in Society of Photo-Optical Instrumentation Engineers (SPIE) Conference Series, Vol. 2569, Wavelet Applications in Signal and Image Processing III, ed. A. F. Laine, M. A. Unser, & M. V. Wickerhauser, 564–574, doi: http://doi.org/10.1117/12.217610
13. CHIME/FRB Collaboration, Amiri, M., Bandura, K., et al. 2018, *ApJ*, 863, 48, doi: http://doi.org/10.3847/1538-4357/aad188
14. CHIME/FRB Collaboration, Andersen, B. C., Bandura, K. M., et al. 2020, *Nature*, 587, 54, doi: http://doi.org/10.1038/s41586-020-2863-y
15. A. Colaiuda, V. Ferrari, L. Gualtieri, J.A. Pons, Mon. Not. R. Astron. Soc. **385**, 2080 (2008). https://doi.org/10.1111/j.1365-2966.2008.12966.x
16. A.A. Costa, R.G. Landim, C.P. Novaes et al., A&A **664**, A20 (2022). https://doi.org/10.1051/0004-6361/202140888
17. Das, S., Anderson, C. J., Ansari, R., et al. 2018, in Society of Photo-Optical Instrumentation Engineers (SPIE) Conference Series, Vol. 10708, Millimeter, Submillimeter, and Far-Infrared Detectors and Instrumentation for Astronomy IX, ed. J. Zmuidzinas & J.-R. Gao, 1070836, doi: http://doi.org/10.1117/12.2313031
18. J. Delabrouille, M. Betoule, J.B. Melin et al., A&A **553**, A96 (2013). https://doi.org/10.1051/0004-6361/201220019
19. T. Delubac, J.E. Bautista, N.G. Busca et al., A&A **574**, A59 (2015). https://doi.org/10.1051/0004-6361/201423969
20. C. Dickinson, Galaxies **6**, 56 (2018). https://doi.org/10.3390/galaxies6020056
21. dos Santos, M. V., Landim, R. G., Hoerning, G. A., et al. 2023, arXiv e-prints, arXiv:2308.06805
22. B.T. Draine, A. Lazarian, ApJ **494**, L19 (1998). https://doi.org/10.1086/311167
23. D.J. Eisenstein, I. Zehavi, D.W. Hogg et al., ApJ **633**, 560 (2005). https://doi.org/10.1086/466512
24. K.S.F. Fornazier, F.B. Abdalla, M. Remazeilles et al., A&A **664**, A18 (2022). https://doi.org/10.1051/0004-6361/202141707
25. K.M. Górski, E. Hivon, A.J. Banday et al., ApJ **622**, 759 (2005). https://doi.org/10.1086/427976
26. T. Gutsunaev, V. Manko, Phys. Lett. A **123**, 215 (1987). https://doi.org/10.1016/0375-9601(87)90063-6
27. A. Hall, C. Bonvin, A. Challinor, Phys. Rev. D **87**, 064026 (2013). https://doi.org/10.1103/PhysRevD.87.064026
28. B.S. Hensley, B.T. Draine, ApJ **836**, 179 (2017). https://doi.org/10.3847/1538-4357/aa5c37
29. E.F. Keane, S. Johnston, S. Bhandari et al., Nature **530**, 453 (2016). https://doi.org/10.1038/nature17140
30. A. Kogut, A.J. Banday, C.L. Bennett et al., ApJ **464**, L5 (1996). https://doi.org/10.1086/310072
31. E.M. Leitch, A.C.S. Readhead, T.J. Pearson, S.T. Myers, ApJ **486**, L23 (1997). https://doi.org/10.1086/310823
32. J. Lesgourgues (2011) arXiv e-prints, arXiv:1104.2932. https://doi.org/10.48550/arXiv.1104.2932
33. A. Lewis, A. Challinor, A. Lasenby, ApJ **538**, 473 (2000). https://doi.org/10.1086/309179
34. D.R. Lorimer, M. Bailes, M.A. McLaughlin, D.J. Narkevic, F. Crawford, Science **318**, 777 (2007). https://doi.org/10.1126/science.1147532
35. A. Macià Escatllar, S.T. Bromley, A&A **634**, A77 (2020). https://doi.org/10.1051/0004-6361/201936419
36. A. Marins, Thesis (2023)
37. Marins, A., Abdalla, F. B., Fornazier, K. S. F., et al. 2022, arXiv e-prints, arXiv:2209.11701
38. F.G. Mertens, M. Mevius, L.V.E. Koopmans et al., Mon. Not. R. Astron. Soc. **493**, 1662 (2020). https://doi.org/10.1093/mnras/staa327
39. M.A. Miville-Deschênes, N. Ysard, A. Lavabre et al., A&A **490**, 1093 (2008). https://doi.org/10.1051/0004-6361:200809484

40. R. Nan, D. Li, C. Jin et al., Int. J. Mod. Phys. D **20**, 989 (2011). https://doi.org/10.1142/S0218271811019335
41. E. Petroff, J.W.T. Hessels, D.R. Lorimer, A&A Rev. **30**, 2 (2022). https://doi.org/10.1007/s00159-022-00139-w
42. Planck Collaboration, Akrami, Y., Arroja, F., et al. 2020a, *A&A*, 641, A10. https://doi.org/10.1051/0004-6361/201833887
43. Planck Collaboration, Aghanim, N., Akrami, Y., et al. 2020b, *A&A*, 641, A6. https://doi.org/10.1051/0004-6361/201833910
44. M.G. Santos, A. Cooray, L. Knox, ApJ **625**, 575 (2005). https://doi.org/10.1086/429857
45. M. Spinelli, I.P. Carucci, S. Cunnington et al., Mon. Not. R. Astron. Soc. **509**, 2048 (2022). https://doi.org/10.1093/mnras/stab3064
46. J.-L. Starck, J. Fadili, F. Murtagh, IEEE Trans. Image Process. **16**, 297 (2007). https://doi.org/10.1109/TIP.2006.887733
47. D. Thornton, B. Stappers, M. Bailes et al., Science **341**, 53 (2013). https://doi.org/10.1126/science.1236789
48. M. Vidal, C. Dickinson, R.D. Davies, J.P. Leahy, Mon. Not. R. Astron. Soc. **452**, 656 (2015). https://doi.org/10.1093/mnras/stv1328
49. F. Villaescusa-Navarro, S. Genel, E. Castorina et al., ApJ **866**, 135 (2018). https://doi.org/10.3847/1538-4357/aadba0
50. Wolfe, A. M., Gawiser, E., & Prochaska, J. X. 2005, , 43, 861, doi: http://doi.org/10.1146/annurev.astro.42.053102.133950
51. H.S. Xavier, F.B. Abdalla, B. Joachimi, Mon. Not. R. Astron. Soc. **459**, 3693 (2016)
52. Z. Yu, F. Deng, S. Sun et al., Res. Astron. Astrophys. **22**, 125007 (2022). https://doi.org/10.1088/1674-4527/ac977c
53. B. Zhang, ApJ **867**, L21 (2018). https://doi.org/10.3847/2041-8213/aae8e3
54. B. Zhang, Nature **587**, 45 (2020). https://doi.org/10.1038/s41586-020-2828-1
55. J. Zhang, P. Motta, C.P. Novaes et al., A&A **664**, A19 (2022). https://doi.org/10.1051/0004-6361/202140887
56. W. Zhu, D. Li, R. Luo et al., ApJ **895**, L6 (2020). https://doi.org/10.3847/2041-8213/ab8e46

BINGO-ABDUS: A Radiotelescope to Unveil the Dark Sector of the Universe

Elcio Abdalla, Alessandro Marins, Filipe Abdalla, Jordany Vieira, Lucas Formigari, Luciano Barosi, Amilcar R. Queiroz, Thyrso Villela, Carlos A. Wuensche, Bin Wang, Chang Feng, Edmar Gurjão, Ricardo Landim, Camila P. Novaes, João R. L. Santos, and Jiajun Zhang

Abstract The *Baryon Acoustic Oscillations from Integrated Neutral Gas Observations* (BINGO) telescope is an international collaboration, led by Brazil and China, aiming to explore the Universe's history through integrated post-reionization 21 cm signals and fast radio emissions. It is concentrated on the important period for understanding the recent Universe dynamics that have shed light on the existence of the Dark Sector. In its first phase, BINGO will explore the southern celestial Hemisphere to observe neutral hydrogen signals from when the Universe was between 8.969 and 12.112 billion years old (These age values are from the cosmology provided by PLANCK satellite [37] assuming a flat-ΛCDM with $H_0 = 67.36$ km/s/Mpc, $\Omega_m = 0.3153$, and $\Omega_\Lambda = 0.6847$.) in an angular resolution of 40 arcmins, and for

E. Abdalla (✉) · A. Marins · F. Abdalla · J. Vieira · L. Formigari
Departamento de Física Geral, Instituto de Física, Universidade de São Paulo, São Paulo, Brazil
e-mail: eabdalla@if.usp.br

A. Marins · C. Feng
Department of Astronomy, School of Physical Science, University of Science and Technology of China, Hefei, China

F. Abdalla
Department of Physics and Astronomy, University College London, Gower Street, London WC1E 6BT, UK

L. Barosi · A. R. Queiroz · E. Gurjão · J. R. L. Santos
Unidade Acadêmica de Física, Universidade Federal de Campina Grande, R. Aprígio Veloso, 58429-900 Campina Grande, Brazil

T. Villela · C. A. Wuensche · C. P. Novaes
Instituto Nacional de Pesquisas Espaciais, Divisão de Astrofísica, Av. dos Astronautas, 1758, São José dos Campos, SP 12227-010, Brazil

T. Villela
Instituto de Física, Universidade de Brasília, Campus Universitário Darcy Ribeiro, Brasília, DF 70910-900, Brazil

B. Wang
School of Aeronautics and Astronautics, Shanghai Jiao Tong University, Shanghai 200240, China

Center for Gravitation and Cosmology, Yangzhou University, Yangzhou 224009, China

© The Author(s), under exclusive license to Springer Nature Switzerland AG 2024

E. Papantonopoulos and N. Mavromatos (eds.), *Compact Objects in the Universe*,
https://doi.org/10.1007/978-3-031-55098-0_11

the next phases reaching far epochs when the Universe was 3.116 billion years old up to 27 arcmin resolution which is poor for identifying individual sources, however important for mapping out the Universe in large scales using the *Intensity Mapping* observational method. For identifying individually fast radio sources, the *Advanced Bingo Dark Universe Studies* (ABDUS) project has been proposed and developed and will combine the current BINGO construction with the main single-dish telescope and stations of phased-array and outrigger, which will reach out up to 2.5 arcsec resolution to pinpoint individual radio sources as Fast Radio Bursts by interferometry.

1 Introduction

From a very early age, curious young people and future explorers are confronted with questions that still have no answers. An experienced explorer recognizes that not having all the answers is acceptable as long as the fire of curiosity continues. Open questions in human understanding are engines for scientific and technological development. In the 1930s, one of the most current meaningful scientific milestones comes up through ignorance about matter distribution and about the behavior of galaxies as a consequence of the existence of a Dark Sector, as we know a posteriori [52, 53]. Dark Matter (DM) emerges with a generic name for a class of candidates for filling in the gap raised in the unseen matter quantity to describe the observed universe. On the other hand, the decade of the 1990s started the running to understand the nature of the Universe's accelerated expansion. Dark Energy (DE) is the name that condenses all fluid candidates to explain the Universe's accelerated behavior within Einstein's gravity. DM and DE together are complex challenges to the current description from elementary known principles since no natural representative arises from Particle Standard Model. DM and DE form the named *Dark Sector* [1], and its nature drives several new projects in current physics and astronomy studies.

However, in contrast to completely new proposals, an explanation of the above-unknown components needs to follow the usual thread of the already-known Universe's history with high accuracy. Or else provide a highly well-grounded new physics. For these, many distinct observational methods and/or tools are used aim-

C. Feng
CAS Key Laboratory for Research in Galaxies and Cosmology, University of Science and Technology of China, Hefei, China

School of Astronomy and Space Science, University of Science and Technology of China, Hefei, China

R. Landim
Technische Universität München, Physik-Department T70, James-Franck-Strasse 1, 85748 Garching, Germany

J. Zhang
Shanghai Astronomical Observatory, Chinese Academy of Sciences, Shanghai 200030, China

ing to elucidate more and more the components from different perspectives. The most promising way to explore the Universe's history and its contents is from the emission lines of neutral hydrogen (HI), the so-called 21 cm signal, also known as hyperfine structure. The former name comes from the size of the wavelength of the emission. To be more precise, atomic neutral hydrogen emits a characteristic signal from the hyperfine splitting of the 1S ground state. Such signal has a wavelength of 21cm, corresponding to the frequency $\nu_{10} = 1420\,\text{MHz}$ in the rest frame. Even with a very low transition probability ($\sim 10^{-15} s^{(-1)}$), in an astrophysical context, HI clouds contain a large amount of HI that can be excited—either by radiative transition or by collision—and emit radiation [18].

Although 21 cm emissions have been used in astronomy for a long time, their use in astrophysics and cosmology is quite recent. Considering as precursor works of the 21 cm for astronomy those of George Field [18–20], and even though the first estimates of the signal are around ~1990–2000, the effective interest in emission in cosmology and astrophysics from the most primordial periods of the Universe begins with the work [41] suggesting the use of 21 cm emission lines to search for spectral or spatial signatures of structures at high redshifts. The first searches for the 21 cm signal in this context took place around the 1980s, but it really gained momentum with the construction of the *Giant Metrewave Radio Telescope* (GMRT) [42], in India in the late 1990s. Since then, several projects have been developed or are under development aiming to use this signal, and due to the HI being a tracer of matter distribution, it can be used for mapping the Universe in *Large Scale Structure* (LSS).

However, 21 cm emission post-reionization ($z < 6$) is a weak signal which makes its identification hard on a large scale [36]. Thus, measuring each source of emission individually is inefficient. Rather, collecting all emissions in the same beam resolution without distinguishing the sources provides a wider covering in a shorter time. Based on this idea, the BINGO project [2], acronym for *Baryon Acoustic Oscillations from Integrated Neutral Gas Observations*, aims to map integrated 21 cm intensity from the southern hemisphere to be the first to identify Baryonic Acoustic Oscillations (BAO) [6] at radio frequency and then to put strong constrain into Dark Sector models. It has been developed to operate at radiofrequency as, firstly, a single-dish with a good resolution and control over the systematic to guarantee the stability of the operation. An approximate prevision for the first pilot tests is for the end of 2023 or the beginning of 2024.

The *Advanced Bingo Dark Universe Studies* (ABDUS) project consists of the next phases of the BINGO project that has been under consideration and should be confirmed soon. The new phases aim to explore the current BINGO configuration together with distant stations of feed horns and phased arrays in order to enable the project to achieve, in addition to improving the coverage and survey speed of the current case, a large amount of data from radio sources such as FRBs. BINGO-ABDUS project will be able to cover half of the sky in a quite precise monitoring and overlapping with important other projects in radioastronomy and also galaxy surveys. Such overlap can enable both a good synergy to realize cross-analysis with galaxy datasets and also enable collaboration with other radiotelescopes in order to achieve unbelievable accuracies for individual source identifications. As an example, a collaboration between BINGO-ABDUS, FAST, and Tianlai has the potential to form one of the largest and most accurate radiotelescopes in the world.

This work aims to provide a brief overview of the BINGO project and its current status, and in addition, gives for the first time some information about the BINGO next step called BINGO-ABDUS, in which the project will be able to be a versatile and in the state-of-the-art of radio operation for looking for transient radio sources.

2 The BINGO Project

The BINGO project consists of a fixed transit telescope covering the south celestial sky by drift scanning mode due to the Earth's rotation. The area coverage is $\sim$5000 deg^2 between $-25°$ and $-10°$ in declination and with an average angular resolution of 40'. The beam resolution in the order of 40' is required to resolve structures of angular scales of $\approx$150 Mpc in the BINGO frequency range and then be possible to identify BAO's feature. The telescope will observe signals within a frequency span of 980–1260 MHz (corresponding to a redshift range of 0.127–0.449).

BINGO has been developed to be an off-axis, crossed-Draconian optical design [15] composed of a parabolic primary mirror and a hyperbolic secondary mirror with $\sim$40 m diameter, and a complex arrange of 28 vertically moving feed horns on the focal plane, as shown in Fig. 1. Its under-illuminated second mirror guarantees reduced spillover effects and a small amplitude of the far beams' sidelobes. Each receiver operates, so far, with a system temperature of 70 K and measures I and V polarization. In addition, its conical corrugated horns provide low cross-polarization levels as well as a small level of optical aberrations. A complete description of the instrumentation and the optical design can be found in [4, 46].

Fig. 1 Artistic view of the BINGO's construction model

Currently, BINGO signed a construction contract with the Chinese company CETC54[1] to manufacture the reflectors. All the feed horns are already in place and in storage, and there is an additional feed horn (called Uirapuru) that is at the Federal University of Campina Grande (UFCG, in Portuguese acronyms) campus, pointing directly to the sky and doing some data collection.

The telescope has been constructed in a far site at the village of Aguiar, in the countryside of the Paraiba state. The *Serra do Urubu* site was chosen after a long time investigating different locations since an abandoned gold mine in Uruguay to the northeast of Brazil, as presented in [34]. The choice followed several criteria as *radio frequency interference* (RFI) level, local topography for supporting the telescope structure, and the ease of assistance and follow-up due to the relative proximity to the UFCG. In Fig. 4 we can see the square focusing on the northeast region of Brazil, where the Paraiba state is located. In this figure, it is possible to see the location of the cost of the country, the capital of the state, and the location of the Campina Grande city.

BINGO telescope aims at using the spatial fluctuations measured from the *Intensity Mapping* observational method to produce tomographic analysis with (auto and cross) angular power spectrum, that is, the statistical two-point function on the harmonic space. The harmonic space is chosen mainly because of the characteristics of observation where there are precise determinations of the redshifts. The tomographic analysis of the two-point functions can be realized in spherical shells.

2.1 Scientific Organization

The international BINGO collaboration has concentrated efforts to develop the physical telescope and, at the same time and in parallel, build a complete and well-structured set-up of pipelines to test and cross-check each process of the telescope and its pre- and post-processing data. Such a process will help achieve the best accuracy in data recovery and obtain strong constraints in the cosmological parameters. The collaboration is organized into four work groups, called "stage", each one in charge of covering a set of specific tasks besides the outreach. For sure, there is no precise separation between the starting point and the end for each stage. Indeed, there is a strong superposition among different stages. A brief description of the arrangement of each stage can be seen as follows:

- **Stage 0:** This stage is responsible for handling calibration, tests, and maintenance of the instrumental parts. All processes, from the interaction between the electric field from the sky and the optical design to the analogic-digital conversion data or even the identification of transient astrophysical phenomena are part of the responsibilities. The construction of the huge reflectors has been developed

[1] 54th Research Institute of the China Electronics Technology Group.

together with CETC54 (China Electronics Technology Group Corporation) company. The development and tests of the feed horns and front-ends (transitions, polarizers, and magic tees) were conducted together with local companies ALLTEC and Metalcard, as presented in [45]. Furthermore, possible *Fast Radio Bursts* (FRBs) [29, 49], pulsars, and *Radiofrequency Interference* (RFI) are first identified in this stage. Machine Learning techniques have been studied to separate chunks of relevant data. FRBs and pulsars phenomena will be monitored in higher precision in the next level of the BINGO, called the ADBUS project, as we will describe soon. A good description of the stage's progress and development can be found in [4, 46].

- **Stage 1:** Stage responsible for the time-ordered data (TOD), map-making, and recording and flagging the possible transient objects. After the data is turned into digital information, it is necessary to record the observed data in a chronological description identifying the main information from the radiometer outputs, pointing locations, and the behavior of the system. Different system effects need to be assumed and understood here in order to achieve precise gain stability and accuracy of the reconstruction of 21 cm data. The TODs will be pre-processed into tomographic images corresponding to BINGO covering by the map-making process. Currently, there are two methods for processing TOD and map-making, one from the University of Manchester and another based on the Hide&Seek[2] software that has been extended and adapted to the BINGO case.[3] One of the main collaboration interests is related to the FRBs, which is a strong astrophysical emission with the origin unknown. A candidate for FRB is not necessarily a reportedly FRB; it is necessary that the candidate fill in a set of criteria to be accepted as an FRB itself. Also, other astrophysical (radio-)transients—such as pulsars—will be monitored by the stage and associated with trained neural networks for identification.
- **Stage 2–3:** Distinct at the beginning, stages 2 and 3 have been merged into the same group due to the increasingly overlapping areas of activity. The group is in charge of simulating the process of generating 21 cm emissions from N-body and lognormal (distribution) mocks [51], simulating possible astrophysical and cosmological sources of contamination [21, 28], and the process of reconstruction the 21 cm signals [13, 30] and BAO identification [32]. Roughly speaking, we can resume this stage as responsible for the data analysis.
- **Stage 4:** The fourth stage is the most theoretical one, interested in investigating the physics behind the data. Remembering that the BINGO concept has been developed aiming to investigate the Dark Sector of the Universe. A first estimation of the quality in constraint some dark energy—from CPL parametrization and interacting dark energy—and modified gravity parameters, as well as the synergy with other projects, have already been published [2, 9] forecasting—in a Fisher matrix context of analysis—the BINGO contribution. Other types of physical interests are also

[2] https://cosmology.ethz.ch/research/software-lab/hide—seek.html.

[3] A complete description about the Hide&Seek adaptation has been writing in a new paper to be published soon and presenting newest map-making results and beam modeling from Zernike polynomials [3].

taken into account in this stage, such as the study of neutrinos, HI parameter density for a range of the Universe's history, and FRB and Pulsar models. In summary, this stage has the purpose, from the project data, to elucidate and constrain the possible explanations to describe the current Universe dynamics observed. The possible constraint on cosmological parameters can have a large impact on the structure of the Dark Sector, especially concerning an internal interaction between DE and DM. The cosmological consequences of the possible localization of FRBs from the outriggers proposed for the project, especially in the case of the ABDUS project, are expected to be very large.

However, the BINGO project is not only about researching scientific issues but also about how to make the data and information produced available for the (civil and scientific) community and how to help society to be closer to the science and technology developed. BINGO outreach has been focusing on efforts to science communication, building a bridge for continuous collaboration between the expert production from the world-class researchers to the social and educational development. The BINGO outreach activities are being publicized on the BINGO website[4] and its social media channels.

3 The ABDUS Project—Advanced Bingo Dark Universe Studies

The ABDUS project honors the great physicist Abdus Salam, and represents the acronyms for *Advanced Bingo Dark Universe Studies*. It is an ambitious Radio Telescope resulting from a collaboration between BINGO (Brazil and China), the Netherlands Institute for Radio Astronomy (ASTRON), and Nançay Observatory. It will be BINGO's next step towards enhancing its resolution and coverage, and mainly to boost BINGO as one of the main telescopes to search for Fast Radio Bursts and strong radio point sources. For that, ABDUS proposes to use detectors in the form of phased arrays, which will allow expressively enlarging the BINGO's capabilities to reach out to $0 \lesssim z \lesssim 2.1$, for the 21 IM scanning and covering up to 50% of the sky (see the left image from Fig. 2) in a resolution between 27'-40', and increasing the current number of beams from 28 up to 200 in a multi-beam electronic synthesis through beamformers. This increase will be conducted by joining the already contracted BINGO structure with outrigger and phased array stations. The farthest station will be $\sim$20 km away from the main telescope site and will make it possible to reach out up to 2.5" angular resolution to pinpoint individual radio sources by interferometry when they are triggered by the main telescope.

Phased arrays (PAs) are a set of closely distributed antenna elements separated by about half wavelength, which can synthesize multiple independent beams and guarantee (Nyquist's field-sampling) requirement for a uniform sensibility coverage.

[4] https://bingotelescope.org/.

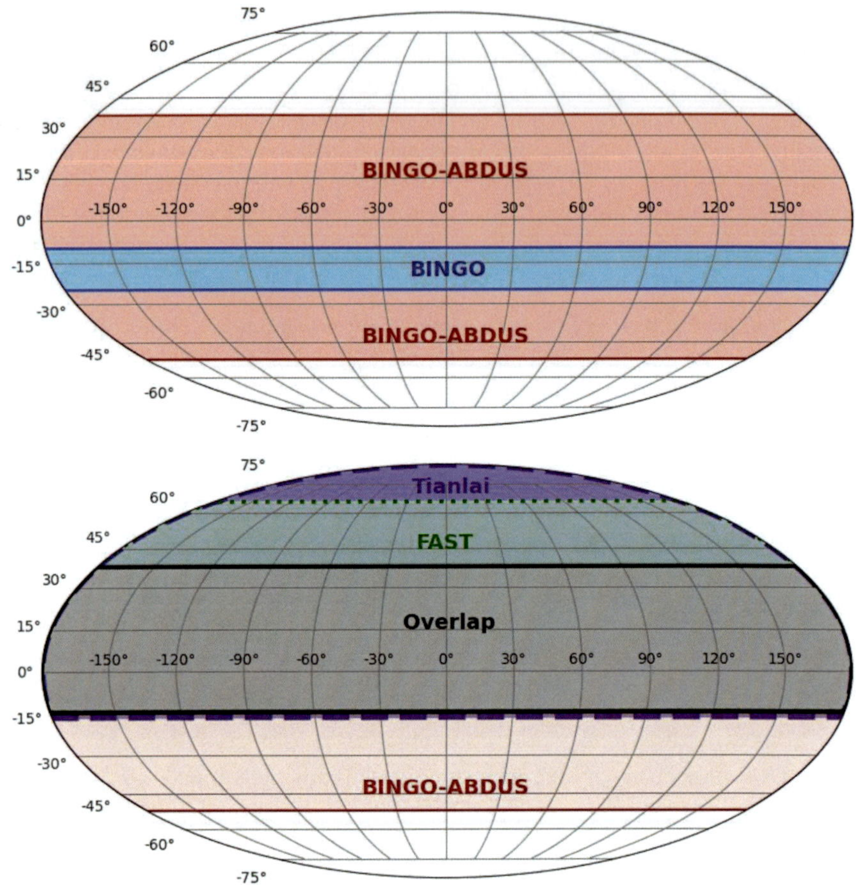

Fig. 2 Unmasked coverage by BINGO, ABDUS, FAST, and Tianlai. On the left, it is possible to check the impressive increase in the region to be observed by the BINGO-ABDUS project. On the right, there is in highlight the Overlap region to be coverage by BINGO, FAST, and Tianlai

In contrast to an interferometric approach, PAs work by imaging "additively" the PA element outputs, and the number of beams simultaneously synthesized depends on the digital processing capacity. Among the advantages of PAs, we can mention good control over the beam patterns, multiple steered beams, the possibility of achieving continuous and uniform sky coverage, an improvement in the FOV due to the increase of the number of beams leading to a higher survey speed, sensitivity optimization in relation to RFI contamination, and wide frequency coverage. Different projects use different approaches to build the PA beams, and an important procedure is through a beamforming technique, which is a process of adding all incoming signals (PA element outputs) weighted by appropriate values (beam weights). The illumination pattern is no longer fixed by the optical design but by the beam weights. Different values of beam weights lead to different directions of pointing, and then beamformers provide a way to observe different locations at the same time, which leads to the advantage of observing a calibrator source simultaneously with the target field [38].

The BINGO's today optical configuration is a reflector system with focal-plane arrays (FPAs); the first extension to the ABDUS setup will be associating phased-array to the BINGO focal plane (so-called ABDUS-1), that is, to create a PA receiver system using Vivaldi antennas (single polarized tapered slot antennas) in two different directions, crossing, providing a two orthogonal polarizations and associated to the beamformers. The PAs will take advantage of the current BINGO optical design and will be placed beside the feed horns' structure, with (an initially proposed) total area of $2.8\,\mathrm{m} \times 20\,\mathrm{m}$, located in a good position of the BINGO's focal plane. At a second moment, there will be outriggers composed of feed horns with mirrors in their mouths and phased-array stations [14] (so-called ABDUS-2). The outriggers and phased array stations will be about 20 km distant from the main telescope to provide sufficient angular resolution through an interferometric approach to pinpoint FRBs and other possible radio sources.

The phased array stations are proposed to cover an area of about $36\,\mathrm{m}^2$ composed of 32 (sub-) tiles[5] of $1.4 \times 0.7\,\mathrm{m}^2$. The ABDUS-2 will be able to achieve an impressive leap in the sky coverage. Assuming the latitude of the stations as that of the current BINGO site (lat 7°2′27.6") and, in a good approximation, the Vivaldi antennas with a great response between $\sim \pm 45°$ in relation to their boresight, the new declination coverage will be within $(-52°, 38°)$ and the sky coverage of $32400\,\mathrm{deg}^2$. This new sky swath to be covered has a huge overlap with other highlighted radiotelescopes, e.g., $\sim$18846 deg^2 considering the FAST and Tianlai region described in the Table 1, between declination $-14.35°$ and $38°$, as shown in Fig. 2. This provided a fantastic opportunity to operate in a Very Large Baseline Interferometry (VLBI) mode with a huge maximum baseline (see Fig. 3).

Several projects are using, or are proposing to use, phased arrays to search Fast Radio Bursts (FRBs), what is crucial due to the low number of current FRB detections ($\gtrsim$800); however, though a 21 cm IM use was first done exploring the Parkes telescope in a crossed analysis with WiggleZ optical survey between 700 and 1084 MHz [26], ABDUS will open an important window of new and accurate data in a huge sky region, in a greater angular resolution, since the use of phased-arrays in the ABDUS range frequency is a gap to be filled in. The first analysis and studies address an increase of the Figure of Merit (FOM) for the Dark Energy equation of state from 19 to (>)3417, and it can achieve up to 12333 (SKA-MID with band1 and band2 can achieve 244 and 3213, respectively). Considering FRBs, as presented in [14] where we assumed only outriggers with different sizes of mirrors coupling them, for two baselines and 4 m mirrors, ABDUS will be able to localize by 20 events per year considering 10 outriggers. In the case of a set of three outriggers in a total of 50 sets (150 outriggers) pointing to the same sky direction, this number leaps to $\sim$180 localized FRBs (by at least 3 baselines) with redshifts up to 1.3.

[5] A tile is composed by four subtitles of $1.4 \times 0.35\,\mathrm{m}^2$ arranged parallel to the largest length.

Table 1 All projects are described here following your 21-cm intensity mapping operation. We will not focus on any other operational frequency bands or suitable characteristics in source monitoring and individual localizing. Any project can possibly have new features not covered here

Project	Type	θ_{FWHM}	Freq. (MHz)	N. of beams	Coverage (deg^2)	Dec.[a] (deg)	T_{sys} (K)
BINGO	Single-dish	40’	980–1260	28	5324	(−25, −10)	70
ABDUS-1	Phased-array	40’	450–1420	44	5324	(−25, −10)	40–60
ABDUS-2	Phased-array	27’	450–1420	200	32400	(−52, 38)	40–60
Tianlai cyl. array[b]	Cylinder interf.	1.8°[c]	700–800	96	10000	(−16, 90)	82–100
Tianlai dish array[d]	Dish interf.	12’	700–800	16	1500[e]	(48, 60)	80
FAST (CRAFTS)[f]	Single-dish	3’	1050–1450	19	20000	(−14, 66)	20
CHIME[g]	Cylinder interf.	40’	400–800	1024	31000	(−21, 65)	55
MeerKAT[h]	Single-dish	1.0°–1.6°	900–1420	64	3000	(−1, 8)	18–22
SKA1-MID[i]	Single-dish	30’-4°	350–1050	64+133	30000	–	25–63

[a] Declination
[b] Value based on [25]
[c] This value assumes the response for the E-W scan direction. For the N-S direction, the cylinder is basically unfocused with >60^0 [25]. Both measures mentioned are assumed at 750 MHz
[d] Data based on [35, 44]
[e] Coverage of 1500 deg^2 after masking. The declination range covers can be tuned to another sky region due to the mobility of the dishes
[f] Values extracted from [22, 48]
[g] Data based on [8]
[h] We are assuming in this table the values provided for L-band, extracted from [10, 43]. MeerKAT will also be able to operate in other modes, and one of the ongoing observations is using a UHF band covering 10000 deg^2
[i] Values assuming only Band1 in an autocorrelation mode as [40]. The declination corresponding to the coverage mentioned in the main papers was not identified

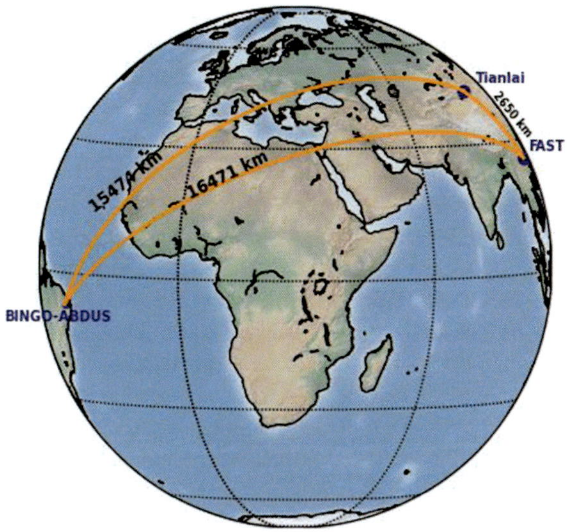

Fig. 3 Triangle formed by BINGO-ABDUS, FAST, and Tianlai and the distance between each on of them in kilometers

4 Other Similar Projects

BINGO is not the only project to measure the 21 cm post-reionization signals from the Universe. Other projects are operating, under development, or being planned. Each project aims to explore a specific frequency span, with a specific sky covering and strategy of observation, developing new instrumental tools and data treatment to achieve the data as precisely as possible within its own propositions. Even though the road to data would be slightly different, the overall goal is the same: helping the scientific community to unveil the nature of the Universe's dynamic behavior and not well-understood astrophysical phenomena.

Considering mainly the post-reionization era, and just to mention some projects, we give below a small summary. Not necessarily these projects below are constrained to the post-reionization, but they have good synergy with the BINGO goals, and all projects to be mentioned are complementary.

- **CHIME:** The Canadian Hydrogen Intensity Mapping Experiment (CHIME) is a radio interferometer located at the Dominion Radio Astrophysical Observatory (DRAO), near Penticton, British Columbia province of Canada, operating as a fixed and transient telescope exploring the drift scan mode covering all northeast celestial sky with a field-of-view (FOV) of $\sim 100°$ in declination. The experiment consists of an array with four $20\,\text{m} \times 100\,\text{m}$ cylindrical telescopes—the shorter dimension has a parabolic curvature—, each instrumented with 256 dual-polarized feeds [8]. The angular resolution is approximately 40' and the frequency resolution is 390 kHz spanning 400–800 MHz (corresponding 21 cm emissions from redshift range 0.8–2.5). In 2022, the CHIME collaboration reported the first detection of a 21 cm signal in cross-correlation with galaxies (luminous red galaxies and emission-line galaxies) and quasars by the eBOSS catalogs [12] through intensity mapping measurements made with an interferometer telescope.

- **MeerKAT:** Located in the arid Upper Karoo region, Northern Cape Province of South Africa, the MeerKAT[6] radio telescope is an array composed of 64 dishes of 13.5 m diameter—each one being an offset Gregorian design—over an area with a diameter of 8 km. 48 dishes are placed within 1 km diameter, and the other 16 antennas are mounted at larger distances up to a 4 km radius from the center. The radio telescope can operate using three possible receivers spanning different frequency bands: UHF-band (580–1015 MHz), L-band (900−1670 MHz), and S-band (1750−3500 MHz). MeerKAT, which will become a part of the final SKA1-Mid (see more above), can operate in either interferometric or single-dish mode. Single-dish observations were carried out with L-band (achieving $z < 0.56$), first in a pilot survey data covering a limited celestial region $153° < RA < 172°$ and $1° < Dec < 8°$ [43]; later, reporting detection of cross-power spectrum [10] with WiggleZ data [16] and the first direct detection of the HI intensity (auto-) power spectrum through intensity mapping at $z \approx 0.32$ and $z \approx 0.44$ [33].
- **SKA:** The conception of the *Square Kilometre Array* (SKA) radio telescope dates back to the 1990s—even without this name—with scientists and engineers from different parts of the world joining efforts to design and build the most sensitive radio telescope to date and only possible to be financed in cooperation on a global scale. The name *Square Kilometre Array* (SKA) was assumed in a radio astronomy conference in Calgary, province of Alberta, Canada, in 1998 [17], combining scientific and technical development to investigate the Universe in a large range of its history. Currently, the radio telescope is the biggest planned radio telescope to be operating in the Intensity Mapping observational method.
  The SKA's 21 cm IM survey operation is split into two phases (SKA1 and SKA2), but here we are commenting only on the most developed one, which is the SKA1.[7] The SKA phase 1 (SKA1) can be separated into two parts due to the receiver frequencies, location, and observational mode: SKA1-LOW and SKA1-MID.
  SKA-Low is hosted in the shire of Murchison, 300 km northeast of Geraldton, Western Australia, located on the traditional lands of the Wajarri Yamaji (Wajarri people). The area comprises 512 stations, with 256 dipole antennas per station, operating in an interferometric mode with a maximum distance between antenna stations of 74 km. Each station has its antennas distributed out within an area of a 35 m radius—designed to support a FOV of $\sim$20 deg^2 at $\sim$100 MHz—, and the core stations—composed of 224 stations and comprising most of the large-scale sensitive—are spread over 500 m, arriving at a maximum baseline size of 1 km [40]. This SKA part is prioritized for imaging and spectral observations of the 21 cm emissions at high redshift. Its stations will be observing in a 35 m (beam

[6] MeerKAT name has double meaning: "more of KAT", where "meer" is the Afrikaans word for "more" and KAT is the acronyms for "Karoo Array Telescope"; and, at the same time, it is a name of a suricate, a member of the mongoose family indigenous to Southern Africa.

[7] The SKA2 (also named as SKA full) configuration has not been decided yet; it has only been traced the main scientific and technical goals [7], achieving a sensibility 10x and a field of view 20x higher than the first phase.

former) station FWHM of $\sim 5°$ (at 100 MHz) and in a frequency range 50–350 MHz (corresponding to $3.06 \lesssim z \lesssim 27.4$, that is, crossing the epoch of reionization and reaching around the Cosmic Dawn epoch when the first lights arose), but only 50–200 MHz ($4.91 \lesssim z \lesssim 27.4$) will be used to 21 cm signals [24].
SKA-MID will be located at the same site as the MeerKAT (Karoo semi-desert plateau) once the latter is an SKA precursor. Its setup is composed of 64 13.5 m dishes from MeerKAT plus 133 (Gregorian offset) dishes with 15 m diameter, displayed in a compact center of 1 km diameter followed by a 3-arm spiral configuration reaching 150 km in the distance between the two extreme dishes. The frequency range is determined through five receiver bands: Band1 350–1050 MHz, Band2 950–1750 MHz, Band3 1650–3000 MHz, Band4 3000–5200 MHz, and 4600–15800 MHz,[8] with only Band1 and 2 being used to the 21 cm signal observations for redshifts ≤ 3 and a covering of $\sim 30000\ \mathrm{deg}^2$ (in a few thousand hours). The maximum baseline, 150 km, at 1.4 GHz, corresponds to an angular resolution $\sim 0.3'$. SKA-MID has the interesting issue of being able to work in both interferometric and single-dish modes. For the single-dish case, each dish can independently operate, and the resulting maps are co-added.

- **FAST:** The *Five-hundred-metre Aperture Spherical Telescope* is the largest and most sensitive (filled-aperture) single-dish radio telescope in the world [31]. The telescope is located in a karst depression called Dawodang in the southern of the Guizhou province, southwest of China. The main FAST reflector is a 500 m paraboloid with an illumination aperture of 300 m in diameter. The telescope reaches the impressive milestone of an effective collection area about 1.6 times larger than that of the Arecibo telescope [5]. FAST reaches, at 1.6 GHz, a beam size of 2.9' (3.5' of Arecibo), a system temperature of 18 K (30 K, Arecibo), spanning a frequency range of 70–3000 MHz, with seven sets of receivers, for $-14.4° <$ Dec $< 65.6°$ [39]. For 21 cm IM observations, there is an L-band 19-beam receiver between 1.05 and 1.45 GHz, with 18 feeds arranged in two concentric hexagonal rings plus one feed in the center. In addition, thanks to its high angular resolution, it is expected to resolve a large number of galaxy clusters, groups, filaments, and voids, reaching the detection of until 6.5×10^5 HI galaxies for $z \sim 0.07$ and a mass threshold of $10^{9.5} M_\odot$. Commissioning and pilot tests were done for the 21 cm IM drift-scan mode (with the L-band 19-beam feed system) in 2019, 2020, and 2021 to investigate frequency resolution to be assumed to allow a low level of radio frequency interference (RFI) and 1/f effects, level of the noise diode power, rotation of the feed arrangement to achieve another declination regions in the sky [23, 27, 50].
- **Tianlai:** This name in Mandarin means "heavenly sound" and represents here the international but Chinese-leaded project of 21 cm signals post-reionization through interferometry Intensity Mapping. Located at Hongliuxia village, Balikun county, in the Xinjiang region, northwest China, the experiment covers the entire northern celestial sky (and also part of the southern one) using the Earth's rotation [11]. At present, the project consists of two radio-interferometers: one is com-

[8] An additional Band6 is desired to cover the frequency range 15–24 GHz.

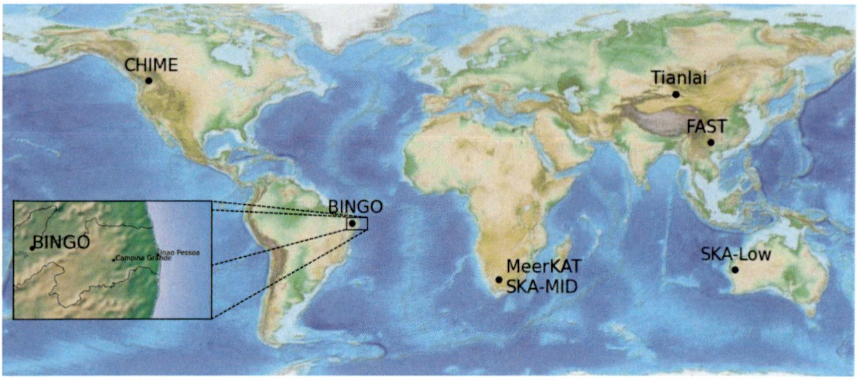

Fig. 4 Some HI Intensity mapping projects displayed around the world. There is a focus on the north of Brazil to show the state of Paraiba, and, in its countryside, the BINGO site located in the village of Aguiar

posed of three arrays of adjustable parabolic cylinder reflectors with 40 m × 15 m size,[9] composed of 96 dual-polarization feeds, and oriented in the N-S direction (called *Tianlai Cylinder Pathfinder Array*). The other radio-interference system (called *Tianlai Dish Pathfinder Array*) is arranged in a near-hexagonal configuration composed of 16 on-axis dishes of 6 m aperture and disposed of in two concentric rings [44]. At this first stage, called Pathfinder, the telescope operates at 700–800 MHz (corresponding to the redshift range 0.78–1.03). The future observations are planned to achieve the frequency range 1330–1430 MHz (for 21 cm emissions, this corresponds to the redshifts <0.068) (Fig. 4).

All the above-cited projects pursue not only 21 cm signals as a scientific goal. They are also important in the investigation of transient phenomena such as pulsars and fast radio bursts (FRB).

References

1. E. Abdalla, A. Marins, Int. J. Modern Phys. D **29**, 2030014 (2020). https://doi.org/10.1142/S0218271820300141
2. E. Abdalla, E.G.M. Ferreira, R.G. Landim et al., A&A **664**, A14 (2022). https://doi.org/10.1051/0004-6361/202140883
3. F.B. Abdalla, A. Marins, J. Alberto, N. Santos, E. Abdalla (2023). (in preparation)
4. F.B. Abdalla, A. Marins, P. Motta et al., A&A **664**, A16 (2022). https://doi.org/10.1051/0004-6361/202141382

[9] The goal is the full-scale Tianlai experiment consist of eight parallel cylinders increasing the length from 40 to 120 m, with a total of 2000 dual-polarization feeds in a frequency range spread between 400 and 1420 MHz [47].

5. D.R. Altschuler, in *Astronomical Society of the Pacific Conference Series*, vol. 278, Single-Dish Radio Astronomy: Techniques and Applications, ed. by S. Stanimirovic, D. Altschuler, P. Goldsmith, C. Salter (2002), pp. 1–24
6. B. Bassett, R. Hlozek, in *Dark Energy: Observational and Theoretical Approaches*, ed. by P. Ruiz-Lapuente (2010), p. 246
7. R. Braun, T. Bourke, J.A. Green, E. Keane, J. Wagg, Advancing astrophysics with the square kilometre array (AASKA14), vol. 174 (2015). https://doi.org/10.22323/1.215.0174
8. CHIME Collaboration, M. Amiri, K. Bandura et al., ApJS **261**, 29 (2022). https://doi.org/10.3847/1538-4365/ac6fd9
9. A.A. Costa, R.G. Landim, C.P. Novaes et al., A&A **664**, A20 (2022). https://doi.org/10.1051/0004-6361/202140888
10. S. Cunnington, Y. Li, M.G. Santos et al., Mon. Not. Roy. Astron. Soc. **518**, 6262 (2023). https://doi.org/10.1093/mnras/stac3060
11. S. Das, C.J. Anderson, R. Ansari et al., *Society of Photo-Optical Instrumentation Engineers (SPIE) Conference Series*, vol. 10708, in *Millimeter, Submillimeter, and Far-Infrared Detectors and Instrumentation for Astronomy IX*, ed. by J. Zmuidzinas, J.-R. Gao (2018), p. 1070836. https://doi.org/10.1117/12.2313031
12. K.S. Dawson, J.-P. Kneib, W.J. Percival et al., AJ **151**, 44 (2016). https://doi.org/10.3847/0004-6256/151/2/44
13. E.J. de Mericia, L.C.O. Santos, C.A. Wuensche et al., A&A **671**, A58 (2023). https://doi.org/10.1051/0004-6361/202243804
14. M.V. dos Santos, R.G. Landim, G.A. Hoerning et al. (2023). arXiv:2308.06805
15. C. Dragone, AT T Tech. J. **57**, 2663 (1978)
16. M.J. Drinkwater, R.J. Jurek, C. Blake et al., Mon. Not. Roy. Astron. Soc. **401**, 1429 (2010). https://doi.org/10.1111/j.1365-2966.2009.15754.x
17. R. Ekers (2012). arXiv:1212.3497. https://doi.org/10.48550/arXiv.1212.3497
18. G.B. Field, Proc. IRE **46**, 240 (1958). https://doi.org/10.1109/JRPROC.1958.286741
19. G.B. Field, ApJ **129**, 525 (1959). https://doi.org/10.1086/146652
20. G.B. Field, ApJ **129**, 536 (1959). https://doi.org/10.1086/146653
21. K.S.F. Fornazier, F.B. Abdalla, M. Remazeilles et al., A&A **664**, A18 (2022). https://doi.org/10.1051/0004-6361/202141707
22. W. Hu, X. Wang, F. Wu et al., Mon. Not. Roy. Astron. Soc. **493**, 5854 (2020). https://doi.org/10.1093/mnras/staa650
23. W. Hu, Y. Li, Y. Wang et al., Mon. Not. Roy. Astron. Soc. **508**, 2897 (2021). https://doi.org/10.1093/mnras/stab2728
24. M.G. Labate, P. Dewdney, R. Braun, M. Waterson, J. Wagg, in *2017 11th European Conference on Antennas and Propagation (EUCAP)* (2017), pp. 2259–2263. https://doi.org/10.23919/EuCAP.2017.7928622
25. J. Li, S. Zuo, F. Wu et al., Sci. China Phys. Mech. Astron. **63**, 129862 (2020). https://doi.org/10.1007/s11433-020-1594-8
26. L.-C. Li, L. Staveley-Smith, J. Rhee, Res. Astron. Astrophys. **21**, 030 (2021). https://doi.org/10.1088/1674-4527/21/2/30
27. Y. Li, Y. Wang, F. Deng et al. (2023). arXiv:2305.06405. https://doi.org/10.48550/arXiv.2305.06405
28. V. Liccardo, E.J. de Mericia, C.A. Wuensche et al., A&A **664**, A17 (2022). https://doi.org/10.1051/0004-6361/202140886
29. D.R. Lorimer, M. Bailes, M.A. McLaughlin, D.J. Narkevic, F. Crawford, Science **318**, 777 (2007). https://doi.org/10.1126/science.1147532
30. A. Marins, F.B. Abdalla, K.S.F. Fornazier et al. (2022). arXiv:2209.11701
31. R. Nan, D. Li, C. Jin et al., Int. J. Modern Phys. D **20**, 989 (2011). https://doi.org/10.1142/S0218271811019335
32. C. Paiva Novaes, J. Zhang, E.J. de Mericia et al. (2022). arXiv:2207.12125. https://doi.org/10.48550/arXiv.2207.12125

33. S. Paul, M.G. Santos, Z. Chen, L. Wolz (2023). arXiv:2301.11943. https://doi.org/10.48550/arXiv.2301.11943
34. M.W. Peel, C.A. Wuensche, E. Abdalla et al., J. Astron. Instrum. **8**, 1940005 (2019). https://doi.org/10.1142/S225117171940005110.48550/arXiv.1811.09464
35. O. Perdereau, R. Ansari, A. Stebbins et al., Mon. Not. Roy. Astron. Soc. **517**, 4637 (2022). https://doi.org/10.1093/mnras/stac2832
36. J.B. Peterson, R. Aleksan, R. Ansari et al., astro2010: the astronomy and astrophysics decadal survey **2010**, 234 (2009). https://doi.org/10.48550/arXiv.0902.3091
37. Planck Collaboration, N. Aghanim, Y. Akrami et al., A&A **641**, A6 (2020). https://doi.org/10.1051/0004-6361/201833910
38. D.C. Price, PASA **39**, e060 (2022). https://doi.org/10.1017/pasa.2022.54
39. L. Qian, R. Yao, J. Sun et al., The Innovation **1**, 100053 (2020). https://doi.org/10.1016/j.xinn.2020.100053
40. SKA Cosmology Science Working Group, D.J. Bacon, R.A. Battye et al., PASA **37**, e007 (2020). https://doi.org/10.1017/pasa.2019.51
41. R. Sunyaev, Y.B. Zeldovich, Astron. Astrophys. **20**, 189 (1972)
42. G. Swarup, Int. Astron. Union Colloq. **131**, 376–380 (1991). https://doi.org/10.1017/S0252921100013671
43. J. Wang, M.G. Santos, P. Bull et al., Mon. Not. Roy. Astron. Soc. **505**, 3698 (2021). https://doi.org/10.1093/mnras/stab1365
44. F. Wu, J. Li, S. Zuo et al., Mon. Not. Roy. Astron. Soc. **506**, 3455 (2021). https://doi.org/10.1093/mnras/stab1802
45. C.A. Wuensche, L. Reitano, M.W. Peel et al., Exp. Astron. **50**, 125 (2020). https://doi.org/10.1007/s10686-020-09666-9
46. C.A. Wuensche, T. Villela, E. Abdalla et al., A&A **664**, A15 (2022). https://doi.org/10.1051/0004-6361/202039962
47. Y. Xu, X. Wang, X. Chen, ApJ **798**, 40 (2015). https://doi.org/10.1088/0004-637X/798/1/40
48. E. Yohana, Y.-Z. Ma, D. Li, X. Chen, W.-M. Dai, Mon. Not. Roy. Astron. Soc. **504**, 5231 (2021). https://doi.org/10.1093/mnras/stab1197
49. B. Zhang, Nature **587**, 45 (2020). https://doi.org/10.1038/s41586-020-2828-1
50. B. Zhang, M. Zhu, Z.-Z. Wu et al., Mon. Not. Roy. Astron. Soc. **503**, 5385 (2021). https://doi.org/10.1093/mnras/stab754
51. J. Zhang, P. Motta, C.P. Novaes et al., A&A **664**, A19 (2022). https://doi.org/10.1051/0004-6361/202140887
52. F. Zwicky, ApJ **86**, 217 (1937). https://doi.org/10.1086/143864
53. Zwicky, General relativity and gravitation **41**, 207 (2009). https://doi.org/10.1007/s10714-008-0707-4

Gravitational Wave Astronomy

Machine Learning Applications in Gravitational Wave Astronomy

Nikolaos Stergioulas

Abstract Gravitational wave astronomy has emerged as a new branch of observational astronomy, since the first detection of gravitational waves in 2015. The current number of $O(100)$ detections is expected to grow by several orders of magnitude over the next two decades. As a result, current computationally expensive detection algorithms will become impractical. A solution to this problem, which has been explored in the last years, is the application of machine-learning techniques to accelerate the detection and parameter estimation of gravitational wave sources. In this chapter, several different applications are summarized, including the application of artificial neural networks and autoenconders in accelerating the computation of surrogate models, deep residual networks in achieving rapid detections with high sensitivity, as well as artificial neural networks for accelerating the construction of neutron star models in an alternative theory of gravity.

1 Introduction

Since 2015, when the first gravitational waves (GWs) from a binary black hole (BBH) system were detected [1], GW detections have become increasingly common, moving closer to the point of being a regular occurrence. After the third observing run (O3), the most recent catalog (GWTC-3, [2]) from the Advanced LIGO [3], Advanced Virgo [4] KAGRA [5, 6] collaboration contained 90 GW events, almost all of which were BBH mergers. The 4th observing run (O4) is currently underway and a larger number of BBH detections are expected [7]. The addition of a fifth interferometer, LIGO-India [8], is expected to significantly enhance both the sensitivity and the sky localization of the network. Moreover, third-generation ground-based detectors such as the Einstein Telescope [9, 10] and Cosmic Explorer [11, 12] are currently being developed and are anticipated to greatly expand our understanding of the astrophysical processes in the Universe [13–15].

N. Stergioulas (✉)
Department of Physics, Aristotle University of Thessaloniki, 54124 Thessaloniki, Greece
e-mail: niksterg@auth.gr

© The Author(s), under exclusive license to Springer Nature Switzerland AG 2024

E. Papantonopoulos and N. Mavromatos (eds.), *Compact Objects in the Universe*,
https://doi.org/10.1007/978-3-031-55098-0_12

The advances in GW astronomy described above were made possible by collaborative efforts in multiple areas. Accurate descriptions of the entire coalescence, including the full inspiral, merger, and ringdown, can be obtained in different ways, with `IMRPhenomXPHM` [16] and `SEOBNRv5PHM` [17] being two examples of waveform models. Recent implementations of these models take into account the spin-induced precession of the binary orbit and contributions from both the dominant and subdominant multipole moments of the emitted gravitational radiation. However, the increased complexity of the waveforms increases their computational cost.

Astronomical observations have enabled a number of attempts to determine the Equation of State (EoS) of Neutron Stars (NSs). These include the NICER mass and radius measurements [18–20], the measurement of tidal deformability through gravitational waves [21–24], as well as joint constraints, e.g., [25–28]. In particular, the detection of the binary NS merger GW170817 [29, 30] has prompted further research in this area.

In recent years, there has been an increase in the utilization of machine learning approaches for the analysis of gravitational wave data (see [31–33] for reviews). This chapter provides a summary of different machine learning applications to gravitational-wave astronomy presented in [34–37].

2 ANN-Accelerated Surrogate Models

Surrogate modeling has been provided to reduce the considerable computational cost of evaluating waveform models [38, 39], which can significantly speed up EOB waveforms (e.g., [38, 40–43]) while still providing high accuracy within its valid parameter range. The SEOBNRv4 model has a three-dimensional parameter space $\boldsymbol{\lambda}$; the mass ratio q between the two black holes and their spins χ_1 and χ_2, assuming that they are aligned with the orbital angular momentum. A surrogate model for this waveform family was presented in [44]. Several machine learning techniques can be used to interpolate or fit the projection coefficients of a reduced basis representation of time-domain waveforms, and the most suitable method depends on the desired accuracy and dimensionality. For low-dimensional parameter spaces, interpolation is a viable option. However, as the dimensionality increases, interpolation becomes difficult due to the large number of data points usually needed. Artificial Neural Networks (ANNs) are proposed as a solution to estimate these coefficients since this approach allows for efficient execution on either a CPU or GPU.

In [34], it was observed that the residual errors after training an ANN to evaluate the coefficients of the surrogate model for the SEOBNRv4 model had a pattern with respect to the input parameters. It was then demonstrated that a second neural network could be trained to model these errors, leading to an improved method, in which the maximum mismatch between SEOBNRv4 waveforms and waveforms generated by the new surrogate model was more than one order of magnitude smaller than the baseline method. Here, we will provide a summary of the steps taken to create the surrogate model and the residual ANN network to accelerate the evaluation of its coefficients, as described in [34].

2.1 Constructing a Surrogate Model

We express the complex gravitational wave strain as $h(t; \boldsymbol{\lambda}) = h_+(t; \boldsymbol{\lambda}) - ih_\times(t; \boldsymbol{\lambda})$, where h_+ and $h_\times$ are the two independent polarizations [45], t is the time, and $\boldsymbol{\lambda}$ is a vector of intrinsic parameters. The SEOBNRv4 model [46] has a three-dimensional parameter space, with each waveform characterized by the mass ratio q (the ratio of the masses of the two black holes) and the dimensionless spins χ_1, χ_2 of the two black holes. Surrogate modeling is a process of approximating given signals using a reduced model, denoted $h_s(t; \boldsymbol{\lambda})$, such that the approximation given by the surrogate model, $h_s(t; \boldsymbol{\lambda})$, accurately reconstructs the actual waveform $h(t; \boldsymbol{\lambda})$ within a preset threshold of error. When considering only the dominant, quadrupole ($l = m = 2$) mode [45], the target becomes $h_s(t; \boldsymbol{\lambda}) \approx h_{2,2}(t; \boldsymbol{\lambda})$ where l, m are the spherical harmonics. To begin the surrogate modeling process, a training set of N waveforms $\{h_i(t; \boldsymbol{\lambda}_i)\}_{i=1}^N$ is created, where $\boldsymbol{\lambda}_i = (q, \chi_1, \chi_2)_i$. The mass ratio is limited to a predetermined interval, such as $1 \leq q \leq 8$, within which the surrogate model is designed to be accurate. The two spins can have values in the range $-0.99 \leq \chi_{1,2} \leq 0.99$.

A *Reduced Order Method* (ROM) basis is constructed from a training set using a greedy algorithm [38]. This is an iterative process that selects $n < N$ waveforms (and their corresponding $\{\boldsymbol{\lambda}_j\}_{j=1}^n$ values, the greedy points) that, after orthonormalization, form the reduced basis $\{e_j\}_{j=1}^n$. Each $\boldsymbol{\lambda}_i$ waveform in the training set is then expressed as a linear combination

$$h(t; \boldsymbol{\lambda}_i) \approx \sum_{j=1}^{n} c_j(\boldsymbol{\lambda}_i) e_j(t), \tag{1}$$

within a given error tolerance, where $\{c_j(\boldsymbol{\lambda}_i)\}_{j=1}^n = \langle h(t; \boldsymbol{\lambda}_i), e_j(t) \rangle$ are the orthogonal projection coefficients.

Next, a new *Empirical Interpolation Method* (EIM) basis $B_k(t)$ is obtained such that a waveform $h(t; \boldsymbol{\lambda}_j)$ can be expressed as a linear combination of the basis, i.e., $h\left(t; \boldsymbol{\lambda}_j\right) = \sum_{k=1}^{n} \alpha_k(\boldsymbol{\lambda}_j) B_k(t)$. The coefficients $\alpha_k(\boldsymbol{\lambda}_j)$ are equal to the waveform at particular times, $\{T_k\}_{k=1}^n$, known as the empirical time nodes, i.e., $\alpha_k(\boldsymbol{\lambda}_j) = h\left(T_k; \boldsymbol{\lambda}_j\right)$. For any other waveform $h\left(t; \boldsymbol{\lambda}_i\right)$ in the training set, the coefficients of the EIM representation are $\alpha_k(\boldsymbol{\lambda}_i) = h\left(T_k; \boldsymbol{\lambda}_i\right)$. This does not require the basis $B_k(t)$, so the coefficients can be computed much faster than the projection coefficients in the ROM basis (which require the projection of the whole waveform).

In the end, a surrogate model is created by interpolating the coefficient matrix $\alpha_k(\boldsymbol{\lambda}_i)$ of the training set to find the coefficients $\hat{\alpha}_k(\boldsymbol{\lambda})$ for any $\boldsymbol{\lambda}$, such that

$$h\left(t; \boldsymbol{\lambda}\right) \approx \sum_{k=1}^{m} \hat{\alpha}_k(\boldsymbol{\lambda}) B_k(t). \tag{2}$$

The complexity of this process increases with the number of parameters in $\boldsymbol{\lambda}$. *Neural networks can be used to speed up this part of the process, as demonstrated in* [44].

In practice, the complex waveform can be expressed in terms of its *amplitude* A and *phase* ϕ, defined through

$$h_+(t;\boldsymbol{\lambda}) - h_\times(t;\boldsymbol{\lambda}) = A(t;\boldsymbol{\lambda})e^{-i\phi(t;\boldsymbol{\lambda})}, \tag{3}$$

which leads to a more compact EIM basis. To construct the ROM and EIM bases, a training set of $N = 2 \times 10^5$ waveforms was randomly sampled in the parameter space of $1 \leq q \leq 8$, $-0.99 \leq \chi_{1,2} \leq 0.99$. The waveforms were aligned in amplitude and initial phase, the phase was unwrapped, and the time series was truncated to a common starting time of $-20000M$, with a total mass of $M = 60M_\odot$. This ensured that all waveforms began with a minimum frequency no larger than 15 Hz, and $100M$ of post-peak ringdown data was kept. The ROM and EIM bases were created using `RomPy` [38, 47]. To evaluate the accuracy of the reconstructed waveforms (after the training is completed), a *validation set* of 3×10^4 SEOBNRv4 waveforms (not included in the training set) was used.

For two waveforms with parameters $\boldsymbol{\lambda}_1$ and $\boldsymbol{\lambda}_2$, the inner product can be defined [48]

$$\langle h(\cdot;\boldsymbol{\lambda}_1), h(\cdot;\boldsymbol{\lambda}_2)\rangle = 4\Re \int_{f_{min}}^{f_{max}} \frac{\tilde{h}(f;\boldsymbol{\lambda}_1)\tilde{h}^*(f;\boldsymbol{\lambda}_2)}{S_n(f)} df, \tag{4}$$

where $\tilde{h}(f;\boldsymbol{\lambda})$ is the Fourier transform of $h(t;\boldsymbol{\lambda})$, $S_n(f)$ denotes the noise power spectral density (PSD) of the GW detector and the star notation stands for the complex conjugate. The inner product can be employed to normalize the Fourier transform of a waveform in the following manner:

$$\hat{h}(f;\boldsymbol{\lambda}) = \frac{\tilde{h}(f;\boldsymbol{\lambda})}{\langle h(\cdot;\boldsymbol{\lambda}), h(\cdot;\boldsymbol{\lambda})\rangle}, \tag{5}$$

Then, the *overlap* between two waveforms is defined as the inner product between normalised waveforms $\hat{h}(\cdot;\boldsymbol{\lambda}_1)$, $\hat{h}(\cdot;\boldsymbol{\lambda}_2)$, maximised over a relative time (t_0) and phase (ϕ_0) shift between the two waveforms:

$$O(\hat{h}(\cdot;\boldsymbol{\lambda}_1), \hat{h}(\cdot;\boldsymbol{\lambda}_2)) = \max_{t_0,\phi_0} \langle h(\cdot;\boldsymbol{\lambda}_1), h(\cdot;\boldsymbol{\lambda}_2)\rangle, \tag{6}$$

and, finally the *mismatch* is given by

$$\mathcal{M}(\hat{h}(\cdot;\boldsymbol{\lambda}_1), \hat{h}(\cdot;\boldsymbol{\lambda}_2)) = 1 - O(\hat{h}(\cdot;\boldsymbol{\lambda}_1), \hat{h}(\cdot;\boldsymbol{\lambda}_2)). \tag{7}$$

The performance of the surrogate model can be evaluated by comparing the waveforms generated by the SEOBNRv4 model with the predictions of the surrogate, using the mismatch defined above.

2.2 Accelerating the Surrogate Model Using ANNs

To construct the surrogate model, an ANN was employed to interpolate the coefficients $\alpha_k(\boldsymbol{\lambda}_i)$ of the training set to find the coefficients $\hat{\alpha}_k(\boldsymbol{\lambda})$ for an arbitrary $\boldsymbol{\lambda}$. The improved model was compared with a baseline model that followed the architecture of [44]. The ANN had four hidden layers with 320 neurons in each. The batch size was 10^3 and the training lasted for 10^3 epochs. The Adam optimizer [49] with a learning rate of 10^{-3} and the ReLU activation function [50] were used for the amplitude network. For the phase network, the Adamax [49] optimizer with a learning rate of 10^{-2} and the softplus activation function [51] were employed. Preprocessing involved using $\log(q)$ as input instead of q, which was then scaled using the `StandardScaler` from `Scikit-Learn` [52]. At the output, the coefficients were used raw for the amplitude network and were scaled using `Scikit-Learn`'s `MinMaxScaler` for the phase network.

The ANN prediction of the EIM coefficients of the training set waveforms will be referred to as $\hat{\boldsymbol{y}}_i \equiv \{\hat{\alpha}_k(\boldsymbol{\lambda}_i)\}_{k=1}^{n}$. During training, the standard mean square error

$$MSE = \frac{1}{N}\sum_{i=1}^{N} \|\hat{\boldsymbol{y}}_i - \boldsymbol{y}_i\|_2^2 \tag{8}$$

was measured and minimized, where the $\|\cdot\|_2$ notation represents the Euclidean norm of a vector. The MSEs were in the range $\sim 10^{-8} - 10^{-7}$.

A second ANN was created to predict the residual errors after establishing the baseline ANN surrogate model. This was done due to the presence of structure in the residuals for some EIM coefficients, as seen in Fig. 1. The final predictions are the sum of the outputs of the two models. For all $\{\boldsymbol{\lambda}_i\}_{i=1}^{N}$ in the training set, one can obtain the corresponding predictions $\{\hat{\boldsymbol{y}}(\boldsymbol{\lambda}_i)\}_{i=1}^{N}$ and calculate the *residual*

$$\boldsymbol{e}_i \equiv \boldsymbol{y}(\boldsymbol{\lambda}_i) - \hat{\boldsymbol{y}}(\boldsymbol{\lambda}_i), \tag{9}$$

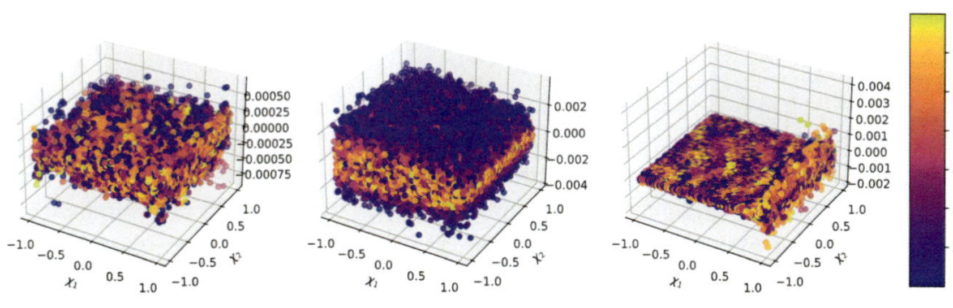

Fig. 1 The residual error for three chosen EIM coefficients for the amplitude is dependent on the input parameters $\boldsymbol{\lambda} = \{\chi_1, \chi_2, q\}$ (the dependence on q is illustrated with a colormap). The example in the left panel shows an unstructured distribution of residuals. In contrast, the example in the center panel reveals a strong dependence on the mass ratio q, while the example on the right displays a large residual error at the highest value of χ_1. Figure from [34]

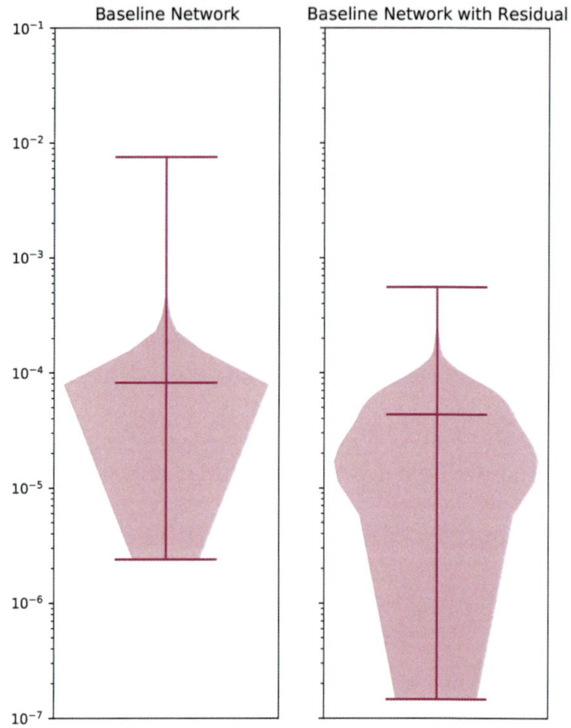

Fig. 2 A comparison of the mismatches (for the validation set) between the baseline network and the case when a second network that models the residual error is added is shown in the violin plots. The median is marked by the middle horizontal line, while the minimum and maximum values are indicated by the extent of the lines. The envelope of each panel is proportional to the density of points, and it is clear that a significant reduction of the mismatch is achieved when the network for the residual error is added. Figure from [34]

where, as already defined, $\boldsymbol{y}$ is the ground truth. The second network was created with the *same input and architecture as the first network*, but this time it was trained on the residuals $\boldsymbol{e}_i$ (which were first scaled using the "MinMaxScaler" from `scikit-learn` [52]) to make predictions for the residual $\hat{\boldsymbol{e}}(\boldsymbol{\lambda})$ at any $\boldsymbol{\lambda}$. When the prediction $\hat{\boldsymbol{e}}$ for the residual is added to the prediction $\hat{\boldsymbol{y}}$ of the first network, an improved prediction is obtained.

$$\tilde{\boldsymbol{y}} \equiv \hat{\boldsymbol{y}} + \hat{\boldsymbol{e}}. \tag{10}$$

Figure 2 illustrates the difference in mismatches (for the validation set) between the baseline network and the case where a second network is added that models the residual error, as a violin plot. The median is marked by the middle horizontal line, whereas the minimum and maximum values are shown by the extent of the lines. The envelope of each panel is proportional to the density of points. The results in Fig. 2 demonstrate that adding a second network to learn the residual errors is beneficial for constructing surrogate models for gravitational waves from BBH inspiral. This strategy is likely to be advantageous for other types of GW template banks, such as binary neutron star inspiral waveforms.

3 Efficient Surrogate Models Using Autoencoders

Autoencoders (AEs) are a type of unsupervised neural network that is trained to reproduce its input by first transforming it into a lower-dimensional representation [53]. Generally, an autoencoder consists of an encoding component that maps the input to a compressed representation and a decoding component that reconstructs the input. Encoding and decoding functions can have symmetrical or asymmetrical architectures and usually comprise multiple layers of fully connected layers, convolutional layers, or recurrent modules. AEs have been studied for a variety of tasks, such as clustering [54, 55], classification [56, 57], and image retrieval [58, 59], due to their ability to extract semantically meaningful representations without labels. A typical AE architecture is shown in Fig. 3, with the input and output layers having the same number of neurons.

In [35] a dataset comprising pairs of mass ratio q_i and corresponding EIM coefficient $\mathbf{a}_i$ ($i = 1, \ldots, N$), created with the EOBNRv2 nonspinning waveform model [60], was used to train an AE, with only the coefficients as input. This unsupervised process revealed a hidden relationship between each mass ratio q_i and the corresponding coefficients, as the mass ratios were unknown to the AE. Specifically, when choosing a two-dimensional intermediate representation, a spiral pattern emerged when visualizing this representation as a function of the mass ratio q, see Fig. 4. Below, we summarize the main steps presented in [35] to add a learnable spiral module to the ANN.

Following [38], a dataset of $N = 1000$ waveforms with mass ratios in the range $1 \leq q \leq 2$ was generated and a surrogate model was built, with a tolerance of 10^{-10}, resulting in a reduced basis of size $n = 11$. Next, a simple symmetric encoder-decoder AE architecture was used, with a two-dimensional hidden representation and two hidden fully-connected layers of 128 neurons on either side. The PReLU non-linearity [61] was used in all layers. The model was built using the PyTorch Deep

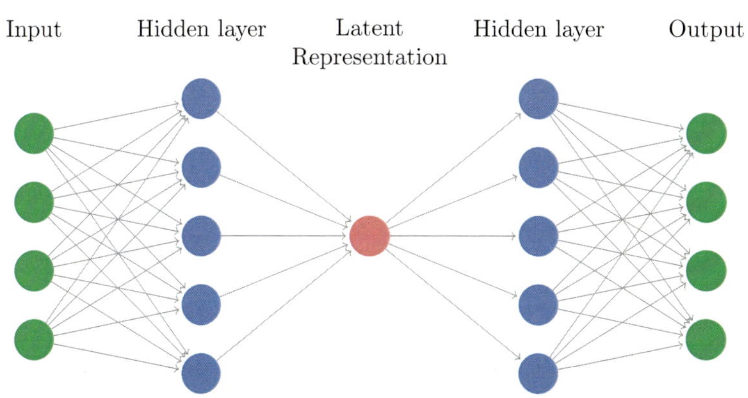

Fig. 3 Single hidden layer architecture of a fully connected Autoencoder. Figure from [35]

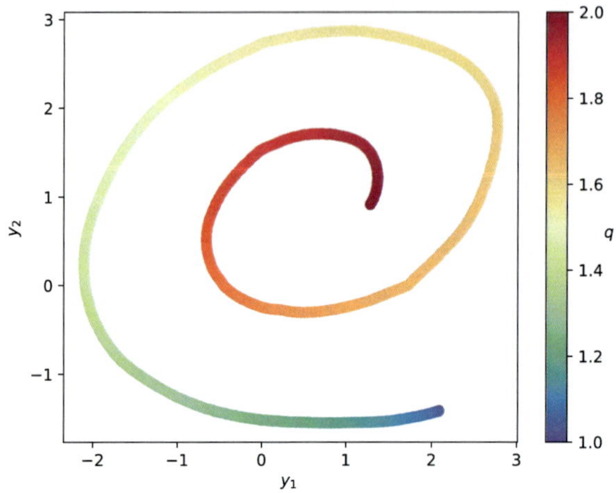

Fig. 4 Hidden representation uncovered by the AE for the empirical interpolation coefficients of a surrogate model of EOBNRv2 waveforms that is valid for $1 \leq q \leq 2$ (the color bar describes different values of the mass ratio q). When the values at the two neurons of the hidden representation are plotted against each other, a spiral structure emerges, along which the mass ratio appears to vary linearly with angle. Figure from [35]

Learning framework [62]. The EIM coefficients were used as input and output for this network. The AE was trained for 100 epochs with an initial learning rate of 0.001 and a batch size of 32. A multi-step multiplicative schedule was used with a gamma value of 0.9 and a step size of 15. The visual representation of the hidden layer is shown in Fig. 4, with the colors indicating the q values for each input coefficient. The spiral manifold in the hidden layer appears to describe a linear relationship between q and the angle θ of the spiral. The mean squared error of the reconstruction is 6.82×10^{-5}.

Based on the spiral pattern that emerged in Fig. 4, a neural spiral module was proposed in [35], which first transforms the input q into an angle θ, defined as

$$\theta := w \cdot q + b, \tag{11}$$

and subsequently maps θ onto a spiral structure of the form

$$\begin{aligned} s_x &:= (\alpha + \beta \cdot \theta) \cdot \cos\theta, \\ s_y &:= (\alpha + \beta \cdot \theta) \cdot \sin\theta, \end{aligned} \tag{12}$$

where w, b, α and β are parameters. These parameters are learnable, since the output is differentiable with respect to each of them. The spiral is fed to multiple, successive fully-connected layers, each with a nonlinear activation function, before reaching the final linear layer. An example of this architecture with two hidden layers is illustrated

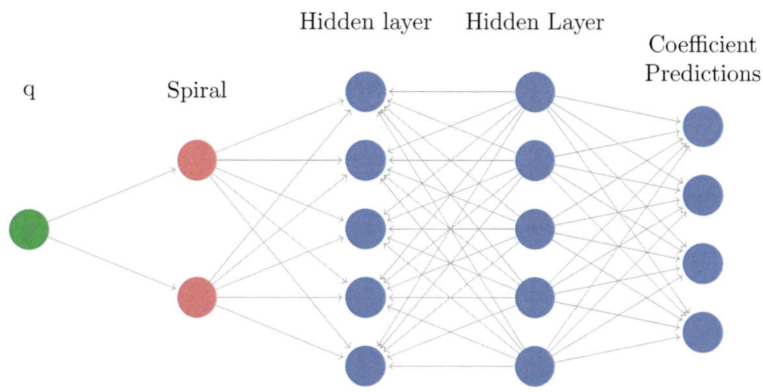

Fig. 5 Fully connected neural network with two hidden layers and the included spiral module. Figure from [35]

in Fig. 5. The inclusion of this module into an ANN accelerates the training process, leading to a significant reduction of the lowest achieved MSE.

The performance of various neural network architectures with fully-connected layers was assessed with and without the spiral module. The metrics used for evaluation were the waveform mismatch, inference speed, and memory requirements, with the maximum batch size that can be processed in a single forward pass on an NVIDIA RTX 2080 Ti GPU. All networks were trained for 2500 epochs with a batch size of 16, using the Adam optimizer [63] and an initial learning rate of 0.001, which was reduced by 0.95 every 150 epochs.

The inclusion of the spiral module significantly improved the mismatch achieved. When only one hidden layer was used, the baseline network with 128 neurons produced waveforms with a very poor mismatch (1.03×10^{-1} median mismatch). However, with the addition of the spiral module, even with only 32 hidden neurons, the median mismatch decreased by about 6 orders of magnitude. The best median and 95th percentile mismatch (9.41×10^{-9} and 3.48×10^{-8}) was achieved by the $\mathcal{S}$-32-64-128-64 network, which was able to generate up to 3.4 million coefficients in a single forward pass on the aforementioned GPU.

Finally, a spiral module was added to a neural network that was trained on a larger dataset of $N = 56000$ waveforms with $1 \leq q \leq 8$, where the q values were equidistant. A validation and a test set were also created, each with 14000 waveforms, and the q values were randomly chosen in the range of $1 \leq q \leq 8$. Figure 6 shows the real and imaginary parts of the first ten coefficients of the EIM basis, $\{a_j(q)\}_{j=1}^{10}$. Despite some modulation of amplitude, each coefficient has a sinusoidal dependence with q (except near $q = 1$, where $dq/da_j = 0$ for all j).

Several neural networks were trained and tested on the dataset. All networks were trained for 5000 epochs, with a batch size of 32, and the Adam optimizer [63] with an initial learning rate of 0.001, which was reduced by 0.9 every 30 epochs. The training and validation loss per epoch for the 32-64-128-64 network and the corresponding architecture with the addition of the spiral is shown in Fig. 7. The

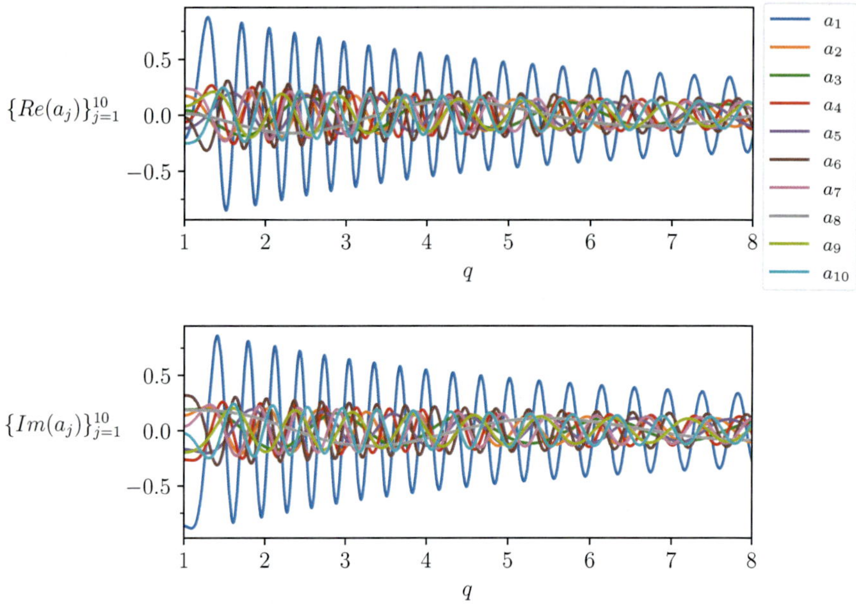

Fig. 6 Real (top) and imaginary parts (bottom) of the empirical interpolation coefficients $a_j(q)$ for a surrogate model of EOBNRv2 waveforms that is valid for $1 \leq q \leq 8$. Figure from [35]

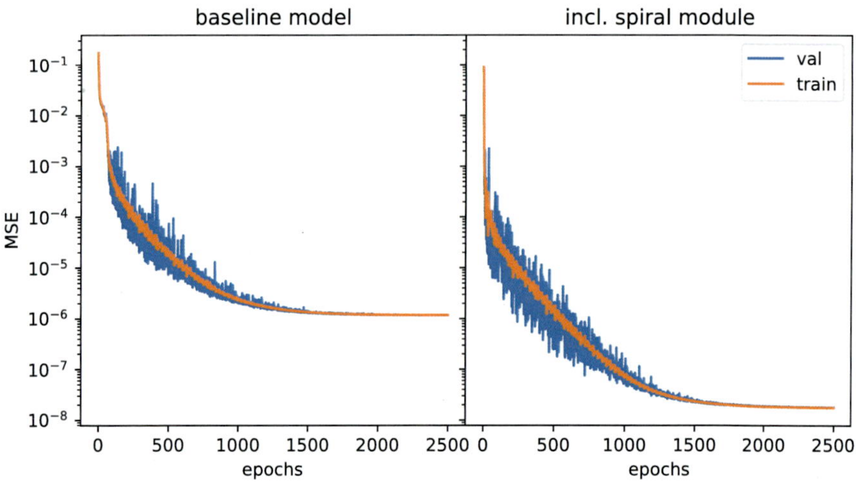

Fig. 7 Training and validation loss per epoch for the 32-64-128-64 network and the corresponding architecture with the addition of the spiral. Figure from [35]

spiral addition resulted in a lower mean squared error, allowing smaller networks to achieve the same accuracy as larger networks, leading to a larger batch size that can be processed in a single forward pass when using a specific GPU card.

4 GW Detection with Deep Residual Networks

The fourth observing run (O4) of gravitational wave detectors, which began in spring of 2023, is expected to result in a significant increase in the number of detections. This increase will be even more pronounced during O5 and the observing runs of the planned third-generation detectors (e.g., Cosmic Explorer [64] and Einstein Telescope [65]). However, the application of traditional matched-filtering techniques to obtain near real-time detection triggers is becoming increasingly costly or even impractical [66], due to both computational efficiency and accuracy. This is especially true for near-threshold systems with random spin directions, which require a much larger parameter space than the aligned-spin case. The situation will become even more difficult if template banks with departures from general relativity (GR) are included. Unmodeled search algorithms, on the other hand, have limited sensitivity, depending on the particular GW source.

Recently, the implementation of machine-learning (ML) methods, such as convolutional neural networks (CNN) or auto-encoders, has been investigated as an attractive solution to the problem of detecting gravitational waves (GWs), see e.g., [31–33, 67–101] for reviews. However, it has been difficult to evaluate the effectiveness of such efforts in a realistic setting. The first Machine-Learning Gravitational-Wave Mock Data Challenge (MLGWSC-1) was completed [102], providing an objective framework for testing the sensitivity and efficiency of ML algorithms on modeled injections in both Gaussian and real O3a detector noise in comparison to traditional algorithms. In [36], the leading ML algorithm in the case of real O3a noise was presented in more detail and it was shown that with further improvements it surpasses, for the first time, the results obtained with standard configurations of traditional algorithms in this specific setting. This was achieved for a component mass range between 7 and 50 $M_{\odot}$ (which corresponds to 70% of the announced events in the cumulative GWTC catalog [103]) and a relatively low false-alarm rate (FAR) as small as one per month.

The AresGW algorithm, described in [36], combines several components that increase the sensitive distance. It is based on a 54-layer one-dimensional deep residual network (ResNet) [104], which is more capable than a simpler CNN. Additionally, the Deep Adaptive Input Normalization (DAIN) [105] was included to address the non-stationary nature of real O3a noise. Furthermore, the dataset was augmented during training to improve the results. The execution speed was increased with the implementation of a framework-specific, module-based whitening layer, which computes the power spectral density (PSD) in batched tensor format. Finally, curriculum learning was used, which allowed the network to learn waveforms with the highest signal-to-noise ratio (SNR) first. The network was created using PyTorch [106] and trained (including validation) on 12 days of data in 31 h on an A6000 GPU (for 14 epochs). The evaluation of one month of test data on the same hardware took less than 2 h. The main findings of [36] are summarized below.

4.1 Training and Test Datasets

The training dataset in [36] spanned a period of 12 days and included real noise from the O3a LIGO run and injections of non-aligned binary black hole waveforms (in accordance with dataset 4 in [102]). Noise was taken from sections of O3a that are available from the Gravitational Wave Open Science Center (GWOSC) [107]. Only segments with a minimum length of 2 h, where both LIGO detectors had good quality data, and excluding 10 s around detections listed in GWTC-2 were included (see [102] for more information). Applying these criteria, the dataset had noise from each of the two aLIGO detectors, Hanford (H1) and Livingston (L1), with a total duration of 11 weeks and a sampling rate of 2048 Hz.

The waveforms injected into the training set were generated using the IMRPhenomXPHM waveform model [108], with a lower-frequency cutoff of 20 Hz. The masses of the individual components, m_1 and m_2, ranged from 7 to 50 solar masses, resulting in a maximum signal duration of 20 s. The signals were uniformly distributed in coalescence phase, polarization, inclination, declination, and right ascension (see [102] for more information). The waveforms were not uniformly distributed in volume, but instead, the chirp distance d_c [102] was sampled (as opposed to the luminosity distance d). This selection increases the number of low-mass systems that can be detected. The spins of the individual components had an isotropically distributed orientation with a magnitude between 0 and 0.99, which means that precession effects were present. All higher-order modes up to (4, −4) available in IMRPhenomXPHM were included. A representative data segment of the training set is shown in Fig. 8.

The training data set was made up of the first 12 days of the 11-week dataset, resulting in 740 k noise segments that formed the background noise. A set of 38k

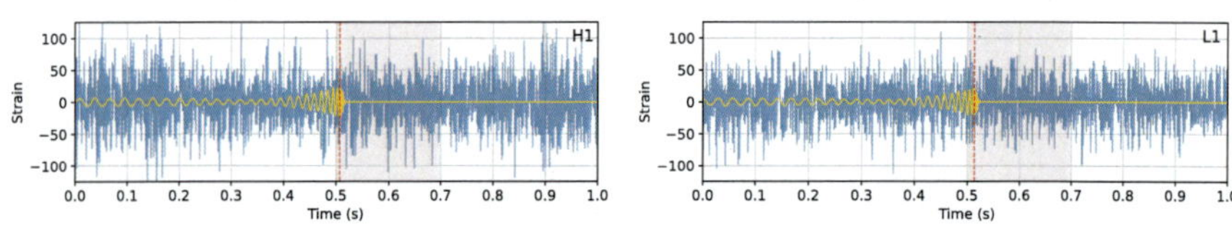

Fig. 8 A 2-channel segment of data from the training set used in [36] is displayed in the left and right panels, respectively, for the Hanford (H1) and Livingston (L1) detectors. The whitened strain of a 1 s segment is shown around the time of coalescence. The coalescence times in the detector frames are within the 0.5–0.7 s range (indicated by the shaded area). The injected waveform is scaled to match the difference between the whitened foreground and background segments. Figure from [36]

different waveforms was randomly injected into around 19 different background segments each, resulting in 740 k foreground segments that contain injections, creating a balanced training set. Additionally, a validation set was created, based on weeks 4 to 7 of the 11-week dataset and with a different random seed for injections. Lastly, the test data set consisted of noise from weeks 8-11 of the 11-week dataset and injections with merger times randomly spaced between 24 and 30 s. For the test dataset the same random seed and offset as in [102] was used.

The training data was first pre-processed using whitening as in [67, 109] and then normalized using the DAIN algorithm [105, 110]. DAIN is trained by backpropagating the network's gradients to its parameter. Furthermore, DAIN can adjust the normalization scheme applied to the input *during inference*, thus allowing it to handle non-stationary data.

4.2 Deep Residual Networks

Residual neural networks [104] employ skip connections to improve training, allowing gradients to better reach the earliest layers of the neural network architecture, thus solving the problem of vanishing gradients [111]. This, in combination with well-crafted training methods [112], leads to more effective training as the number of layers increases. This allows for the training of much deeper networks than simple CNNs.

The deep residual network developed in [36] was based on 1D convolutions for the binary classification of 1 s long (i.e., 2×2048-dimensional) segments into either positive (containing an injection) or negative (pure noise) segments. The network had a depth of 54 layers, which were grouped into 27 blocks containing two convolutional layers with a varying number of filters. Blocks 5, 8, 11, 14, and 17 were 2-strided, which means that the dimensionality was halved and an additional layer was used in the residual connection. Each convolutional layer was followed by batch normalization and a ReLU activation function. The two final individual convolutional layers reduced the output to a binary outcome (noise plus injected waveform vs. noise only). The Adam optimizer [113] was used for backpropagation and regularized binary cross entropy (a variant of the finite cross-entropy loss function [84]) was used as the objective function. During training, dynamic augmentation was also employed. A graphical representation of the network architecture is shown in Fig. 9.

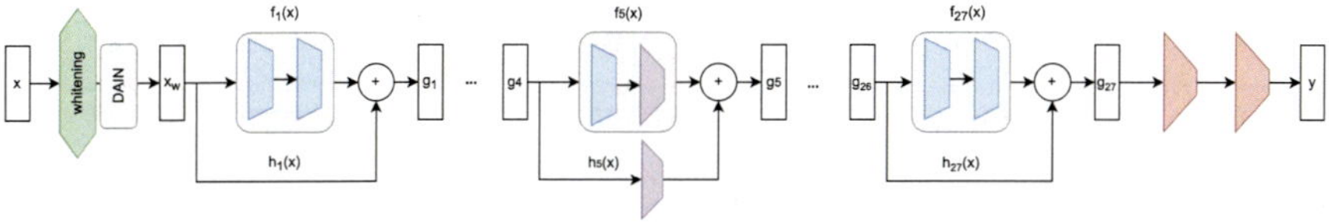

Fig. 9 Description of the residual network architecture in [36]. The input **x** is 2×2048-dimensional (see the text for details). Figure from [36]

A learning strategy was developed such that the network was initially trained on the strongest injections and only later on weaker injections. This is accomplished by using the optimal signal-to-noise ratio (SNR) of the injected signal, which is given by the equation

$$\mathrm{SNR} = 2\sqrt{\int_0^{\infty} df \frac{\tilde{h}(f)^2}{S_n(f)}}, \tag{13}$$

where $\tilde{h}(f)$ is the amplitude of the Fourier transform of the injected signal and $S_n(f)$ is the power spectral density of the detector noise. Instead of using the actual optimal SNR, an empirical relation was created that only depends on the chirp mass, the distance, and the inclination angle ι

$$\mathrm{SNR} = \frac{1261\ \mathrm{Mpc}}{D} \left(\frac{\mathcal{M}_c}{M_\odot}\right)^{5/7} [0.7 + 0.3\cos(2\,\iota)]\,. \tag{14}$$

Figure 10 displays a comparison between the optimal SNR (calculated using Eq. (13) and the PSD of the Hanford detector) for a sample of 10^4 randomly chosen injections and the approximate SNR as determined by Eq. (14). The two distributions are similar; however, the actual SNR is affected by a variety of other factors (such as sky location and spins).

The training process began with signals that were easily recognizable and had a high estimated SNR for the first four epochs. Subsequently, the network was gradually trained on weaker signals, and after the tenth epoch, it learned all the signals in the training set. Figure 10 shows the first eight epochs compared to the SNR distribution. As a result of the chosen learning strategy, initial losses were low. When the network had learned all the signals (after ten epochs), the loss of the training set was equal to that of the validation set.

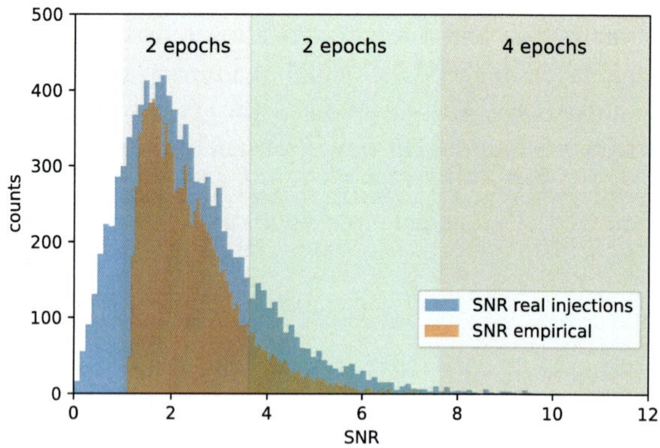

Fig. 10 Comparison of the SNR histogram, calculated with the empirical relation in Eq. (14), to the optimal SNR of 10^4 randomly chosen injections. The shaded areas in the graph represent the limited SNR values used in the first eight epochs of the *learning strategy*. Figure from [36]

4.3 Detection of BBH Injections in Real Noise

The trained network was used to analyze segments of the test dataset, producing a binary output that corresponds to the probability that an injection is present or that the segment only contains noise. The first output was used as a ranking statistic $\mathcal{R}$ (with values between 0 and 1). When $\mathcal{R} > 0.5$, a positive result was recorded. Clustering was applied to any positives that were detected within a time interval of 0.3 s, which were reported as a single detection (see Fig. 11 for an example). After deployment, the output was evaluated every 0.1 s and compared with the known injection times in the test dataset. If a positive output was within 0.3 s of the nominal merger time for a particular injection, it was classified as a true positive and, otherwise, as a false positive.

In order to assess the effectiveness of the search algorithm, one first calculates the false alarm rate as a function of the ranking statistic, that is, the FAR($\mathcal{R}$) function. Subsequently, the sensitivity of the search is determined as a function of the ranking statistic, which produces a relationship between the sensitivity and FAR (see [102] for definitions and the complete methodology).

Figure 12 shows the sensitivity distance as a function of FAR for the best model (ResNet54d+SNR), compared to a simpler setup (ResNet54) and two leading algorithms for GW detection, Coherent WaveBurst (cWB) and PyCBC. cWB is a waveform model-agnostic search pipeline for GW signals based on the constrained likelihood method [114–116], while PyCBC [117] is based on a standard configuration of archival search for compact-binary mergers [118]. The cWB and PyCBC results presented in Fig. 12 are taken from [102], where they were obtained on the same test dataset. cWB uses wavelets, which prevents it from achieving ideal fitting factors. It was recently improved with machine-learning techniques [119]. PyCBC implements matched filtering of waveform templates, but in [102, 118] only aligned-spin templates were used (for more general waveforms, the method could become com-

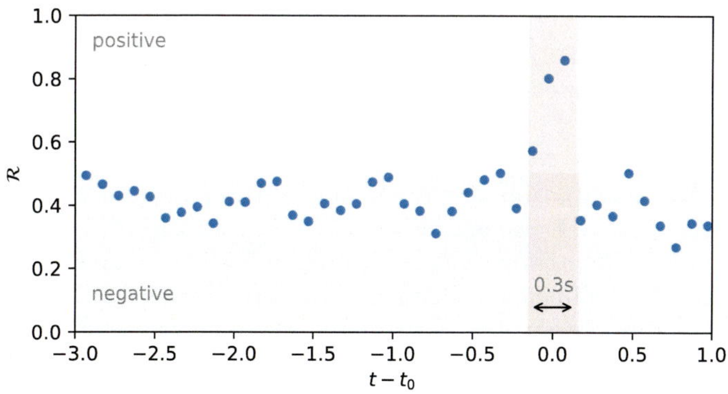

Fig. 11 Representative example of the ranking statistic for overlapping segments (with a stride of 0.1 s). Segments with $\mathcal{R} < 0.5$ are classified as negatives. Segments with $\mathcal{R} > 0.5$ that cluster within 0.3 s of a known injection at time t_0 are reported as a single true positive. Figure from [36]

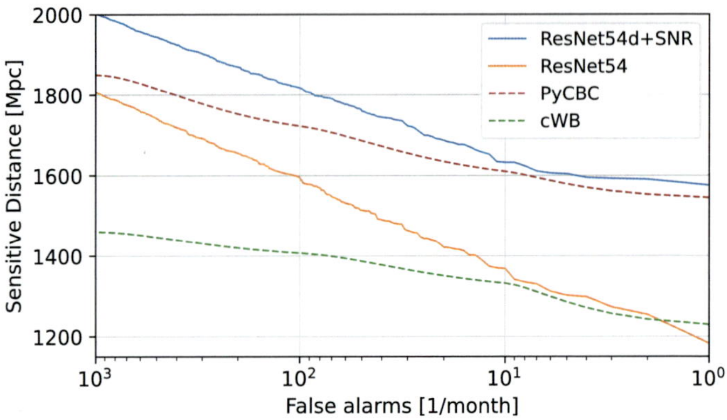

Fig. 12 Comparison of the performance of the best model (ResNet54d+SNR) in [36] with the simpler setup (ResNet54) used in [102] and two leading algorithms for GW detection, Coherent WaveBurst (cWB) and PyCBC (which only used aligned-spin templates, see text). All codes were tested on the same dataset established in [102]. The ResNet54d+SNR model outperformed the other algorithms at all false alarm rates in this setting. Figure from [36]

putationally too expensive). Since the injections in the test dataset of [36] were based on more general waveforms, this particular PyCBC search could not reach ideal fitting factors, leaving room for other algorithms to outperform it. As seen in Fig. 12, the best model presented in [36], which included SNR-based curriculum learning, outperformed the PyCBC results at all FAR and also significantly surpassed the sensitivity of the unmodeled cWB search.

5 ANN Surrogate Models of Neutron Star Mass-Radius Relations in Alternative Theories of Gravity

Several studies have used Artificial Neural Networks (ANNs) to reconstruct the EoS of neutron stars (NSs) based on their observable properties [120–124]. For instance, [125] investigated the use of ANNs with the autoencoder architecture, while [126, 127] employed ANNs to represent the EoS in a model-independent way, utilizing the unsupervised automatic differentiation framework. Other machine-learning techniques have been applied to explore the NS EoS. For example, [128] used a clustering method to identify patterns in mass-radius curves, and [129] investigated correlations among different EoSs of dense matter using unsupervised Machine Learning (ML) techniques. Additionally, attempts have been made to derive nuclear matter characteristics from NS EoS and observations using deep neural networks; see, e.g., [130, 131].

Neutron stars have been the focus of theoretical investigations in alternative theories of gravity (see [132, 133] for reviews and possible tests, and [134] for the particular theory discussed in this work). Bayesian statistics is commonly used to infer the NS EoS from observations of their macroscopic properties. This requires a TOV (Tolman–Oppenheimer–Volkoff) solver to be run multiple times to obtain the final posterior distribution of various parameters. If modified theories of gravity are to be included in these studies, a modified TOV solver, such as the one presented in [134], is needed. However, this algorithm is based on an iterative method for solving a differential equation system, which makes Bayesian inference computationally expensive. Therefore, it would be beneficial to find an alternative way to quickly and accurately predict the macroscopic properties of NSs, given some defining characteristics or other macroscopic properties of each equilibrium model.

Motivated by the need for a faster process, we developed an Artificial Neural Network (ANN) regression for two types of functions, $f_1(\text{EoS}; \alpha, p_c) \rightarrow (M, R)$ and $f_2(\text{EoS}; \alpha, M) \rightarrow R$, which was implemented in [37] and used in a Bayesian inference application in [135]. The EoS is a distinct variable, with each type having one ANN model for each EoS. α is the coupling constant of the theory and p_c is the central pressure of the NS. The first type serves as a surrogate model for the numerical iterative method described in [134], which provides the mass and radius of NSs for a specific EoS and a given pair of α and p_c. The aim is to speed up the process while still meeting strict accuracy requirements. The second type cannot be obtained directly using the iterative method since p_c must be an input. Therefore, implementing a root-finding algorithm would be the only solution, resulting in further time delays. On the other hand, training ANNs to predict R based on (EoS; α, M) offers a more straightforward approach to handling type f_2.

The 4D Horndeski scalar-tensor model, which was studied in [37], was derived from higher-dimensional Einstein-Gauss-Bonnet gravity. The action of this model included several nonlinear terms of a scalar field ϕ, and the mass and radius of the neutron star were affected by the strength of the coupling constant α, which has units of length square. If $\alpha = 0$, then the model is equivalent to Einstein's General Relativity. Further information can be found in [134].

6 Training and Testing

The numerical code from [134] requires two inputs to generate the data sets in [37]: the coupling constant α [km^2] of the theory and the central pressure p_c [10^{35} dyn/cm^2]. These inputs are used to calculate the mass M [$M_\odot$] and radius R [km] of the neutron star. For each type of function, 20 data sets were created, each based on one of the 20 tabulated equations of state.

The type of function $f_1(\text{EoS}; \alpha, \text{p}_\text{c})$ maps a given set of 51 values of α and 200 values of p_c to a unique pair of (M, R) values. The α values were evenly distributed on a linear scale, while the p_c values were logarithmically spaced between 0.1 and 1.2 times the central pressure required for a NS to reach its maximum mass for a specific value of the coupling constant. As an example, Fig. 13 shows the mapping of (α, p_c) pairs to (M, R) pairs for EoS BSk20. The type of function $f_2(\text{EoS}; \alpha, \text{M}) \rightarrow \text{R}$ had

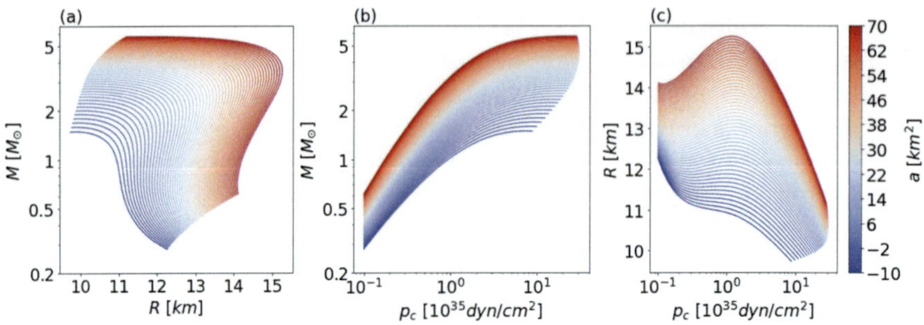

Fig. 13 Data set for EoS BSk20, for modeling the function f_1. Figure from [37]

Table 1 Final network architecture in [37]

Layer	Type f_1	Type f_2
Input layer	(α, p_c)	(α, M)
Hidden layer 1	25-tanh	25-tanh
Hidden layer 2	35-relu	35-relu
Hidden layer 3	25-tanh	25-tanh
Output layer	(M, R)	R

the same size as f_1, but the range of p_c values was chosen differently. The data comprised 51 values of α ranging from -10 to 70, and 200 values of p_c ranging from 0.1 to $p_{\max}$.

The TensorFlow module Keras[1] was employed with a 70:30 train-test ratio. The Mean Square Error (MSE) was selected as the loss function, while the Absolute Relative Error (ARE) was used as the criterion for evaluating the trained models. The ANN architecture for each type is presented in Table 1. The second-order *Broyden-Fletcher-Goldfarb-Shanno optimizer* algorithm (BFGS), as proposed in [136], was implemented as the preferred optimizer.[2]

Figure 14a provides an indication of the training results for f_1, comparing the actual output with the predicted output. Figure 14b displays the training loss per iteration in terms of the Mean Square Error (MSE) and Fig. 14c presents the Absolute Relative Error (ARE) at a given test point i

$$\mathrm{ARE}_i = \frac{1}{m} \sum_{j=1}^{m} \left| \frac{Y_i^j - Y_{i,\,\mathrm{true}}^j}{Y_{i,\,\mathrm{true}}^j} \right| \tag{15}$$

where m is the number of output neurons, Y_i^j is the network output with index j for test point with index i, and $Y_{i,\mathrm{true}}^j$ is the corresponding real output. The Mean ARE

[1] https://www.tensorflow.org/api_docs/python/tf/keras.

[2] BFGS is not included in the list of provided Keras optimizers.

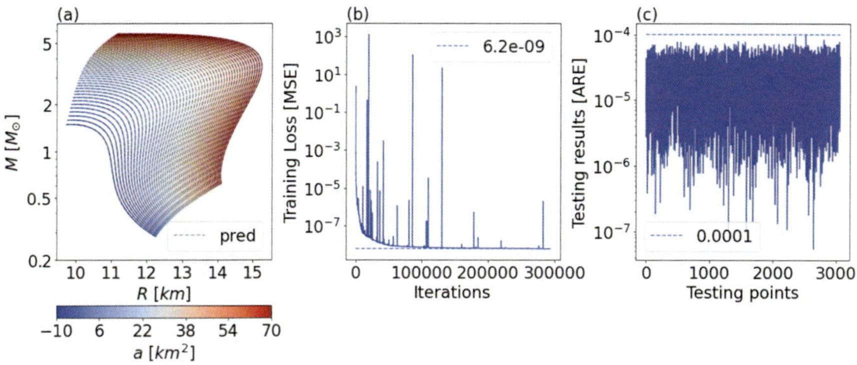

Fig. 14 Training results for f_1 and EoS BSk20. Figure from [37]

over all $n = 3060$ points in the test dataset is denoted as MARE. The MARE ranged between 10^{-5} and 10^{-4} for all EOSs, while the maximum ARE was 6×10^{-4} for f_1 and 9×10^{-3} for f_2, indicating that the absolute relative error never exceeded 1% in the entire domain of the training and test sets.

6.1 Speed-Up

The *speed-up* s when comparing the speed of the trained ANN models of f_1 to the speed of the numerical code, is defined as:

$$s = \frac{\Delta t_{\mathrm{ANN}}}{\Delta t_{\mathrm{num}}}, \tag{16}$$

where Δt_{ANN} is the run time when using an ANN model and Δt_{num} the run time of the numerical code with the iterative numerical scheme. The output of the models can be calculated in three different ways, leading to a different speed-up:

1. `model.predict(`X`)`, with X being one input value,
2. `model(`X`)`, with X being one input value,
3. `model.predict(`$\mathbf{X}$`)`, with $\mathbf{X}$ being an array of input values.

The left panel of Fig. 15 shows the speed up for `model.predict(`X`)`. Panel Fig. 15a displays the speed up using different colors, and the dashed lines with increasing transparency represent six different $M - R$ curves for $a = \{-10, 6, 22, 38, 54, 70\}$ km^2 respectively. Panel Fig. 15b shows the speed-up values for a particular arrangement of the data points, which are sorted in ascending order of α and p_c. Most of the speed-up values were between 10 and 100, with less than 1% outside of this range. Generally, larger input values tend to result in higher speed-ups, with an average speed-up of approximately 25 across all data points.

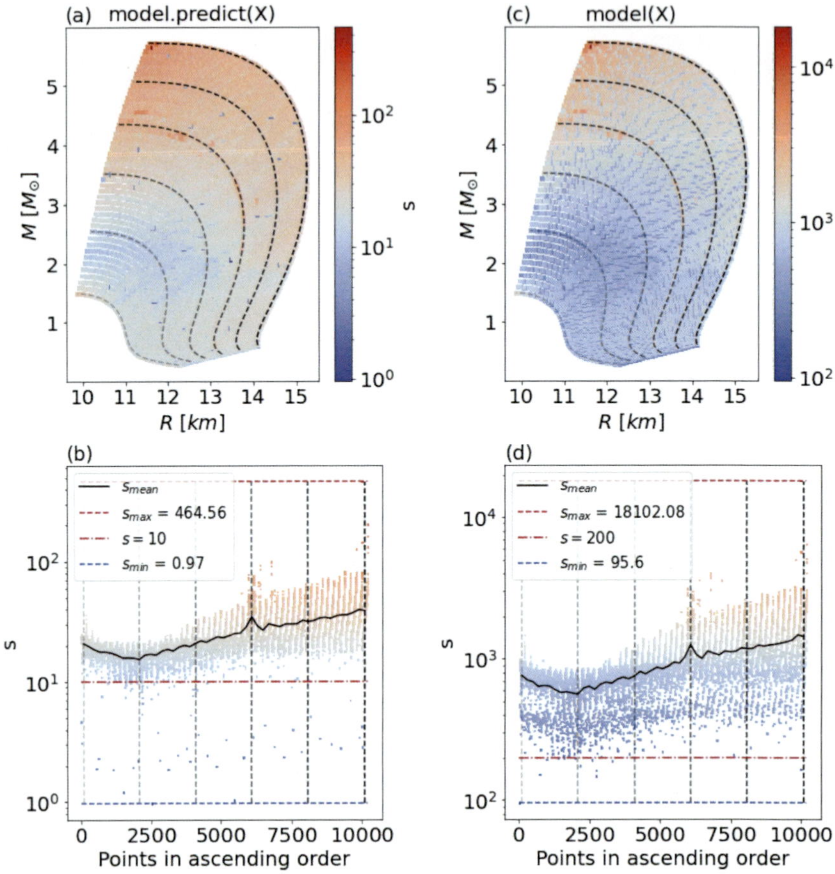

Fig. 15 Speed-up per $M - R$ data point showing how many times faster the trained ANN model is than the numerical code in each case. Figure from [37]

The right panel of Fig. 15 shows the acceleration for the `model(X)` case, which is set up similarly to the left panel. Most of the speed-up values are between 200 and 18000 (less than 0.2% are outside of this range). The average speed-up, taking into account all data points, was greater than 900. For the `model.predict(`**X**`)` case, the 10200 data points were entered as an array of input values **X**. In this case, the *effective run time* Δt_{eff}, can be calculated, which is defined as

$$\Delta t_{eff} = \frac{\Delta t_N}{N}, \tag{17}$$

where Δt_N is the *total run time* for the whole array as input and $N = 10200$. The speed-up ranged from ∼4600 to ∼565000, with a mean value of ∼31000.

7 Conclusions

This comprehensive review has highlighted several advancements in the field of gravitational wave astronomy through the application of machine learning techniques. The exploration of various neural network architectures, including deep residual networks and autoencoders, has demonstrated the potential for improvements in the detection and analysis of gravitational wave signals. The successful integration of machine learning in this domain not only concerns the optimization of data analysis, but also can open new avenues for understanding complex astrophysical phenomena. Moreover, the application of machine learning in neutron star mass-radius relation studies in alternative theories of gravity could be used to break the degeneracy between the equation of state and some alternative theories of gravity.

In conclusion, the integration of machine learning techniques in gravitational wave astronomy and neutron star studies represents a paradigm shift. Not only does it enhance the accuracy and efficiency of existing methodologies, but it also paves the way for novel discoveries in the realm of compact object astrophysics. The continued collaboration between the gravitational wave and machine learning communities is vital for further advancements.

Acknowledgements I am grateful to my collaborators, Theocharis Apostolatos, Stella Fragkouli, Panagiotis Iosif, Alexandra Koloniari, Ioannis Liodis, Paraskevi Nousi, George Pappas, Nikolaos Passalis, Evangelos Smyrniotis, and Anastasios Tefas, for their contributions that led to the main publications summarized in this review. Many thanks to Elena Cuoco for comments on the manuscript. This work was carried out within the framework of the EU COST action No. CA17137. This research has made use of data or software obtained from the Gravitational Wave Open Science Center (gwosc.org), a service of the LIGO Scientific Collaboration, the Virgo Collaboration, and KAGRA. This material is based upon work supported by NSF's LIGO Laboratory which is a major facility fully funded by the National Science Foundation, as well as the Science and Technology Facilities Council (STFC) of the United Kingdom, the Max-Planck-Society (MPS), and the State of Niedersachsen/Germany for support of the construction of Advanced LIGO and construction and operation of the GEO600 detector. Additional support for Advanced LIGO was provided by the Australian Research Council. Virgo is funded, through the European Gravitational Observatory (EGO), by the French Centre National de Recherche Scientifique (CNRS), the Italian Istituto Nazionale di Fisica Nucleare (INFN) and the Dutch Nikhef, with contributions by institutions from Belgium, Germany, Greece, Hungary, Ireland, Japan, Monaco, Poland, Portugal, Spain. KAGRA is supported by Ministry of Education, Culture, Sports, Science and Technology (MEXT), Japan Society for the Promotion of Science (JSPS) in Japan; National Research Foundation (NRF) and Ministry of Science and ICT (MSIT) in Korea; Academia Sinica (AS) and National Science and Technology Council (NSTC) in Taiwan.

References

1. B.P. Abbott et al., Observation of gravitational waves from a binary black hole merger. Phys. Rev. Lett. **116**(6), 061102 (2016)
2. R. Abbott et al., GWTC-3: Compact Binary Coalescences Observed by LIGO and Virgo During the Second Part of the Third Observing Run (2021). arXiv:2111.03606

3. J. Aasi et al., Advanced LIGO. Class. Quant. Grav. **32**, 074001 (2015)
4. F. Acernese et al., Advanced Virgo: a second-generation interferometric gravitational wave detector. Class. Quantum Grav. **32**(2), 024001 (2014)
5. T Akutsu et al., KAGRA: 2.5 generation interferometric gravitational wave detector. Nat. Astron. **3**(1), 35–40 (2019)
6. T. Akutsu et al., Overview of KAGRA: detector design and construction history. Prog. Theor. Exper. Phys. **2021**(5), 05A101 (2020)
7. B.P. Abbott et al., Prospects for observing and localizing gravitational-wave transients with Advanced LIGO, Advanced Virgo and KAGRA. Living Rev. Relat. **23**(1), 3 (2020)
8. M. Saleem et al., The science case for LIGO-India. Class. Quantum Grav. **39**(2), 025004 (2022)
9. M. Punturo et al., The Einstein Telescope: a third-generation gravitational wave observatory. Class. Quantum Grav. **27**(19), 194002 (2010)
10. M. Maggiore et al., Science case for the Einstein telescope. J. Cosmol. Astropart. Phys. **2020**(3), 050 (2020)
11. D. Reitze et al., Cosmic explorer: the U.S. contribution to gravitational-wave astronomy beyond LIGO. Bull. Am. Astron. Soc. **51**, 35 (2019)
12. M. Evans et al., A horizon study for cosmic explorer: science, observatories, and community (2021). arXiv:2109.09882
13. B.P. Abbott et al., Exploring the sensitivity of next generation gravitational wave detectors. Class. Quantum Grav. **34**(4), 044001 (2017)
14. D. Reitze, M. Punturo, P. Couvares, S. Katsanevas, T. Kajita, V. Kalogera, H. Lueck, D. McClelland, S. Rowan, G. Sanders, B.S. Sathyaprakash, D. Shoemaker, J. van den Brand, Expanding the reach of gravitational wave astronomy to the edge of the universe: the gravitational-wave international committee study reports on next generation ground-based gravitational-wave observatories (2021). arXiv:2111.06986
15. V. Kalogera, B.S. Sathyaprakash, M. Bailes, M.-A. Bizouard, et al., The next generation global gravitational wave observatory: the science book (2021). arXiv:2111.06990
16. G. Pratten, C. García-Quirós, M. Colleoni, A. Ramos-Buades, H. Estellés, M. Mateu-Lucena, R. Jaume, M. Haney, D. Keitel, J.E. Thompson, S. Husa, Computationally efficient models for the dominant and subdominant harmonic modes of precessing binary black holes. Phys. Rev. D **103**, 104056 (2021)
17. A. Ramos-Buades, A. Buonanno, H. Estellés, M. Khalil, D.P. Mihaylov, S. Ossokine, L. Pompili, M. Shiferaw, SEOBNRv5PHM: next generation of accurate and efficient multipolar precessing-spin effective-one-body waveforms for binary black holes (2023). arXiv:2303.18046
18. T.E. Riley, A.L. Watts, S. Bogdanov, P.S. Ray, R.M. Ludlam, S. Guillot, Z. Arzoumanian, C.L. Baker, A.V. Bilous, D. Chakrabarty, K.C. Gendreau, A.K. Harding, W.C.G. Ho, J.M. Lattimer, S.M. Morsink, T.E. Strohmayer, A nicer view of PSR j0030+0451: millisecond pulsar parameter estimation. Astrophys. J. **887**(1), L21 (2019)
19. M.C. Miller, F.K. Lamb, A.J. Dittmann, S. Bogdanov, Z. Arzoumanian, K.C. Gendreau, S. Guillot, A.K. Harding, W.C.G. Ho, J.M. Lattimer, R.M. Ludlam, S. Mahmoodifar, S.M. Morsink, P.S. Ray, T.E. Strohmayer, K.S. Wood, T. Enoto, R. Foster, T. Okajima, G. Prigozhin, Y. Soong, Psr j0030+0451 mass and radius from nicer data and implications for the properties of neutron star matter. Astrophys. J. Lett. **887**(1), L24 (2019)
20. M.C. Miller, F.K. Lamb, A.J. Dittmann, S. Bogdanov, Z. Arzoumanian, K.C. Gendreau, S. Guillot, W.C.G. Ho, J.M. Lattimer, M. Loewenstein, S.M. Morsink, P.S. Ray, M.T. Wolff, C.L. Baker, T. Cazeau, S. Manthripragada, C.B. Markwardt, T. Okajima, S. Pollard, I. Cognard, H.T. Cromartie, E. Fonseca, L. Guillemot, M. Kerr, A. Parthasarathy, T.T. Pennucci, S. Ransom, I. Stairs, The radius of psr j0740+6620 from nicer and xmm-newton data. Astrophys. J. Lett. **918**(2), L28 (2021)
21. Eric D. Van Oeveren, John L. Friedman, Upper limit set by causality on the tidal deformability of a neutron star. Phys. Rev. D **95**(8) (2017)
22. T. Hinderer, Tidal Love numbers of neutron stars. Astrophys. J. **677**, 1216–1220 (2008)

23. K. Chatziioannou, Neutron-star tidal deformability and equation-of-state constraints. Gen. Relat. Grav. **52**(11), 109 (2020)
24. T. Dietrich, T. Hinderer, A. Samajdar, Interpreting binary neutron star mergers: describing the binary neutron star dynamics, modelling gravitational waveforms, and analyzing detections. Gen. Relat. Grav. **53**(3), 27 (2021)
25. B. Biswas, Bayesian model selection of neutron star equations of state using multi-messenger observations. Astrophys. J. **926**(1), 75 (2022)
26. T. Dietrich, M.W. Coughlin, P.T.H. Pang, M. Bulla, J. Heinzel, L. Issa, I. Tews, S. Antier, Multimessenger constraints on the neutron-star equation of state and the hubble constant. Science **370**(6523), 1450–1453 (2020)
27. P. Landry, R. Essick, K. Chatziioannou, Nonparametric constraints on neutron star matter with existing and upcoming gravitational wave and pulsar observations. Phys. Rev. D **101**(12), 123007 (2020)
28. G. Raaijmakers, S.K. Greif, K. Hebeler, T. Hinderer, S. Nissanke, A. Schwenk, T.E. Riley, A.L. Watts, J.M. Lattimer, W.C.G. Ho, Constraints on the dense matter equation of state and neutron star properties from NICER's mass-radius estimate of PSR J0740+6620 and multimessenger observations. Astrophys. J. Lett. **918**(2), L29 (2021)
29. B.P. Abbott et al., GW170817: observation of gravitational waves from a binary neutron star inspiral. Phys. Rev. Lett. **119**(16), 161101 (2017)
30. B.P. Abbott et al., Properties of the binary neutron star merger GW170817. Phys. Rev. X **9**(1), 011001 (2019)
31. E. Cuoco, J. Powell, M. Cavaglià, K. Ackley, M. Bejger, C. Chatterjee, M. Coughlin, S. Coughlin, P. Easter, R. Essick, H. Gabbard, T. Gebhard, S. Ghosh, L. Haegel, A. Iess, D. Keitel, Z. Márka, S. Márka, F. Morawski, T. Nguyen, R. Ormiston, M. Pürrer, M. Razzano, K. Staats, G. Vajente, D. Williams, Enhancing gravitational-wave science with machine learning. Mach. Learn.: Scie. Technol. **2**(1), 011002 (2020)
32. V. Benedetto, F. Gissi, G. Ciaparrone, L. Troiano, Ai in gravitational wave analysis, an overview. Appl. Sci. **13**(17) (2023)
33. T. Zhao, R. Shi, Y. Zhou, Z. Cao, Z. Ren, Dawning of a new era in gravitational wave data analysis: unveiling cosmic mysteries via artificial intelligence – a systematic review (2023). arXiv:2311.15585
34. S.-C. Fragkouli, P. Nousi, N. Passalis, P. Iosif, N. Stergioulas, A. Tefas, Deep residual error and bag-of-tricks learning for gravitational wave surrogate modeling. Appl. Soft Comput. **147**, 110746 (2023)
35. P. Nousi, S.-C. Fragkouli, N. Passalis, P. Iosif, T. Apostolatos, G. Pappas, N. Stergioulas, A. Tefas, Autoencoder-driven spiral representation learning for gravitational wave surrogate modelling. Neurocomputing **491**, 67–77 (2022)
36. P. Nousi, A.E. Koloniari, N. Passalis, P. Iosif, N. Stergioulas, A. Tefas, Deep residual networks for gravitational wave detection. Phys. Rev. D **108**(2), 024022 (2023)
37. I. Liodis, E. Smirniotis, N. Stergioulas, A neural-network-based surrogate model for the properties of neutron stars in 4D Einstein-Gauss-Bonnet gravity (2023). arXiv:2309.03991
38. S.E. Field, C.R. Galley, J.S. Hesthaven, J. Kaye, M. Tiglio, Fast prediction and evaluation of gravitational waveforms using surrogate models. Phys. Rev. X **4**(3), 031006 (2014)
39. M. Tiglio, A. Villanueva, Reduced order and surrogate models for gravitational waves (2021). arXiv:2101.11608
40. M. Pürrer, Frequency domain reduced order model of aligned-spin effective-one-body waveforms with generic mass ratios and spins. Phys. Rev. D **93**(6), 064041 (2016)
41. B.D. Lackey, S. Bernuzzi, C.R. Galley, J. Meidam, C. Van Den Broeck, Effective-one-body waveforms for binary neutron stars using surrogate models. Phys. Rev. D **95**(10), 104036 (2017)
42. B.D. Lackey, M. Pürrer, A. Taracchini, S. Marsat, Surrogate model for an aligned-spin effective-one-body waveform model of binary neutron star inspirals using Gaussian process regression. Phys. Rev. D **100**(2), 024002 (2019)

43. Q. Yun, W.-B. Han, X. Zhong, C.A. Benavides-Gallego, Surrogate model for gravitational waveforms of spin-aligned binary black holes with eccentricities. Phys. Rev. D **103**(12), 124053 (2021)
44. S. Khan, R. Green, Gravitational-wave surrogate models powered by artificial neural networks. Phys. Rev. D **103**(6), 064015 (2021)
45. M. Maggiore, *Gravitational Waves Volume 1: Theory and Experiments* (Oxford University Press, 2008)
46. A. Bohé, L. Shao, A. Taracchini, A. Buonanno, S. Babak, I.W. Harry, I. Hinder, S. Ossokine, M. Pürrer, V. Raymond, T. Chu, H. Fong, P. Kumar, H.P. Pfeiffer, M. Boyle, D.A. Hemberger, L.E. Kidder, G. Lovelace, M.A. Scheel, B. Szilágyi, Improved effective-one-body model of spinning, nonprecessing binary black holes for the era of gravitational-wave astrophysics with advanced detectors. Phys. Rev. D **95**(4), 044028 (2017)
47. C.R. Galley, RomPy package (2020). https://bitbucket.org/chadgalley/rompy/
48. C. Cutler, E.E. Flanagan, Gravitational waves from merging compact binaries: how accurately can one extract the binary's parameters from the inspiral waveform? Phys. Rev. D **49**(6), 2658–2697 (1994)
49. D.P. Kingma, J. Ba, Adam: a method for stochastic optimization (2017)
50. V. Nair, G.E. Hinton, Rectified linear units improve restricted boltzmann machines, in *ICML*, pp. 807–814 (2010)
51. H. Zheng, Z. Yang, W. Liu, J. Liang, Y. Li, Improving deep neural networks using softplus units, in *2015 International Joint Conference on Neural Networks (IJCNN)*, pp. 1–4 (2015)
52. scikit learn. 6.3. preprocessing data. https://scikit-learn.org/stable/modules/preprocessing.html
53. P. Vincent, H. Larochelle, Y. Bengio, P.-A. Manzagol, Extracting and composing robust features with denoising autoencoders, in *Proceedings of the 25th International Conference on Machine Learning*, pp. 1096–1103 (2008)
54. J. Xie, R. Girshick, A. Farhadi, Unsupervised deep embedding for clustering analysis, in *International Conference on Machine Learning*, pp. 478–487 (2016)
55. P. Nousi, A. Tefas, Self-supervised autoencoders for clustering and classification. Evol. Syst. 1–14 (2018)
56. P. Nousi, A. Tefas, Deep learning algorithms for discriminant autoencoding. Neurocomputing **266**, 325–335 (2017)
57. P. Nousi, A. Tefas, Discriminatively trained autoencoders for fast and accurate face recognition, in *International Conference on Engineering Applications of Neural Networks*, pp. 205–215 (Springer, 2017)
58. P. Wu, S.C.H. Hoi, H. Xia, P. Zhao, D. Wang, C. Miao, Online multimodal deep similarity learning with application to image retrieval, in *Proceedings of the 21st ACM International Conference on Multimedia*, pp. 153–162 (2013)
59. M.A. Carreira-Perpinán, R. Raziperchikolaei, Hashing with binary autoencoders, in *Proceedings of the IEEE Conference on Computer Vision and Pattern Recognition*, pp. 557–566 (2015)
60. Y. Pan, A. Buonanno, M. Boyle, L.T. Buchman, L.E. Kidder, H.P. Pfeiffer, M.A. Scheel, Inspiral-merger-ringdown multipolar waveforms of nonspinning black-hole binaries using the effective-one-body formalism. Phys. Rev. D **84**(12), 124052 (2011)
61. K. He, X. Zhang, S. Ren, J. Sun, Delving deep into rectifiers: surpassing human-level performance on imagenet classification, in *Proceedings of the IEEE International Conference on Computer Vision*, pp. 1026–1034 (2015)
62. A. Paszke, S. Gross, S. Chintala, G. Chanan, E. Yang, Z. De-Vito, Z. Lin, A. Desmaison, L. Antiga, A. Lerer, Automatic differentiation in pytorch, in *NIPS-W* (2017)
63. D.P. Kingma, J. Ba, Adam: a method for stochastic optimization (2014). arXiv:1412.6980
64. D. Reitze, R.X. Adhikari, S. Ballmer, B. Barish, L. Barsotti, G. Billingsley, D.A. Brown, Y. Chen, D. Coyne, R. Eisenstein, M. Evans, P. Fritschel, E.D. Hall, A. Lazzarini, G. Lovelace, J. Read, B.S. Sathyaprakash, D. Shoemaker, J. Smith, C. Torrie, S. Vitale, R. Weiss, C. Wipf, M. Zucker, Cosmic explorer: the U.S. contribution to gravitational-wave astronomy beyond LIGO. Bull. Amer. Astron. Soc. **51**, 35 (2019)

65. M. Maggiore, C. Van Den Broeck, N. Bartolo, E. Belgacem, D. Bertacca, M.A. Bizouard, M. Branchesi, S. Clesse, S. Foffa, J. García-Bellido, S. Grimm, J. Harms, T. Hinderer, S. Matarrese, C. Palomba, M. Peloso, A. Ricciardone, M. Sakellariadou, Science case for the Einstein telescope. J. Cosmol. Astropart. Phys. **2020**(3), 050 (2020)
66. P. Couvares, I. Bird, Ed Porter, S. Bagnasco, M. Punturo, D. Reitze, S. Katsanevas, T. Kajita, V. Kalogera, H. Lueck, D. McClelland, S. Rowan, G. Sanders, B.S. Sathyaprakash, D. Shoemaker, J. van den Brand, Gravitational wave data analysis: computing challenges in the 3G Era (2021). arXiv:2111.06987
67. H. Gabbard, M. Williams, F. Hayes, C. Messenger, Matching matched filtering with deep networks for gravitational-wave astronomy. Phys. Rev. Lett. **120**, 141103 (2018)
68. D. George, E.A. Huerta, Deep neural networks to enable real-time multimessenger astrophysics. Phys. Rev. D **97**, 044039 (2018)
69. T.D. Gebhard, N. Kilbertus, I. Harry, B. Schölkopf, Convolutional neural networks: a magic bullet for gravitational-wave detection? Phys. Rev. D **100**(6), 063015 (2019)
70. R. Corizzo, M. Ceci, E. Zdravevski, N. Japkowicz, Scalable auto-encoders for gravitational waves detection from time series data. Expert Syst. Appl. **151**, 113378 (2020)
71. M.B. Schäfer, F. Ohme, A.H. Nitz, Detection of gravitational-wave signals from binary neutron star mergers using machine learning. Phys. Rev. D **102**, 063015 (2020)
72. H. Wang, W. Shichao, Z. Cao, X. Liu, J.-Y. Zhu, Gravitational-wave signal recognition of LIGO data by deep learning. Phys. Rev. D **101**(10), 104003 (2020)
73. P.G. Krastev, Real-time detection of gravitational waves from binary neutron stars using artificial neural networks. Phys. Lett. B **803**, 135330 (2020)
74. V. Skliris, M.R.K. Norman, P.J. Sutton, Real-time detection of unmodelled gravitational-wave transients using convolutional neural networks (2020). arXiv:2009.14611
75. Y.-C. Lin, J.-H.P. Wu, Detection of gravitational waves using Bayesian neural networks. Phys. Rev. D **103**(6), 063034 (2021)
76. E.A. Huerta, A. Khan, X. Huang, M. Tian, M. Levental, R. Chard, W. Wei, M. Heflin, D.S. Katz, V. Kindratenko, D. Mu, B. Blaiszik, I. Foster, Accelerated, scalable and reproducible AI-driven gravitational wave detection. Nat. Astron. **5**, 1062–1068 (2021)
77. T. Marianer, D. Poznanski, J.X. Prochaska, A semisupervised machine learning search for never-seen gravitational-wave sources. Mon. Not. R. Astr. Soc. **500**(4), 5408–5419 (2021)
78. W. Wei, A. Khan, E.A. Huerta, X. Huang, M. Tian, Deep learning ensemble for real-time gravitational wave detection of spinning binary black hole mergers. Phys. Lett. B **812**, 136029 (2021)
79. S. Jadhav, N. Mukund, B. Gadre, S. Mitra, S. Abraham, Improving significance of binary black hole mergers in Advanced LIGO data using deep learning: confirmation of GW151216. Phys. Rev. D **104**(6), 064051 (2021)
80. P. Chaturvedi, A. Khan, M. Tian, E.A. Huerta, H. Zheng, Inference-optimized ai and high performance computing for gravitational wave detection at scale. Front. Artif. Intell. **5** (2022)
81. S. Choudhary, A. More, S. Suyamprakasam, S. Bose, SiGMa-Net: deep learning network to distinguish binary black hole signals from short-duration noise transients (2022). arXiv:2202.08671
82. M.B. Schäfer, A.H. Nitz, From one to many: a deep learning coincident gravitational-wave search. Phys. Rev. D **105**, 043003 (2022)
83. F.P. Barone, D. Dell'Aquila, M. Russo, A novel multi-layer modular approach for real-time gravitational-wave detection (2022). arXiv:2206.06004
84. M.B. Schäfer, O. Zelenka, A.H. Nitz, F. Ohme, B. Brügmann, Training strategies for deep learning gravitational-wave searches. Phys. Rev. D **105**(4), 043002 (2022)
85. G. Baltus, J. Janquart, M. Lopez, H. Narola, J.-R. Cudell, Convolutional neural network for gravitational-wave early alert: going down in frequency. Phys. Rev. D **106**, 042002 (2022)
86. M. Andrews, M. Paulini, L. Sellers, A. Bobrick, G. Martire, H. Vestal, DeepSNR: a deep learning foundation for offline gravitational wave detection (2022). arXiv:2207.04749
87. C. Verma, A. Reza, G. Gaur, D. Krishnaswamy, S. Caudill, Can convolution neural networks be used for detection of gravitational waves from precessing black hole systems? (2022) arXiv:2206.12673

88. J. Aveiro, F.F. Freitas, M. Ferreira, A. Onofre, C. Providência, G. Gonçalves, J.A. Font, Identification of binary neutron star mergers in gravitational-wave data using object-detection machine learning models. Phys. Rev. D **106**(8), 084059 (2022)
89. W. Guo, D. Williams, I.S. Heng, H. Gabbard, Y.-B. Bae, G. Kang, Z.-H. Zhu, Mimicking mergers: mistaking black hole captures as mergers. MNRAS **516**(3), 3847–3860 (2022)
90. M. Andrés-Carcasona, A. Menéndez-Vázquez, M. Martínez, L.M. Mir, Searches for mass-asymmetric compact binary coalescence events using neural networks in the LIGO/Virgo third observation period. Phys. Rev. D **107**(8), 082003 (2023)
91. J. Langendorff, A. Kolmus, J. Janquart, C. Van Den Broeck, Normalizing flows as an avenue to studying overlapping gravitational wave signals. Phys. Rev. Lett. **130**(17), 171402 (2023)
92. M. Dax, S.R. Green, J. Gair, M. Pürrer, J. Wildberger, J.H. Macke, A. Buonanno, B. Schölkopf, Neural importance sampling for rapid and reliable gravitational-wave inference. Phys. Rev. Lett. **130**(17), 171403 (2023)
93. S. Bini, G. Vedovato, M. Drago, F. Salemi, G.A. Prodi, An autoencoder neural network integrated into gravitational-wave burst searches to improve the rejection of noise transients. Class. Quantum Grav. **40**(13), 135008 (2023)
94. M. Tian, E.A. Huerta, H. Zheng, Physics-inspired spatiotemporal-graph AI ensemble for gravitational wave detection (2023). arXiv:2306.15728
95. C. Murali, D. Lumley, Detecting and denoising gravitational wave signals from binary black holes using deep learning. Phys. Rev. D **108**(4), 043024 (2023)
96. P. Bacon, A. Trovato, M. Bejger, Denoising gravitational-wave signals from binary black holes with a dilated convolutional autoencoder. Mach. Learn.: Sci. Technol. **4**(3), 035024 (2023)
97. A. McLeod, D. Jacobs, C. Chatterjee, L. Wen, F. Panther, Rapid mass parameter estimation of binary black hole coalescences using deep learning (2022). arXiv:2201.11126
98. R. Qiu, P.G. Krastev, K. Gill, E. Berger, Deep learning detection and classification of gravitational waves from neutron star-black hole mergers. Phys. Lett. B **840**, 137850 (2023)
99. S.-J. Jin, Y.-X. Wang, T.-Y. Sun, J.-F. Zhang, X. Zhang, Rapid identification of time-frequency domain gravitational wave signals from binary black holes using deep learning (2023). arXiv:2305.19003
100. T. Fernandes, S. Vieira, A. Onofre, J.C. Bustillo, A. Torres-Forné, J.A. Font, Convolutional neural networks for the classification of glitches in gravitational-wave data streams. Class. Quantum Grav. **40**(19), 195018 (2023)
101. O.G. Freitas, J.C. Bustillo, J.A. Font, S. Nunes, A. Onofre, A. Torres-Forné, Comparison of neural network architectures for feature extraction from binary black hole merger waveforms (2023). arXiv:2307.16668
102. M.B. Schäfer, O. Zelenka, A.H. Nitz, H. Wang, S. Wu, Z.-K. Guo, Z. Cao, Z. Ren, P. Nousi, N. Stergioulas, P. Iosif, A.E. Koloniari, A. Tefas, N. Passalis, F. Salemi, G. Vedovato, S. Klimenko, T. Mishra, B. Brügmann, E. Cuoco, E.A. Huerta, C. Messenger, F. Ohme, MLGWSC-1: the first machine learning gravitational-wave search mock data challenge (2022). arXiv:2209.11146
103. R. Abbott et al., GWTC-3: compact binary coalescences observed by LIGO and virgo during the second part of the third observing run (2021). arXiv:2111.03606
104. K. He, X. Zhang, S. Ren, J. Sun, Deep residual learning for image recognition, in *Proceedings of the IEEE Conference on Computer Vision and Pattern Recognition*, pp. 770–778 (2016)
105. N. Passalis, A. Tefas, J. Kanniainen, M. Gabbouj, A. Iosifidis, Deep adaptive input normalization for price forecasting using limit order book data, in *IEEE Transactions on Neural Networks and Learning Systems* (2019)
106. A. Paszke, S. Gross, F. Massa, A. Lerer, J. Bradbury, G. Chanan, T. Killeen, Z. Lin, N. Gimelshein, L. Antiga, A. Desmaison, A. Kopf, E. Yang, Z. DeVito, M. Raison, A. Tejani, S. Chilamkurthy, B. Steiner, L. Fang, J. Bai, S. Chintala, Pytorch: an imperative style, high-performance deep learning library, in *Advances in Neural Information Processing Systems 32*, pp. 8024–8035 (Curran Associates, Inc., 2019)

107. R. Abbott et al., Open data from the third observing run of LIGO, Virgo, KAGRA, and GEO. Astrophys. J. Suppl. **267**(2), 29 (2023)
108. G. Pratten, C. García-Quirós, M. Colleoni, A. Ramos-Buades, H. Estellés, M. Mateu-Lucena, R. Jaume, M. Haney, D. Keitel, J.E. Thompson, S. Husa, Computationally efficient models for the dominant and subdominant harmonic modes of precessing binary black holes. Phys. Rev. D **103**, 104056 (2021)
109. S.A. Usman, A.H. Nitz, I.W. Harry, C.M. Biwer, D.A. Brown, M. Cabero, C.D. Capano, T.D. Canton, T. Dent, S. Fairhurst, et al., The pycbc search for gravitational waves from compact binary coalescence. Class. Quantum Grav. **33**(21), 215004 (2016)
110. N. Passalis, J. Kanniainen, M. Gabbouj, A. Iosifidis, A. Tefas, Forecasting financial time series using robust deep adaptive input normalization. J. Signal Process. Syst. **93**(10), 1235–1251 (2021)
111. B. Hanin, Which neural net architectures give rise to exploding and vanishing gradients? in *Advances in Neural Information Processing Systems*, 31 (2018)
112. R. Wightman, H. Touvron, H. Jégou, Resnet strikes back: an improved training procedure in timm (2021). arXiv:2110.00476
113. D.P. Kingma, J.Ba. Adam: a method for stochastic optimization (2014)
114. S. Klimenko, G. Vedovato, M. Drago, F. Salemi, V. Tiwari, G.A. Prodi, C. Lazzaro, K. Ackley, S. Tiwari, C.F. Da Silva, G. Mitselmakher, Method for detection and reconstruction of gravitational wave transients with networks of advanced detectors. Phys. Rev. D **93**, 042004 (2016)
115. M. Drago, S. Klimenko, C. Lazzaro, E. Milotti, G. Mitselmakher, V. Necula, B. O'Brian, G. Prodi, F. Salemi, M. Szczepanczyk, S. Tiwari, V. Tiwari, V. Gayathri, G. Vedovato, I. Yakushin, Coherent waveburst, a pipeline for unmodeled gravitational-wave data analysis. SoftwareX **14**, 100678 (2021)
116. S. Klimenko, G. Vedovato, V. Necula, F. Salemi, M. Drago, R. Poulton, E. Chassande-Mottin, V. Tiwari, C. Lazzaro, B. O'Brian, M. Szczepanczyk, S. Tiwari, V. Gayathri, cwb pipeline library: 6.4.1 (2021)
117. A. Nitz, I. Harry, D. Brown, C.M. Biwer, J. Willis, T.D. Canton, C. Capano, T. Dent, L. Pekowsky, A.R. Williamson, S. De, M. Cabero, B. Machenschalk, D. Macleod, P. Kumar, F. Pannarale, S. Reyes, G.S.C. Davies, dfinstad, S. Kumar, M. Tápai, L. Singer, S. Khan, S. Fairhurst, A. Nielsen, S. Singh, T. Massinger, K. Chandra, Shasvath, Veronica-villa, gwastro/pycbc: v2.0.5 release of pycbc (2022)
118. A.H. Nitz, S. Kumar, Y.-F. Wang, S. Kastha, S. Wu, M. Schäfer, R. Dhurkunde, C.D. Capano, 4-OGC: catalog of gravitational waves from compact-binary mergers (2021). arXiv:2112.06878
119. T. Mishra, B. O'Brien, M. Szczepańczyk, G. Vedovato, S. Bhaumik, V. Gayathri, G. Prodi, F. Salemi, E. Milotti, I. Bartos, S. Klimenko, Search for binary black hole mergers in the third observing run of advanced ligo-virgo using coherent waveburst enhanced with machine learning. Phys. Rev. D **105**, 083018 (2022)
120. Y. Fujimoto, K. Fukushima, K. Murase, Mapping neutron star data to the equation of state using the deep neural network. Phys. Rev. D **101**, 054016 (2020)
121. Y. Fujimoto, K. Fukushima, K. Murase, Methodology study of machine learning for the neutron star equation of state. Phys. Rev. D **98**, 023019 (2018)
122. Y. Fujimoto, K. Fukushima, K. Murase, Extensive studies of the neutron star equation of state from the deep learning inference with the observational data augmentation. J. High Energy Phys. **2021**(3) (2021)
123. M. Ferreira, C. Providência, Unveiling the nuclear matter eos from neutron star properties: a supervised machine learning approach. J. Cosmol. Astropart. Phys. **2021**(07), 011 (2021)
124. P.G. Krastev, Translating neutron star observations to nuclear symmetry energy via deep neural networks. Galaxies **10**(1) (2022)
125. F. Morawski, M. Bejger, Neural network reconstruction of the dense matter equation of state derived from the parameters of neutron stars. Astron. & Astrophys. **642**(A78), 8 (2020)
126. S. Soma, L. Wang, S. Shi, H. Stöcker, K. Zhou, Neural network reconstruction of the dense matter equation of state from neutron star observables. J. Cosmol. Astropart. Phys. **2022**(08), 071 (2022)

127. S. Soma, L. Wang, S. Shi, H. Stöcker, K. Zhou, Reconstructing the neutron star equation of state from observational data via automatic differentiation. Phys. Rev. D **107**(8) (2023)
128. R.V. Lobato, E.V. Chimanski, C.A. Bertulani, Cluster structures with machine learning support in neutron star m-r relations. J. Phys.: Conf. Ser. **2340**(1), 012014 (2022)
129. R.V. Lobato, E.V. Chimanski, C.A. Bertulani, Unsupervised machine learning correlations in eos of neutron stars (2022)
130. M. Ferreira, V. Carvalho, C. Providência, Extracting nuclear matter properties from the neutron star matter equation of state using deep neural networks. Phys. Rev. D **106**(10) (2022)
131. P.G. Krastev, A deep learning approach to extracting nuclear matter properties from neutron star observations (2023)
132. D.D. Doneva, F.M. Ramazanoğlu, H.O. Silva, T.P. Sotiriou, S.S. Yazadjiev, Scalarization (2022). arXiv:2211.01766
133. E. Berti, E. Barausse, V. Cardoso, L. Gualtieri, P. Pani, U. Sperhake, L.C. Stein, N. Wex, K. Yagi, T. Baker, C.P. Burgess, F.S. Coelho, D. Doneva, A. De Felice, P.G. Ferreira, P.C.C. Freire, J. Healy, C. Herdeiro, M. Horbatsch, B. Kleihaus, A. Klein, K. Kokkotas, J. Kunz, P. Laguna, R.N. Lang, T.G.F. Li, T. Littenberg, A. Matas, S. Mirshekari, H. Okawa, E. Radu, R. O'Shaughnessy, B.S. Sathyaprakash, C. Van Den Broeck, H.A Winther, H. Witek, M.E. Aghili, J. Alsing, B. Bolen, L. Bombelli, S. Caudill, L. Chen, J.C. Degollado, R. Fujita, C. Gao, D. Gerosa, S. Kamali, H.O. Silva, J.G. Rosa, L. Sadeghian, M. Sampaio, H. Sotani, M. Zilhao, Testing general relativity with present and future astrophysical observations. Class. Quantum Grav. **32**(24), 243001 (2015)
134. C. Charmousis, A. Lehébel, E. Smyrniotis, N. Stergioulas, Astrophysical constraints on compact objects in 4d einstein-gauss-bonnet gravity. J. Cosmol. Astropart. Phys. **2022**(02), 033 (2022)
135. B. Biswas, E. Smyrniotis, I. Liodis, N. Stergioulas, A Bayesian investigation of the neutron star equation-of-state vs. gravity degeneracy (2023). arXiv:2309.05420
136. E.J. Michaud, Z. Liu, M. Tegmark, Precision machine learning. Entropy **25**(1) (2023)

Features in the Inflaton Potential and the Spectrum of Cosmological Perturbations

Ioannis Dalianis

Abstract Cosmological perturbations, originating in the quantum fluctuations of the fields that drive inflation, are observed to be nearly scale invariant at the largest scales. At smaller scales, however, perturbations are not severely constrained and might be of particular importance if their amplitude is large. They can trigger the creation of primordial black holes (PBHs) or stochastic gravitational waves (GWs). Small-scale perturbations are generated during the later stages of inflation, when possible strong features in the inflaton potential can break scale invariance and leave characteristic imprints on the spectrum. We focus on and review three types of features: inflection points and steep steps in the potential, as well as sharp turns in the inflationary trajectory in field space. We show that such features induce a strong enhancement of the curvature spectrum within a certain wavenumber range. In particular cases, they also generate characteristic oscillatory patterns that are transferred in the spectrum of secondary GWs, which are potentially observable by operating or designed experiments. We demonstrate these effects through the calculation of the primordial power spectrum and the PBH abundance in the context of α-attractors and supergravity (SUGRA) models of inflation.

1 Introduction

Inflationary models that feature strong deviations from the slow-roll (SR) conditions can produce a scale dependent power spectrum of fluctuations. A particularly interesting case is that of an amplified spectrum at wavelengths smaller than that of the CMB scales, where the bounds on the primordial spectrum of curvature perturbations $\Delta^2_{\mathcal{R}}(k)$ are rather weak, see e.g. [1–6]. The motivation to examine such a spectrum is twofold. Firstly, it is an observationally viable possibility with essential cosmological implications, such as the existence of secondary gravitational waves (GWs)

I. Dalianis (✉)
National and Kapodistrian University of Athens, Department of Physics,
Section of Nuclear & Particle Physics, GR–157 84, Athens, Greece
e-mail: idalianis@phys.uoa.gr

© The Author(s), under exclusive license to Springer Nature Switzerland AG 2024
E. Papantonopoulos and N. Mavromatos (eds.), *Compact Objects in the Universe*,
https://doi.org/10.1007/978-3-031-55098-0_13

[7–12] and, possibly, of primordial black holes (PBH) [13–15]. Secondly, it points towards new and challenging inflationary model building directions that go beyond the standard paradigm. Notable inflationary mechanisms that can realize a sizable amplification display the following features:

- step-like transitions in the inflaton potential energy,
- approximate inflection points,
- sharp turns in the multi-field space of the inflaton potential.

These mechanisms are associated with a strong acceleration and deceleration of the inflaton field. In the context of canonical inflation such changes in the velocity are attributed to strong features in the inflationary potential, especially designed and located, see e.g. Ref. [16] for a recent discussion.

A *step-like* feature causes a fast decrease of the potential energy and produces spectral distortions [17–22]: One or more transition points in the inflaton potential energy, at which it jumps from one constant value to another, can result in a sizable enhancement of the spectrum within a certain range of scales. On top of that, characteristic oscillatory spectral patterns are produced. Spectral enhancement due to the *inflection point* mechanism requires that the inflaton traverses a region of its potential V with $V' \sim V'' \sim 0$ [22]. This is a local flattening of the potential that might be deformed so that a shallow minimum and maximum appears. In that region the inflaton gains and loses kinetic energy, which results in an enhancement of the primordial spectrum. Several models implementing this idea have been put forward in the last couple of years, see e.g. [23–31, 38] for some first works. In multi-field inflationary dynamics a different source for intense superhorizon evolution of the curvature perturbations can exist. The mechanism is associated with a *sharp turn* where a slow-roll violation along a direction orthogonal to the inflationary trajectory takes place. This transient phase amplifies the amplitude of the otherwise catatonic isocurvature modes. The amplification of isocurvature perturbations is converted through suitable couplings into an amplification of curvature perturbations [32–39].

2 Beyond Slow-Roll Single-Field Inflation

In the framework of the inflationary universe, the cosmological perturbations naturally emerge from quantum zero-point fluctuations [37]. Fluctuations of the inflaton scalar field ϕ on top of a homogeneous background induce scalar perturbations in the metric, which can be described through the comoving gauge curvature perturbation $\mathcal{R}$. The variance in the fluctuations is quantified by the power spectrum $\Delta^2_{\mathcal{R}}(k)$. Expanding the inflaton-gravity action for canonical single field inflation

$$S = \int d^4x\sqrt{-g}\left[\frac{1}{2}R + \frac{1}{2}(\partial\phi)^2 - V(\phi)\right] \tag{1}$$

up to second order in $\mathcal{R}$, one obtains

$$S_{(2)} = \frac{1}{2} \int \mathrm{d}^4 x \, a^3 \, \frac{\dot{\phi}^2}{H^2} \left[\dot{\mathcal{R}}^2 - \frac{(\partial_i \mathcal{R})^2}{a^2} \right] , \tag{2}$$

where overdots denote derivatives with respect to the coordinate time t. We have chosen units such that the reduced Planck mass is $M_{\mathrm{Pl}} = 1$. After the variable redefinition $v = z\mathcal{R}$ where $z^2 = a^2 \dot{\phi}^2 / H^2 = 2a^2 \epsilon_1$ and switching to conformal time τ, defined by $d\tau = dt/a$, the action is recast into canonical form (ϵ_1 is defined in Eqs. (4)). The evolution of the Fourier modes v_k are described by the Mukhanov-Sasaki equation

$$v_k'' + \left(k^2 - \frac{z''}{z} \right) v_k = 0, \tag{3}$$

where the potential term z''/z is exactly expressed in terms of the Hubble flow functions

$$\epsilon_1 \equiv -\frac{\dot{H}}{H^2} = \frac{\dot{\phi}^2}{2H^2}, \quad \epsilon_2 \equiv \frac{\dot{\epsilon}_1}{H \epsilon_1} = 2\epsilon_1 + 2\frac{\ddot{\phi}}{H\dot{\phi}}, \quad \epsilon_3 \equiv \frac{\dot{\epsilon}_2}{H\epsilon_2}, \tag{4}$$

as

$$\frac{z''}{z} = (aH)^2 \left[2 - \epsilon_1 + \frac{3}{2}\epsilon_2 - \frac{1}{2}\epsilon_1\epsilon_2 + \frac{1}{4}\epsilon_2^2 + \frac{1}{2}\epsilon_2\epsilon_3 \right] . \tag{5}$$

Here the prime denotes derivatives with respect to the conformal time τ. An alternative to ϵ_2 is the second slow-roll parameter $\eta = \epsilon_1 - \epsilon_2/2$ that we will use later (Sect. 3). The super-Hubble evolution of the curvature perturbation, i.e. $k^2 \ll z''/z$, is a linear combination of z and $z \int d\tau / z^2$. Given that $\mathcal{R}_k = v_k / z$ one finds

$$\mathcal{R}_k = c_1(k) + c_2(k) \int \frac{dt}{a^3 \epsilon_1} , \tag{6}$$

where c_1 and c_2 are integration constants. The curvature power spectrum is defined as

$$\Delta_{\mathcal{R}}^2 = \frac{k^3}{2\pi^2} \, |\mathcal{R}_k|^2 \tag{7}$$

and is estimated at a time after the mode exits the horizon and its value freezes.

In the conventional single-field inflationary scenarios based on the slow-roll analysis, the second term in Eq. (6) correspond to a decaying contribution, so that $\mathcal{R}_k$ soon becomes a constant after horizon crossing. For a nearly constant ratio $V'(\phi)/(3H\dot{\phi})$ the first Hubble flow parameter can be found analytically [38] and scales as $\epsilon_1 \propto a^{2\delta}$, where

$$\delta \equiv \frac{\ddot{\phi}}{H\dot{\phi}} = -3 \left(1 + \frac{V'(\phi)}{3H\dot{\phi}} \right) . \tag{8}$$

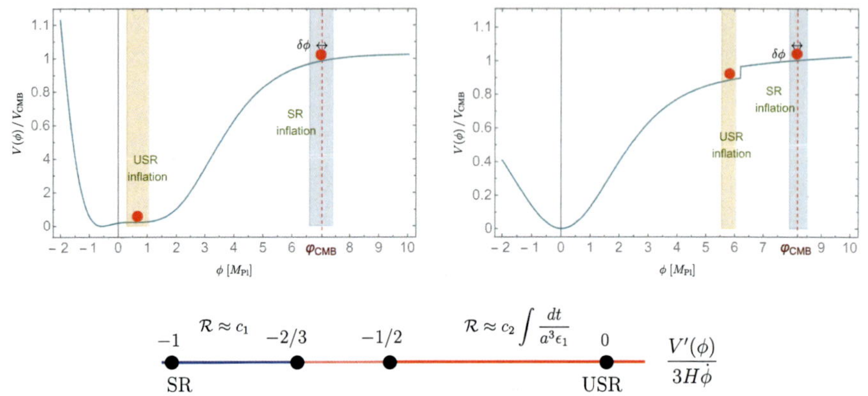

Fig. 1 *Upper panels*: Two examples of inflationary potentials in the framework of α-attractors that realize both a SR and an USR phase: one with an infection point (left panel) and one with a step-like transition (right panel). *Lower:* Intervals of the ratio $V'/(3H\dot{\phi})$ within which a SR or USR phase takes place. The solution for $\mathcal{R}_k$ that characterizes each interval is also shown

For $\epsilon_1 \ll 1$ we have $2\delta \approx \epsilon_2$. During slow-roll inflation we have $\epsilon_2 \ll 1$ and the second term in Eq. (6) is a decaying mode, since $\epsilon_1 \propto a^{\epsilon_2}$ and $\mathcal{R}_k$ is dominated by the constant term. The power spectrum can be described quantitatively well by the expression $\Delta^2_{\mathcal{R}} \approx H^2/(8\pi^2\epsilon_1)$ during the slow-roll regime. The dominant solution is a constant mode and the curvature perturbation is conserved outside the Hubble radius.

The other limit is the ultra-slow roll (USR) inflation [39], which corresponds to a regime in which the effect of the potential is negligible in the equation of motion of the inflaton. In this regime, the Klein-Gordon equation takes the form $\ddot{\phi} \approx -3H\dot{\phi}$, and from Eq. (8) we see that $\epsilon_2 \approx -6$. Therefore, the first Hubble-flow parameter decreases rapidly, i.e. $\epsilon_1 = \dot{\phi}^2/(2M_{\rm Pl}^2 H^2) \propto a^{-6}$, and the solution (6) is dominated by the second mode which grows as $(a^3\epsilon_1)^{-1} \propto a^3$. These two limits[1] are sketched in the graph of Fig. 1.

During USR the scalar field moves on a flat potential. It may initially fast-roll and then decelerate, with the slow-roll approximation breaking down, but it eventually stops because of the Hubble friction at a point that depends on its initial condition. USR inflation has a graceful exit problem and it is also incompatible with the measured tilt of the CMB power spectrum. However, an USR phase is a viable possibility if it begins after the CMB modes exit the horizon, as long as it is guaranteed that the CMB modes do not grow after horizon exit [41, 42]. The USR phase has to be followed by a slow-roll or a constant-roll phase [38]. The motivation to introduce an USR phase is so that it greatly enhances the scalar power spectrum in single-field inflationary models. The enhancement is indicated already by the slow-roll result

[1] Values $-6 \leq \epsilon_2 <\leq -3$ correspond to a phase called non-single-clock inflation [40].

$\Delta^2_{\mathcal{R}} \propto 1/\epsilon_1$, which increases for decreasing ϵ_1 values, but this result is not valid during USR and does not describe accurately the amplification factor or the characteristic wavenumbers. The correct picture is given by the full solution, Eq. (6), that indicates a more subtle superhorizon evolution for the curvature perturbation [43–45].

2.1 Amplification

An USR is phase is driven by a nearly flat part of the potential, where $V'(\phi)$ is negligible compared to the acceleration $\ddot{\phi}$ and the friction term $3H\dot{\phi}$ in the equation of motion. We distinguish the following cases:

- The first case is that of a monotonic potential with a positive derivative, i.e. $V'(\phi) > 0$, where the inflaton continuously rolls down towards smaller potential values. The presence of an USR regime after a slow-roll slope, with strong acceleration followed by deceleration, implies a transition that is connected to the presence of a step in the inflaton potential. In Refs. [17, 18] it was pointed out that the steepness of this step-like transition is critical, with sharp steps enhancing $\Delta^2_{\mathcal{R}}$ significantly.
- The second case is that of a non-monotonic potential with $V'(\phi)$ changing sign. The conventional example is that of an approximate inflection point in the potential. Literally, an inflection point means that there is a field value at which $V'(\phi) = V''(\phi) = 0$. However, a stronger amplification of $\Delta^2_{\mathcal{R}}$ is achieved if the region around the inflection point is deformed into a local minimum and maximum. In this sense, an approximate inflection point indicates that a shallow potential well separates two slow-roll phases and realizes a transient USR regime.

In Fig. 2 a comparison between the aforementioned mechanisms is demonstrated in terms of the ϵ_1 and ϵ_2 parameters. In the left panel of Fig. 2 the inflection point is approximate, i.e. there is a smooth shallow minimum and maximum. In the right panel there is a combination of a step and an inflection point. We display this non-standard inflection point because it is characterized by a sharp minimum and maximum, for which the potential contribution is important. For a step-like feature there is a short phase of very strong acceleration which is followed by a period of deceleration when the inflaton settles onto the second plateau. For the inflection point the most prominent feature is the deceleration.

A third case of $\Delta^2_{\mathcal{R}}$ amplification, which will be discussed in a following section, is that of perpendicular acceleration in multi-field inflation. Also, we comment that relatively slow turns which appear in hybrid inflation models [101] can lead to a change of the inflationary regime and amplification of small scale perturbations [102]. Particular set ups of related mechanisms as well as other model building directions can be found in Refs. [46–48, 103], just to mention a few.

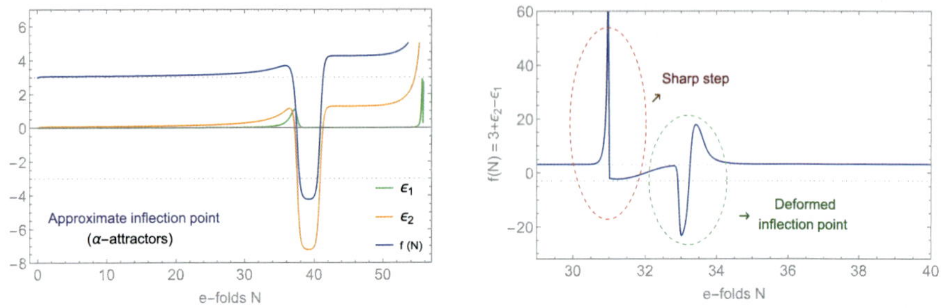

Fig. 2 *Left panel*: The evolution of the first two Hubble flow parameters, ϵ_1 (green), ϵ_2 (orange), and the function $f(N)$ of Eq. (11) for an inflationary model in the context of α-attractors [30]. *Right Panel*: The evolution of the function $f(N)$ for an inflationary model with a sharp step followed by an inflection point [18]. In both plots the dotted grid-lines indicate the critical values ± 3 for the function $f(N)$

2.1.1 An Analytic Calculation

We would like to have a quantitative description of the increase of the amplitude of the curvature perturbation. We start from the equation for the curvature perturbation in Fourier space,

$$\mathcal{R}_k'' + 2\frac{z'}{z}\mathcal{R}_k' + k^2\mathcal{R}_k = 0\,. \tag{9}$$

In terms of efoldings N the equation for the curvature perturbation takes the form

$$R_{k,NN} + f(N)\, R_{k,N} + \frac{k^2}{e^{2N}H^2} R_k = 0\,, \tag{10}$$

where the parameter z reads $z = e^N\, \phi_{,N}$ and the subscript N indicates a derivative. The solutions for $\mathcal{R}_k$ depend on the quantity

$$f(N) = 3 + \frac{2\phi_{,NN}}{\phi_{,N}} - \frac{\phi_{,N}^2}{2M_{\rm Pl}^2} = 3 + \epsilon_2 - \epsilon_1\,. \tag{11}$$

In the slow-roll regime, $f(N)$ acts as a generalized friction term. However, if ϵ_2 becomes negative, it acts in an opposite way and can lead to a dramatic enhancement of the perturbations.

Within the slow-roll approximation, we have $f(N) \approx 3$ and the solution of Eq. (10) can be expressed in terms of the Bessel functions $J_{\pm 3/2}$ as

$$\mathcal{R}_k(N; C_p, C_m, 3) = A\, e^{-\frac{3}{2}N}\left(C_p J_{3/2}\left(e^{-N}\frac{k}{H}\right) + C_m\, J_{-3/2}\left(e^{-N}\frac{k}{H}\right)\right), \tag{12}$$

where we take A to be real without loss of generality. For $-k\tau \gg 1$ we assume the Bunch-Davies vacuum as the initial condition for the evolution of the fluctuations. This selects the values $C_p = 1$, $C_m = i$. The curvature perturbation is $\mathcal{R}_k(N; 1, i, 3) \propto \left(e^{-ik\tau}/\sqrt{k}\right)(1 - i/(k\tau))/a(\tau)$, where the conformal time is $H\tau = -e^{-N} = -1/a$. Within slow-roll and for late times $\tau \to 0^-$, the curvature perturbation approaches a constant value as the mode $\mathcal{R}_k$ moves out of the horizon and freezes, with the absolute value of C_m determining the value of the power spectrum.

This simple picture is modified when the function $f(N)$ deviates from a constant value equal to 3, or equivalently when $|\epsilon_2|$ becomes large. This can occur during an USR phase, where the spectrum might be enhanced by several orders of magnitude and new analytical techniques are needed. Generally, we are interested in one or more transient *non-slow-roll* phases that take place between an initial and a final slow-roll phase. During such a phase the velocity $\varphi_{,N}$ changes fast, by growing or decaying depending on the sign of $f(N) - 3$. In order to obtain an analytical solution and quantitative understanding, we model $f(N)$ through a sequence of square "pulses", each with constant $f(N) = \bar{f}_i \neq 3$. We also approximate the Hubble parameter H as almost constant, which is justified because the change in the Hubble parameter for an inflection point or a step is at most of order 10%, while the enhancement of the spectrum can be several orders of magnitude large. We note that a similar analysis and supplementary results can be found in Refs. [43, 56, 57].

Our starting point is the solution Eq. (12). We assume that a strong feature appears at some moment in time. With the term feature we mean sharp deviations from the minimal and smooth inflationary evolution that have observable consequences, see e.g. Refs. [50–52]. Here we focus on features that can be approximated with a top-hat function that we call a square "pulse". The motivation behind this approximation is the geometrical shape of the function $f(N)$ when a strong feature, such as an inflection point or a step, affects the inflationary trajectory, as depicted in Fig. 2. For $k \to 0$ the evolution of $\mathcal{R}_k$ is given by Eq. (12), resulting in the asymptotic value $|C_m| = 1$ that corresponds to a scale-invariant spectrum, normalized to the CMB value. We are interested in the relative increase of the asymptotic value of $|C_m|$ if at some value of N a "pulse" appears in $f(N)$. For a square "pulse" of height $\bar{f} \neq 3$ that begins at N_1 and ends at N_2, we have

$$\epsilon_2(N) = (\bar{f} - 3)\left[\Theta(N - N_1) - \Theta(N - N_2)\right], \tag{13}$$

where Θ is a step function. The solution involves a linear combination of the Bessel functions $J_{\pm\bar{f}/2}$ and $J_{\pm 3/2}$. The corresponding values of the constants C'_p, C'_m after the pulse is traversed are given by the matrix equation [18]

$$\begin{pmatrix} C'_p \\ C'_m \end{pmatrix} = M_{\text{pulse}}(N_1, N_2, \bar{f}, 3, k)\begin{pmatrix} 1 \\ i \end{pmatrix}. \tag{14}$$

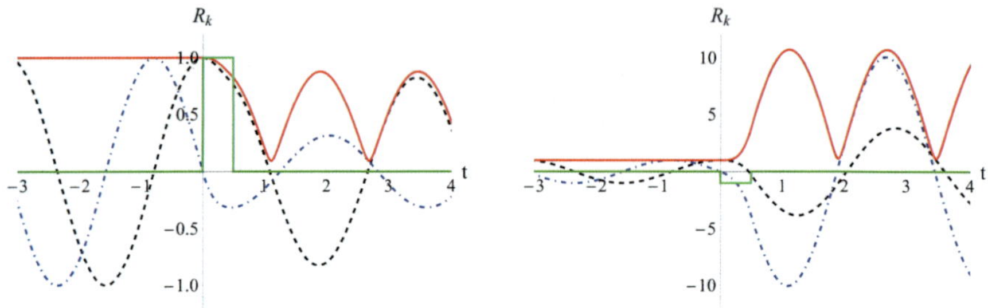

Fig. 3 The real part (dashed curve), imaginary part (dot-dashed curve) and amplitude (red solid curve) of $\mathcal{R}_k(t)$ for the toy-model equation $\mathcal{R}_{k,tt} + \bar{f}\mathcal{R}_{k,t} + k^2\mathcal{R}_k = 0$ for a "pulse" in the interval $0 \leq t \leq 0.5$ and with $k = 2$. We also display the "pulse", with a rescaled maximum $\bar{f}/5$. *Left panel*: $\bar{f} = 5$. *Right panel*: $\bar{f} = -5$ [18]

The matrix M_{pulse} is a product of two square matrices, with each one giving the new coefficients (C_p, C_m) for the transitions $3 \to \bar{f}$ at $N = N_1$ and $\bar{f} \to 3$ at $N = N_2$. The two transitions combined account for the effect of a single square "pulse". The matrix M_{pulse} has elements that depend on the Bessel functions $J_{\bar{f}/2}$, $J_{-\bar{f}/2}$, $J_{1-\bar{f}/2}$, $J_{-1+\bar{f}/2}$ and $J_{3/2}$, $J_{-3/2}$, $J_{1/2}$, $J_{-1/2}$, which can be found by requiring the continuity of the solution and its first derivative at $N = N_1$ and $N = N_2$.

A more general set up involves the evolution of the fluctuation $\mathcal{R}_k$ through a sequence of strong features parameterized by a series of "pulses" in $f(N)$. A product of several M_{pulse} matrices can reproduce the final values of the coefficients (C'_p, C'_m) of the Bessel functions. In the case of a final slow-roll regime after the "pulses", the Bessel functions for large N are $J_{\pm 3/2}$. Clearly, it is possible to reconstruct any smooth function $f(N)$ in terms of short intervals of N during which the function takes constant values. For a single pulse we obtain that for $\bar{f} > 3$ the spectrum is suppressed, while for $\bar{f} < 3$ it is enhanced, see Fig. 3. This behaviour can be also deduced from the expression (6). This approach gives informative analytical expressions and intuitive findings, which we discuss next.

A *first* finding is the size of amplification or the drop of the power spectrum amplitude. For the example of a single pulse that lasts $\Delta N = N_2 - N_1$ efoldings we find that, at late times $N \gg N_2$ and for large wavenumbers k, the power spectrum is scale invariant and amplified by the factor [18]

$$\begin{aligned} \delta\Delta^2_{\mathcal{R}}\big|_{N \gg N_2} &= |C_m|^2 = e^{-(N_2 - N_1)(\bar{f}-3)} \approx e^{-\epsilon_2 (N_2 - N_1)} \\ &= e^{-\int \epsilon_2 \, dN} \end{aligned} \tag{15}$$

relative to the CMB amplitude. In the second line of Eq. (15) the enhancement is written as an integral after taking the continuum limit. We note that analytical expressions for $k \to 0$ are more difficult to obtain because a component of the matrix M_{pulse} scales as $1/k$ in this limit. What is of special interest is that the exponent in the above expression is simply the area of the "pulse" exceeding the value 3. A

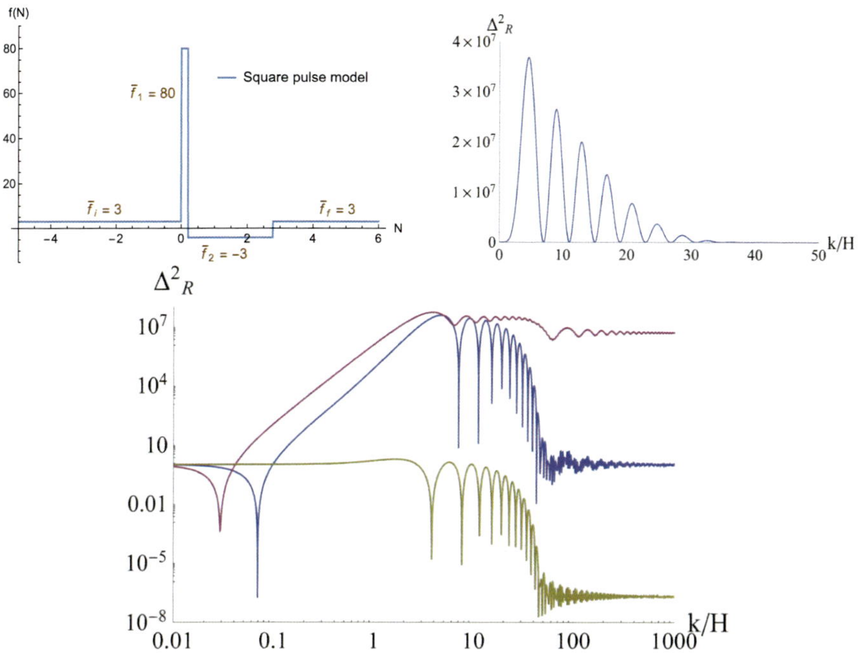

Fig. 4 *Lower panel*: A logarithmic plot of the curvature spectrum resulting from a double "pulse" with $\bar{f}_1 = 80$ between $N_1 = 0$ and $N_2 = 0.2$ and $\bar{f}_2 = -3$ between N_2 and $N_3 = 2.77$ is given by the middle curve depicted with blue color. The form of the "pulses" is shown in the *upper left panel*. The purple upper curve in the main panel is the spectrum resulting from the negative-friction single "pulse". The yellow lower curve is the spectrum resulting from a positive-friction single "pulse". The oscillations of the spectrum are also depicted in a linear plot in the *upper right panel*

general function $f(N)$, that might take positive and negative values, can be broken in infinitesimal "pulses", and the enhancement depends on the integral of $f(N) - 3$ over N. For negative $f(N) - 3$ there is an increase of power, with the negative-friction "pulses" $f(N) < 0$ causing a more prominent amplification, whereas positive $f(N) - 3$ causes a decrease. The positive friction affects most strongly the high-k modes that display a sharp drop to almost zero at a characteristic value of k/H. For positive and negative "pulses" with integrated areas that cancel, we find that the low- and high-k modes are not affected by the presence of the feature. As a result the scale-invariant form of the spectrum is modified only for a finite range of wavenumbers k, see Fig. 4.

A *second* finding is the appearance of oscillatory patterns in the amplitude that we discuss in the following.

2.2 Oscillatory Patterns

The solutions of Eq. (10) have an amplitude that gets suppressed or enhanced, depending on the sign of the friction parameter $f(N)$. Additionally, we observe oscillatory patterns for particular forms of the friction parameter. The origin of oscillations in the amplitude lies in the impact of the strong feature on the phase of the complex quantity $\mathcal{R}_k$. It is known that it is possible to obtain an oscillatory pattern in the spectrum if inflation stops for a certain time interval, so that modes that had exited the horizon reenter, or if there is a change in the sound speed [53]. The analysis with the "pulse" approximation reveals a general pattern: An oscillatory spectrum appears whenever a feature occurs during the evolution of the perturbations that *detunes the relative phase between the real and imaginary parts* of the solution $\mathcal{R}_k$.

A transparent picture of the oscillatory patterns in the spectrum is obtained in the approximation of square "pulses" which, though rough, gives analytic results that make possible the finding of the characteristic frequencies. In particular, the oscillatory behaviour of the solutions can be observed in the matrix M_{pulse}. For a single "pulse" the leading contribution to the power spectrum comes from the (2, 2) component [18],

$$(M_{\text{pulse}})_{22} \approx e^{-\epsilon_2(N_2-N_1)/2} + \epsilon_2 \frac{H}{8k}\left[-2e^{N_1}\sin\left(2e^{-N_1}\frac{k}{H}\right) + 2e^{N_2}\sin\left(2e^{-N_2}\frac{k}{H}\right)\right]. \tag{16}$$

It is the subleading corrections in H/k that introduce oscillatory patterns in the spectrum $|C_m|^2$ with characteristic periods that can be deduced from Eq. (16). The spectrum is expected to vanish at intervals $\delta k/H = e^{N_1}\pi$ and $\delta k/H = e^{N_2}\pi$ and interference patterns appear for $N_1 \simeq N_2$. For a sequence of two or more "pulses" the effect is not simply additive due to the presence of mixing terms.

An example of square "pulses" that approximate the friction function $f(N)$ is depicted in the upper left panel of Fig. 4. Let us discuss it in some detail and compare the exact numerical results displayed in the plots with our analytical results. The feature consists of a positive-friction "pulse" with $\bar{f}_1 = 80$ in the interval between $N_1 = 0$ and $N_2 = 0.2$, followed by a negative-friction "pulse" with $\bar{f}_2 = -3$ in the interval between $N_2 = 0.2$ and $N_3 = 2.77$. In this example we have fine-tuned the integral of $f(N) - 3$ over N to zero. The value of the spectrum for a given value of k/H is equal to $|C'_m|^2$, where (C'_p, C'_m) is given by Eq. (14) with M_{pulse} being a product of three matrices. The resulting curvature power spectrum is depicted by the middle blue curve. The amplification as well as the characteristic modes and the interference patterns between them are visible. In the same figure we also display the spectra that would result from a single "pulse", ignoring the presence of the second. The strong increase of the spectrum results from the effect of the negative-friction "pulse". The full spectrum for the two "pulses" has a maximal value that is comparable to that of the negative-friction single-"pulse" spectrum. A rough estimate can be obtained from the asymptotic value of the spectrum for a single negative "pulse", which according to Eq. (15) is $e^{(N_3-N_2)(3-\bar{f}_2)} = \mathcal{O}(10^7)$. For the positive-friction single-

"pulse" spectrum a sharp drop to almost zero at a characteristic value of k/H is displayed, as expected. Positive friction affects most strongly the high-k modes with the Bessel functions having a zero at a value of their argument roughly equal to $\bar{f}_1/2$. Also the spectra display oscillations with characteristic scales $\delta k/H \approx e^{N_1}\pi = 3.1$, $\delta k/H \approx e^{N_2}\pi = 3.8$ and $\delta k/H \approx e^{N_3}\pi = 50$, consistently with our analysis, see Fig. 4.

The conclusion drawn is that strong features in the background evolution can induce a scale-dependent spectrum which displays, apart from an enhancement that can be several orders of magnitude large, strong oscillatory patterns. This is clearly visible in the example presented. These patterns are expected to become less prominent when the features in $f(N)$ become smoother.

2.3 Step-Like Transitions Versus Approximate Inflection Points

For an inflationary model with an approximate inflection point, the parameter ϵ_2 admits values $\epsilon_2 \lesssim -6$. This is realized when the inflaton crosses a sequence of a shallow local minimum and a local maximum. The inflaton slows down towards the hilltop but it has enough energy to cross the local potential well and transit into a second slow-roll phase. In that region the slope $V'(\phi)$ is negative and ϵ_2 can decrease further producing a very strong enhancement of the curvature power spectrum. In the example presented in Fig. 2 it is apparent that the factor $e^{-\int \epsilon_2 dN}$, which gives the enhancement of the power spectrum, is larger in the region of the inflection point than in the region of the step. A single inflection point, if appropriately tuned (or deformed), suffices to trigger PBH production and source a detectable stochastic GW background.

For a step-like transition the parameter ϵ_2 is bounded from below by the value $\epsilon_2 \gtrsim -6$. According to Eq. (15) the enhancement of the power spectrum is constrained to be smaller than $e^{-\int \epsilon_2 dN} \approx e^{6\,\delta N}$. The duration of the pulse δN is limited by the requirement of graceful exit from the USR phase. Hence, the enhancement of $\Delta^2_{\mathcal{R}}$ is usually insufficient in order to produce a PBH population, and the secondary GW background is not sufficiently strong to be observable by the current detectors. However, an inflationary model with multiple steps, or a combination of features can realize a significant enhancement of the scalar power spectrum that will additionally display a strong oscillatory behaviour, as we stressed above, observable also in the stochastic GW background (see Fig. 7 of Sect. 5).

3 Turns Without Slow-Roll in Two-Field Inflation

The nature of the inflaton field remains elusive. An inflationary phase realized through more than one fields is an attractive scenario and, moreover, a viable possibility, given that multi-field inflation models make their multi-dimensional dynamics manifest at scales smaller than those directly observed on the CMB sky. The phenomenology of interest in multi-field models is related to the fact that the curvature perturbations can evolve even on super-Hubble scales because of the presence of isocurvature perturbations [54–60].

When more than one fields are rolling, one can define an adiabatic perturbation component along the direction tangent to the background classical trajectory, and isocurvature perturbation along the directions orthogonal to the trajectory [59]. Curvature perturbations may be affected by the isocurvature perturbations if the background solution follows a curved trajectory in field space. In the particular case of a sharp turn [61–69], the impact of the isocurvature modes is more prominent, with the resulting curvature spectrum departing significantly from scale invariance. A significant amplification takes place if during the sharp turn the isocurvature modes experience a transition from heavy to light [32, 33, 35, 36].

A turn, even though it is a distinct feature, admits a similar description to that of steps at the level of the slow-roll parameters. This becomes apparent if we decompose, along with the perturbations, the second slow-roll parameter into its tangent $\eta_\parallel$ and orthogonal $\eta_\perp$ components. Sharp steps lead to large positive values of $\eta_\parallel$, while sharp turns result in large $|\eta_\perp|$ values. Allowing both $\eta_\parallel$ and $|\eta_\perp|$ to become large leads to a combination of effects. When the leading effect comes from the rapid change of the $\eta_\parallel$ component, the inflationary dynamics is effectively described by a single-field theory. The $\eta_\perp$ component is rapidly changing when a turn in field space takes place. The isocurvature perturbations get excited and through their coupling to the curvature perturbations, which is proportional to $\eta_\perp$, play a crucial role in shaping the final curvature spectrum.

3.1 Background and Perturbations Evolution

We consider an action for two fields of the form

$$S = \int d^4x \sqrt{-g} \left[\frac{1}{2} R - \frac{1}{2} g^{\mu\nu} \gamma_{ab} \partial_\mu \phi^a \partial_\nu \phi^b - V(\phi) \right], \tag{17}$$

with $a = 1, 2$ and unit Planck mass. γ_{ab} defines a metric in the internal field space parameterized by the coordinates ϕ^a. The Klein-Gordon equation on an expanding, spatially flat background is [65, 70]

$$\frac{D}{dt} \dot{\phi}^a + 3H\dot{\phi}^a + V^a = 0, \tag{18}$$

where $V^a = \gamma^{ab}\partial V/\partial\phi^b$. The covariant derivative is defined as $\frac{D}{dt}\dot{\phi}^a = \ddot{\phi}^a + \Gamma^a_{bc}\dot{\phi}^b\dot{\phi}^c$. The vectors tangent and normal to the path, labeled respectively T^a and N^a, are defined as $T^a = \dot{\phi}^a/\dot{\phi}$ and $N_a = (\det\gamma)^{1/2}\epsilon_{ab}T^b$. They satisfy $T^aT_a = N^aN_a = 1$, $T^aN_a = 0$. Projecting Eq. (18) along T^a, one finds

$$\ddot{\phi} + 3H\dot{\phi} + V_T = 0\,, \tag{19}$$

where $V_T = T^a\partial V/\partial\phi^a$, while $V_N = N^a\partial V/\partial\phi^a$. The first slow-roll parameter is defined as usual: $\epsilon_1 = -\dot{H}/H^2 = \dot{\phi}^2/(2H^2)$. The second slow-roll parameter is defined as

$$\eta^a \equiv -\frac{1}{H\dot{\phi}}\frac{D\dot{\phi}^a}{dt} \tag{20}$$

and can be decomposed into the tangent and normal directions, $\eta^a = \eta_\parallel T^a + \eta_\perp N^a$, where

$$\eta_\parallel = -\frac{\ddot{\phi}}{H\dot{\phi}} \quad \text{and} \quad \eta_\perp = \frac{V_N}{H\dot{\phi}}\,. \tag{21}$$

3.1.1 Perturbations

The dynamics of the scalar field $\delta\phi$ and scalar metric Ψ perturbations can be described by the gauge invariant fields v^T and v^N [65, 70],

$$v^T = aT_a\delta\phi^a + a\frac{\dot{\phi}}{H}\Psi\,, \qquad v^N = aN_a\delta\phi^a\,. \tag{22}$$

A useful parameterization of the perturbations is in terms of curvature and isocurvature fields $\mathcal{R}$ and $\mathcal{S}$,

$$\mathcal{R} = \frac{H}{a\dot{\phi}}v^T\,, \qquad \mathcal{S} = \frac{H}{a\dot{\phi}}v^N\,. \tag{23}$$

The evolution equations for the curvature and isocurvature perturbations in two-field inflation, using the number of efoldings N as independent variable, can be cast in the form

$$\mathcal{R}_{k,NN} + (3+\epsilon_1-2\eta_\parallel)\mathcal{R}_{k,N} + \frac{k^2}{H^2}e^{-2N}\mathcal{R}_k \approx -2\left(\eta_\perp\mathcal{S}_{k,N} + \eta_{\perp,N}\mathcal{S}_k + 3\eta_\perp\mathcal{S}_k\right) \tag{24}$$

$$\mathcal{S}_{k,NN} + (3-\epsilon_1)\mathcal{S}_{k,N} + \frac{k^2}{H^2}e^{-2N}\mathcal{S}_k + \left(\frac{M^2}{H^2} - \eta_\perp^2 + \epsilon_1\,\mathbb{R}\right)\mathcal{S}_k \approx 2\,\eta_\perp\mathcal{R}_{k,N}\,, \tag{25}$$

where $\eta_\parallel = \epsilon_1 - \epsilon_{1,N}/2\epsilon_1$. In the above system of approximate equations we have assumed for simplicity that ϵ_1 is small and roughly constant. $\mathbb{R}$ is the Ricci scalar of the internal manifold spanned by the scalar fields. The Ricci scalar vanishes for a model with standard kinetic terms for the two fields, which is the case of interest to us here. The coefficient in front of $\mathcal{R}_{k,N}$ is the friction term $f(N) = 3 + \epsilon_{2\|} - \epsilon_1$. The mass M of the isocurvature perturbation is given by the curvature of the potential in the direction perpendicular to the trajectory of the background inflaton.

3.2 Enhancement

The friction term is critical for the enhancement of the curvature power spectrum in single field inflation. In the multi-field case the source of amplification can be found in the isocurvature perturbations. We stress that we do not consider the possibility of a curved field manifold, but assume that the fields have standard kinetic terms and we can set $\mathbb{R} = 0$. We are interested in strong deviations from the slow-roll regime during short intervals in N, during which the second slow-roll parameter can grow large. This happens when the evolution of the background inflaton is accelerated either in the direction of the flow through $\eta_{\|}(N)$, or perpendicularly to it through $\eta_{\perp}(N)$. The result is a deviation of the spectrum from scale invariance with a strong enhancement of the curvature perturbations over a range of momentum scales. We list these two main scenaria:

- Suppressed isocurvature modes. This case is implemented for $\eta_{\perp}^2 \ll M^2/H^2$. Hence, the rhs of Eq. (24) vanishes and the system is effectively described by the dynamics of the single field inflation. The curvature mode can be enhanced due to the friction term $f(N)$ that requires large values of the parameter $\eta_{\|} \approx -\epsilon_2/2$. Such values can be attained if the inflaton potential displays an inflection point, or a sharp step [17, 18]. This case corresponds to acceleration in the evolution of the background inflaton in the direction of the flow.
- Excited isocurvature modes. For $\eta_{\perp}^2 \gg M^2/H^2$ the isocurvature contribution in the rhs of Eq. (24) becomes large and acts as a source for the curvature perturbations, leading to their strong enhancement. We assume that this strong feature is effective within a short period. At a later time $\eta_{\perp}$ becomes small and the isocurvature perturbations become suppressed again. This case corresponds to acceleration in the evolution of the background inflaton in the direction perpendicular to the flow.

In the first case, the inflaton potential must feature either a near inflection point or steps finely-tuned with high accuracy in order to achieve strong enhancement. In the second case a single sharp turn in the inflaton trajectory must reach a value of 4π for the enhanced spectrum to have observable consequences [32]. This is possible only if the field manifold is curved, i.e. $\mathbb{R} \neq 0$. For an inflationary set up with canonical kinetic terms an enhancement can be realized if a sequence of turns in field space occurs within a small number of efoldings [34]. Prominent oscillatory patterns appear as well.

3.3 Describing Turns in Field Space

We consider models of a two-component field $\phi^a = (\chi, \psi)$ with vanishing field-manifold curvature. We assume that the potential has an almost flat direction along a curve $\psi = g(\chi)$ with a small inflationary constant slope along this direction and large curvature along the perpendicular direction, so that the flat direction forms a valley. A particular realization of this setup, with $g(\chi) = a^2/\chi$, is given in Ref. [65]. The unit vectors, tangential and normal to the valley at $\phi^a = (\chi, g(\chi))$, are

$$T^a = \frac{1}{\sqrt{1+g'^2(\chi)}}\left(1, g'(\chi)\right), \qquad N^a = \frac{1}{\sqrt{1+g'^2(\chi)}}\left(g'(\chi), -1\right). \tag{26}$$

From the relation $\frac{DT^a}{dt} = -H\eta_\perp N^a$ we deduce the following expression for the component of the η parameter in the normal direction

$$\eta_\perp = \frac{g''(\chi)}{1+g'^2(\chi)}\frac{d\chi}{dN} = \pm\sqrt{2\epsilon_1}\frac{g''(\chi)}{(1+g'^2(\chi))^{3/2}}. \tag{27}$$

Regions in which $g''(\chi) \neq 0$ yield nonzero values for $\eta_\perp$. Practically, the sharp turn is the intersection of two linear parts of the valley causing $\eta_\perp$ to rise and fall sharply from zero. We place the turn near $\chi = \psi = 0$ without loss of generality. The angle of rotation in field space is found from the relation $d\theta/dN = \eta_\perp$, and is given by

$$\Delta\theta = \int_{N_i}^{N_f} \eta_\perp(\chi)dN = \int_{\chi_i}^{\chi_f} \frac{g''(\chi)}{1+g'^2(\chi)}d\chi = \arctan\left(g'(\chi)\right)\Big|_{\chi_i}^{\chi_f}. \tag{28}$$

The maximal angle is $\Delta\theta = \pi$ and can be obtained for $g'(\chi_i) \to -\infty$ before and $g'(\chi_f) \to \infty$ after the turn. The fact that the value of the integral is bounded by π means that the effect of a sharp turn on the amplification of the isocurvature and curvature perturbations is limited. This conclusion changes if multiple turns occur along the flat direction. A sequence of turns with alternating signs can produce a combined effect that can enhance the power spectrum substantially. An analytic quantitative description is possible by approximating the various features through "pulses" [34], in a similar fashion with the analysis done for the case of steps in the previous section.

3.4 The Quantitative Features of the Evolution

We assume that initially the mass of the isocurvature modes M is constant and large ($M/H \gg 1$), so that these modes get suppressed during the linear parts of the inflationary trajectory. In the region of the turn we have $\eta_\perp \gg M/H$ and both the curvature and isocurvature modes get excited. The system is strongly coupled and the evolution becomes non-trivial. The term $\sim \eta_\perp^2$ in the lhs of Eq. (25) acts as a negative mass term, triggering the rapid growth of the field $\mathcal{S}_k$, which subsequently sources through Eq. (24) the curvature perturbation $\mathcal{R}_k$. The resulting growth of $\mathcal{R}_k$ backreacts, generating a source term in the rhs of Eq. (25) and thus moderating the maximal value of the field $\mathcal{S}_k$. The combined effect results in the enhancement of both modes. After the turn, $\eta_\perp$ becomes negligible and $\mathcal{S}_k$ attains a nonzero mass that makes it vanish eventually. During the same period the curvature mode $\mathcal{R}_k$, which has been enhanced within a certain momentum range, freezes. In addition to the enhancement, sharp turns, quite similarly to the case of sharp steps, cause the appearance of distinctive oscillations in the curvature spectrum.

We consider again the "pulse" approximation of the previous section, in which $\eta_\perp$ has the form

$$\eta_\perp(N) = \eta_{\perp 0}\left(\Theta(N - N_1) - \Theta(N - N_2)\right). \tag{29}$$

The parameter $\eta_\perp$ takes a constant value for $N_1 < N < N_2$, and approaches zero very quickly outside this range. The total area of the "pulse" corresponds to the total angle of the turn. During the linear part of the trajectory, i.e. $N < N_1$ and $N > N_2$, the two equations of motion, (24) and (25), decouple

$$\mathcal{R}_{k,NN} + 3\,\mathcal{R}_{k,N} + \frac{k^2}{H^2}e^{-2N}\mathcal{R}_k = 0 \tag{30}$$

$$\mathcal{S}_{k,NN} + 3\,\mathcal{S}_{k,N} + \frac{k^2}{H^2}e^{-2N}\mathcal{S}_k + \frac{M^2}{H^2}\mathcal{S}_k = 0, \tag{31}$$

and the solutions are a combination of Bessel functions:

$$\mathcal{R}_k(N) = e^{-3N/2}\left[C_p J_{\frac{3}{2}}\left(e^{-N}\frac{k}{H}\right) + C_m J_{-\frac{3}{2}}\left(e^{-N}\frac{k}{H}\right)\right] \tag{32}$$

$$\mathcal{S}_k(N) = e^{-3N/2}\left[D_p J_{\frac{1}{2}\sqrt{9-4M^2/H^2}}\left(e^{-N}\frac{k}{H}\right) + D_m J_{-\frac{1}{2}\sqrt{9-4M^2/H^2}}\left(e^{-N}\frac{k}{H}\right)\right]. \tag{33}$$

Initial conditions corresponding to the Bunch-Davies vacuum are obtained for $C_p = 1$, $C_m = i$. For the massive mode $\mathcal{S}_k$ the coefficients are chosen so that they reproduce the free-wave solution at the initial stage of inflation. They have the form $D_p = -\sqrt{2}(1+i)\, f_1(M, H, \varphi)$, $D_m = \sqrt{2}(1+i)\, f_2(M, H, \varphi)$ where the functions f_1 and f_2 can be found in [34].

The evolution in the interval $N_1 < N < N_2$ is complicated because of the coupling between the two modes. The duration of the interval $\Delta N = N_2 - N_1 \lesssim 1$ is short enough so that it is a good approximation to neglect the expansion of space and simplify the evolution equations:

$$\mathcal{R}_{k,NN} + \frac{k^2}{H^2}\mathcal{R}_k + 2\eta_{\perp 0}\mathcal{S}_{k,N} = 0 \tag{34}$$

$$\mathcal{S}_{k,NN} + \left(\frac{k^2}{H^2} + \frac{M^2}{H^2} - \eta_{\perp 0}^2\right)\mathcal{S}_k - 2\eta_{\perp 0}\mathcal{R}_{k,N} = 0. \tag{35}$$

Following Refs. [32, 33], we look for solutions of the form

$$\mathcal{R}_k = Ae^{\omega N}, \quad \mathcal{S}_k = Be^{\omega N}. \tag{36}$$

There are four independent solutions ω_1, ω_2, ω_3 and ω_4. The following relations between A_i and B_i must hold

$$B_{1,2} = h_+\,\omega_{1,2}\,A_{1,2}\,, \qquad B_{3,4} = h_-\,\omega_{3,4}\,A_{3,4}, \tag{37}$$

where h_+ and h_- are functions of $\eta_{\perp 0}$, M/H and k/H [34]. The solutions on either side of N_1 and N_2 can be matched, assuming the continuity of $\mathcal{R}_k(N)$, $\mathcal{S}_k(N)$ and $\mathcal{S}_{k,N}(N)$. The first derivative of $\mathcal{R}_k(N)$ must account for the δ-function arising from the derivative of $\eta_\perp(N)$ at these points. This leads to the conditions

$$\begin{aligned}\mathcal{R}_{k,N}(N_{1-}) &= \mathcal{R}_{k,N}(N_{1+}) + 2\eta_{\perp 0}\mathcal{S}_k(N_1)\\ \mathcal{R}_{k,N}(N_{2-}) &= \mathcal{R}_{k,N}(N_{2+}) - 2\eta_{\perp 0}\mathcal{S}_k(N_2).\end{aligned} \tag{38}$$

The values of the coefficients C_p, C_m, D_p, D_m before the "pulse" , see Eq. (32), are determined by the initial conditions. The coefficients A_i within the "pulse" (37) and eventually the final coefficients of the free solutions C'_p, C'_m, D'_p, D'_m after the "pulse" can be found through the use of the boundary conditions (38). In analogy to the previous section, one can determine the matrix M_{pulse} that links the solutions before and after the "pulse":

$$\begin{bmatrix} C'_p \\ C'_m \\ D'_p \\ D'_m \end{bmatrix} = M_{\text{pulse}}(N_1, N_2, k, M, \eta_\perp) \begin{bmatrix} C_p \\ C_m \\ D_p \\ D_m \end{bmatrix}. \tag{39}$$

The elements of the matrix M_{pulse} are lengthy and we do not present them here explicitly. This matrix facilitates calculations for more general problems with multiple "pulses", occurring when the linear valley of the potential is interrupted by multiple, successive turns.

The impact of the "pulse" on the evolution of the perturbations can be inferred through inspection of the solutions for ω. The solutions $\omega_{3,4}$ are purely imaginary, resulting in oscillatory behaviour. The other two $\omega_{1,2}$ can be either real or imaginary depending on the parameters of the problem. For $\frac{k}{H} \geq \sqrt{\eta_{\perp 0}^2 - \frac{M^2}{H^2}}$ they are imaginary as well, but for $\frac{k}{H} \leq \sqrt{\eta_{\perp 0}^2 - \frac{M^2}{H^2}}$ they become real. It is the real solution that induces an exponential growth or suppression of the curvature power spectrum. In the limit $\Delta N \to 0$, $\eta_{\perp 0} \to \infty$, with the total angle of the turn $\Delta\theta = \eta_{\perp 0}\Delta N$ kept constant, we have the real solutions

$$\omega_{1,2} = \pm\frac{k}{H\sqrt{3}} + \mathcal{O}(\Delta N). \tag{40}$$

For turns that yield real $\omega_{1,2}$ the spectrum is enhanced by a factor [34]

$$\delta\Delta^2_{\mathcal{R}} \sim \exp\left[\frac{2k}{H\sqrt{3}}\Delta N\right] = \exp\left[\frac{2\Delta\theta}{\sqrt{3}}\frac{k/H}{\eta_{\perp 0}}\right]. \tag{41}$$

For inflation models with canonical kinetic terms where $\Delta\theta \lesssim \pi$ the expression (41) implies that the enhancement of the spectrum appears for scales k/H comparable to $\eta_{\perp 0}$. The general solution is a superposition of all independent solutions (36), hence the exponential growth is accompanied by oscillatory behaviour with a characteristic frequency set by the height of the "pulse" $\eta_{\perp 0}$.

The constraint on $\Delta\theta$ implies that a single turn results only in limited growth of the spectrum [32]. However, multiple turns can have an additive, and in particular conditions resonant, effect. In this respect it is important to point out another feature of the solutions. We are interested in the regime $\eta_{\perp 0} \gg M/H$, so that we can neglect the effect of the mass inside the "pulse". For $k/H \ll \eta_{\perp 0}$ the solution $\omega_{1,2}$ is imaginary and strong oscillations, with a frequency set by $\eta_{\perp 0}$, occur within the "pulse". Outside the "pulse" the characteristic frequency of oscillations is set by k or M and is also modulated by the expansion. The continuity of $\mathcal{S}_k(N)$ and its derivative implies that the amplitude of oscillations increases significantly when the perturbation exits the "pulse", see Fig. 5.

In Fig. 5 we present some specific examples of inflationary set-ups with two fields and vanishing internal curvature $\mathbb{R}$ that exhibit trajectories with sharp turns. In the top row the evolution of $\mathcal{S}_k(N)$ (left plot) and $\mathcal{R}_k(N)$ (right plot) is depicted for three "pulses" corresponding to three sharp turns of alternating direction. The pulses are narrow and each turn $|\Delta\theta|$ in field space is smaller than π. The mass of the isocurvature mode is $M/H \approx 5$. The distinctive feature is the strong increase of the amplitude of the isocurvature mode when the perturbation exits the "pulse". The first "pulse" is at the time that we have set $N = 0$ for the number of efoldings and the "pulses" are repeated with a sequence such that a resonance effect occurs, with the amplitude of $\mathcal{S}_k$ being amplified each time a "pulse" is traversed. This effect triggers the growth of the curvature mode.

In the bottom row of Fig. 5 the curvature power spectrum induced by the aforementioned $\eta_\perp(N)$ is shown. Also, the curvature spectra induced by two, three and four turns are displayed. The spectra are normalized to the scale-invariant one and the location of the point $k/H = 1$ is arbitrary. The spectra in the right panel are generated by rather sharp turns and produce an amplified spectrum $\Delta^2_{\mathcal{R}}$ that is possibly testable via the induced stochastic GW background (see Fig. 7 of Sect. 5).

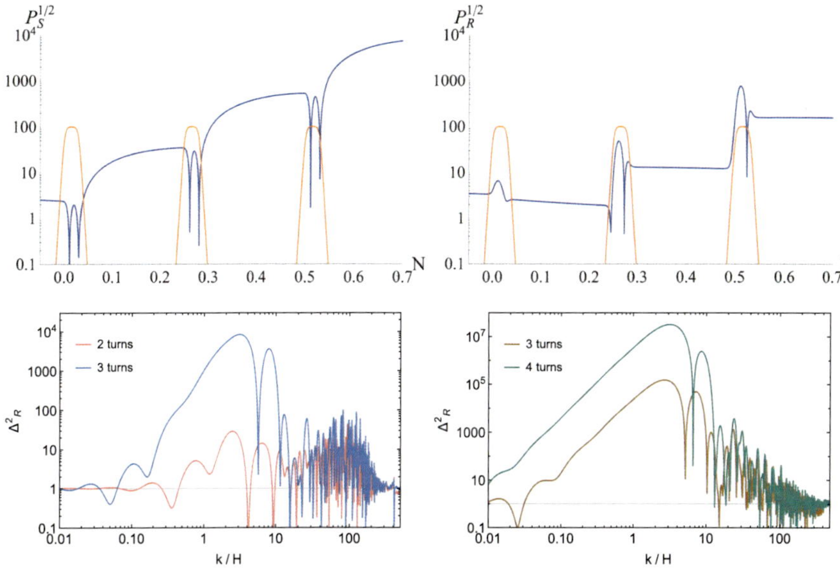

Fig. 5 *Upper panels*: The evolution of the isocurvature mode (left plot) and curvature mode (right plot) are depicted in blue curves, for a time dependent parameter $\eta_\perp(N)$ associated with three sharp turns. The function $|\eta_\perp(N)|$ is depicted in orange curves as a sequence of three "pulses" of alternating sign. *Lower panels*: The corresponding curvature spectrum is depicted in blue (left panel). In pink color we depict the curvature spectrum produced if we keep two out of the three turns. The amplitude of the curvature spectrum is enhanced when we increase the sharpness of the pulses, as was done for three and four turns (right panel)

4 Inflationary Models with Special Features

Embedding models of inflation with strong features in a high energy physics framework is a rather challenging task. A fundamental quantum field theory can be predictive and can reveal aspects of the high energy scale related to the strong feature. In the following we will describe four models: one with step-like transitions and three with an approximate inflection point. We will not present here a model with a sharp turn. A concrete model can be found e.g. in Ref. [71].

4.1 Step-Like Features in the Inflaton Potential

The steps connect regions in which the potential is nearly flat. The basic pattern corresponds to the vacuum energy having one or more transition points at which it jumps from one constant value to another. One can speculate that these points may correspond to filed values at which certain modes decouple very quickly. If these modes have quantum fluctuations that contribute to the vacuum energy a step-like transition may be realized. Decoupling effects become visible when the effective potential is regularized in a mass-sensitive scheme. The renormalization-group equation for the potential can capture the decoupling behavior. However, a concrete quantitative calculation of these effects is elusive due to the fundamental lack of understanding of the nature of vacuum energy or the cosmological constant.

Some intuition can be obtained by considering the role of underlying symmetries. A specific framework, which can be used as a basis, is provided by the α-attractors in supergravity [72, 73]. The theory contains two fields and has conformal invariance and a global SO(1, 1) symmetry. After one of the fields is eliminating through gauge fixing, the following Lagrangian is obtained

$$\mathcal{L} = \sqrt{-g}\left[\frac{1}{2}R(g) - \frac{1}{2}\partial_\mu\phi\partial^\mu\phi - F^2\left(\tanh\frac{\phi}{\sqrt{6\alpha}}\right)\right], \tag{42}$$

where an additional free parameter α has been included. A constant function $F(x)$ preserves the SO(1, 1) symmetry of the initial lagrangian, and corresponds to a cosmological constant in the gauge-fixed version. The value of the cosmological constant is not constrained by the symmetry and is arbitrary. A minimal deformation of the symmetry takes place by assuming that $F(x)$ takes different values over successive ranges of x, with a rapid transition in between. More generally, we can assume that $F(x)$ has the schematic form

$$F(x) \sim x^n + \sum_i A_i\,\Theta(x - x_i), \tag{43}$$

where Θ is a step function. It is understood that each step-function is replaced by a continuous function with a sharp transition at x_i so that the potential varies smoothly. In the framework of α-attractors the dependence of the potential on $\tanh(\phi/\sqrt{6\alpha})$, with α a free parameter, allows for potentials with transitions of arbitrary steepness, see the right panel of Fig. 1. This is an example where the function $F(x)$ gives potentials which are generalizations of the Starobinsky model [74], with the addition of one or more steep steps.

4.2 *Approximate Inflection-Point Inflationary Models*

A primordial scalar spectrum with a maximal enhancement can be realized through inflationary potentials that contain an approximate inflection point. This sort of models feature a shallow potential well (local maximum) that separates potential slopes where the initial and the final slow-roll phases are realized. We will briefly present the models of Refs. [30, 75].

4.2.1 α-attractors

In the context of supergravity and superconformal theory, the general class of α-attractor models have the advantage of flattening the scalar potential for large field values while exhibiting an inflection point for small field values [30]. The necessary conditions the superpotential function $F(x)$ of the scalar Lagrangian (42) should satisfy for the appearance of an approximate inflection point are $V'(\phi_{\text{infl}}) \approx 0$ and $V''(\phi_{\text{infl}}) = 0$. For particular choices of the parameters, the model can be compatible with the CMB constraints [30, 76, 77].

For a polynomial $F(x)$ the simplest choice with one inflection point is the cubic. The corresponding inflationary trajectory for a canonically normalized field is determined by the potential

$$V = V_0 \left[c_0 + c_1 \tanh\left(\frac{\phi}{\sqrt{6\alpha}}\right) + c_2 \tanh^2\left(\frac{\phi}{\sqrt{6\alpha}}\right) + c_3 \tanh^3\left(\frac{\phi}{\sqrt{6\alpha}}\right)\right]^2 , \quad (44)$$

where c_0, c_1, c_2 and c_3 are constants, depicted in the left panel of Fig. 1. Another choice is $F(x) \sim \lambda\Big(x + A\sin(x)\Big)$ and the potential turns out to be

$$V = \lambda^2 \Big[\tanh(\phi/\sqrt{6}) + A \sin\Big(\tanh(\phi/\sqrt{6})/f_\phi\Big)\Big]^2 , \quad (45)$$

having taken $\alpha = 1$ and A, λ, f_ϕ constants. Another example in which a proper inflection point can be generated is found in Ref. [77] where the function $F(x)$ is built in terms of exponential functions.

PBH formation requires particular values for the parameters of the above potentials and these are determined via a subtle numerical process. Firstly, a central PBH mass has to be chosen and this determines the k-position for the $\Delta^2_{\mathcal{R}}(k)$ peak. Secondly, the PBH probability function is exponentially sensitive to the amplitude of the peak and its precise value is found after a delicate selection of the potential parameters. The spectrum $\Delta^2_{\mathcal{R}}(k)$ is reliably found only by solving numerically the Mukhanov-Sasaki equation (3). It goes without saying that consistency with the CMB normalization, the measured n_s and α_s values, as well as adequate efoldings N_* have to be reproduced.

4.2.2 No-Scale Supergravity

A natural framework for formulating models of inflation is supergravity. Potentials that depend on a minimal number of parameters and evade the so-called η problem can be found in no-scale supergravity models [78]. These also emerge as the low energy limit of compactified string models. No-scale models are necessarily multi-field models, and apart from the inflaton, they include additional moduli scalar fields. An inflationary model based on no-scale supergravity [79] can yield a Starobinsky-like potential. By deforming the ordinary $SU(2, 1)/SU(2) \times U(1)$ Kähler potential

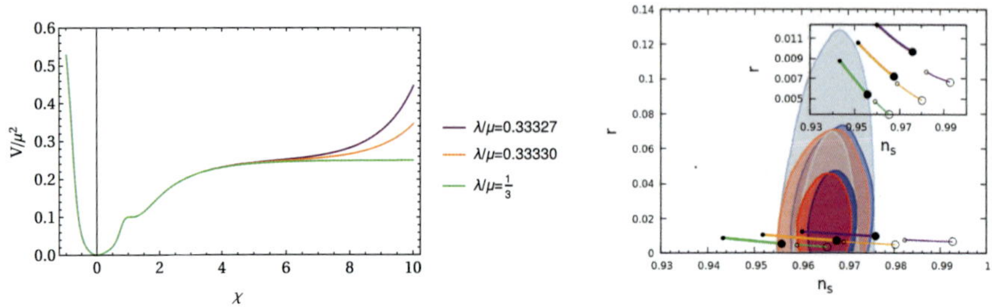

Fig. 6 *Left panel*: The potential given in Eq. (47) as a function of χ, for various values of the ratio λ/μ. *Right panel*: The predictions for the tilt n_s and the tensor-to-scalar ratio r [75]

an inflection point can appear. This is achieved by introducing an exponential term with one extra parameter. All the phenomenological constraints at the CMB scales can be satisfied, also producing PBHs with a significant cosmological population.

Modifications of the Kähler potential that induce an inflection point to the scalar potential have the form

$$K = -3 \ln \left[T + \bar{T} - \frac{\varphi \bar{\varphi}}{3} + a\, e^{-b(\varphi + \bar{\varphi})^2} (\varphi + \bar{\varphi})^4 \right], \tag{46}$$

where T and φ are two chiral superfields that parametrize the noncompact $SU(2,1)/SU(2) \times U(1)$ coset space, and a, b are real numbers. The real part of the field φ plays the role of the inflaton. The simplest superpotential is the Wess-Zumino model, $W = \mu\varphi^2/2 - \lambda\varphi^3/3$, characterized by a mass term μ and a trilinear coupling λ. The scalar potential along the direction $T = \bar{T} = c/2$, $\mathrm{Im}\,\varphi = 0$, is

$$V(\phi) = \frac{3 e^{3b\phi^2 (c\mu^2 - 2\sqrt{\lambda}\mu\phi + 3\lambda^2\phi^2)}}{[-3a\phi^4 + e^{b\phi^2}(-3c + \phi^2)]^2 \, [e^{b\phi^2} - 6a\phi^2(6 + b\phi^2(-9 + 2b\phi^2))]}, \tag{47}$$

where $\phi = \mathrm{Re}\,\varphi$ and c a constant. After a field transformation that puts the kinetic term in canonical form, $\frac{1}{2}\partial_\mu\chi\partial^\mu\chi = K_{\phi\phi}\partial_\mu\phi\partial^\mu\phi$, the above potential can be expressed in terms of the χ field, see Fig. 6.

One can notice that for particular values of the parameters the potential has the required features which ensure that a sizable abundance of PBH is created along with induced GWs [80]. This is achieved via the approximate inflection point, which results from a fine-tuning of the parameters in the modification of the Kähler potential.

5 Secondary GWs

The GW astronomy has started exploring a new range of cosmological phenomena via a network of operating GW detectors. From the cosmological perspective the detection of a cosmic gravitational radiation background would be a milestone in our quest to understand the universe. Stochastic GW backgrounds are thought to originate in several physical sources [81–83]. Among them there is a unambiguous source: the primordial scalar perturbations that source tensor modes at the non-linear level of the cosmological perturbation theory [7–12], the so-called secondary or induced gravitational waves (IGWs).

Once the primordial density perturbations enter the horizon, during the adiabatic expansion of the universe, they start interacting and evolving. The non-linear combinations that act as tensors have an amplitude given roughly by the square of the curvature perturbations. Induced tensors are suppressed by the observed small value of $\delta\rho/\rho$ at the CMB scales. Nevertheless, IGWs can be sizable if the primordial curvature perturbations are enhanced at small scales. In addition, the IGW spectrum is critically influenced by the equation of state of the universe at the time of their birth. Actually, the detection of the relic GW stochastic background can serve as a probe of the very early cosmic history, which is unknown for $t \lesssim 1$ s [84]. In the following we will concentrate on GWs produced during radiation (RD) and early matter domination (EMD) eras.

5.1 GWs in Perturbation Theory

Let us consider a flat Friedmann-Robertson-Walker (FRW) background $g^{(0)}_{\mu\nu} = a^2(\tau)\eta_{\mu\nu}$. In the longitudinal or Newtonian gauge the first-order scalar perturbations (Φ, Ψ) play the role of gravitational potentials, and in the absence of anisotropic stress we have $\Phi = \Psi$. In the second-order Einstein equations, there are combinations of the first-order scalar perturbations Φ, Ψ which act as a source for the tensor modes identified as IGWs. The evolution equation for the IGW tensor $h_{ij}(\tau, \vec{x})$ is

$$h''_{ij} + 2\mathcal{H}h'_{ij} - \partial^2 h_{ij} = \Lambda_{ij,kl} F_{kl} \,. \tag{48}$$

The rhs is the source term and $\Lambda_{ij,kl}$ is the tensor that projects the transverse-traceless part of the $F_{kl}(\tau, \vec{x})$. The solution in Fourier space is given by

$$h_\lambda(\tau, \vec{k}) = \frac{1}{a(\tau)} \int_0^\tau G_k(\tau, \bar{\tau}, w) a(\bar{\tau}) F_\lambda(\bar{\tau}, \vec{k}) d\bar{\tau} \,, \tag{49}$$

where $\lambda = +, \times$ and $e^{(\lambda)}_{ij}$ are the two polarization tensors. $G_k(\tau, \bar{\tau}, w)$ is a Green's function which depends on the background equation of state, $w = p/\rho$. The Fourier mode of the source term is given by (see e.g. Ref. [85])

$$F_{\mathbf{k}} = \int \frac{d^3q}{(2\pi)^{3/2}} e_{ij}(\mathbf{k}) q_i\, q_j \left(2\Phi_{\mathbf{q}}\Phi_{\mathbf{k-q}} + \frac{4}{3(1+w)} (\mathcal{H}^{-1}\Phi_{\mathbf{q}} + \Phi_{\mathbf{q}})(\mathcal{H}^{-1}\Phi_{\mathbf{k-q}} + \Phi_{\mathbf{k-q}}) \right), \tag{50}$$

where $F_{\mathbf{k}}$ is the convolution of two first-order scalar perturbations at different wavenumbers and $\mathcal{H} = aH$. The critical quantity is the gravitational potential Φ that evolves according to the equation (for details see e.g. Ref. [86])

$$\Phi'' + 3(1+c_s^2)\mathcal{H}\Phi' - c_s^2\nabla^2\Phi + \left(2\mathcal{H}' + (1+3c_s^2)\mathcal{H}^2\right)\Phi \propto 4\pi G a^2 \delta S\,. \tag{51}$$

In the absence of entropy perturbations δS it has the exact solution

$$\Phi_k(\tau) = x^{-\ell}\left[C_1(k)J_\ell(x) + C_2(k)Y_\ell(x)\right], \quad x \equiv \sqrt{w}k\tau, \quad \ell \equiv \frac{1}{2}\left(\frac{5+3w}{1+3w}\right), \tag{52}$$

where J_ℓ and Y_ℓ are Bessel functions of order ℓ. For $w > 0$ the potential decays once the perturbations enter the horizon and the IGW are generated mainly at the moment of horizon crossing. In particular, for a RD era ($w = 1/3$) the potential $\Phi(x)$ oscillates with a x^{-2} decaying amplitude in subhorizon scales. For an era of stiff-fluid domination, also called kination era, $w = 1$, the potential decays roughly as $x^{-3/2}$, see e.g. [87–89]. However, for $w = 0$ the potential does not decay and Φ continuously feeds the induced tensors.

The tensor power spectrum is expressed as a double integral involving the power spectrum of the curvature perturbations $\Delta^2_{\mathcal{R}} = \Delta^2_{\Phi}(5+3w)^2/(3+3w)^2$ [85, 90]

$$\Delta^2{}_h(\tau,k) = 4\int_0^\infty dv \int_{|1-v|}^{1+v} du \left(\frac{4v^2-(1+v^2-u^2)^2}{4vu}\right)^2 I^2(v,u,x)\Delta^2{}_{\mathcal{R}}(kv)\Delta^2{}_{\mathcal{R}}(ku) \tag{53}$$

where the dimesionless parameters are $x = k\tau$, $v = k_2/k_1$ and $u = |\mathbf{k}_1 - \mathbf{k}_2|/k$. The function $I(v,u,x)$ includes the time dependence. The energy density parameter of IGW per unit logarithmic frequency interval is given in terms of the tensor power spectrum as $\Omega_{\rm IGW}(\tau,k) \equiv (1/24)\ (k/\mathcal{H}(\tau))^2\ \Delta^2_h(\tau,k)$. As an example, let us assume a Gaussian $\Delta^2_{\mathcal{R}}(k)$ with amplitude $A_{\mathcal{R}}$ and width s. For IGWs produced during RD, the spectral energy density parameter today at the frequency $f_{\rm p}$ where $\Delta^2_{\mathcal{R}}$ is maximal is given by [87]

$$\Omega^{\rm (RD)}_{\rm IGW}(t_0, f_{\rm p}) \sim 5\times 10^{-9}\, s^2 \left(\frac{A_{\mathcal{R}}}{10^{-2}}\right)^2. \tag{54}$$

A detailed analysis can be found in [91]. Amplitudes $A_{\mathcal{R}} \sim 10^{-2}$ can give detectable IGW signals, see e.g. [92] for a recent review.

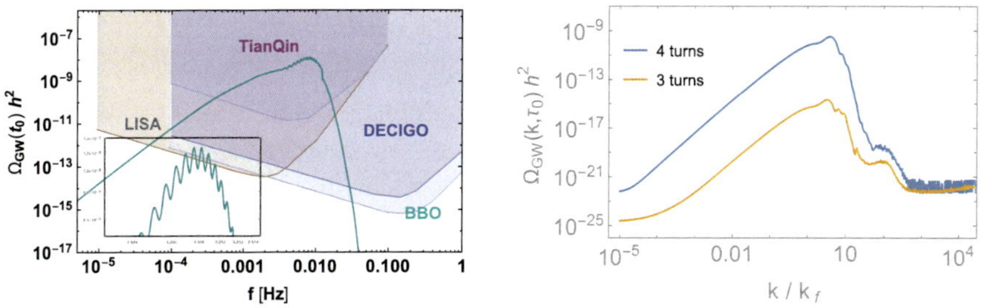

Fig. 7 *Left panel*: The GW density parameter produced by an inflationary model that has two strong features: a step and an inflection. A zoom in about the peak is also depicted. *Right panel*: The GW spectral density produced by inflationary trajectories with 3 and 4 successive turns. The scale k_f is the scale that crosses the horizon at the time when the first feature occurs

5.1.1 Non-linearities

In an EMD era the pressure is negligible, $a \propto \tau^2$ and the linear evolution equation for Φ is given by

$$\Phi_k'' + \frac{6}{\tau}\Phi_k' = 0 \tag{55}$$

with the general solution $\Phi_k(\tau) = c_1(k) + c_2(k)/\tau^5$. The gravitational potential and the density perturbation are related through $\frac{\delta\rho_m}{\rho} = \frac{2}{3\mathcal{H}^2}\nabla^2_{\text{com}}\Phi$, where ∇^2_{com} is the comoving spatial Laplacian. For $\delta\rho/\rho \sim 1$ and working in the Fourier space, one finds a scale k_{NL}^{-1} at which perturbation theory breaks down [93]

$$k_{\text{NL}}(\tau) \sim \frac{\mathcal{H}(\tau)}{(\Delta^2_\Phi)^{1/4}}\,. \tag{56}$$

A perturbation $\delta\rho/\rho$ that has wavelength k^{-1} and power Δ^2_Φ becomes non-linear at about $(\Delta^2_\Phi)^{-1/4}$ Hubble times after horizon entry. Unless reheating is realized fast, one cannot track the evolution of the overdensities during the EMD era in the framework of perturbation theory. This fact restricts the perturbative analysis only to wavenumbers $k < k_{\text{NL}}(\tau_{\text{rh}})$, where τ_{rh} is the conformal time of the cosmic reheating. For example, for $\Delta^2{}_\Phi \sim 10^{-9}$ one can only study scales with wavenumber $k < \mathcal{O}(300)\,k_{\text{rh}}$, while for $\Delta^2{}_\Phi \sim 10^{-2}$ the scales shrink to wavenumbers $k < \mathcal{O}(5)\,k_{\text{rh}}$. This means that the applicability of the perurbative formalism is restricted only within a short period before the reheating of the universe. These theoretical limitations call for a non-perturbative approach.

5.2 GWs Beyond Perturbation Theory and During EMD

A long period of EDM prior to BBN and a modification to the usually assumed RD very early universe is common in many beyond the Standard Model scenarios. For example, moduli with masses around 100 TeV which decay through gravitationaly suppressed couplings lead to a late stage of reheating shortly before the time of BBN. During EMD, IGW production can be described partly with perturbation theory [94, 95]. Beyond the linear regime, a gravitational instability can be understood and traced in terms of the Zel'dovich solution [96]. It describes the growth of inhomogeneities of nonrelativistic matter at scales withing the Hubble horizon and without assuming a spherical symmetry. In the absence of isotropic pressure it is natural to assume that the density perturbations have an initial small deviation from sphericity that we describe by an ellipsoid. In the Zel'dovich approximation the radial coordinate is $r_i = a(t)q_i + b(t)p_i(q_i)$, where p_i is the deviation vector and $b(t)$ is a linearly growing mode in the EMD universe. In an appropriate coordinate system we deal with the diagonal matrices $\partial p_i/\partial q_k = -\text{diag}(\alpha, \beta, \gamma)$ and $D_{i\ell} = \text{diag}(a - \alpha b, a - \beta b, a - \gamma b)$, where $D_{i\ell}$ is the deformation tensor. The (α, β, γ) parameters describe the geometry of different overdensity configurations.

GWs are emitted during the gravitational collapse because of the aspherical and rapid motion of dense matter distributions. The GW emission is described by a multipole expansion of the perturbation $h_{\mu\nu}$ on the background spacetime $g^{(0)}_{\mu\nu}$. To lowest order $h_{ij}(t, d) \propto \ddot{Q}_{ij}(t - d/c)/d$, where Q_{ij} is the mass quadrupole moment and d the distance from the source. This formula is valid for any nearly Newtonian slow-motion source. In terms of the moment of inertia the quadrupole moment is written as

$$Q_{ij} = -I_{ij}(t) + \frac{1}{3}\delta_{ij}\text{Tr}I(t), \tag{57}$$

where I_{ij} is the moment of inertia tensor defined as $I_{ij} = \int d^3r\, \rho(\vec{r})\left(\delta_{ij}|\vec{r}|^2 - r_i r_j\right)$. After choosing the principal axes frame, the moment of inertia tensor is $I_{ij} = \frac{1}{5}M\text{diag}(r_2^2 + r_3^2, r_1^2 + r_3^2, r_1^2 + r_2^2)$ where M is the constant mass enclosed by the ellipsoid. The coordinates of the boundary evolve with time as

$$r_i(t) = \frac{3}{2}t_k^{1/3}t^{2/3}\left(1 - \frac{\xi_i}{2}\left(\frac{t}{t_{\text{max}}}\right)^{2/3}\right), \tag{58}$$

where $\xi_i = \{1, \beta/\alpha, \gamma/\alpha\}$ for the three directions respectively and $t_{\text{max}} = t_{\text{max}}(\alpha, \beta, \gamma, \sigma)$ is the moment of maximum expansion. The variable σ stands for variance of the density perturbations.

The power emitted in the form of GWs from a Hubble patch enclosing a perturbation with a quadrupole tensor Q_{ij} is

$$\frac{dE_e}{dt} = \frac{G}{5c^5}\sum_{ij}\dddot{Q}_{ij}(t)\dddot{Q}_{ji}(t). \tag{59}$$

with G denoting Newton's constant and c the speed of light. The differential energy emitted per logarithmic interval is

$$\frac{dE_e}{d\ln\omega} = \frac{4\pi G}{5c^5}\omega^7 \sum_{ij} |\tilde{Q}_{ij}(\omega)|^2,$$

where $\tilde{Q}_{ij}$ are the Fourier modes in the continuum limit. This expression gives the amount of GW spectral power produced in the EMD era due to a density perturbation with wavelength $q = k^{-1}$ that enters the horizon and evolves non-linearly before the reheating of the universe. Taking the integral over all GW sources, which are all the Hubble patches with size q within our current horizon, and having included a weight function $\mathcal{F}_D$, we find the total energy of the gravitational radiation emitted during the non-perturbative processes of *collapse and halo formation*. The $\mathcal{F}_D$ function is a probability distribution function (PDF) for α, β, and γ for density perturbation with a Gaussian distribution, called Doroshkevich PDF. The differential energy density parameter of the stochastic GW background per observed logarithmic frequency ${\rho_{\rm crit}}^{-1} d\rho_{\rm GW}/d\ln\omega$ is [97]

$$\Omega_{\rm GW}(t,\omega_0) = \frac{1}{\rho_{\rm crit}(t)} \int\int\int_{\mathcal{S}} d\alpha d\beta d\gamma \, \frac{1}{1+z} \frac{4\pi G}{5c^5} \sum_{ij} |\tilde{Q}_{ij}(\omega)|^2 \omega^7 \left(\frac{4\pi}{3} q^3\right)^{-1} \mathcal{F}_{\rm D}(\alpha,\beta,\gamma,\sigma), \tag{60}$$

where $\rho_{\rm crit}$ is the critical energy density of the universe, $\omega = \omega_0(1+z)$, z is the redshift and $f_0 = 2\pi\omega_0$ is the frequency parameter today. GW spectra described by Eq. (60) are displayed in Fig. 8. The integration takes place over an appropriate region $\mathcal{S}$ of the parameter space (α, β, γ).

The central quantity is the quadrupole $Q_{ij}(t)$ and the way it evolves determines the amplitude and the spectral characteristics of the GW signal. The quadrupole can be computed analytically within the Zel'dovich approximation until the moment of maximum expansion and with adequate accuracy until the moment of pancake collapse [97]. The interesting stage of the quadrupole evolution beyond the pancake collapse can be probed by utilizing N-body numerical tools [98]. This method provides valuable quantitative insight on the non-linear regime and illuminates the underlying physical processes. The specification of the spectral details of GWs produced from an EMD enables a reliable test of the very early cosmic history, which is unknown for $t \lesssim 1$ s, through the detection of a relic GW stochastic background.

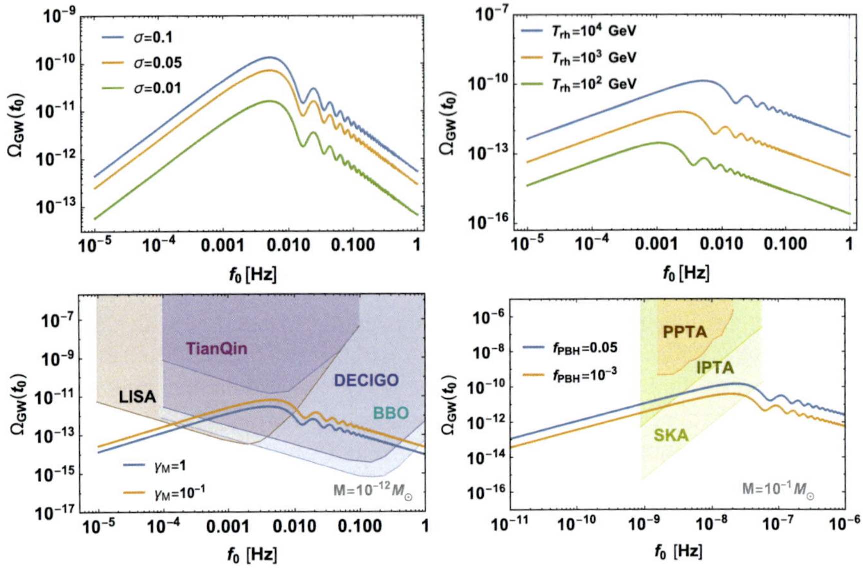

Fig. 8 *Upper panels*: The GW signal, Eq. (60), for a horizon mass $10^{-12} M_{\odot}$ for different σ values and a reheating temperature $T_{\rm rh} = 10^4$ GeV (left panel) and for $\sigma = 0.1$ and three different reheating temperatures (right panel). *Lower panels*: The maximum GW signal in the LISA frequency band for the PBH dark matter scenario and for two σ values (left panel). The GW signal in the PTA frequency range, for a horizon mass $M = 10^{-1} M_{\odot}$ and for two fractional abundances for the associated PBH, 5 and 0.1% respectively, and reheating temperature $\sim$ 80 MeV (right panel)

6 Conclusions

We briefly reviewed basic properties of three inflation mechanisms that transiently realize departures from the slow-roll paradigm. They are characterized by the following features, respectively: step-like transitions, approximate inflection points and sharp turns. Such features can enhance significantly the spectrum of curvature perturbations, possibly triggering PBH production and filling the universe with detectable cosmic GWs. We described the evolution of the perturbations and presented the produced spectra through analytic expressions and complementary numerical means. We highlighted the size of the enhancement and the pattern of oscillations around the peak. We also outlined two particular types of embedding of inflationary models with features into fundamental high-energy frameworks: α-attractors and SUGRA. Finally, we discussed and demonstrated the observational consequences that the excited primordial scalar spectra have on the IGW spectra. We considered the standard RD era as well as the scenario of an EMD era, where calculations beyond the linear regime are required in order to describe the full spectrum of GWs. It is actually the detection of the stochastic IGW background that can be viewed as a direct window to the primordial power spectrum of curvature perturbations at small scales and can be used as a critical discriminator among different inflationary models.

Acknowledgements I would like to thank Vassilis Spanos and Nikolaos Tetradis for collaboration in the work that is reviewed here and for a careful reading of the manuscript. This work was supported by the Hellenic Foundation for Research and Innovation (H.F.R.I.) under the "First Call for H.F.R.I. Research Projects to support Faculty members and Researchers and the procurement of high-cost research equipment grant" (Project Number: 824).

References

1. I. Dalianis, Constraints on the curvature power spectrum from primordial black hole evaporation. JCAP **08**, 032 (2019). (arXiv:1812.09807 [astro-ph.CO])
2. S. Ando, N. Hiroshima, K. Ishiwata, Constraining the primordial curvature perturbation using dark matter substructure. Phys. Rev. D **106**(10), 103014 (2022). (arXiv:2207.05747 [astro-ph.CO])
3. Y. Yang, Constraints on primordial black holes and curvature perturbations from the global 21-cm signal. Phys. Rev. D **102**(8), 083538 (2020). (arXiv:2009.11547 [astro-ph.CO])
4. A.D. Gow, C.T. Byrnes, P.S. Cole, S. Young, The power spectrum on small scales: robust constraints and comparing PBH methodologies. JCAP **02**, 002 (2021). (arXiv:2008.03289 [astro-ph.CO])
5. B. Carr, T. Tenkanen, V. Vaskonen, Primordial black holes from inflaton and spectator field perturbations in a matter-dominated era. Phys. Rev. D **96**(6), 063507 (2017). (arXiv:1706.03746 [astro-ph.CO])
6. B. Carr, K. Kohri, Y. Sendouda, J. Yokoyama, Constraints on primordial black holes. Rept. Prog. Phys. **84**(11), 116902 (2021). (arXiv:2002.12778 [astro-ph.CO])
7. S. Matarrese, O. Pantano, D. Saez, A general relativistic approach to the nonlinear evolution of collisionless matter. Phys. Rev. D **47**, 1311–1323 (1993)
8. S. Matarrese, S. Mollerach, M. Bruni, Second order perturbations of the Einstein-de Sitter universe. Phys. Rev. D **58**, 043504 (1998). arXiv:astro-ph/9707278 [astro-ph]
9. H. Noh, J.C. Hwang, Second-order perturbations of the Friedmann world model. Phys. Rev. D **69**, 104011 (2004)
10. K.N. Ananda, C. Clarkson, D. Wands, The cosmological gravitational wave background from primordial density perturbations. Phys. Rev. D **75**, 123518 (2007). arXiv:gr-qc/0612013 [gr-qc]
11. C. Carbone, S. Matarrese, A Unified treatment of cosmological perturbations from super-horizon to small scales. Phys. Rev. D **71**, 043508 (2005). arXiv:astro-ph/0407611 [astro-ph]
12. D. Baumann, P.J. Steinhardt, K. Takahashi, K. Ichiki, Gravitational wave spectrum induced by primordial scalar perturbations. Phys. Rev. D **76**, 084019 (2007). arXiv:hep-th/0703290 [hep-th]
13. B.J. Carr, S.W. Hawking, Black holes in the early universe. Mon. Not. Roy. Astron. Soc. **168**, 399–415 (1974)
14. P. Meszaros, The behaviour of point masses in an expanding cosmological substratum. Astron. Astrophys. **37**, 225–228 (1974)
15. I.D. Novikov, A.G. Polnarev, A.A. Starobinsky, Ya.. B. Zeldovich, Primordial black holes. Astron. Astroph. **80**, 104 (1979)
16. O. Özsoy, G. Tasinato, Inflation and primordial black holes. Universe **9**(5), 203 (2023). (arXiv:2301.03600 [astro-ph.CO])
17. K. Kefala, G.P. Kodaxis, I.D. Stamou N. Tetradis, Features of the inflaton potential and the power spectrum of cosmological perturbations. arXiv:2010.12483 [astro-ph.CO]
18. I. Dalianis, G.P. Kodaxis, I.D. Stamou, N. Tetradis, A. Tsigkas-Kouvelis, Spectrum oscillations from features in the potential of single-field inflation. Phys. Rev. D **104**(10), 103510 (2021). (arXiv:2106.02467 [astro-ph.CO])

19. A.A. Starobinsky, Spectrum of adiabatic perturbations in the universe when there are singularities in the inflation potential. JETP Lett. **55**, 489–494 (1992)
20. J.A. Adams, B. Cresswell, R. Easther, Inflationary perturbations from a potential with a step. Phys. Rev. D **64**, 123514 (2001). ((arXiv: astro-ph/0102236))
21. K. Inomata, E. McDonough, W. Hu, Amplification of primordial perturbations from the rise or fall of the inflaton. JCAP **02**(02), 031 (2022). (arXiv:2110.14641 [astro-ph.CO])
22. P. Ivanov, P. Naselsky, I. Novikov, Inflation and primordial black holes as dark matter. Phys. Rev. D **50**, 7173–7178 (1994)
23. C. Germani, T. Prokopec, On primordial black holes from an inflection point. Phys. Dark Univ. **18**, 6–10 (2017). (arXiv:1706.04226 [astro-ph.CO])
24. J. Garcia-Bellido, E. Ruiz Morales, Primordial black holes from single field models of inflation. Phys. Dark Univ. **18**, 47–54 (2017). arXiv:1702.03901 [astro-ph.CO]
25. G. Ballesteros, M. Taoso, Primordial black hole dark matter from single field inflation. Phys. Rev. D **97**(2), 023501 (2018). (arXiv:1709.05565 [hep-ph])
26. K. Kannike, L. Marzola, M. Raidal, H. Veermäe, Single field double inflation and primordial black holes. JCAP **09**, 020 (2017). (arXiv:1705.06225 [astro-ph.CO])
27. M.P. Hertzberg, M. Yamada, Primordial black holes from polynomial potentials in single field inflation. Phys. Rev. D **97**(8), 083509 (2018). (arXiv:1712.09750 [astro-ph.CO])
28. M. Cicoli, V.A. Diaz, F.G. Pedro, Primordial black holes from string inflation. JCAP **06**, 034 (2018). (arXiv:1803.02837 [hep-th])
29. O. Özsoy, S. Parameswaran, G. Tasinato, I. Zavala, Mechanisms for primordial black hole production in string theory. JCAP **07**, 005 (2018). (arXiv:1803.07626 [hep-th])
30. I. Dalianis, A. Kehagias, G. Tringas, Primordial black holes from α-attractors. JCAP **01**, 037 (2019). (arXiv:1805.09483 [astro-ph.CO])
31. H. Motohashi, W. Hu, Primordial black holes and slow-roll violation. Phys. Rev. D **96**(6), 063503 (2017). (arXiv:1706.06784 [astro-ph.CO])
32. G.A. Palma, S. Sypsas, C. Zenteno, Seeding primordial black holes in multifield inflation. Phys. Rev. Lett. **125**(12), 121301 (2020). (arXiv:2004.06106 [astro-ph.CO])
33. J. Fumagalli, S. Renaux-Petel, L.T. Witkowski, Oscillations in the stochastic gravitational wave background from sharp features and particle production during inflation. JCAP **08**, 030 (2021). (arXiv:2012.02761 [astro-ph.CO])
34. K. Boutivas, I. Dalianis, G.P. Kodaxis, N. Tetradis, The effect of multiple features on the power spectrum in two-field inflation. JCAP **08**(08), 021 (2022). (arXiv:2203.15605 [astro-ph.CO])
35. J. Fumagalli, S. Renaux-Petel, J.W. Ronayne, L.T. Witkowski, Turning in the landscape: a new mechanism for generating primordial black holes. arXiv:2004.08369 [hep-th]
36. J. Fumagalli, G.A. Palma, S. Renaux-Petel, S. Sypsas, L.T. Witkowski, C. Zenteno, Primordial gravitational waves from excited states. arXiv:2111.14664 [astro-ph.CO]
37. V.F. Mukhanov, G.V. Chibisov, Quantum fluctuations and a nonsingular universe. JETP Lett. **33**, 532–535 (1981)
38. H. Motohashi, A.A. Starobinsky, J. Yokoyama, Inflation with a constant rate of roll. JCAP **09**, 018 (2015). (arXiv:1411.5021 [astro-ph.CO])
39. W.H. Kinney, Horizon crossing and inflation with large eta. Phys. Rev. D **72**, 023515 (2005). arXiv:gr-qc/0503017 [gr-qc]
40. C.T. Byrnes, P.S. Cole, S.P. Patil, Steepest growth of the power spectrum and primordial black holes. JCAP **06**, 028 (2019). (arXiv:1811.11158 [astro-ph.CO])
41. J. Kristiano, J. Yokoyama, Ruling out primordial black hole formation from single-field inflation. arXiv:2211.03395 [hep-th]
42. A. Riotto, "The Primordial Black Hole Formation from Single-Field Inflation is Not Ruled Out," [arXiv:2301.00599 [astro-ph.CO]]
43. S.M. Leach, A.R. Liddle, Inflationary perturbations near horizon crossing. Phys. Rev. D **63**, 043508 (2001). arXiv:astro-ph/0010082 [astro-ph]
44. S.M. Leach, M. Sasaki, D. Wands, A.R. Liddle, Enhancement of superhorizon scale inflationary curvature perturbations. Phys. Rev. D **64**, 023512 (2001). arXiv:astro-ph/0101406 [astro-ph]

45. R. Saito, J. Yokoyama, R. Nagata, Single-field inflation, anomalous enhancement of super-horizon fluctuations, and non-Gaussianity in primordial black hole formation. JCAP **06**, 024 (2008). (arXiv:0804.3470 [astro-ph])
46. G. Ballesteros, M.A.G. García, A.P. Rodríguez, M. Pierre, J. Rey, Primordial black holes and gravitational waves from dissipation during inflation. JCAP **12**, 006 (2022). [arXiv:2208.14978 [astro-ph.CO]
47. G. D'Amico, N. Kaloper, Rollercoaster cosmology. JCAP **08**, 058 (2021). (arXiv:2011.09489 [hep-th])
48. V.C. Spanos, I.D. Stamou, Gravitational waves and primordial black holes from supersymmetric hybrid inflation. Phys. Rev. D **104**(12), 123537 (2021). (arXiv:2108.05671 [astro-ph.CO])
49. P. Carrilho, K.A. Malik D.J. Mulryne, Dissecting the growth of the power spectrum for primordial black holes. Phys. Rev. D **100**(10), 103529 (2019). arXiv:1907.05237 [astro-ph.CO]
50. J. Chluba, J. Hamann, S.P. Patil, Features and new physical scales in primordial observables: theory and observation. Int. J. Mod. Phys. D **24**(10), 1530023 (2015). (arXiv:1505.01834 [astro-ph.CO])
51. A. Slosar, K.N. Abazajian, M. Abidi, P. Adshead, Z. Ahmed, D. Alonso, M.A. Amin, B. Ansarinejad, R. Armstrong, C. Baccigalupi et al., Scratches from the past: inflationary archaeology through features in the power spectrum of primordial fluctuations. Bull. Am. Astron. Soc. **51**(3), 98 (2019). (arXiv:1903.09883 [astro-ph.CO])
52. X. Chen, Primordial non-gaussianities from inflation models. Adv. Astron. **2010**, 638979 (2010). (arXiv:1002.1416 [astro-ph.CO])
53. G. Ballesteros, J. Beltran Jimenez, M. Pieroni, Black hole formation from a general quadratic action for inflationary primordial fluctuations. JCAP **06**(016) (2019). arXiv:1811.03065 [astro-ph.CO]
54. A.A. Starobinsky, J. Yokoyama, Density fluctuations in Brans-Dicke inflation. arXiv:gr-qc/9502002 [gr-qc]
55. M. Sasaki, E.D. Stewart, A General analytic formula for the spectral index of the density perturbations produced during inflation. Prog. Theor. Phys. **95**, 71–78 (1996). arXiv:astro-ph/9507001 [astro-ph]
56. J. Garcia-Bellido, D. Wands, Metric perturbations in two field inflation. Phys. Rev. D **53**, 5437–5445 (1996). arXiv:astro-ph/9511029 [astro-ph]
57. A.D. Linde, V.F. Mukhanov, Nongaussian isocurvature perturbations from inflation. Phys. Rev. D **56**, R535–R539 (1997). arXiv:astro-ph/9610219 [astro-ph]
58. D. Langlois, Correlated adiabatic and isocurvature perturbations from double inflation. Phys. Rev. D **59**, 123512 (1999). arXiv:astro-ph/9906080 [astro-ph]
59. C. Gordon, D. Wands, B.A. Bassett, R. Maartens, Adiabatic and entropy perturbations from inflation. Phys. Rev. D **63**, 023506 (2000). arXiv:astro-ph/0009131 [astro-ph]
60. S. Tsujikawa, H. Yajima, New constraints on multifield inflation with nonminimal coupling. Phys. Rev. D **62**, 123512 (2000). arXiv:hep-ph/0007351 [hep-ph]
61. A. Achucarro, J.O. Gong, S. Hardeman, G.A. Palma, S.P. Patil, Features of heavy physics in the CMB power spectrum. JCAP **01**, 030 (2011). (arXiv:1010.3693 [hep-ph])
62. A. Achucarro, J.O. Gong, S. Hardeman, G.A. Palma, S.P. Patil, Mass hierarchies and non-decoupling in multi-scalar field dynamics. Phys. Rev. D **84**, 043502 (2011). (arXiv:1005.3848 [hep-th])
63. X. Chen, Primordial features as evidence for inflation. JCAP **01**, 038 (2012). (arXiv:1104.1323 [hep-th])
64. G. Shiu, J. Xu, Effective field theory and decoupling in multi-field inflation: an illustrative case study. Phys. Rev. D **84**, 103509 (2011). (arXiv:1108.0981 [hep-th])
65. S. Cespedes, V. Atal, G.A. Palma, On the importance of heavy fields during inflation. JCAP **05**, 008 (2012). (arXiv:1201.4848 [hep-th])
66. A. Avgoustidis, S. Cremonini, A.C. Davis, R.H. Ribeiro, K. Turzynski, S. Watson, Decoupling survives inflation: a critical look at effective field theory violations during inflation. JCAP **06**, 025 (2012). (arXiv:1203.0016 [hep-th])

67. X. Gao, D. Langlois, S. Mizuno, Influence of heavy modes on perturbations in multiple field inflation. JCAP **10**, 040 (2012). (arXiv:1205.5275 [hep-th])
68. A. Achúcarro, J.O. Gong, G.A. Palma, S.P. Patil, Correlating features in the primordial spectra. Phys. Rev. D **87**(12), 121301 (2013). (arXiv:1211.5619 [astro-ph.CO])
69. M. Konieczka, R.H. Ribeiro, K. Turzynski, The effects of a fast-turning trajectory in multiple-field inflation. JCAP **07**, 030 (2014). (arXiv:1401.6163 [astro-ph.CO])
70. A. Achucarro, V. Atal, S. Cespedes, J.O. Gong, G.A. Palma, S.P. Patil, Heavy fields, reduced speeds of sound and decoupling during inflation. Phys. Rev. D **86**, 121301 (2012). (arXiv:1205.0710 [hep-th])
71. S. Bhattacharya, I. Zavala, Sharp turns in axion monodromy: primordial black holes and gravitational waves. JCAP **04**, 065 (2023). (arXiv:2205.06065 [astro-ph.CO])
72. R. Kallosh, A. Linde, JCAP **07**, 002 (2013). (arXiv:1306.5220 [hep-th])
73. S. Ferrara, R. Kallosh, A. Linde, M. Porrati, Phys. Rev. D **88**(8), 085038 (2013). (arXiv:1307.7696 [hep-th])
74. A.A. Starobinsky, A new type of isotropic cosmological models without singularity. Phys. Lett. B **91**, 99–102 (1980)
75. D.V. Nanopoulos, V.C. Spanos, I.D. Stamou, Primordial black holes from no-scale supergravity. Phys. Rev. D **102**(8), 083536 (2020). (arXiv:2008.01457 [astro-ph.CO])
76. L. Iacconi, H. Assadullahi, M. Fasiello, D. Wands, Revisiting small-scale fluctuations in α-attractor models of inflation. JCAP **06**(06), 007 (2022). (arXiv:2112.05092 [astro-ph.CO])
77. I. Dalianis, G. Tringas, Primordial black hole remnants as dark matter produced in thermal, matter, and runaway-quintessence postinflationary scenarios. Phys. Rev. D **100**(8), 083512 (2019). (arXiv:1905.01741 [astro-ph.CO])
78. A.B. Lahanas, D.V. Nanopoulos, The road to no scale supergravity. Phys. Rept. **145**, 1 (1987)
79. J. Ellis, D.V. Nanopoulos, K.A. Olive, No-scale supergravity realization of the starobinsky model of inflation. Phys. Rev. Lett. **111**, 111301 (2013). [erratum: Phys. Rev. Lett. **111**(12), 129902 (2013)]. arXiv:1305.1247 [hep-th]
80. V.C. Spanos, I.D. Stamou, Gravitational waves from no-scale supergravity. Eur. Phys. J. C **83**(1), 4 (2023). (arXiv:2205.05595 [astro-ph.CO])
81. S. Kuroyanagi, T. Chiba, T. Takahashi, Probing the universe through the stochastic gravitational wave background. JCAP **11**, 038 (2018). (arXiv:1807.00786 [astro-ph.CO])
82. N. Christensen, Stochastic Gravitational Wave Backgrounds. Rept. Prog. Phys. **82**(1), 016903 (2019). (arXiv:1811.08797 [gr-qc])
83. C. Caprini, D.G. Figueroa, Cosmological backgrounds of gravitational waves. Class. Quant. Grav. **35**(16), 163001 (2018). (arXiv:1801.04268 [astro-ph.CO])
84. R. Allahverdi, M.A. Amin, A. Berlin, N. Bernal, C.T. Byrnes, M. Sten Delos, A.L. Erickcek, M. Escudero, D.G. Figueroa, K. Freese, et al., The first three seconds: a review of possible expansion histories of the early universe. arXiv:2006.16182 [astro-ph.CO]
85. K. Kohri, T. Terada, Semianalytic calculation of gravitational wave spectrum nonlinearly induced from primordial curvature perturbations. Phys. Rev. D **97**(12), 123532 (2018). (arXiv:1804.08577 [gr-qc])
86. V. Mukhanov, *Physical Foundations of Cosmology* (Cambridge University Press, 2005). ISBN 978-0-521-56398-7
87. I. Dalianis, K. Kritos, Exploring the spectral shape of gravitational waves induced by primordial scalar perturbations and connection with the primordial black hole scenarios. arXiv: 2007.07915 [astro-ph.CO]
88. G. Domènech, Induced gravitational waves in a general cosmological background. Int. J. Mod. Phys. D **29**(03), 2050028 (2020). (arXiv:1912.05583 [gr-qc])
89. I. Dalianis, G.P. Kodaxis, Reheating in runaway inflation models via the evaporation of Mini primordial black holes. Galaxies **10**(1), 31 (2022). (arXiv:2112.15576 [astro-ph.CO])
90. J.R. Espinosa, D. Racco, A. Riotto, A cosmological signature of the SM higgs instability: gravitational waves. JCAP **09**(012) (2018). arXiv:1804.07732 [hep-ph]
91. S. Pi, M. Sasaki, Gravitational waves induced by scalar perturbations with a lognormal peak. JCAP **09**, 037 (2020). arXiv: 2005.12306 [gr-qc]

92. G. Domènech, Scalar induced gravitational waves review. Universe **7**(11), 398 (2021). (arXiv:2109.01398 [gr-qc])
93. H. Assadullahi, D. Wands, Gravitational waves from an early matter era. Phys. Rev. D **79**, 083511 (2009). (arXiv:0901.0989 [astro-ph.CO])
94. K. Inomata, K. Kohri, T. Nakama, T. Terada, Enhancement of gravitational waves induced by scalar perturbations due to a sudden transition from an early matter era to the radiation era. Phys. Rev. D **100**(4), 043532 (2019). (arXiv:1904.12879 [astro-ph.CO])
95. K. Inomata, K. Kohri, T. Nakama, T. Terada, Gravitational waves induced by scalar perturbations during a gradual transition from an early matter era to the radiation era. JCAP **10**, 071 (2019). (arXiv:1904.12878 [astro-ph.CO])
96. Y.B. Zeldovich, Gravitational instability: an Approximate theory for large density perturbations. Astron. Astrophys. **5**, 84–89 (1970)
97. I. Dalianis and C. Kouvaris, Gravitational waves from density perturbations in an early matter domination era. JCAP **07**, 046 (2021). (arXiv:2012.09255 [astro-ph.CO])
98. I. Dalianis and C. Kouvaris, Gravitational waves from collapse of pressureless matter in the early universe. arXiv:2403.15126 [astro-ph.CO]
99. M. Braglia, D.K. Hazra, F. Finelli, G.F. Smoot, L. Sriramkumar, A.A. Starobinsky, Generating PBHs and small-scale GWs in two-field models of inflation. JCAP **08**, 001 (2020). https://doi.org/10.1088/1475-7516/2020/08/001. (arXiv:2005.02895 [astro-ph.CO])
100. M. Braglia, X. Chen, D.K. Hazra, Probing primordial features with the stochastic gravitational wave background. JCAP **03**, 005 (2021). https://doi.org/10.1088/1475-7516/2021/03/005. (arXiv:2012.05821 [astro-ph.CO])
101. A.D. Linde, Hybrid inflation. Phys. Rev. D **49**, 748–754 (1994). https://doi.org/10.1103/PhysRevD.49.748 arXiv:astro-ph/9307002 [astro-ph]
102. J. Garcia-Bellido, A.D. Linde and D. Wands. Density perturbations and black hole formation in hybrid inflation. Phys. Rev. D **54**, 6040–6058 (1996). https://doi.org/10.1103/PhysRevD.54.6040 arXiv:astro-ph/9605094 [astro-ph]
103. M. Braglia, A. Linde, R. Kallosh, F. Finelli, Hybrid α-attractors, primordial black holes and gravitational wave backgrounds. JCAP **04**, 033 (2023). https://doi.org/10.1088/1475-7516/2023/04/033. (arXiv:2211.14262 [astro-ph.CO])
104. S. Choudhury, S. Panda, M. Sami, Galileon inflation evades the no-go for PBH formation in the single-field framework. JCAP **08**, 078 (2023). https://doi.org/10.1088/1475-7516/2023/08/078. (arXiv:2304.04065 [astro-ph.CO])
105. S. Pi, J. Wang, Primordial black hole formation in Starobinsky's linear potential model. JCAP **06**, 018 (2023). https://doi.org/10.1088/1475-7516/2023/06/018. (arXiv:2209.14183 [astro-ph.CO])
106. K.W. Ng, Y.P. Wu, Constant-rate inflation: primordial black holes from conformal weight transitions. JHEP **11**, 076 (2021). https://doi.org/10.1007/JHEP11(2021)076. (arXiv:2102.05620 [astro-ph.CO])

Correction to: Compact Objects in the Universe

Eleftherios Papantonopoulos and Nikolaos Mavromatos

Correction to:
E. Papantonopoulos and N. Mavromatos (eds.), *Compact Objects in the Universe*, https://doi.org/10.1007/978-3-031-55098-0

The original version of the book was inadvertently published incorrect affiliations of the book editors Eleftherios Papantonopoulos and Nikolaos Mavromatos, which have now been corrected. The book has been updated with the changes.

The updated version of the book can be found at
https://doi.org/10.1007/978-3-031-55098-0

© The Author(s), under exclusive license to Springer Nature Switzerland AG 2024

E. Papantonopoulos and N. Mavromatos (eds.), *Compact Objects in the Universe*,
https://doi.org/10.1007/978-3-031-55098-0_14

SPRINGER NATURE

GPSR Compliance

The European Union's (EU) General Product Safety Regulation (GPSR) is a set of rules that requires consumer products to be safe and our obligations to ensure this.

If you have any concerns about our products, you can contact us on ProductSafety@springernature.com

In case Publisher is established outside the EU, the EU authorized representative is:

Springer Nature Customer Service Center GmbH
Europaplatz 3
69115 Heidelberg, Germany

FSC
www.fsc.org
MIX
Papier | Fördert gute Waldnutzung
FSC® C083411

Kolibri 360 GmbH
Ferdinand-Jühlke-Straße 7
99095 Erfurt, Deutschland
produktsicherheit@kolibri360.de